Food and Drug Analysis

Food and Drug Analysis

Special Issue Editor

Ping-Chung Kuo

MDPI • Basel • Beijing • Wuhan • Barcelona • Belgrade • Manchester • Tokyo • Cluj • Tianjin

Special Issue Editor
Ping-Chung Kuo
School of Pharmacy,
College of Medicine,
National Cheng Kung
Taiwan

Editorial Office
MDPI
St. Alban-Anlage 66
4052 Basel, Switzerland

This is a reprint of articles from the Special Issue published online in the open access journal *Molecules* (ISSN 1420-3049) (available at: https://www.mdpi.com/journal/molecules/special_issues/Food_Drug_Analysis).

For citation purposes, cite each article independently as indicated on the article page online and as indicated below:

LastName, A.A.; LastName, B.B.; LastName, C.C. Article Title. *Journal Name* **Year**, *Article Number*, Page Range.

ISBN 978-3-03936-539-5 (Hbk)
ISBN 978-3-03936-540-1 (PDF)

Contents

About the Special Issue Editor vii

Ping-Chung Kuo
Food and Drug Analysis
Reprinted from: *Molecules* **2020**, *25*, 2403, doi:10.3390/molecules25102403 1

Elena Roxana Chiriac, Carmen Lidia Chiţescu, Daniela Borda, Mariana Lupoae, Cerasela Elena Gird, Elisabeta-Irina Geană, Giorgiana-Valentina Blaga and Rica Boscencu
Comparison of the Polyphenolic Profile of *Medicago sativa* L. and *Trifolium pratense* L. Sprouts in Different Germination Stages Using the UHPLC-Q Exactive Hybrid Quadrupole Orbitrap High-Resolution Mass Spectrometry
Reprinted from: *Molecules* **2020**, *25*, 2321, doi:10.3390/molecules25102321 7

Yu-Chi Lin, Jue-Jun Lin, Shu-Rong Chen, Tsong-Long Hwang, Shu-Yen Fang, Michal Korinek, Ching-Yeu Chen, Yun-Sheng Lin, Tung-Ying Wu, Ming-Hong Yen, Chih-Hsin Wang and Yuan-Bin Cheng
Clerodane Diterpenoids from *Callicarpa hypoleucophylla* and Their Anti-Inflammatory Activity
Reprinted from: *Molecules* **2020**, *25*, 2288, doi:10.3390/molecules25102288 27

Ádám Tölgyesi, Luca Kozma and Virender K. Sharma
Determination of *Alternaria* Toxins in Sunflower Oil by Liquid Chromatography Isotope Dilution Tandem Mass Spectrometry
Reprinted from: *Molecules* **2020**, *25*, 1685, doi:10.3390/molecules25071685 37

Chung-Pei Chang, Po-Hsun Hou, Wei-Cheng Yang, Ching-Fen Wu, Chia-Chia Chang, Ming-Yang Tsai, Hsiao-Pei Tsai, Chien-Teng Lin, Yi-Jing Xue, Jiann-Hsiung Wang and Geng-Ruei Chang
Analytical Detection of Sulfonamides and Organophosphorus Insecticide Residues in Fish in Taiwan
Reprinted from: *Molecules* **2020**, *25*, 1501, doi:10.3390/molecules25071501 53

Thanh-Hao Huynh, Lee-Shing Fang, Yu-Hsin Chen, Bo-Rong Peng, You-Ying Chen, Li-Guo Zheng, Yu-Jen Wu, Zhi-Hong Wen, Jih-Jung Chen, Tzu-Chi Lin and Ping-Jyun Sung
Briarenols I—K, New Anti-inflammatory 8,17-Epoxybriaranes from the Octocoral *Briareum excavatum* (Briareidae)
Reprinted from: *Molecules* **2020**, *25*, 1405, doi:10.3390/molecules25061405 67

Emmanouil Tsochatzis, Joao Alberto Lopes, Fabiano Reniero, Margaret Holland, Jenny Åberg and Claude Guillou
Identification of 1-Butyl-Lysergic Acid Diethylamide (1B-LSD) in Seized Blotter Paper Using an Integrated Workflow of Analytical Techniques and Chemo-Informatics
Reprinted from: *Molecules* **2020**, *25*, 712, doi:10.3390/molecules25030712 77

Yongzhe Dong, Jingya Ruan, Zhijuan Ding, Wei Zhao, Mimi Hao, Ying Zhang, Hongyu Jiang, Yi Zhang and Tao Wang
Phytochemistry and Comprehensive Chemical Profiling Study of Flavonoids and Phenolic Acids in the Aerial Parts of *Allium Mongolicum* Regel and Their Intestinal Motility Evaluation
Reprinted from: *Molecules* **2020**, *25*, 577, doi:10.3390/molecules25030577 87

Der-Yen Lee, Yu-Chi Hou, Jai-Sing Yang, Hui-Yi Lin, Tsu-Yuan Chang, Kuo-Hsiung Lee, Sheng-Chu Kuo and Min-Tsang Hsieh
Synthesis, Anticancer Activity, and Preliminary Pharmacokinetic Evaluation of 4,4-Disubstituted Curcuminoid 2,2-bis(Hydroxymethyl)Propionate Derivatives
Reprinted from: *Molecules* **2020**, *25*, 479, doi:10.3390/molecules25030479 107

Rabab Hachem, Gaëtan Assemat, Stéphane Balayssac, Nathalie Martins-Froment, Véronique Gilard, Robert Martino and Myriam Malet-Martino
Comparative Chemical Profiling and Monacolins Quantification in Red Yeast Rice Dietary Supplements by ^{1}H-NMR and UHPLC-DAD-MS
Reprinted from: *Molecules* **2020**, *25*, 317, doi:10.3390/molecules25020317 125

Zheng Pan, Feng Xiong, Yi-Long Chen, Guo-Guo Wan, Yi Zhang, Zhi-Wei Chen, Wen-Fu Cao and Guo-Ying Zhou
Traceability of Geographical Origin in *Gentiana straminea* by UPLC-Q Exactive Mass and Multivariate Analyses
Reprinted from: *Molecules* **2019**, *24*, 4478, doi:10.3390/molecules24244478 159

Hsin-Yi Hung, Chin-Chuan Hung, Jun-Weil Liang, Chin-Fu Chen, Hung-Yi Chen, Po-Chuen Shieh, Ping-Chung Kuo and Tian-Shung Wu
Constituents and Anti-Multidrug Resistance Activity of *Taiwanofungus camphoratus* on Human Cervical Cancer Cells
Reprinted from: *Molecules* **2019**, *24*, 3730, doi:10.3390/molecules24203730 173

Jia-Hao Lee, Hui-Ching Mei, I-Chih Kuo, Tzong-Huei Lee, Yu-Hsin Chen and Ching-Kuo Lee
Characterizing Tyrosinase Modulators from the Roots of *Angelica keiskei* Using Tyrosinase Inhibition Assay and UPLC-MS/MS as the Combinatorial Novel Approach
Reprinted from: *Molecules* **2019**, *24*, 3297, doi:10.3390/molecules24183297 185

Hsien-Tsung Yao, Yu-Hsuan Yang and Mei-Ling Li
Intake of Molecular Hydrogen in Drinking Water Increases Membrane Transporters, *p*-Glycoprotein, and Multidrug Resistance-Associated Protein 2 without Affecting Xenobiotic-Metabolizing Enzymes in Rat Liver
Reprinted from: *Molecules* **2019**, *24*, 2627, doi:10.3390/molecules24142627 197

Xiao-chen Gao, Jing-wei Lv, Chun-nan Li, Nan-xi Zhang, Lin-lin Tian, Xi-ying Han, Hui Zhang and Jia-ming Sun
Screening of the Active Component Promoting Leydig Cell Proliferation from *Lepidium meyenii* Using HPLC-ESI-MS/MS Coupled with Multivariate Statistical Analysis
Reprinted from: *Molecules* **2019**, *24*, 2101, doi:10.3390/molecules24112101 209

Raffaella Colombo and Adele Papetti
Advances in the Analysis of Veterinary Drug Residues in Food Matrices by Capillary Electrophoresis Techniques
Reprinted from: *Molecules* **2019**, *24*, 4617, doi:10.3390/molecules24244617 221

Ming-Shun Wu, Levent Bless B. Aquino, Marjette Ylreb U. Barbaza, Chieh-Lun Hsieh, Kathlia A. De Castro-Cruz, Ling-Ling Yang and Po-Wei Tsai
Anti-Inflammatory and Anticancer Properties of Bioactive Compounds from *Sesamum indicum* L.—A Review
Reprinted from: *Molecules* **2019**, *24*, 4426, doi:10.3390/molecules24244426 239

About the Special Issue Editor

Ping-Chung Kuo Interests: isolation and characterization of natural constituents from plant origin; in-depth study of bioactive molecules using modern analytical techniques; synthesis and structure–activity relationship studies of bioactive molecules; development of the new synthetic methodologies; assessment of bioactivities for natural products. Education: 2004.4 Ph.D. Organic Chemistry, National Cheng Kung University, Tainan, Taiwan. Professional Experiences: 2017/08 to present, Associate Professor, School of Pharmacy, National Cheng Kung University; 2016/02–2017/07, Assistant Professor, School of Pharmacy, National Cheng Kung University; 2008/08–2016/01, Associate Professor, Dept. of Biotechnology, National Formosa University.

molecules

MDPI

Editorial

Food and Drug Analysis

Ping-Chung Kuo

School of Pharmacy, College of Medicine, National Cheng Kung University, Tainan 701, Taiwan; z10502016@email.ncku.edu.tw; Tel.: +886-6-2353535 (ext. 6806)

Received: 19 May 2020; Accepted: 20 May 2020; Published: 21 May 2020

Food can be regarded as functional if it beneficially affects one or more target functions in the body in a way that is relevant to either the state of well-being and health or to the reduction of the risk of a disease [1]. It is also indicated that a functional food is any food that may provide a health benefit beyond the traditional nutrients it contains. An important concern of these functional foods is to provide an appropriate dose of bioactive components in order to have a beneficial rather than a toxic effect on human health [2]. Since the market and consumers have more and more interests in the health-enhancing role of specific foods and physiologically active food components, functional foods have received renewed attentions [3]. Functional foods are usually developed with specific health effects, including but not limited to anti-aging, anti-inflammation, and cardiovascular protection. Therefore, this Special Issue, "Food and drug analysis", wishes to include the various aspects of exploring natural sources as healthy food and characterizing the molecular structures of bioactive principles.

In this issue, Wu et al. [4] reported one review that covered the anti-inflammatory and anticancer properties of bioactive compounds from *Sesamum indicum* L. and provided a common scope of discovery and development of lead compounds from natural sources. The use of foodstuff as natural medicines has already been established through studies demonstrating the pharmacological activities that they exhibit. *Sesamum indicum* L. is composed of lignans such as sesamin, sesamol, sesaminol and sesamolin, and these lignans have been widely studied and are known to possess antiaging, anticancer, antidiabetes, anti-inflammatory and antioxidant properties. Modern chronic diseases, which can transform into clinical diseases, are potential targets of these lignans. The prime example of chronic diseases is rheumatic inflammatory diseases, which affect the support structures and the organs of the body and can also develop into malignancies.

Other articles related to *Callicarpa hypoleucophylla*, *Briareum excavatum*, *Allium mongolicum*, and *Taiwanofungus camphoratus* were also good examples of studies on natural food sources for the possible new drug candidates. Plants of the genus *Callicarpa* are known to possess several medicinal effects. Two new clerodane-type diterpenoids along with seven known compounds were isolated from the leaves and twigs of the Taiwan endemic plant *C. hypoleucophylla* and then characterized. The anti-inflammatory activity of isolated compounds based on the suppression of superoxide anion generation and elastase release was evaluated. Among the isolates, some compounds showed anti-inflammatory activity by suppressing superoxide anion generation and elastase release [5].

Five 8,17-epoxybriaranes, including three new compounds along with two known analogues, were isolated from the octocoral *Briareum excavatum*. Their structures were elucidated by spectroscopic methods, including 1D and 2D nuclear magnetic resonance (NMR) studies and high resolution electrospray ionization mass spectrometry (HRESIMS) analysis. One briarane-type isolate exerted inhibition effects on inducible nitric oxide synthase (*i*NOS) and cyclooxygenase-2 (COX-2) release from RAW 264.7 [6].

Dong et al. [7] designed to isolate and identify the flavonoids and phenols from the aerial parts of *Allium mongolicum* Regel by using various chromatographic and spectrophotometric methods, a bioassay on motility of mouse isolated intestine tissue, as well as qualitative analysis using liquid chromatography/mass spectrometry (LC–MS). The aim of this study is to clarify whether these isolates have the effect of improving gastrointestinal function. As a result, six new flavonoid glycosides and four new phenolic acid glycosides along with twenty-one known compounds were characterized. Among them, eleven flavonoids and three phenolic acids showed significant increase in the height of mouse small intestinal muscle. Furthermore, according to the retention time and the exact mass-to-charge ratio (*m*/*z*), thirty-one compounds were unambiguously identified by comparing to the standard references by using LC–MS. According to these results, a fast analysis method for flavonoids and phenolic acids in *A. mongolicum* was established.

Hung et al. [8] reported some idea against the tumor resistance from natural food sources. Resistance to anti-cancer drugs is one of the main factors of treatment failure resulting in high morbidity. Among the reasons of resistance, the overexpression of efflux pumps leading to multidrug resistance is an important issue that needs to be solved. *Taiwanofungus camphoratus* has been used as a nutritional supplement to treat various cancers. In addition to the four new constituents reported, the major isolates zhankuic acids A–C were evaluated for their P-glycoprotein (P-gp) inhibitory effects and the results showed that zhankuic acid A was the most potent. This study provides support for the use of *T. camphoratus* in the further development of cancer therapy.

Curcumin is the naturally occurring phytochemical from the rhizome of *Curcuma longa* L. It is a polyphenol with a symmetrical structure composed of two ortho-methoxyphenol rings connected to each other through a flexible conjugated hydrocarbon chain. Lee et al. [9] reported a series of derivatives modified from the curcumin di-*O*-2,2-bis(hydroxymethyl)propionate that shows significant in vitro and in vivo inhibitory activity against MDA-MB-231 cells with eight to ten-fold higher potency than curcumin. The established structure–activity relationship and pharmacokinetic outcomes are the guidance for future development of 4,4-disubstituted curcuminoid 2,2-bis(hydroxymethyl)-propionate derivatives as anticancer drug candidates.

Molecular hydrogen (H_2) has been shown to have antioxidant and anti-inflammatory activities that may reduce the development and progression of many diseases. In the study reported by Yao et al. [10], hydrogen-rich water (HRW) was obtained by reacting hybrid magnesium–carbon hydrogen storage materials with water. Then, the effects of intake of the HRW on the activities of xenobiotic-metabolizing enzymes, membrane transporters, and oxidative stress in rats were investigated. The results from this study suggest that the consumption of HRW may not affect xenobiotic metabolism or oxidative stress in liver. However, the intake of HRW may increase the efflux of xenobiotics or toxic substances from the liver into bile by enhancing *p*-glycoprotein and Mrp2 protein expressions.

The liquid chromatography methods including ultrahigh performance liquid chromatography (UHPLC) or coupled with mass spectrometry (LC/MS) provided feasible methods for the analysis of the natural healthy foods or potential medicinal plants. Identification and quantification of polyphenols in plant material are of great interest since they make a significant contribution to its total bioactivity. High-resolution mass spectrometry (HRMS), which is able to provide the accurate mass of unknown compounds, has become an important tool for characterizing chemical components in natural products. In the study reported by Chiriac et al. [11], a UPLC-Orbitrap-MS/MS approach using the variable data acquisition mode was developed and applied for separation, identification, and quantification of the main polyphenolic compounds in *Medicago sativa* L. and *Trifolium pratense* L. sprouts in different germination stages.

Tölgyesi et al. [12] described a liquid chromatography tandem mass spectrometric (LC–MS/MS) method for analyzing five *Alternaria* toxins in sunflower oil. An optimal sample preparation condition was achieved when samples were dissolved in *n*-hexane and extracted with methanol/water mixture, followed by sample pre-concentration with solvent evaporation. This study is focusing only on this lipophilic matrix and in using all corresponding isotopically labeled internal standards (ISTD) to compensate the matrix effect that strongly influences the LC–MS/MS analysis of toxins.

Exposure to residues of antibiotics and insecticides in aquacultured food can adversely affect humans and animals and thus affect public health globally. Hence, Chang et al. [13] used a validated

LC–MS/MS and gas chromatography tandem mass spectrometric (GC–MS/MS) method to examine the levels of residues of 12 sulfonamides as well as 18 organophosphorus insecticides in aquacultured fish in Taiwan. According to the experimental results, the risk of exposure to sulfonamide and organophosphorus insecticide residue by consuming aquacultured fish in Taiwan was thus negligible, signifying no immediate health risk related to the consumption of fish. Residue levels in fish must be continually monitored to further determine the possible effects of these residues on human health.

Pan et al. [14] established the geographical origin traceability in *G. straminea* by analyzing its chemical profiles assisted with a UPLC-Q exactive mass spectrometer, from which 43 compounds were identified by comparing retention times and mass spectrometry. A total of 42 samples from different habitats was determined by a UPLC-Q exactive mass spectrometer and the data were assayed with multivariate statistical analysis. Eight characteristic compounds were identified to determine the geographical origin of the herb. To estimate the key characteristic markers associated with pharmacological function, the inhibiting activities of nitric oxide (NO) production in lipopolysaccharide (LPS)-induced macrophages were also examined. These findings are crucial in the determination of botanical origin and evaluation of the quality of *G. straminea*.

Lee at al. [15] developed an in vitro tyrosinase inhibition assay in combination with ultraperformance liquid chromatography-orbitrap mass spectrometry (UPLC-orbitrap-MS) for the rapid screening and identification of tyrosinase modulators from roots of *Angelica keiskei*. The present study indicated that the combination of in vitro tyrosinase inhibition assay coupled with UPLC–MS/MS could be widely applied to the rapid screening of active substances from various natural resources.

Lepidium meyenii is now widely consumed as a functional food and medicinal product, which is known as an enhancer of reproductive health. However, the specific chemical composition and mechanism of action for improving sexual function are unclear. Gao et al. [16] aimed at screening and determining the potential compounds, which promote mouse leydig cells (TM3) proliferation. The results suggested that three compounds had good activities on the proliferation of TM3 and promoting testosterone secretion, which might be the potential bioactive markers related to the enhancing sexual ability functions of *L. meyenii*. This study also provided the reference for a simple, quick method to screen the promoting Leydig cell proliferation active components in traditional Chinese medicine (TCM).

In addition to the extensively applied chromatographic methods, nuclear magnetic resonance (NMR) spectroscopy is also used in screening for novel bioactive molecules. All these new analytical methods accelerate the research and make the potential targets available in the near future. Hachem et al. [17] reported an easy, rapid and accurate ^{1}H-NMR analytical method to establish the chemical profiles of red yeast rice dietary supplements and to quantify their monacolin contents. The total content of monacolin was close to that measured by UHPLC, as shown by the good linear correlation between the two sets of values.

The rapid dispersion of new psychoactive substances (NPS) presents challenges to customs' services and analytical laboratories, which are involved in their detection and characterization. When the seized material is limited in quantity or of a complex nature, or when the target substance is present in very small amounts, the need to use advanced analytical techniques, efficient workflows and chemo-informatics tools is essential for the complete identification and elucidation of these substances. Tsochatzis et al. [18] described the application of such a workflow in the analysis of a single blotter paper, seized by Swedish customs, that led to the identification of a lysergic acid diethylamide (LSD) derivative, 1-butyl-lysergic acid diethylamide (1B-LSD). Its identification was made possible by comprehensively combining gas chromatography with mass spectrometry detection (GC–MS), liquid chromatography coupled with high-resolution tandem MS (LC–HR-MS/MS), Orbitrap-MS and both 1D and 2D nuclear-magnetic-resonance (NMR) spectroscopy. All the obtained data have been managed, assessed, processed and evaluated using a chemo-informatics platform to produce the effective chemical and structural identification of 1B-LSD in the seized material.

Colombo et al. [19] discussed the advances in the analysis of veterinary drug residues in food matrices by capillary electrophoresis (CE) techniques offered a sensitive and fast analytical technique and guided a new way to the safety concern related to the natural foods. Over the years, the availability of different modes, interfaces, and formats has improved the versatility, sensitivity, and speed of CE techniques. Thus, CE represents a powerful tool for the analysis of a large variety of food matrices and food-related molecules with important applications in food quality and safety. This review focuses the attention of CE applications over the last decade on the detection of different classes of drugs with a potential risk for animal and human health. In addition, considering that the different sample preparation procedures have strongly contributed to CE sensitivity and versatility, the most advanced sample pre-concentration techniques are discussed here.

This issue wishes to provide an intellectual platform for scientists to publish their results covering the topics of bioactive constituents, biological activities, and analytical methodologies in relation to food, drugs, and herbal medicines. The above-mentioned articles form a solid base for future discussions regarding the development of natural healthy foods and the improvement of currently employed analytical methods. Finally, the guest editor wishes to express sincere gratitude to all the authors for their impactful contribution as well as all the reviewers who assisted to ensure high scientific quality of the content.

Conflicts of Interest: The author declares no conflict of interest.

References

1. Rowan, C. Fighting through the functional maze. *Food Eng. Ingred.* **2001**, *15*, 68–81.
2. Falk, M. The impact of regulation on informing consumers about the health promoting properties of functional foods in the U.S.A. *J. Food Sci.* **2004**, *69*, R143–R145. [CrossRef]
3. Hasler, C.M. Functional foods: Their role in disease prevention and health promotion. *Food Technol.* **1998**, *52*, 57–62.
4. Wu, M.S.; Aquino, L.B.B.; Barbaza, M.Y.U.; Hsieh, C.L.; De Castro-Cruz, K.A.; Yang, L.L.; Tsai, P.W. Anti-inflammatory and anticancer properties of bioactive compounds from *Sesamum indicum* L.—A review. *Molecules* **2019**, *24*, 4426. [CrossRef] [PubMed]
5. Lin, Y.C.; Lin, J.J.; Chen, S.R.; Hwang, T.L.; Fang, S.Y.; Korinek, M.; Chen, C.Y.; Lin, Y.S.; Wu, T.Y.; Yen, M.H.; et al. Clerodane diterpenoids from *Callicarpa hypoleucophylla* and their anti-iInflammatory activity. *Molecules* **2020**, *25*, 2288. [CrossRef] [PubMed]
6. Huynh, T.H.; Fang, L.S.; Chen, Y.H.; Peng, B.R.; Chen, Y.Y.; Zheng, L.G.; Wu, Y.J.; Wen, Z.H.; Chen, J.J.; Lin, T.C.; et al. Briarenols I—K, new anti-inflammatory 8,17-epoxybriaranes from the octocoral *Briareum excavatum* (Briareidae). *Molecules* **2020**, *25*, 1405. [CrossRef] [PubMed]
7. Dong, Y.; Ruan, J.; Ding, Z.; Zhao, W.; Hao, M.; Zhang, Y.; Jiang, H.; Zhang, Y.; Wang, T. Phytochemistry and comprehensive chemical profiling study of flavonoids and phenolic acids in the aerial parts of *Allium Mongolicum* Regel and their intestinal motility evaluation. *Molecules* **2020**, *25*, 577. [CrossRef] [PubMed]
8. Hung, H.Y.; Hung, C.C.; Liang, J.W.; Chen, C.F.; Chen, H.Y.; Shieh, P.C.; Kuo, P.C.; Wu, T.S. Constituents and anti-multidrug resistance activity of *Taiwanofungus camphoratus* on human cervical cancer cells. *Molecules* **2019**, *24*, 3730. [CrossRef] [PubMed]
9. Lee, D.Y.; Hou, Y.C.; Yang, J.S.; Lin, H.Y.; Chang, T.Y.; Lee, K.H.; Kuo, S.C.; Hsieh, M.T. Synthesis, anticancer activity, and preliminary pharmacokinetic evaluation of 4,4-disubstituted curcuminoid 2,2-bis(hydroxymethyl)propionate derivatives. *Molecules* **2020**, *25*, 479. [CrossRef] [PubMed]
10. Yao, H.T.; Yang, Y.H.; Li, M.L. Intake of molecular hydrogen in drinking water increases membrane transporters, *p*-glycoprotein, and multidrug resistance-associated protein 2 without affecting xenobiotic-metabolizing enzymes in rat liver. *Molecules* **2019**, *24*, 2627. [CrossRef] [PubMed]
11. Chiriac, E.R.; Chiţescu, C.L.; Borda, D.; Lupoae, M.; Gird, C.E.; Geană, E.-I.; Blaga, G.-V.; Boscencu, R. Comparison of the polyphenolic profile of *Medicago sativa* L. and *Trifolium pratense* L. sprouts in different germination stages using the UHPLC-Q exactive hybrid quadrupole orbitrap high-resolution mass spectrometry. *Molecules* **2020**, *25*, 2321. [CrossRef] [PubMed]

12. Tölgyesi, Á.; Kozma, L.; Sharma, V.K. Determination of *Alternaria* toxins in sunflower oil by liquid chromatography isotope dilution tandem mass spectrometry. *Molecules* **2020**, *25*, 1685. [CrossRef] [PubMed]
13. Chang, C.P.; Hou, P.H.; Yang, W.C.; Wu, C.F.; Chang, C.C.; Tsai, M.Y.; Tsai, H.P.; Lin, C.T.; Xue, Y.J.; Wang, J.H.; et al. Analytical detection of sulfonamides and organophosphorus insecticide residues in fish in Taiwan. *Molecules* **2020**, *25*, 1501. [CrossRef] [PubMed]
14. Pan, Z.; Xiong, F.; Chen, Y.L.; Wan, G.G.; Zhang, Y.; Chen, Z.W.; Cao, W.F.; Zhou, G.Y. Traceability of geographical origin in *Gentiana straminea* by UPLC-Q exactive mass and multivariate analyses. *Molecules* **2019**, *24*, 4478. [CrossRef] [PubMed]
15. Lee, J.H.; Mei, H.C.; Kuo, I.C.; Lee, T.H.; Chen, Y.H.; Lee, C.K. Characterizing tyrosinase modulators from the roots of *Angelica keiskei* using tyrosinase inhibition assay and UPLC–MS/MS as the combinatorial novel approach. *Molecules* **2019**, *24*, 3297. [CrossRef] [PubMed]
16. Gao, X.C.; Lv, J.W.; Li, C.N.; Zhang, N.X.; Tian, L.L.; Han, X.Y.; Zhang, H.; Sun, J.M. Screening of the active component promoting Leydig cell proliferation from *Lepidium meyenii* using HPLC-ESI-MS/MS coupled with multivariate statistical analysis. *Molecules* **2019**, *24*, 2101. [CrossRef] [PubMed]
17. Hachem, R.; Assemat, G.; Balayssac, S.; Martins-Froment, N.; Gilard, V.; Martino, R.; Malet-Martino, M. Comparative chemical profiling and monacolins quantification in red yeast rice dietary supplements by ^{1}H-NMR and UHPLC-DAD-MS. *Molecules* **2020**, *25*, 317. [CrossRef] [PubMed]
18. Tsochatzis, E.; Lopes, J.A.; Reniero, F.; Holland, M.; Åberg, J.; Guillou, C. Identification of 1-butyl-lysergic acid diethylamide (1B-LSD) in seized blotter paper using an integrated workflow of analytical techniques and chemo-informatics. *Molecules* **2020**, *25*, 712. [CrossRef] [PubMed]
19. Colombo, R.; Papetti, A. Advances in the analysis of veterinary drug residues in food matrices by capillary electrophoresis techniques. *Molecules* **2019**, *24*, 4617. [CrossRef] [PubMed]

molecules

MDPI

Article

Comparison of the Polyphenolic Profile of *Medicago sativa L.* and *Trifolium pratense L.* Sprouts in Different Germination Stages Using the UHPLC-Q Exactive Hybrid Quadrupole Orbitrap High-Resolution Mass Spectrometry

Elena Roxana Chiriac [1,2], Carmen Lidia Chiţescu [2,*], Daniela Borda [3], Mariana Lupoae [2], Cerasela Elena Gird [1], Elisabeta-Irina Geană [4], Giorgiana-Valentina Blaga [3] and Rica Boscencu [1]

1. Faculty of Pharmacy, "Carol Davila" University of Medicine and Pharmacy of Bucharest, 37 Dionisie Lupu Street, Sector 2, 020021 Bucharest, Romania; roxana.chiriac10@yahoo.com (E.R.C.); cerasela.gird@umfcd.ro (C.E.G.); rica.boscencu@umfcd.ro (R.B.)
2. Faculty of Medicine and Pharmacy, "Dunarea de Jos" University of Galaţi, 35 A.I. Cuza Str., 800010 Galaţi, Romania; mariana.lupoaie@gmail.com
3. Faculty of Food Science and Engineering, "Dunarea de Jos" University of Galaţi, Str. Domnească 111, 800201 Galaţi, Romania; daniela.borda@ugal.ro (D.B.); giorgianablaga89@ymail.com (G.-V.B.)
4. National Research &Development Institute for Cryogenics and Isotopic Technologies (ICSI Rm. Valcea), 4th Uzinei Street, 240050 Râmnicu Vâlcea, Romania; irina.geana@icsi.ro

* Correspondence: carmen.chitescu@ugal.ro

Academic Editor: Ping-Chung Kuo

check for updates

Received: 19 April 2020; Accepted: 14 May 2020; Published: 15 May 2020

Abstract: Identification and quantification of polyphenols in plant material are of great interest since they make a significant contribution to its total bioactivity. In the present study, an UPLC-Orbitrap-MS/MS approach using the variable data acquisition mode (*v*DIA) was developed and applied for rapid separation, identification, and quantification of the main polyphenolic compounds in *Medicago sativa L.* and *Trifolium pratense L.* sprouts in different germination stages. Based on accurate MS data and fragment ions identification strategy, a total of 29 compounds were identified by comparing their accurate masses, fragment ions, retention times, and literatures. Additionally, a number of 30 compounds were quantified by comparing to the reference standards. Data were statistically analysed. For both plant species, the sprouts of the third germination day are valuable sources of bioactive compounds and could be used in phytotherapy and nutrition. Although *Trifolium pratense L.* (Red Clover) is considered to be a reference for natural remedies in relieving menopause disorders, alfalfa also showed a high level of biological active compounds with estrogenic activity.

Keywords: germination; bioactive compounds; UPLC-Orbitrap-MS/MS; MS fragmentation pattern; identification

1. Introduction

The consumption of sprouts, common in Asia, has been growing in western countries in the last decade once they were found to possess a broad spectrum of biologically active properties such as antioxidant, anti-inflammatory, allelopathic, and viewed as a valuable dietary supplement [1–4]. *Medicago sativa L* and *Trifolium pratense* sprouts are commonly consumed worldwide. They belong to the *Fabaceae* family, generally known by their commonly edible seeds [4–6].

Due to the increase in the use of sprouts in the human diet, there has been an expansion in the scientific literature regarding their phytochemical contents and biological proprieties. Nutritional properties of *Fabaceae* have been investigated in numerous studies [1,7].

The most interesting secondary metabolite classes found in *Medicago species* are the triterpene saponins and the polyphenolic compounds. The extraction, profiling, and identification of *M. sativa* saponins were extensively studied [8–11]. In contrast to the saponins, *Medicago* polyphenols are less genus-specific and generally encountered in many legumes. Nonetheless, alfalfa contains specific potentially valuable flavonoids with phyto-estrogenic abilities, which makes it particularly interesting for chemical characterization and pharmacological studies [12].

Red clover (*Trifolium pratense* L) sprouts are recognised as a source of phytoestrogens with high biological activity and as a dietary supplement reducing menopausal symptoms [13,14]. However, most studies focused only on certain classes of phenolic compounds such as isoflavones compounds with phytoestrogenic activity and their glycosides [13,15,16]. Among them, other polyphenolic compounds were identified by high-performance liquid chromatography HPLC in *Trifolium pratense L.*: glycitein, pratensein, pseudobaptigenin, and prunetin [16]. The comprehensive profile of phenolic compounds in aerial parts of *Trifolium pratense L.* extracts was obtained by the HPLC-tandem mass spectrometry HPLC-MS/MS technique [17,18].

Fewer reports have developed methods for qualitative and/or quantitative analysis of polyphenols in alfalfa. Flavonoids content in alfalfa was analysed in aerial parts [19,20] or in commercial sprouts by HPLC [5,6]. Ferulic acid, luteolin, myricetin, and *p*-cumaric acid were quantified in alfalfa sprouts by Oh and Rajashekar (2009) [1] using an HPLC system. However, due to the limitation of applied instrument methods, only high-level components were studied in previous studies. Moreover, a comprehensive overview of the polyphenols content variation in red clover and alfalfa during germination has not yet been reported.

High resolution mass spectrometry (HRMS), which is able to provide the accurate mass of unknown compounds, has become an important tool for characterizing chemical components in a natural product [21,22]. In the present work, we describe a comparative study conducted on alfalfa and red clover seeds and sprouts during different germination stages. Extensive characterisation of polyphenolic compounds was done by quantitative and qualitative analysis performed using ultra high-performance liquid chromatography-Q Exactive hybrid quadrupole-orbitrap high resolution accurate mass spectrometry (UHPLC-Q-Orbitrap HRMS). Thus, 29 compounds were tentatively identified without a reference standard, based on their retention times, mass spectra in a full scan mode (MS), and fragmentation patterns observed in MS-MS mass spectra. A number of 30 major compounds were unambiguously identified and quantified by comparing with reference standards.

A fully non-targeted approach of data acquisition with and without fragmentation in one single run was developed. A full scan acquisition event without fragmentation at 70,000 full width at half-maximum FWHM of resolving power was followed by five consecutive fragmentation events at a resolving power of 35,000 FWHM (variable data independent acquisition, *v*DIA) where all ions from the full scan range are fragmented. Data Independent Acquisition is an advanced option to perform untargeted fragmentation, where the entire full scan mass range is segmented in a number of subsequent fixed m/z precursor ion ranges, which are fragmented subsequently. Thus, the fragment can be restricted to the masses in a certain fragmentation event. Zomer and Mol (2015) [23] and Elmiger (2018) [24] used this approach in the analyses of small molecules as pesticides and drugs. Compared to all-ion fragmentation, vDIA can improve selectivity because product ions result from a smaller range of precursor ions [23,24].

The data obtained were subjected to statistical processing using multivariate analysis (PCA) and hierarchical clustering analysis (HCA). The present study results could represent a novel opportunity for food science and health promotion so that only certain classes of phenolic compounds in alfalfa and red clover plants/sprouts were described in previous works [13,15,18,19]. Based on our knowledge, this is the first comparison study on the chemical profile of polyphenolic compounds in sprouts of red clover and alfalfa in different germination stages.

2. Results and Discussions

Both the extraction and the instrumental method used were optimized within the present study. Three different extraction procedures are tested and compared: tinctures, microwave-assisted (MAE), and ultrasound assisted extraction (UAE). The details regarding the extraction optimisation are presented in the Supplementary materials along with details on the optimisation of the electrospray ionization parameters and mobile phase selection. According to the experimental results (Supplementary Figure S1), the UAE method was subjected for further method validation.

2.1. Identification of Phenolic Compounds in Alfalfa and Red Clover Sprouts

Identification and quantification of polyphenols in plant material are of great interest as they make a significant contribution to its total bioactivity. A specific non-target UHPLC-Q-Orbitrap HRMS method for rapid identification of the samples components was developed, optimized, and validated. A total of 59 polyphenolic compounds were simultaneously identified including nine phenolic acids, 22 isoflavones and glucoside derivatives, 11 flavone, six flavanone, and nine flavonols. Among them, 30 major compounds were unambiguously quantified by comparing with reference standards. The retention time, compound name, formula, *m/z* values of adduct ions, and MS/MS fragment ions in negative ESI mode, mass error, and accurate molecular mass are shown in Table 1.

For the compounds without available references, the structures were presumed based on high-accuracy analysis of deprotonated precursors and fragment ions of specific components. The chemical elemental composition for each peak was assigned within a mass error of 2 ppm. Based on literature [4,12,13,17,18,20,25], a self-built chemical database of known polyphenolic compounds in seeds and sprouts was achieved. A total of 29 compounds were identified in the analysed extracts (Table 2, Supplementary Figure S2).

The fragmentation pattern of polyphenols in negative electrospray ionization has been extensively studied [26–31]. The retro-Diels–Alder (rDA) reaction, loss of a methyl radical, and the mechanism of eliminating CO, CH_2CO, and CO_2 at ring C are followed by successive specific fragmentations that were previously described. MS ^{2} data obtained in the present study was consistent of literature sources [31–33].

A number of 13 isoflavones and two isoflavones glycosides were identified and listed in Table 2. In the MS-MS spectra of the methoxylated isoflavones, the loss of the $[M\text{-}H\text{-}CH_3]^-$ radical anion, loss of a hydrogen atom in these radical anions, and the neutral losses of CO and CO_2 were commonly observed [31].

Four isomeric peaks displayed the same $[M-H]^-$ at *m/z* 283.06 ($C_{16}H_{12}O_5$) in the extracted ion chromatogram of extracts of alfalfa and red clover spouts (Figure 1) and were assigned as biochanin A, prunetin, calycosin, and glycitein after the comparison with the fragmentation pattern in the mentioned databases based on the presence of some key fragments (Table 3, Supplementary Figure S3). In both alfalfa and red clover sprouts corresponding to the germination days 3 to 5, a fifth peak at *m/z* 283.06 was displayed at 16.58 min. (mass error 1.3) and it was assigned to 5,7-dihydroxy-2′-methoxyflavone based on the pattern fragmentation.

Glycitein, which is available as a reference standard, was identified by comparing with the retention time and fragmentation pattern of the reference solution. The fragment ion *m/z* 147.00 corresponding to $[M\text{-}H\text{-}CH_3\text{-}CO\text{-}B\text{-}ring]^-$ is a characteristic fragment ion of glycitein, which can be used to differentiate glycitein from its isomers and was identified only in glycitein spectra. The rest of the compounds were deduced based on the presence of diagnostic fragment ions. The loss of $CH_3\cdot$ (*m/z* 268) followed by loss of a hydrogen atom (*m/z* 267) was characteristic of all four compounds. The loss of CO· moiety from demethoxylated precursors generated the ion *m/z* 240 and the loss of CO_2 generated fragment *m/z* 224. Loss of a hydrogen atom was observed in the fragments *m/z* 240 and *m/z* 224, which both gave peaks at *m/z* 239 and *m/z* 223. Both *m/z* 240 and *m/z* 239 ions further lose CO·and CO_2 to produce *m/z* 212 and *m/z* 211 ions, respectively, in biochanin and glycitein. In an attempt to differentiate biochanin, prunetin and isoprunetin Frański (2018) [31] showed that biochanin and prunetin cannot be differentiated by MS/MS experiments in which fragmentation occurs in a collision chamber. In the previously mentioned study, by recording full scan mass spectra at high cone voltages, differences in ions' abundance were obtained. We tried to overlap that information on our data but the differences were not consistent and were not repeatable in real vegetal samples.

Table 1. The 30 compounds identified from alfalfa and red clover sprouts by UHPLC-Q-Exactive with structures confirmed by comparison with reference standards.

Compound Name	R.T. (min)	Formula	Exact Mass	Error (ppm)	Adduct Ion (*m/z*)	MS ^{2} Fragments (*m/z*)
Flavonoids (flavan-3-ols, flavone, flavonols, flavonone, flavone glucoside)						
Catechin	7.8	$C_{15}H_{14}O_6$	290.07904	1.47	289.07176	245.08192; 203.07088; 151.03908; 125.02320; 109.02821
Epicatechin	10.19	$C_{15}H_{14}O6$	290.07904	1.25	289.07176	245.08192; 203.07088; 151.03908; 109.02821
Quercitin	16.59	$C_{15}H_{10}O_7$	302.04265	0.86	301.0354	245.04601; 178.99809; 273.04059; 121.02814
Rutin (quercetin3-rutinoside)	14.20	$C_{27}H_{30}O_{16}$	610.15338	0.5	609.14613	300.02777; 271.02505; 255.02995; 243.02980; 165.01841;151.00258
Apigenin	17.54	$C_{15}H_{10}O_5$	270.05282	1.18	269.04502	227.03389; 181.06430; 151, 00194; 149.002266; 117.03271
Kaempferol	17.06	$C_{15}H_{10}O_6$	286.04774	0.57	285.04049	255.02977; 201.01866; 151.00262; 107.01250; 92.9266
Isorhamnetin	13.20	$C_{16}H_{12}O_7$	316.0583	1.35	315.05105	300.0271; 227.03508; 163.00369; 151.00264; 107.01190;
Naringenin	19.69	$C_{15}H_{12}O_5$	272.06847	0.5	271.06122	253.05055; 151.00269; 119.04903; 107.01258
Naringin	14.11	$C_{27}H_{32}O_{14}$	580.1792	2.02	579.17185	356.99371; 255.02995; 119.04884
Hesperitin	16.83	$C_{16}H_{14}O_6$	302.07904	1.34	301.07179	283.06204; 267.06650; 252.04286; 151.00266; 125.02319
Pinostrobin	17.40	$C_{16}H_{14}O_4$	270.08921	1.89	269.08196	254.05864; 210.06839; 177.05495; 148.01559
Pinocembrin	18.24	$C_{15}H_{12}O_4$	256.07356	1.04	255.06631	239.0713; 237.0557; 227.0713; 179.0349; 147.0451
Chrysin	17.63	$C_{15}H_{10}O_4$	254.05791	1.65	253.05066	208.96011; 151.03899; 107.04897; 89.04897; 65.03819
Myricetin	10.42	$C_{15}H_{10}O_8$	318.03757	1.37	317.03032	178.9986; 164.92636; 151.00368; 137.02442; 107.01258
Galangin	19.98	$C_{15}H_{10}O_5$	270.05282	1.48	269.04557	239.03345; 227.03389; 225.05580; 211.03877; 169.06425
Hyperoside (quercetin 3-galactoside)	13.98	$C_{21}H_{20}O_{12}$	464.09548	1.03	463.08768	300.02771; 355.02985; 271.02491; 243.02969; 178.99773; 151.00262
Isoflavone						
Genistin	14.77	$C_{21}H_{20}O_{10}$	432.10565	1.45	431.09837	311.05637; 269.04590; 271.05133; 181.06580
Genistein	18.07	$C_{15}H_{10}O_5$	270.05282	1.24	269.04502	159.04420; 133.02835; 201.05527; 181.06546; 107.01257
Daidzin	11.42	$C_{21}H_{20}O_9$	416.11073	1.49	415.10348	252.0451; 251.0349; 224.0487; 223.0398
Daidzein	16.50	$C_{15}H_{10}O_4$	254.05791	0.87	253.05066	226.05887; 224.04649; 209.06091; 197.06055; 135.00686; 117.03333
Ononin	26.14	$C_{22}H_{22}O_9$	430.12638	2.06	429.11913	355.0969; 341.1109; 267.1028; 252.00778
Formononetin	18.74	$C_{16}H_{12}O_4$	268.07356	1.19	267.06631	252.04298; 223.03986; 195.04466; 132.02049
Glycitein	16.33	$C_{16}H_{12}O_5$	284.06847	1.18	283.06122	268.0375; 240.0483; 211.03979; 196.05252; 167.02063
Phenolic Acid						
Gallic Acid	1.73	$C_7H_6O_5$	170.02152	0.16	169.01427	125.02318; 141.01823
Chlorogenic Acid	8.20	$C_{16}H_{18}O_9$	354.09508	0.24	353.08783	192.05876; 191.05544; 173.04474; 127.03876; 85.02806
Caffeic Acid	8.71	$C_9H_8O_4$	180.04226	0.37	179.03501	135.04390; 107.04881
Ferulic Acid	14.98	$C_{10}H_{10}O_4$	194.05791	0.62	193.05066	178.02635; 149.05974; 134.03615; 106.0424
Ellagic Acid	14.44	$C_{14}H_6O_8$	302.00627	1.62	300.99899	185.02349; 283.98961; 229.01391; 157.01006
Abscinic Acid	15.73	$C_{15}H_{20}O_4$	264.13616	1.42	263.12891	263.12854; 219.13864; 204.11502; 153.09126; 136.05162
p-coumaric Acid	10.77	$C_9H_8O_3$	164.04734	0.18	163.03954	211.0764; 135.00754; 119.0502; 17.0332; 116.0267
Syringic Acid	15.38	$C_9H_{10}O_5$	198.05282	0.41	197.04555	182.02049; 123.00697; 166.99693

Table 2. The 29 compounds identified from alfalfa and red clover sprouts by UHPLC-Q-Exactive for which the structures were presumed based on high-accuracy analysis of deprotonated precursors and fragment ions of specific components.

Compound Name	R.T. (min)	Formula	Exact Mass	Error (ppm)	Adduct Ion (*m/z*)	MS ^{2} Fragments (*m/z*)
coumestrol	18.22	$C_{15}H_8O_5$	268.03717	1.2	267.0299	266.0373; 239.0487; 211.04058; 167.10689
coumestrol 3-*O*-glucoside	12.46	$C_{21}H_{18}O_{10}$	430.0900	1.5	429.08274	417.23584; 387.22531; 367.11652; 345.13385; 267.03012
biochanin A	20.09	$C_16H_{12}O_5$	284.06847	1.27	283.06122	269.04132; 268.03809; 267.03015; 239.03487; 224.04756; 212.04695; 211.03928; 195.13850; 154.06250; 132.02031
Sissotrin (biochanin A7-*O*-β-Dglucoside)	22.02	$C_{22}H_{22}O_{10}$	446.1213	1.49	445.11404	269.04587; 283.06130; 268.03778; 166.92354; 131.94290
Prunetine (*O*-methyl genistein)	18.11	$C_{22}H_{12}O_5$	284.06847	1.39	283.06122	269.04132; 268.03809; 267.03015; 240.04300; 239.03487;211.03928
5,7-dihydroxy-2′-methoxyflavone	16.58	$C_{22}H_{12}O_5$	284.06847	1.39	283.06122	269.04584; 268.03799; 267.03015; 240.04300; 239.03473; 223.03952; 211.03963; 148.01559
calycosin (3′-hydroxy-formononetin)	17.52	$C_{16}H_{12}O_5$	284.06847	1.68	283.06122	269.06122; 268.03784; 225.0554; 226.03493; 151.00259; 157.08203; 117.03323; 107.01257
irilone	17.21	$C_{16}H_{10}O_6$	298.04774	0.96	297.04049	269.04590; 252.04297; 178.9951; 133.02837
baptigenin	11.74	$C_{15}H_{10}O_6$	286.04774	1.44	285.04046	269.04565; 240.04242; 136.01556; 109.0282
pseudobaptigenin	21.29	$C_{16}H_{10}O_5$	282.05282	1.36	281.04557	253.05089; 254.05385; 255.14954; 223.02847; 224.04770
pratensein	19.70	$C_{16}H_{12}O_6$	300.06339	0.96	299.05614	284.03284; 283.02505; 257.04111; 135.00754; 211.03932
afrormosin	18.38	$C_{17}H_{14}O_5$	298.08412	1.34	297.07687	282.05362; 283.06802; 267.03021; 253.04797; 167.04965
tectorigenin	18.20	$C_{16}H_{12}O_6$	300.06339	1.41	299.05611	284.03293; 255.03006; 227.03448
alfalone	14.80	$C_{17}H_{14}O_5$	298.08412	1.06	297.07687	281.0450; 269.04120; 211.03958; 135.00395
irisolidone	16.52	$C_{17}H_{14}O_6$	314.07904	1.44	313.07179	298.04849; 269.04581; 255.02989; 211.03957; 165.01836
medicarpin	16.59	$C_{16}H_{14}O_4$	270.08921	1.48	269.08196	254.054408; 253.14426; 141.10812; 117.03334
liquiritigenin	15.30	$C_{15}H_{12}O_4$	256.07356	1.23	255.06631	211.0764; 135.00761; 119.04889; 117.03323
isoliquiritigenin	18.19	$C_{15}H_{12}O_4$	256.07356	1.04	255.06631	211.0764; 135.00760; 119.04889; 117.03323
kaempferol-3-*O*-rutinoside	18.24	$C_{27}H_{30}O_{15}$	594.15847	1.3	593.15122	299.05615; 284.03281; 255.02997; 227.0341; 229.05032; 133.02834
kaempferol-*O*-glucoside	13.62	$C_{21}H_{20}O_{11}$	448.10056	1.42	447.09331	284.04077; 284.03299; 255.02995; 243.02979
Isorhamnetin3-*O*-glucoside	15.37	$C_{22}H_{22}O_{12}$	478.1111	1.43	477.10381	315.04871; 314.04370; 271.02518; 243.03003; 285.04083; 300.02777; 151.00262
ethyl gallate	15.38	$C_9H_{10}O_5$	198.05282	0.42	197.04557	181.04961; 169.01326; 151.0031; 121.02814; 107.01214; 83.01234
Luteolin7-glucoside	15.19	$C_{21}H_{20}O_{11}$	448.10056	1.59	447.09331	287.0359; 286.04431; 285.04077; 227.0359; 199.03964; 151.00264
vitexin (apigenin8-*C*-glucoside)	14.79	$C_{21}H_{20}O_{10}$	432.10565	1.62	431.09839	341.05179; 339.14789; 269.04587; 240.04268; 225.05542; 197.06077
apigetrin (apigenin-7-glucoside)	14.29	$C_{21}H_{20}O_{10}$	432.10565	1.78	431.09839	269.0478; 267.03009; 257.08200; 151.00267
chryosoeriol	15.24	$C_{16}H_{12}O6$	300.06339	1.29	299.05614	284.03284; 269.04572; 256.0357; 255.05562; 207.08704; 151.00266
chryosoeriol7-glucoside	17.32	$C_{22}H_{22}O_{11}$	462.11621	1.78	461.10893	446.23111; 289.04852; 283.06140; 255.02997
tricin	18.17	$C_{17}H_{14}O_7$	330.07395	0.68	329.06668	299.05634; 284.03290; 243.03044; 227.03470; 161.02370
azelaic acid	15.11	$C_9H_{16}O_4$	188.10486	1.05	187.09761	169.08600; 143.10655; 125.09581; 123.08015; 97.06589

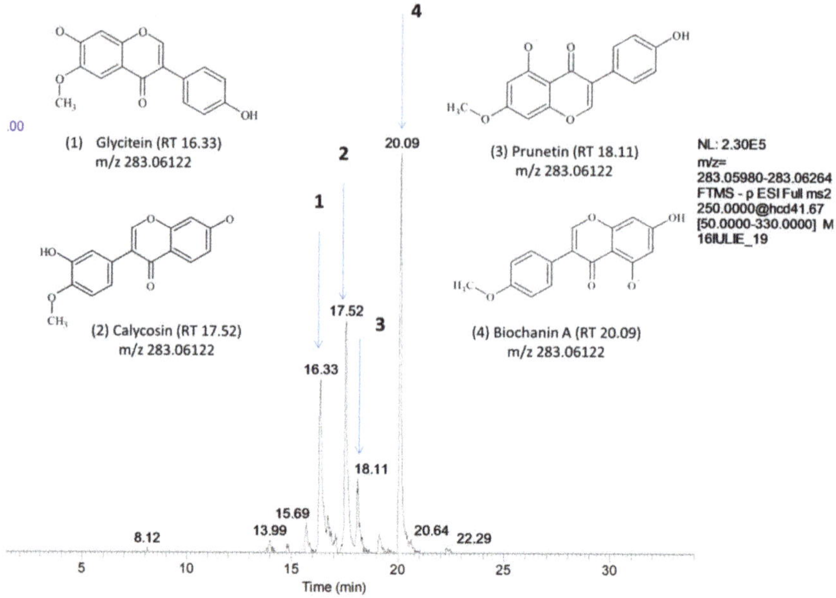

Figure 1. Extracted chromatograms for $[M-H]^-$ 286.06, full MS ^{2} (50–330 *m/z*). The isomeric peaks were assigned to **(1)** glycitein, **(2)** calycosin, **(3)** prunetin, and **(4)** biochanin A.

Table 3. Key ions in HRMS-MS spectra (*m/z* with relative abundances (%) in parenthesis) for the identification of biochanin A and isomers.

Ions	Biochanin A (20.09 min)	Calycosin (17.52 min)	Prunetine (18.11 min)	Glycitein (16.33min)	5,7-dihydroxy-2′-methoxy-flavone (16.58 min)
$[M\text{-}H]^-$	283.06	283.06	283.06	283.06	283.06
$[M\text{-}H\text{-}CH_3]^-$	268.03 (97)	268.03 (100)	268.03 (100)	268.03 (100)	268.03 (100)
$[M\text{-}H\text{-}OH]^-$	267.06 (10)	267.06 (4)	267.06 (11)	267.06 (11)	267.06 (18)
$[M\text{-}H\text{-}CO_2]^-$	-	239.07 (13)	239.07 (6)	239.07 (20)	239.07 (35)
$[M\text{-}H\text{-}CO]^-$	255.06 (25)	255.06 (9)	255.06 (42)	255.06 (12)	255.06 (56)
$[M\text{-}H\text{-}CH_3\text{-}CO]^-$	240.04 (5)	240.04 (10)	240.04(44)	240.04 (29)	240.04 (14)
$[M\text{-}H\text{-}CH_3\text{-}CO_2]^-$	224.04 (22)	224.04 (100)	224.04 (10)	-	-
$[M\text{-}H\text{-}CH_3\text{-}C_2H_2O]^-$	226.04 (12)	226.04 (10)	-	-	-
$[M\text{-}H\text{-}CH_3\text{-}H\text{-}CO]^-$	239.03 (10)	239.03 (10)	239.03 (17)	-	239.03 (10)
$[M\text{-}H\text{-}CH_3\text{-}H\text{-}CO_2]^-$	223.04 (16)	223.04 (12)	-	-	-
$[M\text{-}H\text{-}CH_3\text{-}2CO]^-$	212.02 (59)	-	-	212.02 (18)	-
$[M\text{-}H\text{-}CH_3\text{-}CO\text{-}CO_2]$	-	-	196.05 (15)	196.05 (62)	
$[M\text{-}H\text{-}CH_3\text{-}CO\text{-}H\text{-}CO_2]^-$	195.13 (17)	-	-	-	
$[M\text{-}H\text{-}CO\text{-}C\text{-}ring]^-$	-	193.05 (18)			
$[M\text{-}H\text{-}CH_3\text{-}CO\text{-}H\text{-}CO]^-$	211.03 (10)	-	211.03 (6)		-
$[M-H-CH_3-CO-CO_2-CO]^-$	-	-	-	168 (12)	-
$[M\text{-}H\text{-}CO\text{-}B\text{-}ring]^-$	-	-	167.03 (23)	167.03 (23)	-
$[M\text{-}H\text{-}CH_3\text{-}CO\text{-}B\text{-}ring]^-$	147.04 (41)	147.04 (27)	-	147.04 (18)	147.04 (48)
[A-ring fragment]$^-$	135.08 (74)	-	-	-	-
[B-ring fragment]$^-$	132.02 (48)	132.02 (12)	-	-	-

Among possible differences between those four compounds, the mass spectral decompositions in the MS2 mode concerning the retrocyclization cleavages are of special interest. The ions of type $[M\text{-}H\text{-}CH_3\text{-}CO\text{-}CO_2\text{-}CO]^-$(*m/z* 168) characteristic for glycitein, $[M\text{-}H\text{-}CO\text{-}B\text{-}ring]^-$(*m/z* 167.03) characteristic for prunetin and glycitein and $[M\text{-}H\text{-}CH_3\text{-}CO\text{-}B\text{-}ring]^-$(*m/z* 147) were used in tentative compound identification (Table 3). Ions [A-ring fragment]$^-$(*m/z* 135.00) and [B-ring fragment]$^-$(*m/z* 132.02) were characteristic for biochanin and were not detected in the rest of the isomers.

The peak at the retention time of 18.22 displayed at *m/z* 267.03 ($C_{15}H_8O_5$) was deduced as coumestrol based on the presence of diagnostic fragment ions *m/z* 266.03 produced by loss of H and *m/z* 239.03, 223.04 and 211.01 (mzCloude^{TM}). Biachi (2016) suggested the fragmentation pattern for coumestrol [34]. The loss of CO· moiety from the precursor ion generated the ion *m/z* 239.03 and the loss of CO2, the ion *m/z* 223. The *m/z* 239.03 ion further lose CO·and CO_2 to produce *m/z* 211.04 and *m/z* 167.10 ions, respectively. All fragments were identified in the alfalfa extracts samples.

At 12.46-min and 22.02-min hexose glycoside of coumestrol and, respectively, biochanin A were identified. The loss of sugar moieties was observed, producing a fragment ion $[M\text{-}H\text{-}Glc]^-$ corresponding to the aglycone form. For apigenin glucoside, two isomers were observed and assigned as vitexin (apigenin8-C-glucoside) and apigetrin (apigenin-7-glucoside). The fragmentation of each isomer was predicted using MS Fragmenter software. Due to their structural differences, fragment *m/z* 340.05 $[M\text{-}C_3H_8O_3]^-$ was specific only for apigetrin (Figure 2).

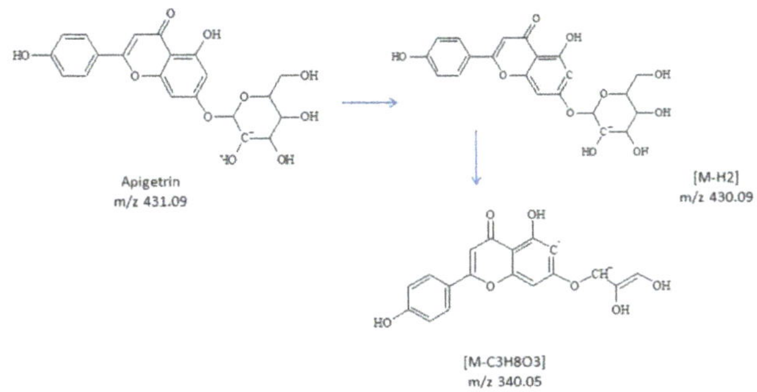

Figure 2. Proposed pathway for fragment *m/z* 340.05, specific for apigetrin.

Peaks observed at the retention time 15.30 and 18.1921 minutes displayed the same exact mass *m/z* 256.0735 and was assigned as liquiritigenin and isoliquiritigenin ($C_{15}H_{12}O_4$) in reference to the mass spectral database and the previously reported study [33,34]. The diagnostic ions were *m/z* 135.00, 119.04, which were consistent with typical $[^{1,3}A - H]^-$ and $[^{1,3}B - H]^-$ fragments [33].

Three isomeric peaks with the same $[M - H]^-$ ion at *m/z* 299.06 ($C_{16}H_{12}O_6$) displayed at 15.24, 18.20, and 19.70 min in the extracted ion chromatogram of alfalfa samples and were assigned to chryosoeriol, tectorigenin, and pratensein based on literature [12,15] and fragmentation pattern (Supplementary Figure S4). The fragments at m/z 284.03 formed by demethoxylation and m/z 256.06 generated by the future loss of CO were characteristic for all three compounds (Table 4). Fragment ions at *m/z* 267.03 $[M\text{-}H\text{-}HO\text{-}CH_3]^-$ and 255.06 $[M\text{-}H\text{-}CO_2]^-$ are also detected in the MS2 spectrum of tectorigenin. The minor fragment ion at *m/z* 135 $[M\text{-}H\text{-}Aring]^-$ correspond to the cleavage of the B ring pathway. The results were consistent with previous studies [35,36]. The formation of an ion at *m/z* 151.04 comprising the ring A of the flavone skeleton was also c haracteristic for the mass spectral decomposition in tectorigenin.

The extracted chromatogram for *m/z* 285.04 $[M - H]^-$ in the red clover spout extracts revealed two peaks at 11.74 min 17.06 min. The last one was identified as kaempferol according to the retention time and fragmentation pattern of the reference standard. According to the fragment interpretation using MS Fragmenter software and the published data, the first peak was identified as baptigenin. In the MS/MS, spectrum were detected signals at *m/z* 269.04 corresponding to the $[M\text{ -}OH]^-$ fragment, *m/z* 240.04 as $[M\text{-}H\text{-}CO_2\text{-}H]^-$, *m/z* 136.01 as $[M\text{-}B\text{-}ring\text{-}C_2H]^-$, and 109.02 corresponding to the $[B\ ring]^-$.

Peck at $[M - H]^-$ *m/z* 297.04 (RT 17.21) was identified in the red clover sprouts samples as irilone, which is in agreement with other studies [25,37]. The diagnostic ions were: *m/z* 269.04 corresponding to $[M\text{-}H\text{-}CO]^-$, *m/z* 252.04 $[M\text{-}H\text{-}CO_2\text{-}H]^-$, *m/z* 178.99 $[M\text{-}H\text{-}B\ ring\text{-}CO]^-$, *m/z* 133.02 $[M\text{-}C_8H_5O_4]^-$. Isoflavone pseudobaptigenin $[M - H]^-$ *m/z* 281.04 (RT 21.29) was identified in both alfalfa and red clover sprout samples based on the diagnostic ions *m/z* 285.05 $[M\text{-}H\text{-}CO]^-$, *m/z* 251.03 $[M\text{-}H\text{-}CO\text{-}H_2]^-$, and *m/z* 135.00 $[M\text{-}C_9H_7O_2]^-$. Azelaic acid, $[M - H]^-$ at *m/z* 187.09 (RT 15.11), which is naturally occurring in wheat, rye, barley, oat seeds, and sorghum, was identified in all analysed samples based on the diagnostic ions *m*/*z* 171.10 [M-OH]$^-$, *m*/*z* 125.09 [M-CH_3O_4]$^-$, and *m*/*z* 123.08 [M-CH_4O_3]$^-$, according to the MS Fragmenter Software.

The other compounds listed in Table 2 were identified according to their molecular mass, formula, MS/MS fragments, and related literature by using the same approach. Regarding those compounds, distribution in the analysed samples and the variation during germination is shown in Supplementary Figure S5.

Table 4. Key ions in HRMS-MS spectra (*m*/*z* with relative abundances (%) in parenthesis) for identifying tectorigenin A and isomers.

Ions	Tectorigenin (RT 18.20 min)	Chryosoeriol (RT 15.24 min)	Pratensein (RT 19.70 min)
[M−H]$^-$	299.05	299.05	299.05
[M-H-CH_3]$^-$	284.03 (57)	284.03 (21)	284.03 (100)
[M-H-CO]$^-$	-	-	271.06 (21)
[M-H-CO-H]$^-$	270.05 (100)	-	-
[M-H-CH_3-OH]$^-$	267.02(13)	-	267.02 (48)
[M-CH_3O]$^-$	269.04 (65)	269.04 (100)	-
[M-H-CH_3-CO]$^-$	256.03 (5)	-	256.03 (100)
[M−H-CO_2]$^-$	255.06(24)	255.06 (15)	255.06 (24)
[M-H-$C_9H_6O_2$]$^-$	153.01 (25)	-	-
[M-H-CH3-CO-B-ring]$^-$	151.00 (31)	-	-
[M-$C_9H_9O_3$]$^-$	135.00 (34)	135.00 (10)	-

The isoflavone aglycones biochanin A, pseudobaptigenin, calycosin, prunetin, and pratensein previously reported in the red clover aerial part [16,18,19] were detected in both red clover and alfalfa seeds and sprouts. Comestrol, tricin, vitexin (apigenin 8-C-glucoside), and tectorigenin were found only in alfalfa sprout samples, while afrormosin, baptigenin, and irilone were detected only in red clover samples. Afrormosin and irilone appear due to various modifications of the isoflavones A-ring and have been reported in aerial parts of plant form *Fabaceae* family [25].

Although biochanin A, formononetin, genistein, daidzein, and their glycosides are commonly determined in red clover as phytoestrogens of interest [18,38,39], other compounds identified in the present study that may contribute to the significant estrogenic activity of red clover include medicarpin, liquiritigenin, and isoliquiritigenin [19]. All of these compounds were found both in alfalfa and red clover sprouts.

Chrysoeriol and its glycoside, previously reported in the aerial part of alfalfa [20,38], were identified in both species of spouts. Medicarpin and chrysoeriol show significant antiangiogenic and cytotoxic activity [40] suggesting an interesting potential in cancer therapy and justify further studies.

Coumestrol was identified only in alfalfa sprouts. However, alfalfa sprout was recognised as a major source of coumestrol [19]. Tricin was detected in all samples of alfalfa seeds and sprouts, which is consistent with other studies reporting tricin identification in the alfalfa aerial part [12]. Luteolin-7-*O*-glucoside, kaempferol-3-*O*-glucoside, kaempferol-3-*O*-rutinoside previously reported in red clover, alfalfa, and mung bean sprouts [6] were identified in both species in different germination stages.

The compounds identified in sprouts' samples exhibit a wide range of biological effects, including antioxidant, antimicrobial, phytoestrogenic effects, and anticarcinogenic activity [13,15,19,40]. Our study findings confirm that both alfalfa and red clover sprouts are important sources of isoflavones besides soy and soy-derived products. Furthermore, according to the literature, while widely used soy is a source of poorly absorbed isoflavones glycosides, red clover and alfalfa contains easily-absorbed free aglycones forms [5].

2.2. Quantification Result

The developed UPLC-HRMS/MS method was applied for the routine determination of 30 compounds in the extracts of seeds and sprouts of alfalfa and red clover identified in Table 1a. The method was validated according to the section "Quantitative Method validation." Validation parameters are provided in Supplementary Table S1. The result of the quantification of the target compounds in alfalfa and red clover seeds and sprouts were shown in Tables 5 and 6. The analysis was conducted in duplicate. The obtained data were expressed as mean values ± standard deviations.

Among the quantified compounds, ononin, naringin, and epicatechin levels were below quantification limits. The results of the quantitative analysis are consistent with other studies [6,18,39]. Germination is a process known to be accompanied by a spectrum of significant changes in metabolites. Phenolic compounds are already present in the earliest plant stages and have crucial functions in plants' evolution and adaptation. While the total polyphenols' content varies, there are some kinetic transformations of the individual components that are directed towards reducing or increasing the activity over time to support specific metabolic pathways [25].

The quantitative results showed a high content in catechin, hesperetin, and quercetin in red clover seeds. The content in catechin sharply decreased after the first day of germination to a value almost six-fold lower and it continued to be reduced over time, up 20-fold on the fifth day of red clover germination. Quercetin concentration was four times lower on the fifth day of germination when compared to red clover seeds. The concentration of hesperetin in the red clover seeds linearly decrease 3.47-fold from the first to the fifth day of germination. Rather, rutin concentration increased during germination, reaching the maximum concentration on the fifth day of germination.

Respecting the alfalfa seeds and sprouts, lower variation of compounds' concentrations was registered for the sprouts in the second, third, and fourth day of germination. However, clear differences were observed between the seeds and spouts. For example, the considerably higher concentrations of kaempferol, quercetin, syringic acid, and hyperoside were measured in the seeds. Hesperitin dynamics during germination was reversed when compared to red clover, registering a marked increase in the third day of germination, which is followed by a slight decrease.

The dynamics of ferulic acid and *p*-coumaric acid concentration during germinations was similar in alfalfa to the one in red clover, with higher concentrations of *p*-coumaric acid and lower concentrations of ferulic acid in alfalfa than in red clover during germination. *p*-coumaric acid can be converted into caffeic acid via hydroxylase activity during plant germination [25]. According our results, *p*-coumaric acid showed a similar decrease during germination for both species. In alfalfa sprouts, almost double the amount of *p*-coumaric acid was measured when compared to red clover. However, caffeic acid was detected only in red clover samples, which suggests different metabolic pathways. The highest concentrations of caffeic acid were measured in red clover seeds and in sprouts on the third day of germination.

The profile of the isoflavones with estrogenic activity varied greatly with the phenological stage along the germination. In the red clover sprouts, aglycones daidzein and genistein concentrations increased from the first to the fourth day of germination, with a maximum on the third day, while, on the fifth day, the concentration was considerably lower. The concentration of formononetin in red clover increased during germination 12-fold up to day 4 when compared to the seeds. Glycitein was found in the highest concentration on the third day of germination.

Formonetin and glycitein concentrations in alfalfa sprouts on the third germination day were comparable with red clover sprouts in the same germination stage, while the concentrations in genistein, genistin, daidzein, and daidzin were considerably lower. However, other compounds that manifest an estrogenic effect include myricetin and apigenin [41]. Apigenin was found only in alfalfa sprouts with an increasing concentration up to three times higher in sprouts on the third day when compared to seeds. Apigenin was previously reported in alfalfa aerial parts [20,42]. Myricetin concentration in the alfalfa sprouts on the third germination day was about 10-fold higher than in red clover sprouts.

Table 5. The results of the quantitative analysis for alfalfa sprouts in μg per g of dried weight vegetal material (alfalfa sprouts samples coded as: ALF – alfalfa, day 1-first day of germination, day 2-second day of germination, day 3-third day of germination, day 4-fourth day of germination, and day 5-fith day of germination).

	μg/g DW	ALF Seeds	ALF Day 1	ALF Day 2	ALF Day 3	ALF Day 4	ALF Day 5
1	catechin	2.16 ± 0.16	2.5 ± 0.18	5.89 ± 0.14	7.53 ± 0.51	NF *	NF
2	caffeic acid	NF	NF	NF	NF	NF	NF
3	myricetin	117 ± 4.6	109.5 ± 6.05	113.8 ± 3.6	214 ± 5.6	62.2 ± 1.3	56.3 ± 2.45
4	*p*-cumaric acid	39.24 ± 2.5	30.18 ± 1.2	24.2 ± 0.8	23.2 ± 1.4	22.72 ± 0.9	15.2 ± 0.75
5	syringic acid	25.63 ± 1.04	4.48 ± 0.17	3.87 ± 0.22	3.88 ± 0.09	3.52 ± 0.16	2.8 ± 0.14
6	genistin	2.12 ± 0.06	6.15 ± 0.32	9.28 ± 0.45	4.8 ± 0.05	3.24 ± 0.15	1.04 ± 0.7
7	chlorogenic acid	2.24 ± 0.11	2.3 ± 0.12	2.36 ± 0.09	2.32 ± 0.06	2.32 ± 0.14	2.48 ± 0.07
8	ferulic acid	82.5 ± 4.2	63.9 ± 3.1	39.8 ± 2.04	43.03 ± 2.14	38.29 ± 1.54	50.10 ± 4.01
9	hyperoside	1209.2 ± 10.7	NF	NF	NF	NF	NF
10	isohamnetin	18.58 ± 0.85	26.12 ± 1.2	35.32 ± 1.62	36.12 ± 4.2	40.24 ± 2.87	15.24 ± 1.45
11	rutin	4.36 ± 0.28	3.85 ± 0.15	4.36 ± 0.32	6.96 ± 0.7	5.88 ± 0.14	3.92 ± 0.25
12	gallic acid	NF	NF	NF	NF	NF	NF
13	ellagic acid	7.8 ± 0.55	8.2 ± 0.17	7.15 ± 0.48	7.8 ± 0.84	7.17 ± 0.12	6.90 ± 0.21
14	formononetin	NF	NF	2.04 ± 0.07	133.5±6.2	12.04±0.17	2.24±0.06
15	pinocembrin	0.52 ± 0.02	1.08 ± 0.04	2.48 ± 0.04	2.6 ± 0.02	2.68 ± 0.07	5.12 ± 0.11
16	apigenin	8.35 ± 0.67	10.29 ± 1.62	13.12 ± 0.31	26.57 ± 2.83	19.33 ± 1.63	33.43 ± 1.44
17	pinstrobin	1.52 ± 0.32	1.6 ± 0.4	1.68 ± 0.7	1.76 ± 0.02	2.24 ± 0.014	6.92 ± 0.41
18	kaempferol	328 ± 9.2	162.14 ± 6.8	15.32 ± 1.87	9.2 ± 0.11	8.48 ± 0.15	6.72 ± 0.1
19	hesperetin	14.79 ± 1.02	26.74 ± 1.26	39.2 ± 1.42	366.91 ± 12.3	209.2 ± 11.8	75.9 ± 2.1
20	genistein	NF	20.15 ± 0.95	41.36 ± 1.12	105.8 ± 3.2	27.4 ± 1.05	10.96 ± 0.42
21	naringenin	0.32 ± 0.01	0.41 ± 0.02	0.52 ± 0.02	0.2 ± 0.01	0.4 ± 0.014	0.52 ± 0.011
22	quercitin	1108.64 ± 9.5	836.1 ± 9.9	725 ± 11.2	393.7 ± 7.9	299.6 ± 7.3	138.96 ± 5.3
23	glycitein	NF	10.01 ± 0.32	21.85 ± 1.4	43.69 ± 5.1	5.2 ± 0.23	6.1 ± 0.12
24	daidzin	NF	NF	10.9±0.32	5.2±0.12	NF	NF
25	daidzein	NF	NF	34.44 ± 2.6	53.96 ± 3.45	72.92 ± 2.9	12.44 ± 0.47
26	crysin	1.53 ± 0.4	1.5 ± 0.08	1.34 ± 0.4	1.67 ± 0.07	1.8 ± 0.09	2.0 ± 0.08
27	abiscisic acid	0.61 ± 0.3	0.48 ± 0.2	0.38 ± 0.15	0.32 ± 0.17	0.47 ± 0.15	NF
	Σ polyphenols	2974.60 ± 1.32	1327.97 ± 3.35	1156.48 ± 3.22	1496.27 ± 3.52	846.92 ± 4.94	455.40 ± 5.01

* NF – not found

Table 6. The results of the quantitative analysis for red clover sprouts in μg per g of dried weight (DW) vegetal material (red clover sprouts samples coded as: RCV - red clover, day 1-first day of germination, day 2-second day of germination, day 3-third day of germination, day 4-fourth day of germination, day 5-fith day of germination).

	μg/g DW	RCV Seeds	RCV Day 1	RCV Day 2	RCV Day 3	RCV Day 4	RCV Day 5
1	catechin	134.6 ± 5.8	23.17 ± 3.7	17.57 ± 2.6	10.39 ± 1.55	7.57 ± 0.70	7 ± 1.06
2	caffeic acid	16 ± 1.76	NF *	NF	11.88 ± 1.2	NF	NF
3	myricetin	199.5 ± 6.5	67.41 ± 3.9	30.41 ± 1.5	26.98 ± 4.03	44.22 ± 3.2	16.81 ± 1.06
4	*p*-cumaric acid	19.48 ± 1.086	16.48 ± 1.44	14.68 ± 0.95	12.6 ± 0.6	11.44 ± 0.55	11.6 ± 0.28
5	syringic acid	16.11 ± 1.14	8.68 ± 1.5	6.37 ± 0.24	5.48 ± 1.14	3.17 ± 0.17	2.19 ± 0.04
6	genistin	NF	0.76 ± 0.04	12.2 ± 1.04	20.38 ± 1.28	19.31 ± 2.09	11.9 ± 1.10
7	chlorogenic acid	NF	NF	NF	NF	NF	NF
8	ferulic acid	175.48 ± 6.8	67.56 ± 4.6	99.64 ± 7.5	82 ± 5.2	97.12 ± 8.3	101.48 ± 6.9
9	hyperoside	750.9 ± 10.02	196.2 ± 4.6	12.6 ± 0.6	NF	NF	NF
10	isohamnetin	28.6 ± 0.52	56.04 ± 0.12	43.6 ± 0.34	30.52 ± 0.47	88.48 ± 1.28	38.16 ± 0.9
11	rutin	16 ± 0.9	67.78 ± 4.6	71.14 ± 5.02	71.56 ± 3.95	122.68 ± 7.21	140.2 ± 6.95
12	gallic acid	2 ± 0.04	1.2 ± 0.02	1.04 ± 0.01	NF	1.56 ± 0.02	1.64 ± 0.03
13	ellagic acid	9.08 ± 0.06	10.20 ± 1.14	8.21 ± 1.09	7.8 ± 1.11	7.16 ± 1.15	NF
14	formononetin	14.8 ± 1.03	35.68 ± 3.6	134.2 ± 6.2	172.76 ± 8.05	180.12 ± 6.12	141.36 ± 4.8
15	pinocembrin	NF	0.2 ± 0.03	0.2 ± 0.02	0.24 ± 0.02	0.48 ± 0.03	0.28 ± 0.01
16	apigenin	NF	NF	NF	NF	NF	NF
17	pinstrobin	1.6 ± 0.14	1.6 ± 0.12	1.92 ± 0.2	2.4 ± 0.04	3.24 ± 0.15	4.52 ± 0.04
18	kaempferol	78.8 ± 3.8	5.76 ± 2.05	5.48 ± 2.12	5.48 ± 1.5	6.48 ± 1.9	5.48 ± 2.4
19	hesperetin	2824.8 ± 8.5	759.86 ± 7.3	424.83 ± 6.8	343.62 ± 10.05	203.44 ± 6.5	173.95 ± 4.6
20	genistein	28.76 ± 2.06	435.44 ± 4.7	593.44 ± 5.2	607.2 ± 3.9	499.92 ± 4.8	NF
21	naringenin	1.32 ± 0.7	0.16 ± 0.01	0.24 ± 0.03	0.32 ± 0.03	0.28 ± 0.04	NF
22	quercitin	6714 ± 9.54	2105 ± 6.25	1840.84 ± 5.07	1169.17 ± 5.10	1406.05 ± 3.9	1633.61 ± 3.04
23	glycitein	2.49 ± 1.05	1.9 ± 1.01	17.43 ± 0.95	44.54 ± 2.45	43.74 ± 1.90	27 ± 1.05
24	daidzin	NF	94.1 ± 4.3	78.3 ± 3.7	12.4 ± 2.4	NF	NF
25	daidzein	NF	NF	185.2 ± 6.8	220.2 ± 4.2	263.1 ± 3.7	114.5 ± 4.6
26	crysin	1.09 ± 0.1	1.09 ± 0.2	NF	1.10 ± 0.07	1.11 ± 0.09	1.10 ± 0.4
27	abiscisic acid	0.98 ± 0.4	0.17 ± 0.01	0.17 ± 0.02	0.52 ± 0.25	1.00 ± 0.51	0.92 ± 0.32
	Σ polyphenols	11036.40 ± 2.18	2056.45 ± 3.07	3599.72 ± 1.45	2859.55 ± 3.68	3011.68 ± 2.98	2433.72 ± 4.21

* NF – not found

In contrast to the red clover sprouts, which contain genistein as the major isoflavons, in alfalfa sprouts, the most abundant isoflavone was formononetin, which is consistent with other studies [37]. Genistein content in alfalfa and red clover spouts, reaching the maximum level on the third day of germination (607.2 μg/g for red clover and 105.8 μg/g for alfalfa), might be considered high, even compared to soybean in which genistein was reported as the major isoflavone, ranging from 84 μg/g to 583 μg/g DW along the reproductive stages [42]. According to the quantitative analysis, the sum of the isoflavones with estrogenic activity (daidzein, genistein, glycosides, apigenin, formononetin, myriceitin, and glycitein) was 906.1 μg/g DW for red clover and 587.5 μg/g DW for alfalfa sprouts on the third day of germination. Due to the different compounds' bioavailability [43], consideration of their biological activity could be speculative.

While rutin concentrations in all red clover samples were significantly higher than in alfala, comparable concentration of naringenin, syringic acid, ellagic acid, isorhamnetin, and pinostrobin were measured in both species during germination. For isorhamentin, the same nonlinear variation was observed for both species with the maximum concentration reached on the fourth day of germination. Isorhamnetin 3-*O*-glucoside was also identified in both species. Isorhamnetin has been previously detected, but not quantified in alfalfa aerial parts [38].

Abscisic acid, which is a plant hormone that has important roles in seed development and maturation [44], was present in relatively small amounts in both red clover and alfalfa in seeds and during germination.

In conclusion, consider the quantitative results for both plants species. Sprouts on the third day of germination could be considered as valuable sources of bioactive polyphenols with a potential health impact. The important health promoting potential of polyphenols resides in various biological activities, including antioxidant, estrogenic, anti-carcinogenic, and vasodilatory [16,25,37]. Different studies described the interactions of phenolic derivatives with intracellular receptors and signaling pathways to induce adaptive responses and regulation of apoptotic genes and mitochondrial function [3,15]. They suggest that *Fabaceae* sprout consumptions may also reduce the risk of osteoporosis, help alleviate menopausal symptoms, and prevent cardiovascular disease, hypertension, and hormone-dependent tumours [5,41].

2.3. Multivariate Data Analysis

Unsupervised classification by PCA and HCA were used in order to show the grouping of the investigated samples. PCA explained 59.68% of the total variation using principal components with a higher contribution brought by PC1 (33.26%) when compared to PC2 (26.42%), (Figure 3). Along PC1, the germinated alfalfa and red clover seeds from the initial seeds can be discriminated. Along the PC2 axis, two clusters including the firs cluster including alfalfa seeds and germinated alfalfa seeds located on the left side and cluster of two grouping the germinated red clover seeds located on the right side are observed. Non-germinated red clover seeds are clearly discriminated from the other seeds. PCA analysis revealed the correlations among the polyphenols' composition of different germinated seeds. Our results showed that ellagic and *p*-coumaric acids, kaempferol, and myricetin represent polyphenol markers of alfalfa seeds, while chlorogenic acid, chrysin, apigenin, and pinocembrin are representative for germinated alfalfa seeds. Red clover seeds were characterised by catechin, naringenin, hyperoside, and syringic acid, while gallic and ferulic acids, quercetin, hesperitin, and abscisic acid characterise the red clover seeds on the first and second day of germination and caffeic acid, pinostrobin, glycitein, rutin, isorhamnetin, genistein, formonetin, daidzein, and genistin are representative of the germinated red clover seeds on day 3, day 4, and day 5.

The Ward's hierarchical clustering method with Euclidean distances as measures of dissimilarity based on the phenolic compounds' profile was applied. The dendrogram shows the clustering of the samples in four separate groups (Figure 4). At a dissimilarity level of 1200000, class 1 refers to the initial alfalfa seeds (ALF seeds) (C1), while class 2 (C2) include the germinated alfalfa seeds in the five days of germination and red clover seeds on the first day of germination. Class 3 (C3) refers to the initial red clover seeds (RCV seeds) and class 4 (C4) correspond to the germinated red clover seeds, which reveal a polyphenolic composition similar to the initial alfalfa seeds.

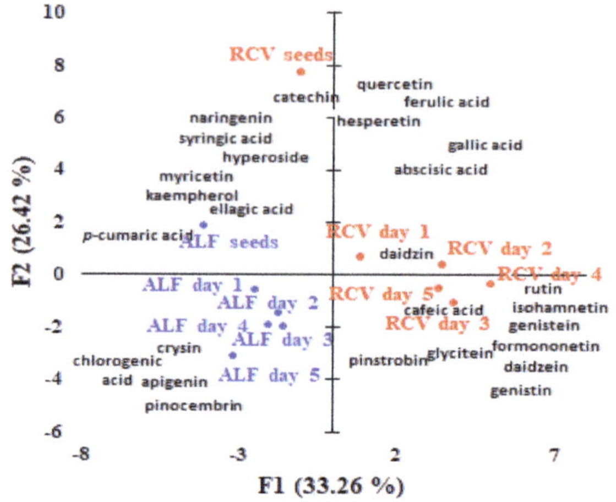

Figure 3. Bi-plot of the principal components PC1 and PC2 resulted from the PCA analysis with normalized Quatrimax rotation data of the polyphenols' concentrations quantified and the samples (alfalfa and red clover sprouts samples coded as: RCV — red clover, ALF — alfalfa, s — seeds, day 1-first day of germination, day 2-second day of germination, day 3-third day of germination, day 4-fourth day of germination, and day 5-fifth day of germination).

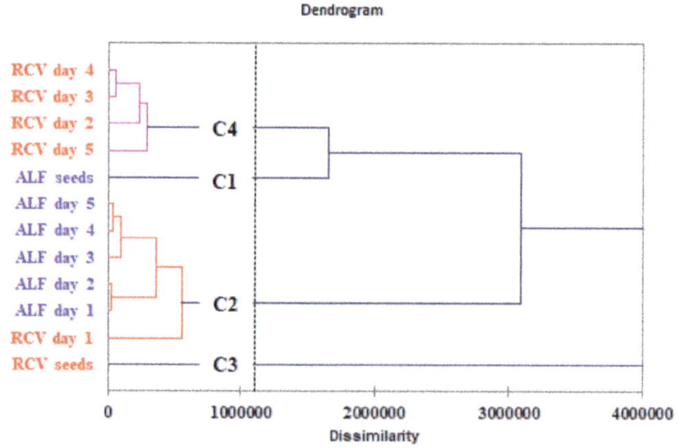

Figure 4. Dendrogram of the 12 objects (alfalfa and red clover sprout samples) represented by phenolic compounds' profile obtained by Ward's hierarchical clustering method (hierarchical cluster analysis).

The second PCA-analysis (Figure 5) was performed for the results of the qualitative analysis: identified polyphenols obtained from the screening HRMS/MS [2] and indicated with a present / absent decision.

PCA analysis revealed the correlations among the polyphenols' composition of different germinated seeds. Thereby, coumestrol, tricin, vitexin, tectorigenin, and sissotrin represent polyphenol markers of alfalfa seeds, while isoliquiritigenin, apigetrin, irisolidone kaempferol-*O*-glucoside, medicarpin, irilone, alfalone, and afrormosin are representative for germinated alfalfa seeds. Azaleic acid, isorhamnetin, luteolin-7-glucoside, liquitrigenin, pratensein, pseudobaptigenin, chrysoeriol,

chrysoeriol-7-glucoside, and kaempferol-3-rutinoside were characteristic for alfalfa in the five days of germination, while biochanin A, calycosin, prunetin, and baptigenin are characteristic for red clover.

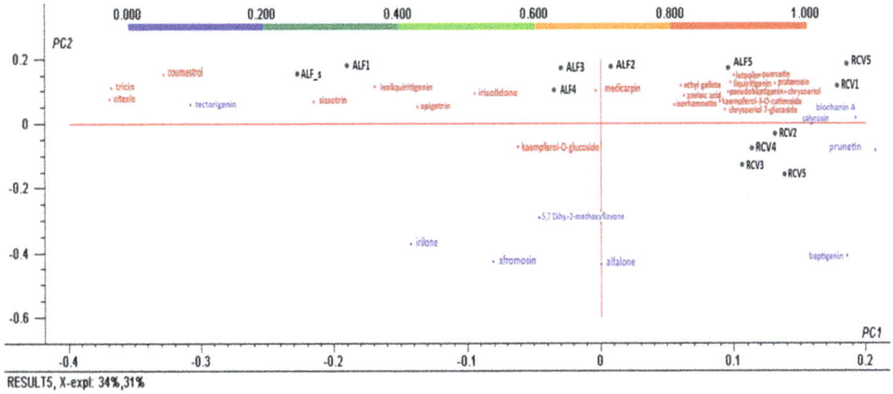

Figure 5. Bi-plot of the principal components PC1 and PC2 resulting from the PCA analysis with normalized Quatrimax rotation data of the tentatively identified polyphenols and the samples.

3. Materials and Methods

3.1. Reagents

The reference standards of 30 compounds (bold in Table 1) were purchased from Sigma–Aldrich (Aquator, Iasi, Romania). Organic solvents' methanol and ethylic alcohol, HPLC grade, were purchased from Merck Romania. Formic acid (98%) was ultrapure water (LC-MS grade) and was purchased from Merck (Merck Romania, Bucharest, Romania). For the calibration of the mass spectrometer, the Pierce^{TM} LTQ Velos electrospray ionization (ESI) positive and negative ion calibration solutions (Thermo Fisher Scientific) were used.

3.2. Stock Solutions

The stock standard solutions of the reference standards were dissolved in methanol with a concentration of 1.0 mg/mL for each compound, respectively. A series of working standard solutions (concentrations ranged from 0.05 to 1.0 µg/mL) were prepared by the successive dilution of the mixture of standard solutions with 20% methanol. All the solutions were stored at 4 °C before use.

3.3. Plant Samples – Germination

The seeds of *Trifolium pratense* and *Medicago sativa* were purchased from Agrosem, Targu Mures, Romania. Approximately 50 g of seed of each plant species were washed in a plastic container with 20 °C sterile distilled water for 30 min and then transferred to a growth chamber (automat sprout germinator, Biovita model GE-1, Cluj-Napoca, Romania) controlled at 25 °C and 80% humidity in the dark condition. Seed hydration was automatically controlled as follows. The seeds were soaked for 7 h. Then they were irrigated with water every 5 min for 7 h, and then they were irrigated for 1 min every 4 h. For both plant species, we obtained sprouts at 24 h, 48 h, 72 h, 96 h, and 120 h. The entire sprouts were dried for 4.5 h at 40 ° C in a fruit dryer (Zilan model- ZLN-96451). Average humidity loss was 90% for alfalfa and 89% for red clover. After drying, the twins were kept in closed containers, away from light. Before analysis, aliquots of 5 g of each sample were ground.

3.4. Extraction

Three extraction procedures were compared in the present work, maceration, ultrasound assisted extraction (UAE), and microwave assisted extraction (MAE). The extracts were prepared in the same condition for both plant species.

A tincture was prepared according to EU Pharmacopeia using 70% ethanol (1:10 g DW/ g). The solvent was initially heated at 60 °C. The maceration continued for 10 days in dark conditions.

An ultrasonic assisted extraction (UAE) method optimised by response surface methodology (RSM) [45] was adapted for the present study. The extraction conditions were a ratio of liquid to solid of 1:10 g DW/g 70% ethanol, 60 °C, 60 min, at 60 kHz.

A microwave assisted extraction (MAE) procedure was adapted after Zhang, 2008 [26]. An amount of 0.5 g of each sprout sample were extracted with 15 mL and 50% ethanol (1:25 g DW/g) at 50 °C (10 min. gradient with 5 min. maintaining), and microwave power at 300 W.

For all procedures, the extracts were filtered through Whatman No. 2 filter paper and a 0.20 nm Millipore MF syringe filter. Dilution 1:3 with water:methanol (80:20) before instrumental analysis was completed.

3.5. Instrumentation

3.5.1. LC Parameters

A Thermo Scientific Dionex Ultimate 3000 Series RS pump coupled with a Thermo Scientific Dionex Ultimate 3000 Series TCC-3000RS column compartments and a Thermo Fisher Scientific Ultimate 3000 Series WPS-3000RS autosampler controlled by Chromeleon 7.2 Software (Thermo Fisher Scientific, Waltham, MA and Dionex Softron GMbH Part of Thermo Fisher Scientific, Germany) were used for analysis.

The application of a 35-min gradient over an ultra-performance Accucore U-HPLC Column C18 (150 × 2.1 mm, 2.6 μm), (Thermo Scientific) was applied. The column temperature was set at 40 °C. The mobile phase consisted of: eluent A, ultrapure water containing 500 μL/L formic acid (pH 2.5), and eluent B, methanol. The step gradient was as follows: 0–1 min 100% A, 1–10 min linear increase to 30% B, 10–26 linear increased to 100% B and held for 4.0 min, 30–32.5 decreasing to 0% B. The initial conditions were obtained again at the 35th min with an equilibration time of 2.5 min. The run was performed at 0.4 mL/min for a total of 35 min.

3.5.2. MS Parameters and Data Processing

A HESI (Heated Electrospray) ion source was used for the ionization. The HESI parameters were optimized as follows. Nitrogen as sheath and auxiliary gas flow rate was set at 8 and respectively 6 units. The source heater temperature was set at 300 °C. The capillary temperature was set at 300 °C. The aux gas heater temperature was set at 300 °C. The electrospray voltage was 2800 V. The S lens RF level was 50.

Detection of the compounds was performed using an Q-Exactive mass spectrometer. Full scan data in negative mode was acquired at a resolving power of 70,000 FWHM at *m/z* 200. For the compounds of interest, a scan range of *m/z* 100–1000 Da was chosen. The automatic gain control (AGC) was set at 3e6 and the injection time was set to 200 ms. The scan rate was set at 2 scan/sec. External calibration was performed by calibration solution in a positive and a negative mode.

A total of six scan events were combined including one full scan event with mentioned parameters and five MS-MS events. In the MS^2 scan events, the precursor ion ranges were *m/z* 95–205, 195–305, 295–405, 395–505, and 500–10005, which were consecutively selected, fragmented in an higher-energy collisional dissociation cell HCD, and measured in five separate Orbitrap scans at a resolving power of 35,000 FWHM. The fragmentation events were performed at 30, 60, and 80 NCE (normalised collision energy). The C-trap parameters for all scan events were the following: Automatic Gain Control (AGC) target 1e6 and the injection time of 100 ms.

Data were evaluated by the Quan/Qual Browser Xcalibur 2.3 (Thermo Fisher). The mass tolerance window was set to 5 ppm for the two analysis modes. For the MS/MS analysis, detection of at least two fragment ions with the appropriate ion-ratio was performed by comparing the reference standards.

For those compounds without available references, the most reasonable molecular formula with a lower mass error was sought in the chemical Chemspider database (www.chemspider.com). Considering that the flavones, isoflavones, and phenolic acids had the same skeleton, the fragment ions from MS-MS analysis were used to further confirm the chemical structure with the aid of NORMAN MassBank (https://massbank.eu/MassBank/), mzCloude^{TM} Advanced Mass Spectral Database (https://www.mzcloud.org/), and PubChem (https://pubchem.ncbi.nlm.nih.gov/). ACDLabs MS Fragmenter 2019.2.1 software was used to generate a fragmentation pattern of the identified compounds for a comparison analysis.

3.5.3. Quantification Method Validation

In the full scan mode, the accurate mass of the precursor ion was used for quantification. Five-point calibration curves were obtained with a mixed standard solution in the concentration range of 25 to 2500 ng/mL. Linearity was evaluated using the coefficients of correlation (R^2). The LOD (limit of detection) and LOQ (limit of quantification) were determined as 3.3 times and 10 times, respectively. The standard deviation of the y-intercept divided by the slope of the calibration curve (ICH Validation of analytical procedures: text and methodology Q2(R1), International Conference on Harmonization, 2005) [46]. The UHPLC-Q-Orbitrap mass spectrometry was validated with respect to specificity, linearity, sensitivity, and reproducibility (ICH guidelines [46]).

The precision of the established method was evaluated by intra-day and inter-day variability, and the relative standard deviations (RSD) were taken as a measure. The RSDs were used as a measure and the acceptance criterion should be within 5.0%. The intra-day and inter-day variability were measured to assess the precision of the developed method using samples of alfalfa and red clover sprouts extract in the second day of germination. The intra-day precision was evaluated by analyzing six replicates prepared from the mentioned samples, and the inter-day precision was examined over three consecutive days with six samples per day. The repeatability was determined by injection of six samples prepared by following the same procedure.

3.5.4. Multivariate Data Analysis

Principal component analysis (PCA) is a multidimensional scale analysis that enables transformation of the variables into new ones, called principal components. The role of principal components is to explain the maximum amount of variance with the fewest number of components. PCA was performed for the quantitative and qualitative matrix containing 12 samples and 25 phenolic compounds. The PCA was applied for the quantitative correlation matrix using unit vector normalization and to the binary matrix for the qualitative analysis. Quatrimax rotation was performed and the bi-plots were selected for the visualization of the results.

Additionally, the Ward's hierarchical clustering method with Euclidean distances as measures of dissimilarity were applied.

4. Conclusions

In this paper, qualitative and quantitative analyses were combined together for the integrated characterisation and comparative analysis of the polyphenolic profile of *Medicago sativa L.* and *Trifolium pratense L.* sprouts in different germination stages. A variable data independent acquisition (vDIA) approach was used, which improved both selectivity and sensitivity for the fragment ions. This was beneficial for screening performance and identification capabilities.

By comparing MS/MS fragmentation patterns of reference compounds and the systematic identification strategy, a total of 59 polyphenolic compounds including isoflavones, flavones, flavonones, flenolic acids, and flavonols were identified in the alfalfa and red clover sprout extracts. A quantitative determination method had been validated and applied for the quantification of 30 compounds. Three extraction methods were optimised and compared.

The 29 phenolic compounds that have been identified in sprout extracts are: isoflavones with estrogeninc action as biochanin A, coumestrol, prunetin, isoflavones as irilone, pratensein, pseudobaptigenin, flavone as tricin, chriosoeriol, and phenolic acid as ethyl gallate. Glucosides of apigenin, kaempherol, and coumestrol or isorhamnetine were also identified. For both plant species, sprouts in the third and fourth germination days were found to contain higher quantities of biologically active isoflavones as genistin, daidzein, formononetin, glycitein, apigenin, hesperetin, quercetin, ferulic acid, and *p*-coumaric acid.

The method presented in this paper has been demonstrated as an effective pathway for analysing the bioactive compounds in a complex sample from a natural resource as sprouts of alfalfa and red clover. This study also demonstrated the feasibility and advantage of the *v*DIA strategy on untargeted screening. The development of advanced methods for analysis of individual, biologically-active compounds will enable future understanding of their mechanisms of action on human organisms.

Despite the well-known medicinal properties of *M. sativa* and wide consumption of alfalfa sprouts, only a few reports on biological activity of single compounds have been published. This study provides an important scientific basis for further study on clinical application and functional food of alfalfa and red clover sprouts.

Supplementary Materials: The following are available online. Figure S1. Influence of different procedures on the extraction yield of the main active compounds. Results are presented as mean (n = 3) values ± STDev. Superscripts with different letters indicate significant differences ($p < 0.05$). Figure S2. Total ions current TIC and the extracted chromatograms of the main identified compounds in alfalfa extract on the third day of germination (the chromatograms were extracted from TIC using a 5 ppm mass accuracy window, negative ion mode, full scan, base peak in the range 150–1000 *m*/*z*). Figure S3. Influence of different procedures on the extraction yield of the main active compounds. Results are presented as mean (n = 3) values ± STDev. Superscripts with different letters indicate significant differences ($p < 0.05$). Figure S4. Extracted ion chromatogram for *m*/*z* 299.05 in alfalfa sprout on the third day of germination (**A**) and MS-MS spectra of the diagnostic ions of chryosoeriol (**B**), tectorigenin (**C** and **D**), and pratensein (**E**). Figure S5: Results of the calitative screening: variation of the compounds in the samples (alfalfa and red clover sprout samples coded as flow: RCV-red clover, ALF-alfalfa, s-seeds, day 1-first day of germination, day 2-s day of germination, day 3-third day of germination, day 4-fourth day of germination, day 5-fith day of germination). Table S1: UHPLC-MS/MS method validation parameters.

Author Contributions: Conceptualization, E.R.C., and C.L.C. Methodology, C.L.C., M.L., and E.-I.G. Software, D.B. Validation, E.R.C., C.L.C., and G.-V.B. Formal analysis, C.E.G. Writing—original draft preparation, E.R.C. and C.L.C. Writing— E.R.C. and C.L.C. Visualization, all authors. Supervision, R.B. All authors have read and agreed to the published version of the manuscript.

Funding: This research was not funded by a research project.

Acknowledgments: The authors thank the Romanian Center for Modeling in Recirculating Aquaculture Systems for the infrastructure provided by the project POS 2.2.1, No 622/11.03.2014.

Conflicts of Interest: The authors declare no conflict of interest.

References

1. Oh, M.M.; Rajashekar, C.B. Antioxidant content of edible sprouts: Effects of environmental shocks. *J. Sci. Food Agric.* **2009**, *89*, 2221–2227. [CrossRef]
2. Brajdes, C.; Bahrim, G.; Dinica, R.; Vizireanu, C. Phenolics composition and their biochemical stability confirmation by in vitro gastrointestinal conditions simulation, for a new functional fermented beverage based on sprouted buckwheat. *Rom. Biotechnol. Lett.* **2013**, *18*, 7232–7242.
3. Goszcz, K.; Duthie, G.G.; Stewart, D.; Leslie, S.J.; Megson, I.L. Bioactive polyphenols and cardiovascular disease: Chemical antagonists, pharmacological agents or xenobiotics that drive an adaptive response? *Br. J. Pharm.* **2017**, *174*, 1209–1225. [CrossRef] [PubMed]
4. Benincasa, P.; Falcinelli, B.; Lutts, S.; Stagnari, F.; Galieni, A. Sprouted grains: A comprehensive review. *Nutrients* **2019**, *11*, 421. [CrossRef]

5. Plaza, L.; de Ancos, B.; Cano, P.M. Nutritional and health-related compounds in sprouts and seeds of soybean (Glycine max), wheat (Triticum aestivum L.) and alfalfa (Medicago sativa) treated by a new drying method. *Eur. Food Res. Technol.* **2003**, *216*, 138–144. [CrossRef]
6. Silva, L.R.; Pereira, M.J.; Azevedo, J.; Gonçalves, R.F.; Valentão, P.; de Pinho, P.G.; Andrade, P.B. Glycine max (L.) Merr., Vigna radiata L. and Medicago sativa L. sprouts: A natural source of bioactive compounds. *Food Res. Int.* **2013**, *50*, 167–175. [CrossRef]
7. Grela, E.R.; Kiczorowska, B.; Samolińska, W.; Matras, J.; Kiczorowski, P.; Rybiński, W.; Hanczakowska, E. Chemical composition of leguminous seeds: Part I—content of basic nutrients, amino acids, phytochemical compounds, and antioxidant activity. *Eur. Food Res. Technol.* **2017**, *243*, 1385–1395. [CrossRef]
8. Witkowska, H.E.; Biały, Z.; Jurzysta, M.; Waller, G.R. Analysis of saponin mixtures from alfalfa (*Medicago sativa L.*) roots using mass spectrometry with MALDI techniques. *Nat. Prod. Com.* **2008**, *3*, 1395–1410. [CrossRef]
9. Oleszek, W.; Stochmal, A. Cap 25. High performance Liquid Chromatography of Triterpenes (including saponins). In *High Performance Liquid Chromatography in Phytochemical Analysis*; Waksmundzka-Hajnos, M., Sherma, J., Eds.; CRC Press, Taylor & Francis group: New York, NY, USA, 2010; p. 649.
10. Abbruscato, P.; Tosi, S.; Crispino, L.; Biazzi, E.; Menin, B.; Picco, A.M.; Pecetti, L.; Avato, P.; Tava, A. Triterpenoid glycosides from *Medicago sativa* as antifungal agents against Pyricularia oryzae. *J. Agric. Food Chem.* **2014**, *62*, 11030–11036. [CrossRef]
11. Lei, Z.; Watson, B.S.; Huhman, D.; Yang, D.S.; Sumner, L.W. Large-Scale Profiling of Saponins in Different Ecotypes of *Medicago truncatula*. *Front. Plant Sci.* **2019**, *10*, 850. [CrossRef]
12. Rafińska, K.; Pomastowski, P.; Wrona, O.; Górecki, R.; Buszewski, B. Medicago sativa as a source of secondary metabolites for agriculture and pharmaceutical industry. *Phytochem. Lett.* **2017**, *20*, 520–539.
13. Kolodziejczyk-Czepas, J. Trifolium species – The latest findings on chemical profile, ethnomedicinal use and pharmacological properties. *J. Pharm. Pharm.* **2016**, *68*, 845–861. [CrossRef] [PubMed]
14. Rietjens, I.M.; Louisse, J.; Beekmann, K. The potential health effects of dietary phytoestrogens. *Br. J. Pharm.* **2017**, *174*, 1263–1280. [CrossRef] [PubMed]
15. Budryn, G.; Grzelczyk, J.; Pérez-Sánchez, H. Binding of red clover isoflavones to actin as a potential mechanism of anti-metastatic activity restricting the migration of cancer cells. *Molecules* **2018**, *23*, 2471. [CrossRef] [PubMed]
16. Spanguolo, P.; Rasini, E.; Luini, A.; Legnaro, M.; Luzzani, M.; Casareto, E.; Carreri, M.; Paracchini, S.; Marino, F.; Cosentino, M. Isoflavone content and estrogenic activity of different batches of red clover (Trifolium pretense L.) extracts: An in vitro study in MCF-7 cells. *Fitoterapia* **2014**, *94*, 62–69.
17. Klejdus, B.; Vitamvásová-Štěrbová, D.; Kubáň, V. Identification of isoflavone conjugates in red clover (Trifolium pratense) by liquid chromatography–mass spectrometry after two-dimensional solid-phase extraction. *Anal. Chim. Acta* **2001**, *450*, 81–97. [CrossRef]
18. Vlaisavljević, S.; Kaurinović, B.; Popović, M.; Vasiljević, S. Profile of phenolic compounds in Trifolium pratense L. extracts at different growth stages and their biological activities. *Int. J. Food Prop.* **2017**, *20*, 3090–3101.
19. Hong, Y.H.; Wang, S.C.; Hsu, C.; Lin, B.F.; Kuo, Y.H.; Huang, C.J. Phytoestrogenic compounds in alfalfa sprout (Medicago sativa) beyond coumestrol. *J. Agric. Food Chem.* **2011**, *59*, 131–137. [CrossRef]
20. Goławska, S.; Łukasik, I.; Kapusta, I.; Janda, B. Do the contents of luteolin, tricin, and chrysoeriol glycosides in alfalfa (Medicago sativa L.) affect the behavior of pea aphid (Acyrthosiphon pisum)? *Pol. J. Environ. Stud.* **2012**, *21*.
21. Wang, Z.; Qu, Y.; Wang, L.; Zhang, X.; Xiao, H. Ultra-high performance liquid chromatography with linear ion trap-Orbitrap hybrid mass spectrometry combined with a systematic strategy based on fragment ions for the rapid separation and characterization of components in Stellera chamaejasme extracts. *J. Sep. Sci.* **2016**, *39*, 1379–1388. [CrossRef]
22. Sun, Z.; Zuo, L.; Sun, T.; Tang, J.; Ding, D.; Zhou, L.; Zhang, X. Chemical profiling and quantification of XueBiJing injection, a systematic quality control strategy using UHPLC-Q Exactive hybrid quadrupole-orbitrap high-resolution mass spectrometry. *Sci. Rep.* **2017**, *7*, 1–15. [CrossRef] [PubMed]
23. Zomer, P.; Mol, H.G. Simultaneous quantitative determination, identification and qualitative screening of pesticides in fruits and vegetables using LC-Q-Orbitrap™-MS. *Food Addit. Contam. Part A* **2015**, *32*, 1628–1636. [CrossRef] [PubMed]

24. Elmiger, M.P.; Poetzsch, M.; Steuer, A.E.; Kraemer, T. Parameter Optimization for Feature and Hit Generation in a General Unknown Screening Method—Proof of Concept Study Using a Design of Experiment Approach for a High Resolution Mass Spectrometry Procedure after Data Independent Acquisition. *Anal. Chem.* **2018**, *90*, 3531–3536. [CrossRef] [PubMed]
25. Gholami, A.; De Geyter, N.; Pollier, J.; Goormachtig, S.; Goossens, A. Natural product biosynthesis in Medicago species. *Nat. Prod. Rep.* **2014**, *31*, 356–380. [CrossRef]
26. Zhang, Y.M.; Yan, S.J.; Cao, Z.Z.; Shi, S.L. Methodological Study for Total Flavonoid Extraction from Alfalfa by Microwave Assistance. *Acta Agrestia Sin.* **2008**, *16*, 76–80.
27. Kang, J.; Hick, L.A.; Price, W.E. A fragmentation study of isoflavones in negative electrospray ionization by MSn ion trap mass spectrometry and triple quadrupole mass spectrometry. *Rapid Commun. Mass Spectrom.* **2007**, *21*, 857–868. [CrossRef]
28. Ben Said, R.; Hamed, A.I.; Mahalel, U.A.; Al-Ayed, A.S.; Kowalczyk, M.; Moldoch, J.; Stochmal, A. Tentative characterization of polyphenolic compounds in the male flowers of Phoenix dactylifera by liquid chromatography coupled with mass spectrometry and DFT. *Int. J. Mol. Sci.* **2017**, *18*, 512. [CrossRef]
29. Raju, K.S.R.; Kadian, N.; Taneja, I.; Wahajuddin, M. Phytochemical analysis of isoflavonoids using liquid chromatography coupled with tandem mass spectrometry. *Phytochem. Rev.* **2015**, *14*, 469–498. [CrossRef]
30. Schmidt, J. Negative ion electrospray high-resolution tandem mass spectrometry of polyphenols. *J. Mass Spectrom.* **2016**, *51*, 33–43. [CrossRef]
31. Frański, R.; Gierczyk, B.; Kozik, T.; Popenda, Ł.; Beszterda, M. Signals of diagnostic ions in the product ion spectra of $[M-H]^-$ ions of methoxylated flavonoids. *Rapid Commun. Mass Spectrom.* **2019**, *33*, 125–132.
32. Ablajan, K. A study of characteristic fragmentation of isoflavonoids by using negative ion ESI-MSn. *J. Mass Spectrom.* **2011**, *46*, 77–84. [CrossRef] [PubMed]
33. Zhao, X.; Zhang, S.; Liu, D.; Yang, M.; Wei, J. Analysis of Flavonoids in Dalbergia odorifera by Ultra-Performance Liquid Chromatography with Tandem Mass Spectrometry. *Molecules* **2020**, *25*, 389. [CrossRef] [PubMed]
34. Liu, J.; Luo, L.; Zhang, H.; Jia, B.; Lu, J.; Li, P.; Chen, J. Rapid screening for novel antioxidants in Glycyrrhiza inflata using high-resolution peak fractionation. *J. Funct. Foods* **2015**, *16*, 40–49. [CrossRef]
35. Bhat, G.; Shawl, A.S.; Shah, Z.; Tantry, M. HPLC-DAD-ESI-MS/MS identification and characterization of major constituents of Iris crocea, Iris germanica and Iris spuria growing in Kashmir Himalayas, India. *J. Anal. Bioanal. Tech.* **2014**, *5*, 1. [CrossRef]
36. Zhang, W.D.; Qi, L.W.; Yang, X.L.; Huang, W.Z.; Li, P.; Yang, Z.L. Identification of the major metabolites of tectorigenin in rat bile by liquid chromatography combined with time-of-flight and ion trap tandem mass spectrometry. *Rapid Commun. Mass Spectrom.* **2008**, *22*, 2677–2684. [CrossRef]
37. Barreira, J.C.; Visnevschi-Necrasov, T.; Nunes, E.; Cunha, S.C.; Pereira, G.; Oliveira, M.B.P. Medicago spp. as potential sources of bioactive isoflavones: Characterization according to phylogenetic and phenologic factors. *Phytochemistry* **2015**, *116*, 230–238. [CrossRef]
38. Stochmal, A.; Simonet, A.M.; Macias, F.A.; Oleszek, W. Alfalfa (Medicago sativa L.) flavonoids. 2. Tricin and chrysoeriol glycosides from aerial parts. *J. Agric. Food Chem.* **2001**, *49*, 5310–5314. [CrossRef]
39. Mattioli, S.; Dal Bosco, A.; Martino, M.; Ruggeri, S.; Marconi, O.; Sileoni, V.; Falcinelli, B.; Castellini, C.; Benincasa, P. Alfalfa and flax sprouts supplementation enriches the content of bioactive compounds and lowers the cholesterol in hen egg. *J. Funct. Foods* **2016**, *22*, 454–462. [CrossRef]
40. Gatouillat, G.; Alabdul Magid, A.; Bertin, E.; Okiemy-Akeli, M.G.; Morjani, H.; Lavaud, C.; Madoulet, C. Cytotoxicity and apoptosis induced by alfalfa (Medicago sativa) leaf extracts in sensitive and multidrug-resistant tumor cells. *Nutr. Cancer* **2014**, *66*, 483–491. [CrossRef]
41. Barlas, N.; Özer, S.; Karabulut, G. The estrogenic effects of apigenin, phloretin and myricetin based on uterotrophic assay in immature Wistar albino rats. *Toxicol. Lett.* **2014**, *226*, 35–42. [CrossRef]
42. Kumar, V.; Rani, A.; Dixit, A.K.; Bhatnagar, D.; Chauhan, G.S. Relative changes in tocopherols, isoflavones, total phenolic content, and antioxidative activity in soybean seeds at different reproductive stages. *J. Agric. Food Chem.* **2009**, *57*, 2705–2710. [CrossRef] [PubMed]
43. Beck, V.; Rohr, U.; Jungbauer, A. Phytoestrogens derived from red clover: An alternative to estrogen replacement therapy? *J. Steroid Biochem. Mol. Biol.* **2005**, *94*, 499–518. [CrossRef] [PubMed]
44. Chen, K.; Li, G.J.; Bressan, R.A.; Song, C.P.; Zhu, J.K.; Zhao, Y. Abscisic acid dynamics, signaling and functions in plants. *J. Integr. Plant Biol.* **2019**, *62*, 25–54. [CrossRef] [PubMed]

45. Jing, C.L.; Dong, X.-F.; Tong, J.-M. Optimization of ultrasonic-assisted extraction of flavonoid compounds and antioxidants from alfalfa using response surface method. *Molecules* **2015**, *20*, 15550–15571. [CrossRef]
46. ICH. Harmonised Tripartite Guideline. Validation of Analytical Procedures: Text and Methodology Q2 (R1). In Proceedings of the International Conference on Harmonization, London, UK, 6 November 2005; Available online: http://www.ich.org/products/guidelines/quality/quality-single/article/validation-of-analytical-procedures-text-and-methodology.html (accessed on 6 May 2019).

Sample Availability: Samples of the compounds are not available from the authors.

molecules

Article

Clerodane Diterpenoids from *Callicarpa hypoleucophylla* and Their Anti-Inflammatory Activity

Yu-Chi Lin [1], Jue-Jun Lin [1], Shu-Rong Chen [1], Tsong-Long Hwang [2,3,4], Shu-Yen Fang [2], Michal Korinek [2,3], Ching-Yeu Chen [5], Yun-Sheng Lin [6], Tung-Ying Wu [6,7], Ming-Hong Yen [8], Chih-Hsin Wang [9,*] and Yuan-Bin Cheng [1,10,11,*]

1. Graduate Institute of Natural Products, Center for Natural Product Research and Development, College of Pharmacy, Kaohsiung Medical University, Kaohsiung 80708, Taiwan; m8952612@hotmail.com (Y.-C.L.); conk4wu0g3@gmail.com (J.-J.L.); highshorter@hotmail.com (S.-R.C.)
2. Graduate Institute of Natural Products, Graduate Institute of Biomedical Sciences, College of Medicine, Chang Gung University, Taoyuan 33302, Taiwan; htl@mail.cgu.edu.tw (T.-L.H.); susan850903@gmail.com (S.-Y.F.); mickorinek@hotmail.com (M.K.)
3. Research Center for Industry of Human Ecology, Research Center for Chinese Herbal Medicine, and Graduate Institute of Health Industry Technology, College of Human Ecology, Chang Gung University of Science and Technology, Taoyuan 33302, Taiwan
4. Department of Anesthesiology, Chang Gung Memorial Hospital, Taoyuan 33302, Taiwan
5. Department of Physical Therapy, Tzu-Hui Institute of Technology, Pingtung 92641, Taiwan; chingyeu1971@yahoo.com.tw
6. Department of Biological Science and Technology, Meiho University, Pingtung 912009, Taiwan; x00010106@meiho.edu.tw (Y.-S.L.); kuma0401@gmail.com (T.-Y.W.)
7. Department of Food Science and Nutrition, Meiho University, Pingtung 912009, Taiwan
8. School of Pharmacy, College of Pharmacy, Kaohsiung Medical University, Kaohsiung 80708, Taiwan; yen@kmu.edu.tw
9. Division of Pharmacy, Zuoying Branch of Kaohsiung Armed Forces General Hospital, Kaohsiung 81342, Taiwan
10. Department of Marine Biotechnology and Resources, National Sun Yat-sen University, Kaohsiung 80424, Taiwan
11. Department of Medical Research, Kaohsiung Medical University Hospital, Kaohsiung 80708, Taiwan

* Correspondence: hsin610255@gmail.com (C.-H.W.); jmb@kmu.edu.tw (Y.-B.C.); Tel.: +886-0906-7288-62 (C.-H.W.); +886-7312-1101 (ext. 2197) (Y.-B.C.)

Academic Editor: Ping-Chung Kuo
Received: 23 April 2020; Accepted: 12 May 2020; Published: 13 May 2020

Abstract: Plants of the genus *Callicarpa* are known to possess several medicinal effects. The constituents of the Taiwan endemic plant *Callicarpa hypoleucophylla* have never been studied. Therefore, *C. hypoleucophylla* was selected for our phytochemical investigation. Two new clerodane-type diterpenoids, named callihypolins A (**1**) and B (**2**), along with seven known compounds were isolated from the leaves and twigs of the Lamiaceae plant *C. hypoleucophylla* and then characterized. The structures of compounds **1** and **2** were elucidated by spectroscopic data analysis, specifically, two-dimension nuclear magnetic resonance (NMR). The anti-inflammatory activity of compounds **1–9** based on the suppression of superoxide anion generation and elastase release was evaluated. Among the isolates, compounds **2–4** showed anti-inflammatory activity (9.52–32.48% inhibition at the concentration 10 μm) by suppressing superoxide anion generation and elastase release. Our findings not only expand the description of the structural diversity of the compounds present in plants of the genus *Callicarpa* but also highlight the possibility of developing anti-inflammatory agents from *Callicarpa* endemic species.

Keywords: *Callicarpa hypoleucophylla*; clerodane diterpenoid; anti-inflammatory activity

1. Introduction

Callicarpa (family Lamiaceae) is a genus of about 190 species of herbaceous plants. The plant is geographically found throughout east and southeast Asia, Australia, Madagascar, southeast North America, and South America [1]. Folkloric usage of various parts of *Callicarpa* includes preparations used as fish poisons [2,3], insect repellents [1], and for some medical indications [3]. The phytochemical investigation of this genus has resulted in the identification of diterpenoids, phenylethanoids, phenypropanoids, and flavonoids. These components display various biological effects, such as anti-inflammatory [4–6], anti-platelet aggregation [7], hemostatic [8], antioxidative [9,10], cytotoxic [6,11,12], and neuroprotective [13], antitubercular [14], hepatoprotective [15,16], antimicrobial [17], anti-arthritic [18], as well as analgesic properties [19]. From the above-mentioned phytochemical and biological studies, we know this genus may offer a rich supply of bioactive phytochemicals. Because the phytochemical profile of the Taiwanese endemic plant *Callicarpa hypoleucophylla* has never been analyzed, we carried out an investigation of the constituents and bioactivity of *C. hypoleucophylla*. A meticulous separation of an ethanolic extract of *C. hypoleucophylla* led to the isolation of two new clerodane-type diterpenoids that we named callihypolins A and B (**1** and **2**), together with seven known analogues (**3–9**). The anti-inflammatory evaluation of these isolates is also presented in this paper.

2. Results and Discussion

The leaves and twigs of *C. hypoleucophylla* were extracted with 95% ethanol; the yielded extracts were suspended in H_2O and extracted with ethyl acetate (EtOAc). The EtOAc-soluble part was further partitioned with hexanes/methanol (MeOH)/H_2O (4:3:1) to obtain a MeOH layer. The MeOH layer was subjected to extensive chromatography by normal- and reversed-phase HPLC, using a normal-phase silica gel open column and a Sephadex LH-20 resin column, supplying callihypolins A and B (**1** and **2**) as well as seven known compounds (4a*R*,5*S*,6*R*,8a*R*)-5-[2-(2,5-dihydro-5-methoxy-2-oxofuran-3-yl)ethyl]-3,4,4a,5,6,7,8,8a-octahydro-5,6,8a-trimethylnaphthalene-1-carboxylic acid (**3**) [20], patagonic acid (**4**) [21], limbatolide F (**5**) [22], limbatolide A (**6**) [23], methyl (4a*R*,5*S*,6*R*,8*S*,8a*R*)-3,4,4a,5,6,7,8,8a-octahydro-8-hydroxy-5,6,8a-trimethyl-5-[2-(2-oxo-2,5-dihydrofuran-3-yl)ethyl]naphthalene-1-carboxylate (**7**) [24], clerodermic acid (**8**) [25], and visclerodol acid (**9**) [26] (Figure 1).

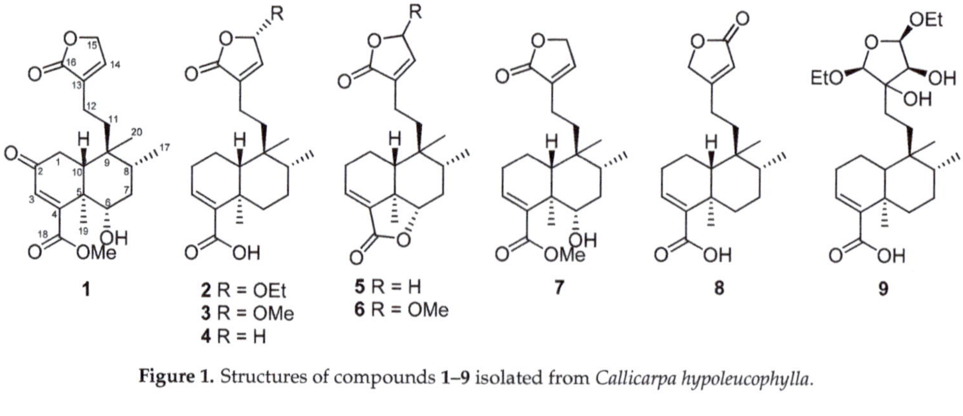

Figure 1. Structures of compounds **1–9** isolated from *Callicarpa hypoleucophylla*.

The molecular formula of compound **1** was established to be $C_{21}H_{28}O_6$ on the basis of the $[M + Na]^+$ peak at *m/z* 399.17785 (calcd. 399.17781 for $C_{21}H_{28}O_6Na$) obtained from high-resolution electrospray ionization mass spectrometry (HRESIMS) (Figure S13). The IR absorption bands of compound **1** indicated the presence of hydroxy (3451 cm^{-1}), α,β-unsaturated-γ-lactone (1739 cm^{-1}), and carboxyl (1678 cm^{-1}) functionalities. The ^{13}C and distortionless enhancement by polarization transfer (DEPT)-135 NMR data (Figure S2) showed the presence of 21 carbons divided into 7 quaternary carbons (including 3 carbonyls), 5 methines, 5 methylenes, and 4 methyls. The ^1H (Figure S1) and ^{13}C NMR signals of compound 1 showed some characteristic peaks such as an olefinic methine singlet (δ_H 5.96, δ_C 126.6, C-3), two tertiary methyls (δ_H 1.33, δ_C 14.0, Me-19; δ_H 0.84, δ_C 17.3, Me-20), a secondary methyl (δ_H 0.90, J = 6.8 Hz, δ_C 15.3, Me-17), as well as a butenolide unit (δ_C 134.0, C-13; δ_H 7.09, 1 H, quin, J = 1.7 Hz, δ_C 143.9, C-14; δ_H 4.77, 1H, dd, J = 3.8, 1.7 Hz, δ_C 70.2, C-15; δ_C 174.0, C-16). The above NMR data indicated that the structure of compound **1** was similar to that of dichrocephnoid E [27], a clerodane diterpenoid, except for a methylene corresponding to C-6 that was replaced by an oxymethine (δ_H 3.84, δ_C 72.5) and an additional methoxy (δ_H 3.81, δ_C 52.8) present in compound **1**. The whole structure of compound 1 was then determined, starting from characteristic signals, by means of correlation spectroscopy (COSY), heteronuclear single quantum correlation (HSQC), and heteronuclear multiple bond correlation (HMBC) NMR correlations (Figures S3–S5). The COSY spectrum (Figure 2) showed cross-peaks with signals at H-1 (δ_H 2.43, 2.56)/H-10 (δ_H 2.00); H-6 (δ_H 3.84)/H-7 (δ_H 1.61, 1.70)/H-8 (δ_H 1.76)/Me-17 (δ_H 0.90); H-14 (δ_H 7.09)/H-15 (δ_H 4.77). Moreover, the key HMBC correlations (Figure 2) of H-1 with C-2, H-3 with C-1, C-4, C-5, and C-18; Me-19 with C-4, C-5, C-6, and C-10, Me-17 with C-7, C-8, and C-9, Me-20 with C-9, C-10, and C-11, and methoxy proton with C-18 led to the construction of the decalin core of compound **1**, including a hydroxy group at C-6 and a methyl ester substituted at C-4. The linkage between C-12 and butanolide via C-13 was established by comparing the corresponding NMR data with those of similar analogues and confirmed by mass spectrometry analysis [22,24,27]. The planar structure of compound **1** is represented in Figure 2. The relative stereochemistry of compound **1** was deduced from nuclear overhauser effect spectroscopy (NOESY) correlations (Figure 2 and Figure S6) and by comparison of its spectroscopic data with those of clerodane analogues. The NOESY experiment showed correlations of H-6 (δ_H 3.84)/H-10 (δ_H 2.00)/H-8 (δ_H 1.76), which indicated protons located on the β face of the molecule. On the other hand, Me-20 presented NOESY correlations with Me-19 and Me-17, but neither Me-19 nor Me-20 correlated with H-10, suggesting that compound **1** is an *ent*-clerodane-type molecule with *trans*-decalin core [28]. The *trans* A/B ring junction was also evidenced by the carbon chemical shifts of C-19 (δ_C 14.0) and C-20 (δ_C 17.3) [29–31]. Thus, these correlations indicated that the hydroxy group at C-6 had an α-configuration, as confirmed by the coupling constants of H-6 with H-7α (J = 12.6 Hz) and H-7β (J = 4.4 Hz) [32,33]. All the spectral data appeared thus to be in agreement with the structure and stereochemistry of compound **1**.

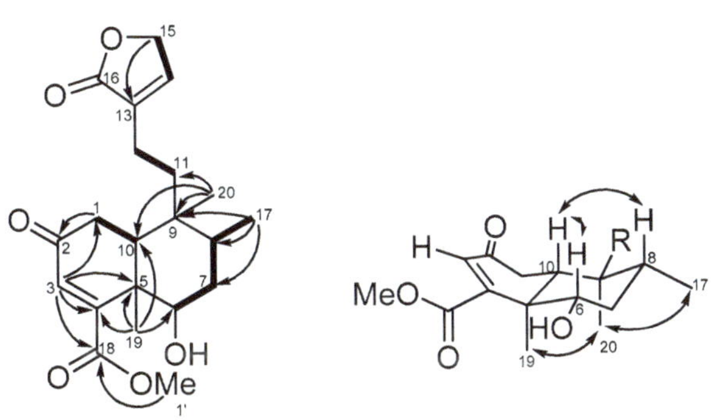

Figure 2. COSY (bold bond), selected HMBC (arrow), and NOESY (left-right arrow) correlations of compound **1**.

Callihypolin B (**2**) was isolated as a yellow oil. It possesses the molecular formula $C_{22}H_{32}O_5$, corresponding to seven indices of hydrogen deficiency, as determined by the HRESIMS ion at *m/z* 399.21419 $[M + Na]^+$ (calcd. 399.21420) (Figure S14) and ^{13}C NMR data. The IR spectrum revealed the presence of ester (1768 cm^{-1}) and conjugated carbonyl (1682 cm^{-1}) groups. The ^1H NMR data of compound **2** (Table 1, Figure S7) demonstrated the presence of one ethoxy [δ_H 3.94 (m) and 3.74 (m); 1.27 (t, *J* = 7.1 Hz)], one secondary methyl [δ_H 0.81 (d, *J* = 6.2 Hz)], two tertiary methyls (δ_H 0.76 and 1.23), and two olefinic methines [δ_H 6.85 (m), and 6.76 (d, *J* = 1.2)], together with one hemiacetal methine [δ_H 5.79 (brd, *J* = 1.2)]. The ^{13}C NMR and DEPT spectra (Table 1, Figure S8) of compound **2** showed the presence of 22 carbon signals ascribable to 4 methyls, 7 methylenes (of which one was oxygenated), 2 olefinic methines, 3 aliphatic methines, 2 aliphatic quaternary carbons, 2 olefinic quaternary carbons, and 2 carbonyl carbons. Two carbonyls and two C=C double bonds accounted for four indices of hydrogen deficiency, so the remaining three indices suggested that compound **2** was a tricyclic compound. In the ^1H-^1H COSY spectrum (Figure S9), the correlations of H_2-1/H_2-2/H_2-3, H_2-6/H_2-7/H-8/Me-17, H_2-11/H_2-12, H-14/H-15, and H_2-1′/Me-2′ were used to establish the presence of five fragments, as shown in Figure 3. In the HMBC spectrum (Figure 3, Figure S11), the cross-peaks of H-3 with C-4 and C-18; of Me-19 with C-4, C-5, C-6, and C-10; and of H-10 with C-1 and C-5 revealed the presence of a cyclohexene ring (ring A), in which a carboxyl group and Me-19 were attached to C-4 and C-5, respectively. The presence of a cyclohexane ring (ring B) with Me-20 attached at C-9 was elucidated by the HMBC correlations of Me-20 to C-8, C-9, and C-10, as well as of H-10 to C-9. Additionally, both H_3-20 and H-10 showed correlations with C-11 and indicated the linkage between ring B and C-11 via C-9. The HMBC cross-peaks of H-14 to C-13 (δ_C 139.0) and C-16 (δ_C 171.5); H-15 (δ_H 5.79) to C-16 and C-1′ (δ_C 66.0), as well as H_2-12 to C-13 and C-16, revealed the presence of an α,β-unsaturated γ-lactone ring with an ethoxy group located at C-15. Thus, the planar structure of compound **2** could be established. The stereochemistry of compound **2** was determined by its NOESY spectrum, relative NMR data, and circular dichroism spectrum. The NOESY experiments (Figure 3 and Figure S12) carried out on compound **2** showed correlations of Me-19/Me-20/Me-17, and H-6β (δ_H 2.44)/H-10/H-8, whereas no correlation was revealed between H-10 and Me-19. These data, as well as the carbon chemical shift of Me-19 at δ_C 20.5 [29], indicated that compound **1** is characterized by a type TC clerodane skeleton under a chair conformation of ring B [34], a *trans* relationship between rings A and B, α-orientations of Me-17, Me-19, and Me-20, and β-orientation of H-10. The ethoxy group attached at C-15 in the butenolide moiety was assigned to the α-face by comparison with the circular dichroism (CD) data of known butenolides and by applying the octant rule. The CD spectrum showed a negative Cotton effect near 243 nm (π-π*) and supported the *S* configuration of C-15 [31,35,36]. Thus, the structure and stereochemistry of compound **2** were clearly determined.

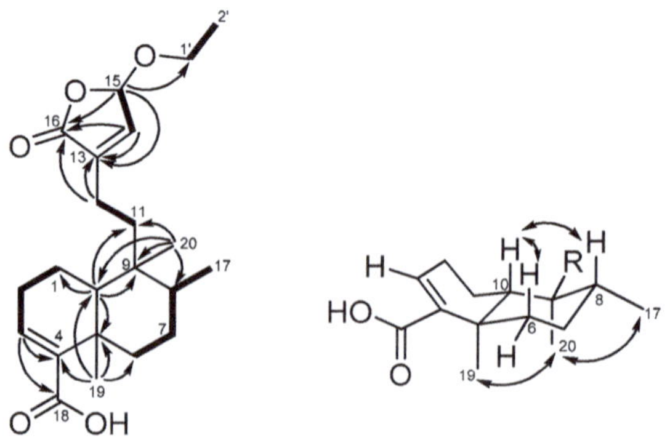

Figure 3. COSY (bold bond), selected HMBC (arrow), and NOESY (left-right arrow) correlations of compound **2**.

Table 1. ^1H and ^{13}C NMR Data of compounds **1** and **2** in $CDCl_3$.

Position	**1** [a]		**2** [b]	
	δ_H Mult. (J in Hz)	δ_C, Type	δ_H Mult. (J in Hz)	δ_C, Type
1	2.56 (dd, 18.0, 14.2) 2.43 (dd, 18.0, 3.5)	34.4, CH_2	1.46 (m) 1.66 (m)	17.3, CH_2
2		199.0, C	2.27 (m)	27.4, CH_2
3	5.96 (s)	126.6, CH	6.85 (m)	140.3, CH
4		160.4, C		141.2, C
5		45.5, C		37.5, C
6	3.84 (dd, 12.6, 4.4)	72.5, CH	2.44 (m) 1.14 (m)	35.7, CH_2
7	1.70 (dt, 12.6, 4.4) 1.61 (m)	36.1, CH_2	1.46 (m) 1.42 (m)	27.2, CH_2
8	1.76 (m)	34.5, CH	1.50 (m)	36.2, CH
9		38.5, C		38.7, C
10	2.00 (dd, 14.2, 3.5)	44.9, CH	1.32 (d, 11.6)	46.6, CH
11	1.61 (m) 1.47 (m)	34.8, CH_2	1.50 (m) 1.66 (m)	35.7, CH_2
12	2.19 (m) 2.00 (tm, 13.0)	18.7, CH_2	2.20 (m) 2.03 (m)	18.9, CH_2
13		134.0, C		139.0, C
14	7.09 (quin, 1.7)	143.9, CH	6.76 (quin, 1.2)	141.4, CH
15	4.77 (dd, 3.9, 1.7)	70.2, CH_2	5.79 (brd, 1.2)	101.6, CH
16		174.0, C		171.5, C
17	0.90 (d, 6.8)	15.3, CH_3	0.81 (d, 6.2)	15.9, CH_3
18		169.8, C		172.3, C
19	1.33 (s)	14.0, CH_3	1.23 (s)	20.5, CH_3
20	0.84 (s)	17.3, CH_3	0.76 (s)	18.2, CH_3
1′	3.81 (s)	52.8, CH_3	3.94 (m) 3.74 (m)	66.0, CH_2
2′			1.27 (t, 7.1)	15.0, CH_3

[a] ^1H and ^{13}C-NMR were measured at 600 and 150 MHz. [b] ^1H and ^{13}C-NMR were measured at 400 and 100 MHz.

Compounds **1–9** were evaluated for their inhibitory activities on superoxide anion generation and elastase release in formyl-methionyl-leucyl-phenylalanine (fMLF)/cytochalasin (CB)-induced human neutrophils. The formyl peptide fMLF in combination with the priming agent CB serves as a stimulator that mimics the over-activation of neutrophils by a pathogen or an immune system reaction [37]. As shown in Table 2, compounds **2–4** exerted anti-inflammatory activity by suppressing superoxide anion generation and elastase release. The positive control genistein, which acts via inhibition of protein tyrosine kinases, showed a profound effect on the respiratory burst (89% inhibition of superoxide generation) and only a mild effect on degranulation (22.8% inhibition of elastase release). Among the tested samples, the new compound **2** showed the best activity, suppressing 32.2% of superoxide generation and 17.6% of elastase release. To exclude possible toxicity to the cells, the lactate dehydrogenase (LDH) release assay was employed, and none of the tested clerodane diterpenoids resulted toxic to human neutrophils (Figure 4). Clerodane diterpenes with an open lactone ring at C16 were previously reported to exert inhibitory effects on the function of neutrophils activated by fMLF/CB, including respiratory burst [38] and degranulation [39]. Thus, our results well correlate with the anti-inflammatory effects of previously isolated clerodane diterpenes and indicate the potential of the new compounds for the development of anti-inflammatory drugs targeting neutrophils.

Table 2. Inhibitory effects of compounds **1–9** on superoxide anion generation and elastase release in formyl-methionyl-leucyl-phenylalanine (fMLF)/ cytochalasin (CB)-induced human neutrophils.

Compound	Superoxide Anion Inh %		Elastase Release Inh %	
1	20.28 ± 5.98	*	8.26 ± 3.72	
2	32.19 ± 6.92	**	17.55 ± 2.64	***
3	31.19 ± 5.99	**	12.15 ± 2.38	***
4	32.88 ± 4.41	***	13.57 ± 1.48	***
5	23.65 ± 7.67	*	7.33 ± 1.56	**
6	8.44 ± 6.40		10.50 ± 3.23	*
7	7.93 ± 5.86		9.30 ± 2.91	*
8	15.23 ± 6.37		11.80 ± 3.55	*
9	18.80 ± 7.82		16.30 ± 3.74	**
Genistein [a]	89.00 ± 3.00	***	22.79 ± 2.25	***

Percentage of inhibition (Inh %) at 10 μM concentration. Results are presented as mean ± S.E.M. (n = 4–5); * $p < 0.05$, ** $p < 0.01$, *** $p < 0.001$ compared with the control (solvent). [a] Genistein served as a positive control.

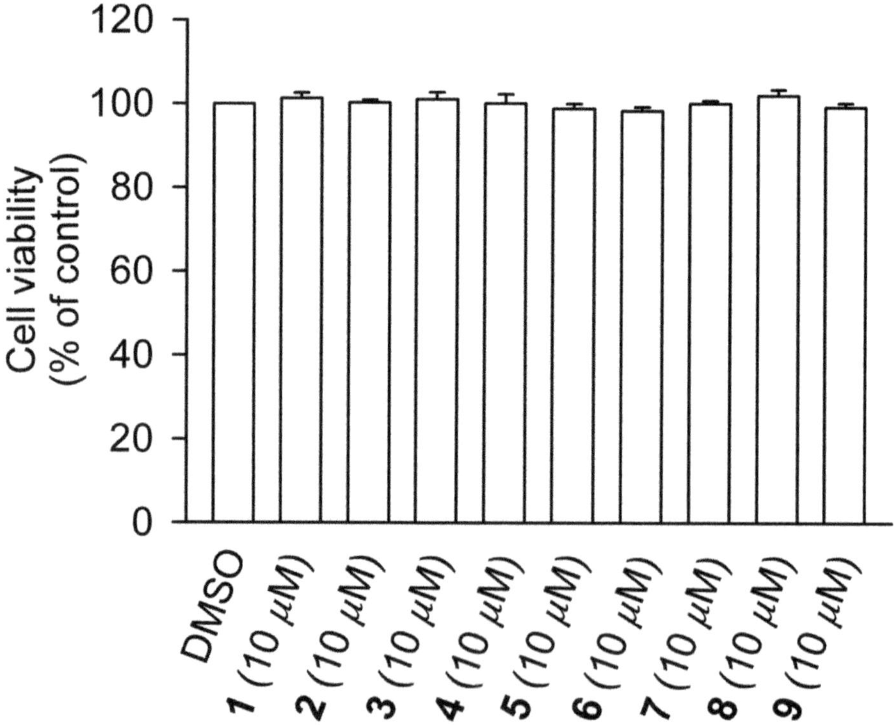

Figure 4. Compounds **1–9** do not cause LDH release in human neutrophils. Human neutrophils were incubated with DMSO (as a control) or compounds **1–9** (10 μM) for 15 min. Cytotoxicity was evaluated by LDH release. All data are presented as the means ± S.E.M. (n = 3).

3. Experimental

3.1. General

Silica gel 60 (Merck) was used for open-column chromatography (CC). Luna C_{18} (5 m, 250 × 10 mm, Phenomenex), Luna CN (5 m, 250 10 mm, Phenomenex), and Luna phenyl-hexyl (5 m, 250 × 10 mm, Phenomenex) semi-preparative columns were used for high-performance liquid chromatography

(HPLC). HPLC used a Shimadzu LC-10AT pump with an SPD-20A UV-Vis detector. The UV spectra were obtained by using a Jasco UV-530 ultraviolet spectrophotometer (Jasco, Tokyo, Japan), whereas the IR spectra were obtained on a Jasco FT-IR-4600 spectrophotometer (Jasco, Tokyo, Japan). Optical rotations were measured with a Jasco P-1020 digital polarimeter (Jasco, Tokyo, Japan). NMR spectra were obtained using JEOL JNM ECS 400 MHz (JEOL, Tokyo, Japan) and Varian 600 MHz NMR spectrometers (Varian, Palo Alto, CA, USA). ESI–MS data were collected on a VG Biotech Quattro 5022 mass spectrometer (VG Biotech, Altrincham, UK). High-resolution ESI–MS data were obtained with a Bruker APEX II spectrometer (Bruker, Bremen, Germany). Circular dichroism spectra were recorded on a JASCO J-810 spectrophotometer (Jasco, Tokyo, Japan).

3.2. Plant Material

The plant samples of *C. hypoleucophylla* were collected in Kaohsiung city, Taiwan, in May 2018. The plant material was identified by one of the authors, Dr. Ming-Hong Yen. A voucher sample (specimen code: CH001) was deposited at the Graduate Institute of Natural Products, College of Pharmacy, Kaohsiung Medical University, Kaohsiung, Taiwan.

3.3. Extraction and Isolation

Air-dried leaves and twigs of *C. hypoleucophylla* (17.0 kg) were extracted three times with 95% ethanol at room temperature for 72 h each time. The extract was evaporated under reduced pressure to get a crude extract (3.6 kg). Next, the ethanol extract of *C. hypoleucophylla* was suspended and dissolved in H_2O and then partitioned with ethyl acetate to obtain an ethyl acetate layer (118.3 g). The the ethyl acetate layer was further partitioned between hexanes and 75% MeOH to acquire hexanes and MeOH layers, respectively.

Due to the results of the cytotoxic assay, the MeOH layer (45.6 g) was selected for further isolation. At first, it was loaded on a normal-phase silica gel open column and was eluted by stepwise hexanes with ethyl acetate (1:0~0:1) followed by stepwise ethyl acetate with methanol (1:0~0:1) to obtain seven subfractions (CH1~7), according to TLC analysis. The third sub-fraction, CH3, was isolated on Sephadex LH-20 and eluted with MeOH to afford four subfractions (CH3-1–4). Then, repeated column chromatography isolation on CH3-3 yielded CH3-3-1–5 fractions. CH3-3-2 (500.3 mg) was separated by silica gel CC (dichloromethane/MeOH, 100:1→0:1) to afford more subfractions (CH3-3-2-1–6). Fr. CH3-3-2-6 was purified by normal-phase HPLC using a Phenomenex Luna-CN column (hexane/dichloromethane/methanol, 30:10:1, 1.5 mL/min) to give compounds **2** (33.1mg), **3** (7.2 mg), **4** (62.7 mg), and 5 (26.7 mg). Fr. CH3-3-2-4 was isolated by reverse-phase HPLC using a CN column and gave compounds **1** (1.9 mg) and 7 (0.7 mg). Fr. CH3-3-3 was subjected to silica gel CC (CH_2Cl_2/MeOH, 1:0→0:1) followed by NP-CN HPLC and elution with (hexane/dichloromethane/methanol, 40:10:1, 2.0 mL/min) to obtain compound **8** (9.8 mg). In addition, Fr. CH3-2 was separated by normal-phase silica gel CC with hexane/dichloromethane/methanol (100:40:1→0:0:1) to afford Frs. CH3-2-1–5. Fr. CH3-2-5 was purified by silica gel CC (CH_2Cl_2/MeOH, 1:0→0:1) followed by RP-phenyl-hexyl HPLC (methanol/H_2O, 65/35, 2.0 mL/min) to give compounds **6** (2.5 mg) and **9** (7.0 mg).

3.4. Spectroscopic Data

Callihypolin A (**1**) yellow oily, $[\alpha]_D^{26}$ −1.0° (*c* 0.05, MeOH); IR (neat) ν_{max} 3452, 2956, 1768, 1682, 1376, 1342, 1202, 1141, 1018 cm^{-1}; ^{1}H-NMR and ^{13}C-NMR ($CDCl_3$, 600/150 MHz) see Table 1; HRESIMS *m/z* 399.17785 (calcd for $C_{21}H_{28}O_6Na$, 399.17781).

Callihypolin B (**2**) yellow oily, $[\alpha]_D^{26}$ −47.6° (*c* 0.05, MeOH); IR (neat) ν_{max} 3451, 2930, 1739, 1678, 1450, 1253, 1072 cm^{-1}; ^{1}H-NMR and ^{13}C-NMR ($CDCl_3$, 400/100 MHz) see Table 1; HRESIMS *m/z* 399.21419 (calcd for $C_{22}H_{32}O_5Na$, 399.21420).

3.5. Superoxide Anion Generation and Elastase Release Assays by Human Neutrophils

Human neutrophils were obtained from the venous blood of healthy adult volunteers (20–30 years old), following a reported procedure [37]. Superoxide anion generation by fMLF (0.1 μM)/CB (1μM)-activated neutrophils was evaluated based on the reduction of ferricytochrome c, as previously described [37,40]. Elastase release by the fMLF (0.1 μM)/CB (0.5 μM)-activated neutrophils was determined using N-methoxysuccinyl-Ala-Ala-Pro-Val-p-nitroanilide as the elastase substrate, according to a previous protocol [37,40]. The concentration was 10 μM for compounds **1–9**. Genistein was used as a positive control.

3.6. Cytotoxicity Test

A lactate dehydrogenase (LDH) assay kit (Promega, Madison, WI, USA) was utilized to evaluate the cytotoxicity of the samples in human neutrophils. Human neutrophils were treated with DMSO or compounds **1–9** for 15 min at 37 °C. Cell-free supernatants were collected, and the amount of LDH was evaluated [37].

4. Conclusions

The first phytochemical investigation of the leaves and twigs of the Taiwanese endemic plant *Callicarpa hypoleucophylla* has resulted in the isolation of nine clerodane-type diterpenoids, compounds **1–9**, including two new compounds designated callihypolins A and B (compounds **1** and **2**). All isolates from *C. hypoleucophylla* possess a TC *ent*-clerodane skeleton, which is different from that of the phyllocladane and labdane diterpenoids that were identified as major components of the other well-studied species *Callicarpa macrophylla* Vahl, which is recorded in the Pharmacopoeia of the People's Republic of China. These results reflect the unique properties of *C. hypoleucophylla* from the perspective of chemotaxonomy. Moreover, the anti-inflammatory activity of the isolated compounds highlights the potential of clerodane-type diterpenoids for further pharmaceutic development.

Supplementary Materials: The NMR spectra of compounds **1** and **2** are available online.

Author Contributions: Y.-B.C., C.-Y.C. and C.-H.W. conceived and designed the experiments; Y.-C.L. and J.-J.L. carried out the plant extraction and isolation of the compounds; T.-L.H., S.-Y.F., and M.K. conducted the biological studies; M.-H.Y. collected and identified the material; C.-Y.C., Y.-S.L., and T.-Y.W. assisted with the interpretation of various data; Y.-C.L., S.-R.C., Y.-B.C. contributed to the writing of the manuscript. All authors have read and agreed to the published version of the manuscript.

Funding: This work was funded by grants from the Ministry of Science and Technology of Taiwan (MOST-107-2628-B-037-001 and MOST-108-2320-B-037-013-MY3 awarded to Prof. Yuan-Bin Cheng; MOST 106-2320-B-255-003-MY3 and MOST 108-2320-B-255-003-MY3 awarded to Prof. Tsong-Long Hwang). This work was also supported by grants from Zuoying Branch of Kaohsiung Armed Forces General Hospital (KAFGH-ZY-A-109030).

Conflicts of Interest: The authors declare no conflict of interest.

References

1. Tu, Y.; Sun, L.; Guo, M.; Chen, W. The medicinal uses of *Callicarpa*, L. in traditional Chinese medicine: An ethnopharmacological, phytochemical and pharmacological review. *J. Ethnopharmacol.* **2013**, *146*, 465–481. [CrossRef] [PubMed]
2. Kawazu, K.; Inaba, M.; Mitsui, T. Studies on fish-killing components of *Callicarpa candicans*. *Agr. Biol. Chem.* **1967**, *31*, 494–497.
3. Cantrell, C.L.; Klun, J.A.; Bryson, C.T.; Kobaisy, M.; Duke, S.O. Isolation and identification of mosquito bite deterrent terpenoids from leaves of American (*Callicarpa americana*) and Japanese (*Callicarpa japonica*) beautyberry. *J. Agric. Food Chem.* **2005**, *53*, 5948–5953. [CrossRef] [PubMed]
4. Zhang, L.; Dong, L.; Huang, J.; Liu, M.; Li, G.; Zhang, C.; Zhang, K.; Wang, J. 3,4-*seco*-Labdane diterpenoids from the leaves of *Callicarpa nudiflora* and their inhibitory effects on nitric oxide production. *Fitoterapia* **2013**, *89*, 218–223. [CrossRef]

5. Dong, L.; Zhang, L.; Zhang, X.; Liu, M.; Wang, J.; Wang, Y. Two new 3,4-*seco*-labdane diterpenoids from *Callicarpa nudiflora* and their inhibitory activities against nitric oxide production. *Phytochem. Lett.* **2014**, *10*, 127–131. [CrossRef]
6. Cheng, H.H.; Cheng, Y.B.; Hwang, T.L.; Kuo, Y.H.; Chen, C.H.; Shen, Y.C. Randainins A–D, based on unique diterpenoid architectures, from *Callicarpa randaiensis*. *J. Nat. Prod.* **2015**, *78*, 1823–1828. [CrossRef]
7. Zhou, Z.; Wei, X.; Fu, H.; Luo, Y. Chemical constituents of *Callicarpa nudiflora* and their anti-platelet aggregation activity. *Fitoterapia* **2013**, *88*, 91–95. [CrossRef]
8. Wu, A.Z.; Zhai, Y.J.; Zhao, Z.X.; Zhang, C.X.; Lin, C.Z.; Zhu, C.C. Phenylethanoid glycosides from the stems of *Callicarpa peii* (hemostatic drug). *Fitoterapia* **2013**, *84*, 237–241. [CrossRef]
9. Luo, Y.H.; Zhou, Z.Q.; Ma, S.C.; Fu, H.Z. Three new antioxidant furofuran lignans from *Callicarpa nudiflora*. *Phytochem. Lett.* **2014**, *7*, 194–197. [CrossRef]
10. Cai, H.; Xie, Z.; Liu, G.; Sun, X.; Peng, G.; Lin, B.; Liao, Q. Isolation, identification and activities of natural antioxidants from *Callicarpa kwangtungensis* Chun. *PLoS ONE* **2014**, *9*, e93000. [CrossRef]
11. Jones, W.P.; Lobo-Echeverri, T.; Mi, Q.; Chai, H.B.; Soejarto, D.D.; Cordell, G.A.; Swanson, S.M.; Kinghorn, A.D. Cytotoxic constituents from the fruiting branches of *Callicarpa Americana* collected in southern Florida. *J. Nat. Prod.* **2007**, *70*, 372–377. [CrossRef] [PubMed]
12. Mei, W.L.; Han, Z.; Cui, H.B.; Zhao, Y.X.; Deng, Y.Y.; Dai, H.F. A new cytotoxic iridoid from *Callicarpa nudiflora*. *Nat. Prod. Res.* **2010**, *24*, 899–904. [CrossRef] [PubMed]
13. Xu, J.; Sun, Y.; Wang, M.; Ren, Q.; Li, S.; Wang, H.; Sun, X.; Jin, D.Q.; Sun, H.; Ohizumi, Y.; et al. Bioactive diterpenoids from the leaves of *Callicarpa macrophylla*. *J. Nat. Prod.* **2015**, *78*, 1563–1569. [CrossRef] [PubMed]
14. Chen, J.J.; Wu, H.M.; Peng, C.F.; Chen, I.S.; Chu, S.D. *seco*-Abietane diterpenoids, a phenylethanoid derivative, and antitubercular constituents from *Callicarpa pilosissima*. *J. Nat. Prod.* **2009**, *72*, 223–228. [CrossRef] [PubMed]
15. Huang, B.; Fu, H.Z.; Chen, W.K.; Luo, Y.H.; Ma, S.C. Hepatoprotective triterpenoid saponins from *Callicarpa nudiflora*. *Chem. Pharm. Bull.* **2014**, *62*, 695–699. [CrossRef] [PubMed]
16. Luo, Y.H.; Fu, H.Z.; Huang, B.; Chen, W.K.; Ma, S.C. Hepatoprotective iridoid glucosides from *Callicarpa nudiflora*. *J. Asian. Nat. Prod. Res.* **2016**, *18*, 274–279. [CrossRef]
17. Chung, P.Y.; Chung, L.Y.; Navaratnam, P. Potential targets by pentacyclic triterpenoids from *Callicarpa farinosa* against methicillin-resistant and sensitive. *Staphylococcus aureus*. *Fitoterapia* **2014**, *94*, 48–54. [CrossRef]
18. Gupta, S.K.; Gupta, A.; Gupta, A.K.; Pakash, D.V. *In vitro* anti-arthritic activity of ethanolic extract of *Callicarpa Macrophylla* flower. *Int. Res. J. Pharm.* **2013**, *4*, 160–162. [CrossRef]
19. Yadav, V.; Jayalakshmi, S.; Singla, R.K.; Patra, A.; Khan, S. Assessment of anti-inflammatory and analgesic activities of *Callicarpa macrophylla* Vahl. roots extracts. *Webmed Cent. Pharmacol.* **2012**, *3*, WMC003366.
20. Ahmad, V.U.; Farooq, U.; Abbaskhan, A.; Hussain, J.; Abbasi, M.A.; Nawaz, S.A.; Choudhary, M.I. Four new diterpenoids from *Ballota limbata*. *Helv. Chim. Acta.* **2004**, *87*, 682–689. [CrossRef]
21. Pinto, M.E.F.; Silva, M.S.D.; Schindler, E.; Filho, J.M.B.; El-Bachá, R.D.S.; Castello-Branco, M.V.S.; Agra, M.D.F.; Tavares, J.F. 3′,8″-Biisokaempferide, a cytotoxic biflavonoid and other chemical constituents of *Nanuza plicata* (Velloziaceae). *J. Braz. Chem. Soc.* **2010**, *21*, 1819–1824. [CrossRef]
22. Farooq, U.; Khan, A.; Ahmad, V.U.; Kousar, F.; Iqbal, S. Limbatolide F and G: Two new *trans*-clerodane diterpenoids from *Otostegia limbata*. *Pol. J. Chem.* **2005**, *79*, 1757–1762. [CrossRef]
23. Ahmad, V.U.; Khan, A.; Farooq, U.; Kousar, F.; Khan, S.S.; Nawaz, S.A.; Abbasi, M.A.; Choudhary, M.I. Three new cholinesterase-inhibiting *cis*-clerodane diterpenoids from *Otostegia limbata*. *Chem. Pharm. Bull.* **2005**, *53*, 378–381. [CrossRef] [PubMed]
24. Iqbal Choudhary, M.; Mohammad, M.Y.; Musharraf, S.G.; Onajobi, I.; Mohammad, A.; Anis, I.; Shah, M.R.; Atta-Ur-Rahman. Biotransformation of clerodane diterpenoids by *Rhizopus stolonifer* and antibacterial activity of resulting metabolites. *Phytochemistry* **2013**, *90*, 56–61. [CrossRef] [PubMed]
25. Raha, P.; Das, A.K.; Adityachaudhuri, N.; Majumder, P.L. Cleroinermin, *aneo*-clerodane diterpenoid from *Clerodendron inermi*. *Phytochemistry* **1991**, *30*, 3812–3814. [CrossRef]
26. Huang, Z.; Jiang, M.Y.; Zhou, Z.Y.; Xu, D. Two new clerodane diterpenes from *Dodonaea viscosa*. *Z. Naturforsch.* **2010**, *65b*, 83–86. [CrossRef]
27. Song, B.; Ding, G.; Tian, X.H.; Li, L.; Zhou, C.; Zhang, Q.B.; Wang, M.H.; Zhang, T.; Zou, Z.M. Anti-HIV-1 integrase diterpenoids from *Dichrocephala benthamii*. *Phytochem. Lett.* **2015**, *14*, 249–253. [CrossRef]

28. Rustaiyan, A.; Simozar, E.; Ahmadi, A.; Grenz, M.; Bohlmann, F. A hardwickiic acid derivative from *Pulicaria gnaphalodes*. *Phytochemistry* **1981**, *20*, 2772–2773. [CrossRef]
29. Heymann, H.; Tezuka, Y.; Kikuchi, T.; Supriyatna, S. Constituents of *Sindora sumatrana* M_{IQ}. III. new *trans*-clerodane diterpenoids from the dried pods. *Chem. Pharm. Bull.* **1994**, *42*, 1202–1207. [CrossRef]
30. García, A.; Ramírez-Apan, T.; Cogordan, J.A.; Delgado, G. Absolute configuration assignments by experimental and theoretical approaches of *ent*-labdane- and *cis-ent*-clerodane-type diterpenes isolated from *Croton glabellus*. *Can. J. Chem.* **2006**, *84*, 1593–1602. [CrossRef]
31. Chang, F.R.; Huang, S.T.; Liaw, C.C.; Yen, M.H.; Hwang, T.L.; Chen, C.Y.; Hou, M.F.; Yuan, S.S.; Cheng, Y.B.; Wu, Y.C. Diterpenes from *Grangea maderaspatana*. *Phytochemistry* **2016**, *131*, 124–129. [CrossRef] [PubMed]
32. Calderón, C.; De Ford, C.; Castro, V.; Merfort, I.; Murillo, R. Cytotoxic clerodane diterpenes from *Zuelania guidonia*. *J. Nat. Prod.* **2014**, *77*, 455–463. [CrossRef] [PubMed]
33. Oberlies, N.H.; Burgess, J.P.; Navarro, H.A.; Pinos, R.E.; Fairchild, C.R.; Peterson, R.W.; Soejarto, D.D.; Farnsworth, N.R.; Kinghorn, A.D.; Wani, M.C.; et al. Novel bioactive clerodane diterpenoids from the leaves and twigs of *Casearia sylvestris*. *J. Nat. Prod.* **2002**, *65*, 95–99. [CrossRef] [PubMed]
34. Tokoroyama, T. Synthesis of clerodane diterpenoids and related compounds-stereoselective construction of the decalin skeleton with multiple contiguous stereogenic centers. *Synthesis* **2000**, *5*, 611–633. [CrossRef]
35. Wu, T.H.; Cheng, Y.Y.; Liou, J.R.; Way, T.D.; Chen, C.J.; Chen, Y.H.; Kuo, S.C.; El-Shazly, M.; Chang, F.R.; Wu, Y.C.; et al. Clerodane diterpenes from *Polyalthia longifolia* var. *pendula* protect SK-N-MC human neuroblastoma cells from β-amyloid insult. *RSC Adv.* **2014**, *4*, 23707–23712. [CrossRef]
36. Itokawa, H.; Morita, H.; Katou, I.; Takeya, K.; Cavalheiro, A.J.; de Oliveira, R.C.; Ishige, M.; Motidome, M. Cytotoxic diterpenes from the rhizomes of *Hedychium coronarium*. *Planta Med.* **1988**, *54*, 311–315. [CrossRef]
37. Liu, F.C.; Yu, H.P.; Chen, P.J.; Yang, H.W.; Chang, S.H.; Tzeng, C.C.; Cheng, W.J.; Chen, Y.R.; Chen, Y.L.; Hwang, T.L. A novel NOX2 inhibitor attenuates human neutrophil oxidative stress and ameliorates inflammatory arthritis in mice. *Redox Biol.* **2019**, *26*, 101273. [CrossRef]
38. Chang, F.R.; Hwang, T.L.; Yang, Y.L.; Li, C.E.; Wu, C.C.; Issa, H.H.; Hsieh, W.B.; Wu, Y.C. Anti-inflammatory and cytotoxic diterpenes from formosan *Polyalthia longifolia* var. pendula. *Planta Med.* **2006**, *72*, 1344–1347. [CrossRef]
39. Chang, H.L.; Chang, F.R.; Chen, J.S.; Wang, H.P.; Wu, Y.H.; Wang, C.C.; Wu, Y.C.; Hwang, T.L. Inhibitory effects of 16-hydroxycleroda-3,13(14)*E*-dien-15-oic acid on superoxide anion and elastase release in human neutrophils through multiple mechanisms. *Eur. J. Pharmacol.* **2008**, *31*, 332–339. [CrossRef]
40. Yang, S.C.; Chung, P.J.; Ho, C.M.; Kuo, C.Y.; Hung, M.F.; Huang, Y.T.; Chang, W.Y.; Chang, Y.W.; Chan, K.H.; Hwang, T.L. Propofol inhibits superoxide production, elastase release, and chemotaxis in formyl peptide-activated human neutrophils by blocking formyl peptide receptor 1. *J. Immunol.* **2013**, *190*, 6511–6519. [CrossRef]

Sample Availability: Samples of the compounds are available from the authors.

molecules

Article

Determination of *Alternaria* Toxins in Sunflower Oil by Liquid Chromatography Isotope Dilution Tandem Mass Spectrometry

Ádám Tölgyesi [1,*], Luca Kozma [1] and Virender K. Sharma [2,*]

1 KERMI Department, ÉMI-TÜV SÜD Ltd., Dugonics utca 11, 1043 Budapest, Hungary; kozma.luca@emi-tuv.hu

2 Program for the Environment and Sustainability, Department of Environmental and Occupational Health, School of Public Health, Texas A&M University, 212 Adriance Lab Rd., 1266 TAMU, College Station, TX 77843, USA

* Correspondence: tolgyesi.adam@emi-tuv.hu (Á.T.); vsharma@tamu.edu (V.K.S.)

Academic Editor: Ping-Chung Kuo

Received: 27 February 2020; Accepted: 3 April 2020; Published: 7 April 2020

Abstract: *Alternaria* toxins have gained attention as a potential health risk and can be classified as emerging mycotoxins. As a result, they are candidates to be regulated by the European Commission. This paper describes a liquid chromatography tandem mass spectrometric (LC-MS/MS) method for analyzing five *Alternaria* toxins in sunflower oil, which is a rather different type of sample to those matrices investigated in earlier published papers. An optimal sample preparation condition was achieved when samples were dissolved in *n*-hexane and extracted with methanol/water mixture, followed by sample pre-concentration with solvent evaporation. This study is the first focusing only on this lipophilic matrix and in using all corresponding isotopically labeled internal standards (ISTD) to compensate the matrix effect that strongly influences the LC-MS/MS analysis of toxins. Target compounds were separated on Zorbax Extend C-18 column enabling the analysis at alkaline pH of 8.8 that was necessary to obtain appropriate peak shape of tenuazonic acid and to separate the analytes at baseline. The method was validated according to the EU 2002/657/EC Decision and all the analytical performance characteristics met the requirements. The recovery was between 74% and 122% in fortified sunflower oil samples and the precision varied from 9% to 22%. The method was successfully demonstrated for sunflower seed quality check (QC) samples. Finally, 16 different sunflower oil samples were measured; and tenuazonic acid and tentoxin toxins were detected at levels close to LOQ concentrations.

Keywords: *Alternaria* toxins; LC-MS/MS; isotope dilution; sunflower oil; validation; real sample analysis

1. Introduction

Clear evidence of animal and human illness and death caused by fungal metabolites have been reported worldwide since the 1970s. The secondary metabolites of fungi growing on agricultural commodities, called mycotoxins, are still considered as a major health concern [1]. Analytical methods, therefore, have been developed and subsequently validated to determine mycotoxins in different food and feed samples [2]. These methods are used in the monitoring laboratories to screen and confirm the samples that are contaminated with mycotoxins. In the European Union (EU), the maximum levels (ML) of regulated mycotoxins in food are in force [3]. For some mycotoxins without any ML, the European Food Safety Authority (EFSA) has published opinions, often indicating the need of harmonized and suitable methods to make sound exposure assessments. According to the EFSA, cereals, vegetables, and oilseeds frequently contain *Alternaria* mycotoxins, which could cause mutagenic, genotoxic, fetotoxic, and teratogenic effects [4]. These agricultural commodities are mostly infected by *Alternaria* species (e.g., *Alternaria alternata*) that produce more than 70 secondary metabolites from which the five most important ones are tenuazonic acid (TEA), altenuene (ALT), alternariol (AOH), tentoxin (TEN), and alternariol monomethyl ether (AME) (Figure 1) [4]. Currently, these *Alternaria* toxins are not regulated because there has not been enough information available to establish ML (i.e., risk assessment). Additionally, the development of a standard method is imperative, which is not available yet [4]. Hence, a few control laboratories have analyzed them on a regular basis using LC-MS/MS technique, which is the most suitable technique according to the EFSA [4]. However, interlaboratory comparisons (ILC) have already been organized for these toxins in order to support the legislation [5,6].

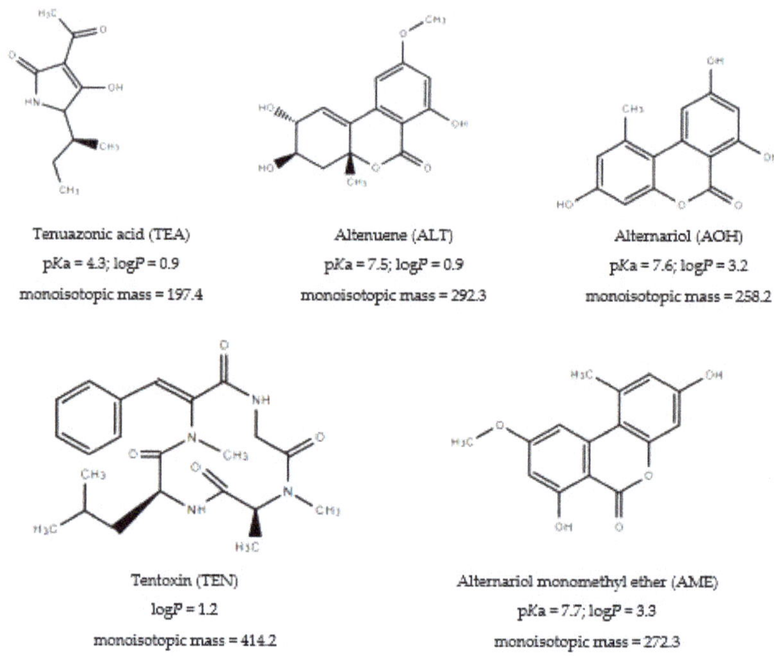

Figure 1. Structure and physical-chemical properties of five toxins analyzed in this study.

The first ILC was a proficiency test (PT) for *Alternaria* toxins in tomato juice, organized by the Federal Institute for Risk Assessment (BfR, Berlin, Germany) in 2014 [5]. Afterwards, the European Commission Joint Research Centre, EU Reference Laboratory for Mycotoxins (JRC, Geel, Belgium), performed the method validation study (MVS) to find a candidate LC-MS/MS method as a possible basis for drafting a standard method for *Alternaria* toxins in 2015 [6]. The MVS was started with a pre-trial and included tomato juice as test samples, followed by the final trial with tomato juice, cereals and sunflower seed samples. The reproducibility of the candidate method did not fulfill the requirements of the European Committee for Standardization (CEN). Due to lack of isotopically labeled internal standards (ISTDs), the candidate method showed low interlaboratory precision of some compounds (i.e., TEA and AME) [6]. In 2018, the ISTDs for five *Alternarias* mentioned above became commercially available. Therefore, the MVS was repeated by utilizing the LC-MS/MS determination with isotope dilution (LC-ID-MS/MS) and the obtained results met the requirements [7]. This MVS showed that the application of isotope dilution is critically important for analyzing *Alternaria* toxins in food samples using LC-MS/MS method.

Furthermore, Liu and Rychlik [8] published the advantage of using isotopically labeled TEN and its derivates for quantification of toxins in various food samples. They reported the synthesis and application of TEN-d3, DH-TEN-d3 (dihydrotentoxin-d3), and isoTEN-d3 (isotentoxin-d3) for quantifying native toxins in cereal-, vegetable-, and fruit-based samples and included different types of oils as well [8]. In another research performed by Liu and Rychlik, the biosynthesis of ^{13}C-labeled ISTDs for seven *Alternaria* toxins was described [9]. The application of AOH-^{13}C14, ALT-^{13}C15, and AME-^{13}C15 in the future can further enhance the quantification of toxins in food since the ^{13}C-labeled ISTDs have advantages over the deuterated ISTDs. Namely, there is no retention time difference between the native target compound and its corresponding ^{13}C-labeled analogue. This enables the total compensation of matrix effect (ME) during LC-MS/MS analysis. Furthermore, there will not be substantial overlap between ISTD signals and the isotopic signals of analyte if the molecular mass of the isotopologue is more than 5 mass units. However, the authors reported the isotope effect between AME and AME-^{13}C15 when acetonitrile/2-propanol mixture was used as the organic modifier in the eluent [9]. Therefore, only the deuterated AOH and AME are commercially available so far.

This paper describes the use of a LC-ID-MS/MS method for analyzing five toxins mentioned above in sunflower oil samples for the first time. Even though the high contamination of sunflower seeds with *Alternaria* toxins (TEA: LOQ—5400 µg/kg; AOH: LOQ—1200 µg/kg; TEN: LOQ - 880 µg/kg; AME: LOQ—440 µg/kg) have been recently reported worldwide [4,6,10–12], the existing methods (Table S1) involved mainly the vegetable-, cereal-, fruit-based, and oilseed samples [8–36]. The oil samples got less attention, so far only three studies have included the analysis of this matrix [8,11,12]. The reason for excluding this sample matrix could be the different sample manipulations needed for this lipophilic sample. Thus, this matrix is now in the focus of the current study. The aims of the work presented here were to: (i) set up LC-ID-MS/MS separation process for five *Alternaria* toxins without chemical derivatization; (ii) develop a sample preparation approach, which is suitable for sunflower oil; (iii) fine-tune the LC-ID-MS/MS method to achieve the quantification limit as low as possible; (iv) perform inhouse validation of the method to meet the requirements set by EU; and (v) apply the method for real samples and also naturally contaminated and spiked sunflower seed QC samples.

2. Results

2.1. General Conditions for LC-ID-MS/MS Separation

The fine-tuning of ion transitions in the MS/MS instrument was carried out with individual standard solutions (2 µg/mL) in methanol and employing electrospray (ESI) source in negative ion mode according to our previous paper [13]. The isotopically labeled ISTDs were also tuned using the ISTD mixture (Section 4.1). The two most intense ion transitions of target compounds used for the MS/MS detection are detailed in Table 1. The ALT, AOH, and AME toxins are weak acidic molecules (Figure 1) and show an appropriate retention on C-18 HPLC columns and high sensitivity during MS or MS/MS detection [9–36]. TEN can be considered as a neutral molecule but can be also measured by LC-MS/MS method with negative ionization and fit-for-purpose sensitivity [8,10–13]. In the case of TEA, however, special HPLC conditions are necessary due to its different isomer forms appearing in aqueous phase at acidic pH [13,19,20]. Therefore, chemical derivatization with 2,4-dinitrophenylhidrazine was introduced for TEA that makes it a suitable compound for HPLC analysis [13,19,20]. The drawback of the derivatization approach is the longer sample preparation time, lower selectivity, and increased noise of analysis [13]. Recently, HPLC separations at alkaline pH conditions on C-18 column were reported for TEA separation in food matrices [6,21,22,30,31,33]. A pH above 8.0 results in reproducible retention time and peak shape for TEA but decreases its retention time due to the deprotonated hydrophilic form of TEA at alkaline pH (Figure 1).

Table 1. MS/MS detection parameters for *Alternaria* toxins detected in APCI and ESI ionization modes employing negative or positive ion mode. The quantifier ion transition is highlighted with bold.

Compounds	Ionization Mode	Precursor Ion (*m/z*)	Product Ion (*m/z*)	Dwell Time (ms)	Declustering Potential (V)	Entrance Potential (V)	Cell Exit Potential (V)	Collision Energy (V)	Collision Cell Exit Potential (V)
TEA	negative	196	83	50	−70	−9	−12	−32	0
			139	50				−26	0
TEA-^{13}C2		198	**141**	50	−70	−9	−12	−26	0
ALT		291	161	50	−80	−10	−22	−52	0
			203	50				−40	0
ALT-d6		296	**203**	50	−80	−10	−22	−40	0
AOH		257	**147**	50	−65	−6	−22	−46	0
			213	50				−30	−5
AOH-d3		260	**218**	50	−65	−6	−22	−30	−5
TEN		413	141	50	−80	−5	−14	−28	0
			271	50				−20	−2
TEN-d3		416	**274**	50	−80	−5	−14	−20	−2
AME		271	228	50	−60	−2	−16	−36	−2
			256	50				−30	−2
AME-d3		274	**259**	50	−60	−2	−16	−30	−2
TEA	positive	198	139	50	66	10	12	19	4
			153	50				17	4
TEA-^{13}C2		200	**155**	50	66	10	12	17	4
ALT		293	139	50	61	12	16	79	4
			257	50				19	4
ALT-d6		299	**262**	50	61	12	16	19	4
AOH		259	**128**	50	116	9	14	57	4
			185	50				14	4
AOH-d3		262	**131**	50	116	9	14	57	4
TEN		415	119	50	91	8	20	23	4
			132	50				53	4
TEN-d3		418	**135**	50	91	8	20	53	4
AME		273	115	50	126	9	16	69	4
			128	50				71	4
AME-d3		276	**131**	50	126	9	16	71	4

This condition, however, does not require derivatization, hence, it was tested in the present study with a HPLC column suitable for separation at above pH 8.0. A Zorbax Extend C-18 column allows separation at pH up to 11.5. The alkaline condition (Section 4.7) resulted in baseline separation for the five toxins and appropriate retention for TEA on this column (Figure 2). The pH of the mobile phase was tested between 8.0 and 9.0. Retention time shift and difference in sensitivity were not observed. The apparent retention factor (k′) calculated for TEA was higher than 2.0 under all conditions. However, the ESI source did not result in enough sensitivity for the analysis, and the instrumental limit of quantification (LOQ) was not lower than 50 ng/mL, but the aim was to detect AME and AOH below 10 ng/mL [6]. Consequently, the ESI positive ionization mode was also tested (Table 1) under acidic pH condition (Section 4.7), but a better sensitivity could not be achieved. Moreover, the peak shape of TEA was irreproducible under acidic condition using this HPLC column mentioned above. Therefore, the atmospheric pressure chemical ionization (APCI) source was also tested in negative ion mode and under alkaline HPLC condition. It should be pointed out that the same ion transitions were used for performing the detections in both ESI and APCI modes. Only different polarities (positive or negative) resulted in various ion traces. This instrument gave increased sensitivity for these toxins with APCI source using negative ionization. The instrumental LOQ could be lowered at least with one order of magnitude for all compounds in comparison to those values obtained with ESI probe. The best LC-MS/MS conditions were obtained using alkaline pH condition for HPLC separation at pH 8.8 and employing APCI source with negative ionization mode (Figure 2).

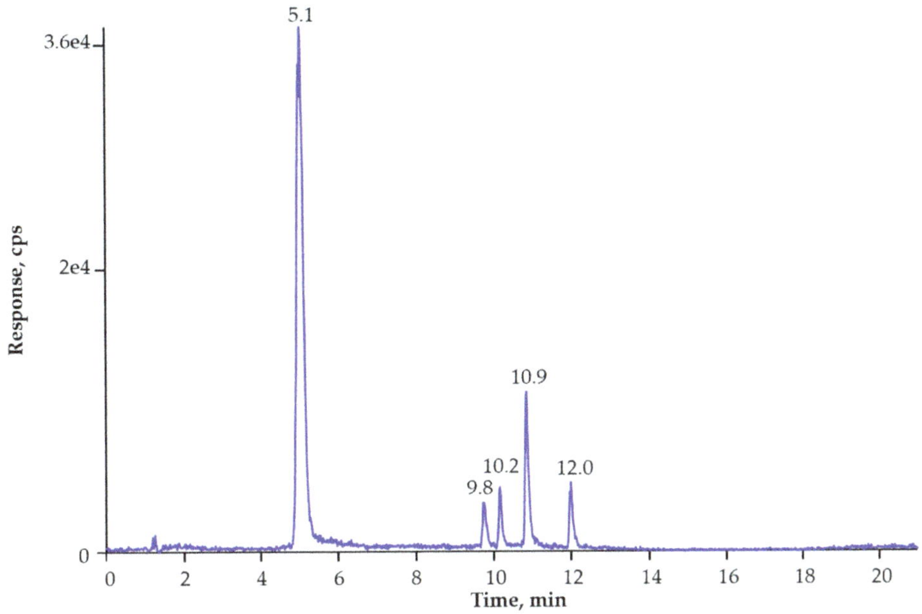

Figure 2. Total ion current chromatogram of five toxins at 10 µg/kg using LC-APCI(-)-MS/MS separation at pH 8.8. Compounds: TEA (5.1 min); ALT (9.8 min); AOH (10.2 min); TEN (10.9 min); and AME (12.0 min). The concentrations of ISTDs were: TEA-13C2 (83 µg/kg), ALT-d6 (33 µg/kg), AOH-d3 (17 µg/kg), TEN-d3 (17 µg/kg), and AME-d3 (17 µg/kg).

2.2. *Development of Sample Preparation*

2.2.1. Sample Preparation without SPE Clean-Up

Contrary to the vegetable-, fruit- or cereal-based food samples investigated frequently for *Alternarias* earlier (Table S1), the sunflower oil is a very lipophilic matrix and needs a unique sample preparation approach. Therefore, an experimental design was carried out to obtain the appropriate accuracy in different types of sunflower oil samples. A central composition design (CCD) has been planned with the statistical software R, version 3.0.2 for Windows. Two grams of the sample was used for sample extraction with methanol/water mixture; and *n*-hexane was applied for the elimination of lipophilic matrix constituents. This sample weight and these solvents have been found suitable for the *Alternaria* analysis (Table S1). The factors and levels were the following: (I) sample-to-hexane ratio: 1.0, 1.5, or 2.0; (II) methanol content in the extraction medium: 70%, 80%, or 90%; and (III) sample-to-extraction solvent ratio: 2.0, 4.0, or 6.0. A naturally contaminated sunflower oil was used for the experimental design that contained TEA (7.1 µg/kg) and TEN (12.8 µg/kg). An oil sample containing the toxins in much higher concentrations than LOQ would have been better for the CCD, but a sample with greater natural contamination could not be found. An optimal condition was achieved with the sample-to-hexane ratio of 1.0; 80% (*v*/*v*) methanol for extraction, and sample-to-extraction solvent ratio of 4.0 (Figure 3).

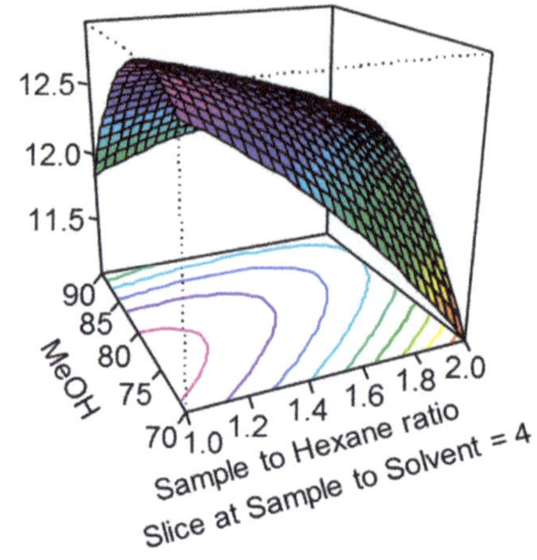

Figure 3. Response surface of TEN: slice at sample to extraction solvent ratio of 4.0.

In this case, the sample dilution was 4-fold that increased the LOQ value, and AOH could not detected below 10 μg/kg with a signal-to-noise ratio (SNR) higher than 10. Therefore, an aliquot of methanolic extract (6 mL, equal to 1.5 g sample) was evaporated, and the final sample volume was adjusted to 0.5 mL with water that ended up with 3-fold sample pre-concentration (Section 4.2. and Section 4.3.). In this case, all compounds were detected with appropriate SNR at the desired levels. The other aim of sample evaporation was to lower the methanol content of the injection solution, and consequently, the deformation of TEA peak on the chromatogram could be avoided.

One aim of sample preparation is to reduce the ME of LC-MS/MS analysis [37]. ME is caused by the co-eluting matrix constituents and strongly influences the quantification [37]. The ME was studied with the optimal sample preparation conditions and was evaluated using the general approach [37]. Three matrix-matched calibrations were prepared from blank samples (i.e., three different sunflower oils). Blank extracts were spiked with standard mixture: the fortification levels were 10, 20, 30, 40, and 50 μg/kg for TEA, ALT, and TEN; and were 5, 10, 15, 20, and 25 μg/kg for AOH and AME. Low concentration levels were set due to the naturally low contamination of oils with these toxins reported in previous studies (see Table S1). AOH and AME are considered more toxic [4,6], therefore, twice lower levels were set for these two compounds. Calibrants in neat (matrix-free) solvent were also prepared and analyzed. The slopes of matrix-matched calibrations were compared to the slope of neat calibration in order to calculate the absolute ME. ME (%) = (slope in matrix-matched calibration/slope in neat calibration-1) × 100. ME < 0% means ion suppression, and ME > 0% indicates ion enhancement. The relative standard deviation (RSD%) of slopes obtained from the matrix-matched calibrations ($n = 3$) was calculated and evaluated as the relative ME [37]. Therefore, the relative ME means the precision of slopes in different matrix-matched calibrations.

The results obtained without ISTD correction indicated that the high matrix suppression influences the signal of AOH (20–48% ion suppression) and AME (75–88% ion suppression) (Table 2). For TEA, ALT, and TEN, a moderate ME could be seen. The relative ME was also considerable for AOH and AME (22–42%). The ME, however, could be compensated with isotope dilution (Table 2). The calibration evaluated by the ISTD method showed that the ME is greatly compensated, mainly for those two compounds (i.e., AOH and AME) that are considerably influenced by the co-eluting matrix constituents. The relative ME was also improved with ISTD correction. The high ion suppression for AOH and AME, however, indicated that considerable losses of the analytes occurred in the ion source. The reason for high ME is the remaining impurities (i.e., phospholipids) after extraction. Therefore, the SPE clean-up was tested for reducing the number and concentration of matrix constituents, which may lead to lower ME.

Table 2. The matrix effect (ME%) and relative matrix effect (RSD% of slopes) evaluated under different sample preparation and evaluation conditions. ME% < 0 means ion suppression, and ME% > 0 means ion enhancement.

	TEA	ALT	AOH	TEN	AME
	Preparation without SPE clean-up and evaluation without ISTD correction				
ME% (sample 1)	8	−7	−20	4	−75
ME% (sample 2)	−3	−3	−40	−13	−86
ME% (sample 3)	−5	−18	−48	−14	−88
Relative ME%	7	9	22	11	42
	Preparation without SPE clean-up and evaluation with ISTD correction				
ME% (sample 1)	11	−4	11	10	−6
ME% (sample 2)	4	−6	10	0	−11
ME% (sample 3)	5	−19	31	2	5
Relative ME%	4	9	10	5	5
	Preparation with mixed-mode SPE clean-up and evaluation without ISTD correction				
ME% (sample 1)	2	10	−45	10	−57
ME% (sample 2)	−10	−8	−57	−15	−50
ME% (sample 3)	−13	3	−46	−18	−62
Relative ME%	9	11	14	17	14
	Preparation with mixed-mode SPE clean-up and evaluation with ISTD correction				
ME% (sample 1)	8	1	2	9	6
ME% (sample 2)	−6	−7	4	−12	−3
ME% (sample 3)	−6	3	−18	−4	−8
Relative ME%	8	5	13	11	7

2.2.2. Sample Preparation with Mixed-mode SPE Clean-Up

A sample clean-up utilizing mixed-mode SPE purification was tested. An aliquot (5 mL) of methanolic sample extracts (Section 4.2.) was diluted with 1% (*v*/*v*) acetic acid in water (35 mL) to lower the methanol content of the sample solvent. Diluted extracts were subjected to SPE clean-up (Section 4.4) using mixed-mode polymeric strong cation exchange cartridges (Strata-XL-C). Under acidic condition, this cartridge could selectively adsorb the toxins and basic matrix constituents on the reversed-phase and the cation exchange part of the cartridge, respectively [13]. Hence, the basic matrix solutes could be eliminated from the samples. The ME was studied after SPE clean-up. The SPE purification did not improve the absolute ME considerably. The signals of AOH and AME were still considerably suppressed (~ 50%) in the ion source, and only a slight improvement could be seen for AME (Table 2), however, the relative ME (14–17%) was significantly enhanced. This was an advantage of SPE clean-up, but the response correction with ISTD was also necessary after SPE clean-up. Again, the ISTD dilution could well compensate the ME (Table 2).

2.2.3. Sample Preparation With Normal-Phase SPE Clean-Up

Oil samples can be easily dissolved in hexane that allow testing the normal-phase (NP) SPE clean-up (Section 4.5) with silica cartridges (Strata-Si-1). The washing solvent was a mixture of *n*-hexane and ethyl acetate, while the elution solvent was a mixture of methanol and acetonitrile. An optimization was based on an experimental design using CCD. The factors and levels were the following: (I) ethyl acetate content of the washing solvent: 10%, 20%, and 30% (*v*/*v*); (II) acetonitrile content of the elution solvent: 0%, 25%, and 50% (*v*/*v*); and (III) sample pre-concentration: 3-fold, 4-fold, and 6-fold. The same naturally contaminated sunflower oil was used for the experimental design mentioned above (12.8 μg/kg TEA and 7.1 μg/kg TEN). The results showed that there is no significant difference in concentrations obtained for TEN under different conditions. However, the NP SPE considerably lowered the accuracy of TEA. The recovery of TEA was around only 10% in all settings. The TEA could not be eluted from the silica cartridge with solvent containing only organic phase. Therefore, we added water into the elution solvent and tested the methanol/water mixture for sample elution with 10%, 20%, 30%, and 40% (*v*/*v*) water in methanol. Ten percent water in the elution solvent already resulted in ~ 70% recovery for TEA, which did not improve with a higher percentage of water. The drawback of having water in the elution solvent was that AME could not be eluted from the NP cartridge due to its lipophilic character. Hence, the NP SPE could not be used for all toxins involved.

2.3. Method Validation

The method was validated in accordance with the Commission Decision 2002/657/EC decision [38] and CEN/TR 16059:2016 guidelines [39]. The fortification levels were 10, 20, and 30 μg/kg for TEA, ALT, and TEN, respectively; and were 5, 10, and 15 for AOH and AME, respectively. These levels were set in line with the validation ranges used in the MVS [6]. Investigations at higher concentration levels were not needed because natural contamination of oils was reported at low μg/kg levels only. Six parallel samples were analyzed at each level that are in accordance with the EU guideline (Table S2). Measurements were carried out over 3 days, and all 54 samples were analyzed (3 levels × 6 samples × 3 days). The performance characteristics were as follows: selectivity, identification, linearity, recovery, precision, and limit of quantification (LOQ).

Blank samples were spiked and analyzed using the optimized method (Sections 4.2 and 4.3). The chromatograms obtained from the blank samples were free of any interfering peak. For identification, the ion ratios (IAs) were calculated for all compounds in both neat standard solutions and samples. IAs were all within the tolerance ranges for all toxins (Table S2). The selectivity and identification met the criteria of EU guidelines [38]. Five-point calibration curves were performed to evaluate the linearity. Concentration levels, determination coefficients (R^2), and equations are given in Table S2.

The requirement for recovery has been obtained between 70% and 120% at spiking levels used for validation [38,39]. The recovery varied from 73.6% to 95% at levels between 5 μg/kg and 15 μg/kg for AOH and AME. In the case of TEA, ALT, and TEN, the recovery was between 92.4% and 122% at the concentration range of 10–30 μg/kg. Only one value (122%) exceeded the acceptable ranges. Below the concentration of 100 μg/kg, the precision should be as low as possible [38], normally, RSD ≤ 30% [39]. The within-laboratory precision varied from 10.1% to 22.2% (Table S2). The LOQ was calculated from the SNR and evaluated as 10 times of SNR. The LOQ was checked by fortifying blank samples ($n = 6$) with standard solution to obtain the individual LOQ levels, and samples were analyzed. The SNR was above 10 in each sample and the IAs were in the acceptable ranges.

2.4. Analysis of Sunflower Oil Samples

Sixteen different brands and lots of sunflower oil samples were collected and analyzed for the five toxins mentioned above. Three samples were contaminated at low levels, in which only TEA and TEN were detected. One sample (cold pressed oil) contained both TEA (12.8 μg/kg) and TEN (7.1 μg/kg). The other two samples (refined oils) contained TEN at concentrations between 4.5 μg/kg and 5.0 μg/kg.

2.5. Analysis of Sunflower Seed QC Samples

In lack of sunflower oil QC samples, sunflower seed QC samples were measured. The method optimized for sunflower oil had to be modified to obtain the suitable method for sunflower seeds (Section 4.6). Both spiked (C08 SP and Q25 SP) and naturally contaminated (W52 NC) samples were tested. The samples were leftovers from MVS performed by JRC in 2018. The detected values and reference concentrations are given in Table S2. Even though the method presented herein was developed for sunflower oil samples, the concentrations detected in sunflower seed samples were not considerably different to the reference values. The method could not detect ALT at all, since the reference concentrations were all below the LOQ (10 μg/kg). Also, AOH and AME were not found in

C08 SP due to the same reason. The accuracy of the method for sunflower seed samples was between 72% and 129% (Table S2).

3. Discussion

3.1. Method Development for LC-MS/MS Separation

The MS/MS detection of *Alternaria* toxins can be carried out in both positive and negative ionization modes (Table S1). The choice of polarization mode can be instrument dependent, but the negative mode usually results in a higher sensitivity for these toxins (Table S1) due to their weak acidic characteristics. We also observed considerable enhancement in sensitivity when negative ionization was applied. In addition to the ionization mode, the choice of ion source can also influence the sensitivity of MS/MS detection of *Alternarias*. Zwickel et al. [30] tested three ion sources (ESI, APCI, and atmospheric pressure photo ionization) for these toxins and found that the ESI was the most suitable one. In general, the ESI was employed (see Table S1), but Prelle et al. [16] reported three times higher responses for TEA when APCI was used, while the rest of the toxins had similar sensitivity in both ESI and APCI modes. Even though the ESI source of the applied LC-MS/MS instrument enabled appropriate sensitivity for the compounds other than *Alternarias*, the sensitivity for *Alternaria* toxins was quite a bit lower than those reported in earlier methods utilizing other types of instruments (Table S1). This led to the application of an APCI probe that significantly improved the instrumental LOQ for all compounds. One participant in the MVS 2018 used the same instrument as in our study and also applied the APCI source [7].

The mobile phase pH was set at 8.8 due to the negative ionization mode and the chromatographic separation of TEA. In the existing methods, the alkaline pH was used when the detection was carried out in negative ionization mode; and the acidic or neutral eluent pH was utilized if the positive ion mode or polarity switching was employed (Table S1). Even though the alkaline mobile phase pH is not usual in LC-MS/MS separation, it is feasible for *Alternaria* toxins due to the chromatographic problem with TEA at acidic pH condition. Moreover, the alkaline pH enhanced the sensitivity in the negative ion mode. The acidic pH condition did not result in the appropriate peak shape for TEA and also lowered the sensitivity of MS/MS detection in negative ionization mode. The chemical derivatization, suggested in some papers [13,19,20], was not tested. Although this approach enabled the simultaneous separation of *Alternaria* toxins [13], it could have further increased the LOQ and the preparation time and costs.

3.2. Method Development for Sample Preparation

In this study, we focused on sunflower oil samples only and developed a LC-ID-MS/MS method involving a unique sample preparation approach suitable for this kind of lipophilic matrix. The goal was to develop a dilute-and-shoot method that is frequently used in toxin analysis by LC-MS/MS method [40]. The non-polar matrix constituents of oil were eliminated with hexane that could easily dissolve the oil. The toxins have low solubility in hexane; hence, the target compounds could be extracted into a non-miscible solvent such as water, methanol or acetonitrile. Even though some studies have reported the use of the general acetonitrile-based mycotoxin extraction solvent mixture (acetonitrile/water/acetic or formic acid) [41] for *Alternarias* (Table S1), we did not prefer the acetonitrile as a solvent due the lower solubility of *Alternarias* in acetonitrile. Methanol is a more suitable solvent for these toxins, and therefore, methanol/water/acetic acid mixture has been utilized for extraction in the candidate method for standardization [6]. In the case of cereal samples, the extraction medium should contain water due to the starch content of samples; and the aqueous methanolic solvent in our case was needed to obtain better solvent separation between the hexane layer and the extraction medium. Also, water can enhance the extraction of TEA with polar characteristics. The experimental design showed that 80% (*v*/*v*) methanol in water gave the best extraction from the naturally contaminated oil. In other types of samples (e.g., tomato, cereals, and oilseeds), ~ 80% methanol also resulted in the optimal extraction solvent composition (Table S1). The HPLC separation was carried out at alkaline pH, so acid was not added into the extraction solvent to avoid the large pH difference between the injection solvent and the mobile phase, which could deform the chromatographic peak. The experimental design also indicated that the optimal sample-to-solvent ratio was 4.0, which is a general ratio in mycotoxin analysis based on the dilute-and-shoot approach [40,41].

High ME (mainly ion suppression) usually influences the mycotoxin analysis based on the LC-MS/MS method [40,41] that is also true for *Alternarias* [8,9,13,18,19,23,28,29,31,42]. The lower sensitivity of our instrument and the high ME for AOH and AME increased the LOQ. Hence, sample pre-concentration with evaporation and reconstitution was necessary to obtain appropriate LOQ (≤ 10 μg/kg) for all compounds. It should be pointed out that an instrument with higher sensitivity would allow further dilution of the extracts that could decrease the preparation time and the ME of analysis. The elimination of co-eluting matrix constituents was tested with SPE clean-up on mixed-mode cation exchange cartridges. Although the mixed-mode SPE and subsequent reversed-phase HPLC measurements enabled an orthogonal separation approach, considerable improvement in ME could not be seen (Table 2), and only the relative ME was enhanced. However, the sample preparation time and overall costs were also increased. In conclusion, the mixed-mode SPE clean-up did not improve the overall analytical process since it is time-consuming and more expensive compared to the dilute-and-shoot approach. The NP SPE clean-up was alternatively tested since this approach requires only sample dissolution in hexane and the dissolved samples can be directly subjected to NP SPE. The NP SPE clean-up would be a simpler clean-up approach, but using this process, we lost either the TEA or the AME, depending on the elution solvent composition. Overall, the dilute-and-shoot approach was the most suitable sample preparation method.

The need of isotope dilution for *Alternaria* toxin analysis by LC-MS/MS method has been strongly suggested by Asam and Rychlik [42] in 2015. Accordingly, isotopically labeled analogues were necessary for the analysis. In line with that, an important conclusion of the ME study was that all corresponding isotopically labeled analogues were necessary for the analysis. This study is the first in using all ISTDs for five *Alternarias* analyzed. While a moderate ion suppression influences the signal of TEA, ALT, and TEN, the ME for AOH and AME is much higher. This means that AOH-d3 and AME-d3 cannot compensate the ME of other analytes, and the ISTDs are not exchangeable. The differences in ME among the compounds analyzed can originate from the retention time differences between toxins and from the various structures of analytes. Even though ALT has similar structure to AOH, the 1.2 min of retention time difference (Figure 2) resulted in considerable difference in the ion suppression (Table 2). On the other hand, there was a significant difference in slopes of matrix-matched calibrations of AME obtained from three different oils. The relative ME was evaluated from the precision of slopes in the matrix-matched calibrations and showed that the matrix-matched calibration could strongly influence the quantification of AME. Hence, the isotope dilution method is needed for appropriate quantification. The relative ME values for AME were significantly improved with the ISTD evaluation (Table 2). In general, the relative ME values were improved for all compounds under ISTD evaluation. It means that the slopes of three different matrix-matched calibrations were close to each other, and they were nearly free of ME.

3.3. Real Sample Analysis

Chulze et al. [43] has already reported the high (30 μg/kg AME—15.796 μg/kg TEA) and frequent (85%) contamination of sunflower seeds with *Alternaria* toxins in 1995. Under processing sunflower oil from the oilseeds, the *Alternaria* toxins may appear in the oil product due to the contamination of sunflower seeds with these toxins. The high natural contamination of sunflower seeds with *Alternarias* reported recently worldwide [4,6,10–12] indicates that cross contamination with toxins can occur in the final sunflower oil products. Even though Chulze et al. [43] has described the decrease of *Alternaria* toxins during the processing of sunflower seeds to oil, the TEA and AME contamination of raw seeds were still detectable in lower concentrations in the oil after processing [43]. Due to the non-polar character of AME, the occurrence of this toxin in lipophilic oil matrix is more likely, as reported previously [43]. Accordingly, the polar characteristics of TEA inhibit its accumulation in oil that was also proven in another study [43], while the AOH contamination of oilseeds could not be detected in the oil product at all. It should be pointed out that the method used by Chulze et al. [43] had a LOQ of 50 μg/kg (AOH), but the recent methods have much lower analytical limits. To the best of our knowledge, no other newer studies have dealt with the decrease of *Alternaria* during the process of oil from sunflower seeds or other types of oilseeds. Since there is a great consumption of sunflower oil worldwide, the need for involving this sample in toxin analysis is to support the legislation. The analysis on sunflower oil was performed by Liu and Rychlik [8] in 2013. López et al. [11] has conducted studies involving other types of foods as well (Table S1). In 2016, López et al. [11] found relatively high (up to 1350 μg/kg) and frequent (80%) contamination of sunflower seeds with TEA, but sunflower oils contaminated with TEA above LOQ (5 μg/kg) were not found [11]. Only AME was detected at 17 μg/kg in one of 11 oil samples, and other toxins were all below the LOQ [11]. Liu and Rychlik [8] also investigated several types of refined and cold-pressed oil samples like pumpkin seed oil, rapeseed oil, sunflower oil, and thistle oil. They detected TEN in three refined (up to 3.95 μg/kg) and three cold-pressed (up to 6.73 μg/kg) sunflower oils, and also in a rapeseed cold pressed oil (up to 0.64 μg/kg). Furthermore, the dihydrotentoxin could be detected (up to 4.48 μg/kg) in three cold-pressed sunflower oil [8].

We have analyzed 16 sunflower oil samples: one sample was a cold pressed sample, and the others were refined ones. In agreement with Chulze et al. [43] and López et al. [11], AOH was not detected and a low concentration of TEA was found (12.8 μg/kg), but only in the cold pressed oil. In three samples, TEN was detected between 4.5 μg/kg and 7.1 μg/kg, similar to those reported by Liu and Rychlik [8]. We have also found that the cold-pressed oil is more likely to be contaminated than the refined samples. Comparatively, Chuzle et al. [43] did not investigate the TEN and López et al. [11] did not find TEN in sunflower oil. Both TEA and TEN have the least toxicity [4] and the detected concentrations are below the validation range suggested by CEN [6], therefore, these contaminations may not cause any risk to human health.

The focus of our study was on sunflower oil since ML would be set for sunflower in near future.

4. Materials and Methods

4.1. Standards, Reagents, Equipment, Samples

Tenuazonic acid (TEA), altenuene (ALT), alternariol (AOH), tentoxin (TEN), and alternariol monomethyl ether (AME) analytical standards were obtained from Romer Labs (Tulln, Austria) and individual 100 μg/mL stock solutions in methanol were prepared and then kept at −18 °C for a year. The isotopically labeled analogues (ISTDs) were purchased from ASCA GmbH (Berlin, Germany). An ISTD mixture containing TEA-^{13}C2 (2.5 μg/mL), ALT-d6 (1 μg/mL), AOH-d3 (0.5 μg/mL), TEN-d3 (0.5 μg/mL), and AME-d3 (0.5 μg/mL) in methanol was prepared and stored at −18 °C for a half year. Methanol, acetonitrile, *n*-hexane, ethyl acetate, ammonia (25%), acetic acid, and ammonium acetate were either of LC-MS or HPLC grade, purchased from the Merck-Sigma group (Schnelldorf, Germany). The PTFE syringe filters (13 mm, 0.45 μm), Strata-XL-C mixed-mode polymeric strong cation exchange SPE cartridges (3 mL, 200 mg), and Strata-Si-1 silica SPE cartridges (6 mL, 500 mg) were acquired from Gen-lab Ltd. (Budapest, Hungary). The LC-MS/MS analysis was carried out by an Agilent 1100 HPLC pump (Agilent; Waldbronn, Germany), which was coupled to an AB Sciex 3200 QTRAP triple quad MS detector, equipped with a Turbo Ion Spray APCI or ESI sources (Sciex; Warrington, Cheshire, UK). Data acquisition and evaluation were performed using the Analyst software version 1.5.2. (Sciex; Warrington, Cheshire, UK). Sample shaking and centrifugation were done using horizontal shaker SM 30 B (Edmund Bühler, Bodelshausen, Germany) and Jouan B4i centrifuge (Thermo Fisher Scientific, Budapest, Hungary), respectively. Sunflower oil samples of different brands and lots were purchased at local shops and originated from the EU. Three sunflower quality check (QC) samples (i.e., W54 NC, C08 SP, and Q25 SP) were leftovers from MVS organized by JRC in 2018. The stability of toxins in both food samples and sample extracts was studied by JRC during the MVS in 2016 and 2018 [6,7]. *Alternaria* toxins are stable at least up to 4 months in samples and they do not degrade in the autosampler in the aqueous injection solution during validation.

4.2. Sample Extraction

Sunflower oil samples (2.00 g) were weighed in polypropylene (PP) centrifuge tubes, followed by the addition of 2 mL *n*-hexane. The oils were completely dissolved in the tubes by vortex-mixing for 5 s. The sample-to-hexane ratio was 1.0. Then, 8 mL methanol–water (80/20, *v*/*v*) mixture was added to the samples and the tubes were capped, followed by vortex-mixing for 5 s. The sample-to-extraction solvent ratio was 4.0. Afterwards, the samples were shaken for 45 min at 180 rpm at ambient temperature. Then, the extracts were centrifuged at 4000× *g* for 2 min at ambient temperature and the hexane layer was discarded.

4.3. Sample Pre-Concentration

An aliquot (6.0 mL) of the extracts (equal to 1.5 g sample) was transferred into glass tubes and evaporated at 45 °C under a gentle stream of nitrogen to ~0.2 mL. Then, 50 µL ISTD mixture (Section 4.1) and water were added into the tubes to obtain 0.5 mL of volume, followed by sample reconstitution by vortex-mixing for 20 s. Finally, samples were filtered through the PTFE syringe filters into the HPLC vials and analyzed by the LC-ID-MS/MS method. In this case, the sample pre-concentration was 3-fold. The concentrations of ISTDs were: TEA-^{13}C2 (83 µg/kg), ALT-d6 (33 µg/kg), AOH-d3 (17 µg/kg), TEN-d3 (17 µg/kg), and AME-d3 (17 µg/kg).

4.4. Sample Clean-Up on Mixed-Mode Cation Exchange Cartridges

An aliquot (5.0 mL) of the extracts was diluted with 35 mL 1% (*v*/*v*) acetic acid in water in new PP centrifuge tubes, which were capped. After homogenization by handshaking for 10 s, the diluted samples were subjected to SPE clean-up on Strata-XL-C cartridges (3 mL, 200 mg). Cartridges were conditioned with 3.0 mL methanol, followed by 3.0 mL water and 3.0 mL 1% (*v*/*v*) acetic acid in water. Diluted samples (40 mL) were passed the cartridges through the dropwise method. Then, cartridges were washed with 3.0 mL water, followed by 3.0 mL *n*-hexane. Afterwards, the cartridges were dried under vacuum for 1.0 min and the samples were eluted with 5.0 mL methanol into glass tubes. Samples were then evaporated and reconstituted as written in Section 4.3.

4.5. Sample Clean-Up on Silica Cartridges

Sunflower oil samples (3.00 g) were weighed into the PP centrifuge tubes, followed by the addition of 6.0 mL *n*-hexane. The oils were completely dissolved in the tubes by vortex-mixing for 5 s. Samples were subjected to SPE clean-up on Strata-Si-1 cartridges (6 mL, 500 mg). Cartridges were conditioned with 6.0 mL methanol, followed by 6.0 mL *n*-hexane. The samples dissolved in *n*-hexane were passed the cartridges through dropwise method. Then, cartridges were washed with 6.0 mL *n*-hexane. Afterwards, the cartridges were dried under vacuum for 1 min and samples were eluted with 6.0 mL methanol into glass tubes. Samples were then evaporated and reconstituted as written above.

4.6. Sample Preparation for Sunflower Seed Samples

Sunflower seed QC samples (2.00 g) were extracted with 8 mL methanol/water (80/20, *v*/*v*) mixture, then the extracts were de-fattened with 2.0 mL *n*-hexane (Section 4.2). The hexane layer was discarded, and 0.5 mL extract was diluted with ISTD mixture (50.0 µL). Afterwards, extracts were filtered through the PTFE syringe filters into HPLC vials and analyzed by LC-ID-MS/MS.

4.7. LC-ID-MS/MS Separation

Toxins were separated on a Zorbax Extend C-18 (150 mm × 3 mm, 5 µm) HPLC column (Agilent; Waldbronn, Germany) using a binary gradient elution. Solvent A contained 5.0 mM ammonium acetate in water (pH adjusted to 8.8 with ammonium hydroxide) and solvent B was methanol. The mobile phase consisted of 10% B at 0 min; 10% B at 1 min; 100% B at 10 min; 100% B at 14.0 min; 10% B at 14.1 min. Stop time was 21 min. The flow rate was 0.5 mL/min. The column thermostat was maintained at 30 °C. The injection volume was 10.0 µL. Compounds were detected in APCI negative ionization mode and using multiple reaction monitoring (MRM) scan type in the triple quadrupole MS/MS instrument. The ion transitions are given in Table 1. The LC-ID-MS/MS analysis was carried out using all corresponding isotopically labeled ISTDs. The ISTDs were employed to compensate the signal suppression/enhancement in the ion source (matrix effect) caused by the co-eluting matrix constituents.

During the method development, the positive ionization mode (Table 1) with both APCI and ESI sources and the ESI negative ionization mode were also tested. When the positive ionization was employed in the ion source, the separation was performed with 0.3% (*v*/*v*) acetic acid in water (mobile phase A) and 0.3% (*v*/*v*) acetic acid in methanol (mobile phase B) eluent composition using the same gradient elution as written above.

The ion source settings were as follows: nebulizer current (only with APCI ion source): −4 (negative ion mode) or 4 (positive ion mode); drying gas temperature: 600 °C; nebuliser pressure: 30 unit; drying gas flow: 30 unit; curtain gas: 20 unit; capillary voltage: −4200 V (negative ion mode) or + 5000 V (positive ion mode); collision gas (N_2): medium unit; interface heater: on.

The optimal ion transitions are given in Table 1, and the optimal ionization mode was APCI negative. These detection parameters were used during validation and real sample analysis.

Supplementary Materials: The following are available online. Table S1: Existing LC-MS/MS methods for *Alternaria* toxins; Table S2: The validation results for Alternaria toxins in sunflower oil samples and the results of analyzing sunflower seed QC samples by the modified method.

Author Contributions: Methodology, Á.T. and L.K.; validation, L.K. and Á.T.; formal analysis, L.K. and Á.T.; investigation, L.K. and Á.T.; writing—Original draft preparation, Á.T., V.K.S. and L.K.; writing—Review and editing, Á.T., V.K.S. and L.K. All authors have read and agreed to the published version of the manuscript.

Funding: This research received no external funding.

Acknowledgments: Authors like to thank Carlos Gonçalves for providing the QC samples. We thank Mingbao Feng for his comments on the revised manuscript. We thank the anonymous reviewers for their comments, which improve the manuscript greatly.

Conflicts of Interest: The authors declare no conflict of interest.

References

1. De Ruyck, K.; De Boevre, M.; Huybrechts, I.; De Saeger, S. Dietary mycotoxins, co-exposure, and carcinogenesis in humans: Short review. *Mutat. Res. Rev. Mutat. Res.* **2015**, *766*, 32–41. [CrossRef] [PubMed]
2. Zöllner, P.; Mayer-Helm, B. Trace mycotoxin analysis in complex biological and food matrices by liquid chromatography–atmospheric pressure ionisation mass spectrometry. *J. Chromatogr. A* **2006**, *1136*, 123–169. [CrossRef] [PubMed]
3. European Commission. Commission Regulation (EC), No 1881/2006 of 19 December 2006 setting maximum levels for certain contaminants in foodstuffs. *Off. J. Eur. Union* **2006**, *364*, 5.
4. EFSA, European Food Safety Authority. Scientific Opinion on the risks for public and animal health related to the presence of *Alternaria* toxins in food and feed. *EFSA J.* **2011**, *9*, 1–97.
5. Zwickel, T.; Fauhl-Hassek, C.; Genrich, C.; Klaffke, H.; Pydde, E.; Rychlik, M. Report on the Proficiency Test about *Alternaria* Toxins in Food. In Proceedings of the Poster Presentation at the 36th Mycotoxin Workshop, Göttingen, Germany, June 2014.
6. Tölgyesi, Á.; Stroka, J. Collaborative Study Report: Determination of *Alternaria* toxins in Cereals, Tomato Juice and Sunflower Seeds by Liquid Chromatography Tandem Mass Spectrometry. 2016. Available online: https://ec.europa.eu/jrc/en/publication/collaborative-study-report-determination-alternaria-toxins-cereals-tomato-juice-and-sunflower-seeds (accessed on 26 July 2019).
7. Goncalves, C.; Joint Research Centre, Geel, Belgium. Personal communication, 2019.

8. Liu, Y.; Rychlik, M. Development of a Stable Isotope Dilution LC–MS/MS Method for the *Alternaria* Toxins Tentoxin, Dihydrotentoxin, and Isotentoxin. *J. Agric. Food Chem.* **2013**, *61*, 2970–2978. [CrossRef]
9. Liu, Y.; Rychlik, M. Biosynthesis of seven carbon-13 labeled *Alternaria* toxins including altertoxins, alternariol, and alternariol methyl ether, and their application to a multiple stable isotope dilution assay. *Anal. Bioanal. Chem.* **2015**, *407*, 1357–1369. [CrossRef]
10. López, P.; Venema, D.; de Rijk, T.; de Kok, A.; Scholten, J.M.; Mol, H.G.J.; de Nijs, M. Occurrence of *Alternaria* toxins in food products in The Netherlands. *Food Control* **2016**, *60*, 196–204. [CrossRef]
11. López, P.; Venema, D.; Mol, H.; Spanjer, M.; de Stoppelaar, J.; Pfeiffer, E.; de Nijs, M. *Alternaria* toxins and conjugates in selected foods in the Netherlands. *Food Control* **2016**, *69*, 153–159. [CrossRef]
12. Hickert, S.; Bergmann, M.; Ersen, S.; Cramer, B.; Humpf, H.-U. Survey of *Alternaria* toxin contamination in food from the German market, using a rapid HPLC-MS/MS approach. *Mycotoxin Res.* **2016**, *32*, 7–18. [CrossRef]
13. Tölgyesi, Á.; Stroka, J.; Tamosiunas, V.; Zwickel, T. Simultaneous analysis of *Alternaria* toxins and citrinin in tomato: An optimised method using liquid chromatography-tandem mass spectrometry. *Food Addit. Contam.* **2015**, *32*, 1512–1522. [CrossRef]
14. Di Mavungu, J.D.; Monbaliu, S.; Scippo, M.-L.; Maghuin-Rogister, G.; Schneider, Y.-J.; Larondelle, Y.; Callebaut, A.; Robbens, J.; Van Peteghem, C.; De Saeger, S. LC-MS/MS multi-analyte method for mycotoxin determination in food supplements. *Food Addit. Contam.* **2009**, *26*, 885–895. [CrossRef] [PubMed]
15. Scott, P.M.; Lawrence, G.A.; Lau, B.P.Y. Analysis of wines, grape juices and cranberry juices for *Alternaria* toxins. *Mycotoxin Res.* **2006**, *22*, 142–147. [CrossRef] [PubMed]
16. Prelle, A.; Spadaro, D.; Garibaldi, A.; Gullino, M.L. A new method for detection of five *Alternaria* toxins in food matrices based on LC–APCI-MS. *Food Chem.* **2013**, *140*, 161–167. [CrossRef] [PubMed]
17. Magnani, R.; De Souza, G.D.; Rodrigues-Filho, E. Analysis of alternariol and alternariol monomethyl ether on flavedo and albedo tissues of tangerines (*Citrus reticulata*) with symptoms of alternaria brown spot. *J. Agric. Food Chem.* **2007**, *55*, 4980–4986. [CrossRef] [PubMed]
18. Asam, S.; Konitzer, K.; Schieberle, P.; Rychlik, M. Stable isotope dilution assays of alternariol and alternariol monomethyl ether in beverages. *J. Agric. Food Chem.* **2009**, *57*, 5152–5160. [CrossRef]
19. Asam, S.; Liu, Y.; Konitzer, K.; Rychlik, M. Development of a stable isotope dilution assay for tenuazonic acid. *J. Agric. Food Chem.* **2011**, *59*, 2980–2987. [CrossRef]
20. Siegel, D.; Rasenko, T.; Koch, M.; Nehls, I. Determination of the *Alternaria* mycotoxin tenuazonic acid in cereals by high-performance liquid chromatography–electrospray ionization iontrap multistage mass spectrometry after derivatization with 2,4-dinitrophenylhydrazine. *J. Chromatogr. A* **2009**, *1216*, 4582–4588. [CrossRef]
21. Puntscher, H.; Cobankovic, I.; Marko, D.; Warth, W. Quantitation of free and modified *Alternaria* mycotoxins in European food products by LC-MS/MS. *Food Control* **2019**, *102*, 157–165. [CrossRef]
22. Gambacorta, L.; Darra, N.E.; Fakhoury, R.; Logrieco, A.F.; Solfrizzo, M. Incidence and levels of *Alternaria* mycotoxins in spices and herbs produced worldwide and commercialized in Lebanon. *Food Control* **2019**, *106*, 106724. [CrossRef]
23. Gotthardt, M.; Asam, S.; Gunkel, K.; Moghaddam, A.F.; Baumann, E.; Kietz, R.; Rychlik, M. Quantitation of Six *Alternaria* Toxins in Infant Foods Applying Stable Isotope Labeled Standards. *Front. Microbiol.* **2019**, *10*, 109. [CrossRef]
24. De Berardis, S.; De Paola, E.L.; Montevecchi, G.; Garbini, D.; Masino, F.; Antonelli, A.; Melucci, D. Determination of four *Alternaria* alternata mycotoxins by QuEChERS approach coupled with liquid chromatography-tandem mass spectrometry in tomato-based and fruit-based products. *Food Res. Int.* **2018**, *106*, 677–685. [CrossRef] [PubMed]
25. Dong, H.; Xian, Y.; Xiao, K.; Wu, Y.; Zhu, L.; He, J. Development and comparison of single-step solid phase extraction and QuEChERS clean-up for the analysis of 7 mycotoxins in fruits and vegetables during storage by UHPLC-MS/MS. *Food Chem.* **2019**, *274*, 471–479. [CrossRef] [PubMed]
26. Nguyen, T.T.T.; Kim, J.; Jeon, S.J.; Lee, C.W.; Magan, N.; Lee, H.B. Mycotoxin production of *Alternaria* strains isolated from Korean barley grains determined by LC-MS/MS. *Int. J. Food Microbiol.* **2018**, *268*, 44–52. [CrossRef]
27. Ediage, E.N.; Van Poucke, C.; De Saeger, S. A multi-analyte LC–MS/MS method for the analysis of 23 mycotoxins in different sorghum varieties: The forgotten sample matrix. *Food Chem.* **2015**, *177*, 397–404. [CrossRef] [PubMed]

28. Walravens, J.; Mikula, H.; Rychlik, M.; Asam, S.; Devos, T.; Njumbe Ediage, E.; Diana Di Mavungu, J.; Jacxsens, L.; Van Landschoot, A.; Vanhaecke, L.; et al. Validated UPLC-MS/MS methods to quantitate free and conjugated *Alternaria* toxins in commercially available tomato products and fruit and vegetable juices in Belgium. *J. Agric. Food Chem.* **2016**, *64*, 5101–5109. [CrossRef]
29. Walravens, J.; Mikula, H.; Rychlik, M.; Asam, S.; Ediage, E.N.; Di Mavungu, J.D.; Van Landschoot, A.; Vanhaecke, L.; De Saeger, S. Development and validation of a ultra-high performance liquid chromatography tandem mass spectrometric method for the simultaneous determination of free and conjugated *Alternaria* toxins in cereal-based foodstuffs. *J. Chromatogr. A* **2014**, *1372*, 91–101. [CrossRef]
30. Zwickel, T.; Klaffke, H.; Richards, K.; Rychlik, M. Development of a high performance liquid chromatography tandem mass spectrometry based analysis for the simultaneous quantification of various *Alternaria* toxins in wine, vegetable juices and fruit juices. *J. Chromatogr. A* **2016**, *1455*, 74–85. [CrossRef]
31. Rico-Yuste, A.; Walravens, J.; Urraca, J.L.; Abou-Hany, R.A.G.; Descalzo, A.B.; Orellana, G.; Rychlik, M.; De Saeger, S.; Moreno-Bondi, M. Analysis of alternariol and alternariol monomethyl ether in foodstuffs by molecularly imprinted solid-phase extraction and ultra-high-performance liquid chromatography tandem mass spectrometry. *Food Chem.* **2018**, *243*, 357–364. [CrossRef]
32. Rodríguez-Carrasco, Y.; Mañes, J.; Berrada, H.; Juan, C.; Wang, M. Development and Validation of a LC-ESI-MS/MS Method for the Determination of *Alternaria* Toxins Alternariol, Alternariol Methyl-Ether and Tentoxin in Tomato and Tomato-Based Products. *Toxins* **2016**, *8*, 328. [CrossRef]
33. Zhao, K.; Shao, B.; Yang, D.; Li, F. Natural occurrence of four Alternaria mycotoxins in tomato- and citrus-based foods in China. *J. Agric. Food Chem.* **2015**, *63*, 343–348. [CrossRef]
34. Wei, D.; Wang, Y.; Jiang, D.; Feng, X.; Li, J.; Wang, M. Survey of *Alternaria* Toxins and Other Mycotoxins in Dried Fruits in China. *Toxins* **2017**, *9*, 200. [CrossRef] [PubMed]
35. Wang, C.; Fan, Y.; He, W.; Hu, D.; Wu, A.; Wu, W. Development and Application of a QuEChERS-Based Liquid Chromatography Tandem Mass Spectrometry Method to Quantitate Multi-Component *Alternaria* Toxins in Jujube. *Toxins* **2018**, *10*, 382. [CrossRef] [PubMed]
36. Houissa, E.; Lasram, S.; Sulyok, M.; Šarkanj, B.; Fontana, A.; Strub, C.; Krska, R.; Schorr-Galindo, S.; Ghorbel, A. Multimycotoxin LC-MS/MS analysis in pearl millet (*Pennisetum glaucum*) from Tunisia. *Food Control* **2019**, *106*, 106738. [CrossRef]
37. Matuszewski, B.K.; Constanzer, M.L.; Chavez-Eng, C.M. Strategies for the assessment of matrix effect in quantitative bioanalytical methods based on HPLC–MS/MS. *Anal. Chem.* **2003**, *75*, 3019–3030. [CrossRef]
38. European Commission. Commission Decision of 12 August 2002 implementing Council Directive 96/23/EC concerning the performance of analytical methods and the interpretation of results. 2002/657/EC. *Off. J. Eur. Union* **2002**, *221*, 8.
39. CEN/TR 16059:2016 Food Analysis. Performance Criteria for Single Laboratory Validated Methods of analysis for the Determination of Mycotoxins. Available online: infostore.saiglobal.com/preview/98697665123.pdf?sku=874908_saig_nsai_nsai_2079941, (accessed on 1 November 2010).
40. Malachova, A.; Sulyok, M.; Beltran, E.; Berthiller, F.; Krska, R. Multi-Toxin Determination in Food—The Power of "Dilute and Shoot" Approaches in LC–MS–MS. *LCGC Eur.* **2015**, *28*, 542–555.
41. Sulyok, M.; Berthiller, F.; Krska, R.; Schuhmacher, R. Development and validation of a liquid chromatography/tandem mass spectrometric method for the determination of 39 mycotoxins in wheat and maize. *Rapid Commun. Mass Spectrom.* **2006**, *20*, 2649–2659. [CrossRef]
42. Asam, S.; Rychlik, M. Recent developments in stable isotope dilution assays in mycotoxin analysis with special regard to *Alternaria* toxins. *Anal. Bioanal. Chem.* **2015**, *407*, 7563–7577. [CrossRef]
43. Chulze, S.N.; Torres, A.M.; Dalcero, A.M.; Etcheverry, M.G.; Ramirez, M.L.; Farnochi, M.C. *Alternaria* mycotoxins in sunflower seeds: Incidence and distribution of the toxins in oil and meal. *J. Food Prot.* **1995**, *58*, 1133–1135. [CrossRef]

Sample Availability: Samples of the compounds are not available from the authors.

molecules

MDPI

Article

Analytical Detection of Sulfonamides and Organophosphorus Insecticide Residues in Fish in Taiwan

Chung-Pei Chang [1,†], Po-Hsun Hou [2,3,†], Wei-Cheng Yang [4], Ching-Fen Wu [5], Chia-Chia Chang [6], Ming-Yang Tsai [7,8], Hsiao-Pei Tsai [9], Chien-Teng Lin [5], Yi-Jing Xue [5], Jiann-Hsiung Wang [5] and Geng-Ruei Chang [5,9,10,*]

1 Department of Anesthesiology, Show Chwan Memorial Hospital, 1 Section, 542 Chung-Shan Road, Changhua 50008, Taiwan; fentanylpei@yahoo.com.tw
2 Department of Psychiatry, Taichung Veterans General Hospital, 4 Section, 1650 Taiwan Boulevard, Taichung 40705, Taiwan; peterhopo2@yahoo.com.tw
3 Faculty of Medicine, National Yang-Ming University, 2 Section, 155 Linong Street, Beitou District, Taipei 11221, Taiwan
4 Department of Veterinary Medicine, School of Veterinary Medicine, National Taiwan University, 4 Section. 1 Roosevelt Road, Taipei 10617, Taiwan; yangweicheng@ntu.edu.tw
5 Department of Veterinary Medicine, National Chiayi University, 580 Xinmin Road, Chiayi 60054, Taiwan; cfwu@mail.ncyu.edu.tw (C.-F.W.); vet540423@gmail.com (C.-T.L.); asdzxc0523@gmail.com (Y.-J.X.); jhwang@mail.ncyu.edu.tw (J.-H.W.)
6 Animal Drugs Inspection Branch, Animal Health Research Institute, Council of Agriculture, 21 Muchang, Ciding Village, Zhunan Township, Miaoli County 35054, Taiwan; bccchia@mail.nvri.gov.tw
7 Animal Industry Division, Livestock Research Institute, Council of Agriculture, Executive Yuan, 112 Muchang, Xinhua Dist, Tainan 71246, Taiwan; mytsai@mail.tlri.gov.tw
8 Graduate Institute of Bioresources, National Pingtung University of Science and Technology, 1 Shuefu Road, Neipu, Pingtung 91201, Taiwan
9 Veterinary Teaching Hospital, National Chiayi University, 580 Xinmin Road, Chiayi 60054, Taiwan; tsaibelle@mail.ncyu.edu.tw
10 Ph.D. Program of Agriculture Science, National Chiayi University, 300 Syuefu Road, Chiayi 60004, Taiwan
* Correspondence: grchang@mail.ncyu.edu.tw; Tel.: +886-5-2732946
† These authors contributed equally to this work.

Academic Editor: Ping-Chung Kuo
Received: 26 February 2020; Accepted: 23 March 2020; Published: 25 March 2020

check for updates

Abstract: Exposure to residues of antibiotics (e.g., sulfonamides) and insecticides (e.g., organophosphorus insecticides) in aquacultured food can adversely affect humans and animals and thus affect public health globally. Here, using a validated method, we examined the levels of residues of 12 sulfonamides as well as 18 organophosphorus insecticides in aquacultured fish in Taiwan. A total of 52 fish samples (i.e., 20 tilapia, 16 milk fish, and 16 perch samples) were obtained from Taiwanese aquafarms from June 2018 to October 2019. We detected 0.02 and 0.03 mg/kg of sulfamethazine (a sulfonamide) in one tilapia and one milk fish, respectively, and 0.02, 0.05, and 0.03 mg/kg of chlorpyrifos (an organophosphorus insecticide) in one tilapia, one milk fish, and one perch, respectively; thus, among the samples, 3.85% and 5.77% contained sulfonamides and organophosphorus insecticide residues, respectively. Furthermore, we assessed human health risk based on the estimated daily intakes (EDIs) of these residues: EDIs of sulfonamide and organophosphorus insecticide residues were <1.0% of the acceptable daily intake recommended by the Joint Food and Agriculture Organization of the United Nations/World Health Organization Expert Committee on Food Additives. The risk of exposure to sulfonamide and organophosphorus insecticide residue by consuming aquacultured fish in Taiwan was thus negligible, signifying no immediate health risk related to the consumption of fish. Our findings can constitute a reference in efforts geared toward ensuring food safety and monitoring veterinary drug and insecticide residue levels in aquacultured organisms. Residue levels in fish must be continually monitored to further determine possible effects of these residues on human health.

Keywords: sulfonamide; organophosphorus insecticide; residue; fish; risk assessment

1. Introduction

The aquaculture industry is of great economic importance and has been growing faster than other animal-farming industry worldwide [1]. Taiwan's environment and geographic location are highly appropriate for aquaculture advancement, with the country's aquaculture history spanning more than three centuries [2]. In Taiwan, more than 35 major and candidate aquatic species are commercially aquacultured [3]. Over 2010–2018, Taiwan's revenue derived from land aquaculture was up to US$1 billion on average [3]; in addition, during the 2010s, Taiwan's estimated annual land aquaculture production was 300,000 t [4]. Moreover, in Taiwan, major developments in aquaculture occurred over the 1960s to the 1990s [2], mainly because of the strong support of the government since the 1960s [5]. In particular, the government strongly supported the farming of fish, such as tilapia, milkfish, perch, eels, and groupers. From the beginning of tilapia farming in 1978 until 2018, Taiwan's average annual tilapia production increased to 60,000 t [3].

Taiwan has a subtropical climate and a limited amount of usable land; thus, aquaculture farms in Taiwan are located close to agricultural and residential areas, predisposing the cultured organisms to fungal, bacterial, parasitic, and viral infections. This thus necessitates applying prophylactic and therapeutic veterinary drugs, such as antibiotics [6]. Nevertheless, excessively applying these drugs could result in the cultured organisms retaining drug residues at harvest, potentially exposing consumers to substances toxic for human consumption [2]. We previously detected chloramphenicol in hard clam [6] and shrimp [2] and quinolones in shrimp [2] aquafarms in Taiwan. Moreover, these aquafarms use several insecticides to control ectoparasites and endoparasite growth [7] and several pesticides to limit the growth of aquatic weeds in bodies that include canals, fishponds, and lakes [6]. These activities lead to the introduction of these chemicals into water and soil and consequently contaminate aquaculture regions. Aquaculture techniques that are usually executed in Taiwan's inner regions include polyculturing with waterfowls and mixed breeding with different aquatic products; this may also introduce chemical pollution to the soil as well as water environments and thus negatively affect aquacultured organisms [8]. In aquaculture, excessively applying chemicals may engender public health concerns as well as ecological impacts; for example, bacterial resistance could be induced and chemical residues may be added to the environment, thus increasing allergy and cancer risks. Therefore, research should be executed on whether aquacultured fish contain veterinary drug and insecticide residues—possibly bioaccumulated in their edible parts.

The recent continual decrease in the availability of wild-caught fish has led to a rise in aquaculture fish production as well as a rise in their consumption. However, in Taiwan, major food safety incidents over the past decade have been related to aquatic product consumption [9] and economic losses have been incurred because the import of some Taiwanese aquacultured food products has been banned by several countries worldwide [10]. Consequently, Taiwan's government is paying increasing attention toward aquacultured product safety and quality assessment and management. Sulfonamides, synthetic derivatives of sulfonic acid, are used broadly because they are low-cost, are effective against some bacterial infections, and improve animal performance. In the Asia Pacific region, sulfonamides are commonly used against bacterial infections and other diseases in aquaculture [1]. In China, for instance, sulfur drugs were detected in all fish samples from typical marine aquaculture regions [11]. However, sulfonamides have several side effects; for example, they have been demonstrated to engender reduced filet palatability of aquacultured products, increased kidney damage and infertility risk, and physiological and immunological response disruptions [1,12]. Massive and repeated illegal use of organophosphorus insecticides to control parasitic diseases in aquacultured organisms has also been noted in Taiwan [4]; these insecticides mainly inhibit acetylcholinesterase activity and disrupt nerve function in humans [13].

The Taiwan Food and Drug Administration (TFDA) has yet to establish the maximum residue limits (MRLs) for sulfa drugs (except for sulfadimethoxine and sulfamonomethoxine) or organophosphorus insecticides in fish. However, in 2019, the TFDA defined the MRLs of sulfa drugs in livestock, chicken, milk, and eggs and completely banned the use of sulfa drugs in fish aquaculture. Therefore, in this study, we determined the concentrations and accumulation levels of sulfonamides and organophosphorus insecticides in major aquacultured fish in Taiwan. In addition, we examined these contaminants' estimated daily intake (EDI) through seafood consumption in Taiwanese adults. Our findings provide information that can be potentially useful during the development of effective measures for safe aquaculture and aquacultured product consumption. Moreover, our results highlight the potential phthalate burden imposed on consumers due to excessive plastic material use in Taiwan.

2. Results

2.1. Detected Levels as Well as Rates of Residues of Sulfonamides in Analyzed Fish Samples

In total, 52 fish samples were analyzed. Table 1 presents the detected levels of banned sulfonamide residues in different fish samples. In all fish samples, the predominant sulfonamide residue was sulfamethazine (Figure S1): 5% in tilapia (1/20; concentration: 0.03 mg/kg) and 6.25% in milk fish (1/16; concentration: 0.02 mg/kg). The data for the remaining 19 tilapia, 15 milk fish, and 16 perch samples, which did not demonstrate detectable levels of sulfonamides, are not shown. Moreover, on average, 0.002, 0.001, and 0 mg/kg sulfonamides were detected in all tilapia, milk fish, and perch samples, respectively. Finally, only 2 (3.85%) of all 52 samples contained sulfonamides (i.e., sulfamethazine).

Table 1. Detected levels of residues of prohibited sulfonamides in fish samples collected over June 2018 to October 2019.

Fish	Targets Detected	Surveyed Samples	No. of Residue	Detected Residues [1] (mg/kg)	Average [2] (mg/kg)	Residual Ration [3] (%)
Tilapia	sulfamethazine	20	1	0.03 ± 0.0003	0.002 ± 0.0015	5
Milk fish	sulfamethazine	16	1	0.02 ± 0.0005	0.001 ± 0.0010	6.25
Perch	undetected	16	0	0	0	0
Total	sulfamethazine	52	2	0.02–0.03	0.001 ± 0.0006	3.85

[1] Values are given as means ± SEM. [2] Estimated for all samples (detected and not detected to have sulfonamide residues). [3] Samples with residual concentrations lower than the limit of quantification were considered to have undetectable concentrations.

2.2. Detection Levels and Rates of Organophosphorus Insecticide Residues in Fish Samples

Table 2 presents the detected levels of the residues of prohibited organophosphorus insecticides in different fish samples. In all fish samples, the predominant organophosphorus insecticide residue was of chlorpyrifos (Figure S2): 5% in tilapia (1/20; concentration: 0.02 mg/kg), 6.25% in milk fish (1/16; concentration: 0.05 mg/kg), and 6.25% in perch (1/16; concentration: 0.03 mg/kg). The data of the remaining 19 tilapia, 15 milk fish, and 16 perch samples, which did not demonstrate detectable levels of organophosphorus insecticides, are not shown. That is, in general, the detected levels of chlorpyrifos were between 0.02 and 0.05 mg/kg. Moreover, on average, 0.001, 0.003, and 0.002 mg/kg chlorpyrifos were detected in all tilapia, milk fish, and perch samples, respectively. Finally, only 3 (5.77%) of all 52 samples contained sulfonamides (i.e., sulfamethazine).

Table 2. Detected levels of residues of organophosphorus insecticides in fish samples obtained over June 2018 to October 2019.

Fish	Targets Detected	Surveyed Samples	No. of Residue	Detected Residues [1] (mg/kg)	Average [2] (mg/kg)	Residual Ration [3] (%)
Tilapia	Chlorpyrifos	20	1	0.02 ± 0.0003	0.001 ± 0.0010	5.0
Milk fish	Chlorpyrifos	16	1	0.05 ± 0.0006	0.003 ± 0.0030	6.25
perch	Chlorpyrifos	16	1	0.03 ± 0.0003	0.002 ± 0.0018	6.25
Total	Chlorpyrifos	52	3	0.02–0.05	0.002 ± 0.0011	5.77

[1] Values are given as means ± SEM. [2] Estimated for all samples (detected and not detected to have organophosphorus insecticide residues). [3] Samples with residual concentrations lower than the limit of quantification were considered to have undetectable concentrations.

2.3. *EDI Levels of Sulfonamide Residues Through Fish in Taiwanese Adults*

Average EDI levels extrapolated from average sulfamethazine levels in Taiwanese women and men were 0.001 and 0.002 μg/kg body weight/day, respectively (Table 3). The acceptable daily intake (ADI) level of sulfamethazine residues through food is 0.05 mg/kg, as stipulated by the Joint Food and Agriculture Organization of the United Nations/World Health Organization Expert Committee on Food Additives (JECFA) [1,14]. The obtained EDI levels were determined to be substantially lower than the JECFA-recommended ADI levels of sulfamethazine (Table 3). Moreover, the EDI levels expressed as the percentages of the ADIs were 0.004% and 0.002% in men and women respectively. Taken together, fish consumption led to a low risk assessment for sulfamethazine, with EDIs that are <0.1% of the ADIs in both men as well as women.

Table 3. Estimated daily intake (EDI) levels of sulfonamide residues in Taiwanese adults.

Sulfonamides	EDI (μg/kg Body Weight/Day) Weight/Day Weight/Day)		EDI% of ADI		ADI (JECFA) (mg/kg Body Weight/Day)
	Male	Female	Male	Female	
sulfamethazine	0.002	0.001	0.004	0.002	0.05

2.4. *EDI Levels of Residues of Organophosphorus Insecticides Ingested Through Fish in Taiwanese Adults*

The average EDI levels extrapolated from the average chlorpyrifos levels in Taiwanese women and men were 0.002 and 0.003 μg/kg body weight/day, respectively (Table 4). The EDI levels were noted to be substantially lower than the JECFA-recommended chlorpyrifos ADI level of 0.01 mg/kg (Table 4) [4,15]. Moreover, the EDI levels expressed as the percentages of the ADIs were 0.03% and 0.02% in men and women, respectively. Taken together, fish consumption led to a low risk assessment for chlorpyrifos, with EDIs that are <0.1% of the ADIs in men as well as women.

Table 4. EDI levels of organophosphorus insecticide residues in Taiwanese adults.

Insecticides	EDI (μg/kg Body Weight/Day) Weight/Day Weight/Day)		EDI% of ADI		ADI (JECFA) (mg/kg Body Weight/Day)
	Male	Female	Male	Female	
Chlorpyrifos	0.003	0.002	0.03	0.02	0.01

3. Discussion

We analyzed 52 fish samples from aquaculture regions in Taiwan for the residues of 12 sulfonamides (i.e., sulfamerazine, sulfaethoxypyridazine, sulfathiazole, sulfadiazine, sulfamethoxypyridazine, sulfapyridine, sulfadoxine, sulfamethazine, sulfadimethoxine, sulfamethoxazole, sulfamonomethoxine, and sulfameter,) and 18 organophosphorus insecticides (i.e., chlorfenvinphos, chlorpyrifos, diazinon, fenamiphos, fenitrothion, fenthion, formothion, iprobenfos, malathion, methacrifos, methamidophos, methidathion, phoxim, profenophos, prothiofos, pyrazophos, triazophos, and trichlorfon). The TFDA allows the presence of only sulfadimethoxine and sulfamonomethoxine in aquacultured products at total residual levels of <0.1 mg/kg and prohibits the use of any other sulfa drugs and organophosphorus insecticides in fish aquaculture. These compounds' residues existing in fish were thus identified; moreover, the identified levels of the residues were noted to adequately demonstrate the level of legal compliance concerning the application of these products.

The limit of quantification (LOQ) of all sulfonamides in fish samples was 10 ng/g—identical to the LOQ recommended by the TFDA for sulfonamide contamination in edible livestock, chicken, milk, and aquacultured foods and half of that for sulfonamide contamination in animal viscera [16]. However, the TFDA has not recommended a strict LOQ for organophosphorus insecticide contamination in aquacultured foods. Nevertheless, the TFDA-recommended LOQ is 10 ng/g for organophosphorus insecticide contamination in livestock (e.g., pork) and chicken muscles [17]. Previous methods developed for analyzing organophosphorus insecticide contamination in bivalves, crustaceans, fish, and cephalopods have involved the use of various apparatuses. For instance, gas chromatography flame ionization revealed the LOQs of chlopyriphos, ethion, ethoprophos, fensulfothion, isoxathion, and parathion to be 2–50 ng/g [18]; gas chromatography–mass spectrometry revealed the LOQs of propetamphos, diazinon, disulfoton, malathion, fenthion, and triazophos to be 7–15.2 ng/g [19]; and high-performance liquid chromatography (HPLC)–tandem mass spectrometry (LC-MS/MS) coupled with gel permeation chromatography revealed the LOQs of profenofos, chlorpyrifos, malathion, phosmet, triazophos, trichlorphon, and dimethoate to be 0.05–0.2 ng/g [20]. Compared with the aforementioned LOQs, our present analytical method detected a lower LOQ (i.e., 5 ng/g) for organophosphorus insecticide multiresidues, indicating the usefulness of our method in detecting trace organophosphorus insecticide residues.

Here, we adopted the analytical approach for sulfonamide residues suggested by the TFDA [16] and that for organophosphorus insecticide residues developed by the European Committee for Standardization, the QuEChERS (quick, easy, cheap, effective, rugged, and safe) method [4,21]—both of which are suitable for detecting trace chemical residues. For the validation of its analytic method, the TFDA [2,22] recommends an acceptable recovery rate of 70–120% [relative standard deviation (RSD) < 15%], 70–120% (RSD < 20%), and 50–125% (RSD < 35%) for chemical residues in food matrices detected in the 0.1–1, 0.001–0.01, and <0.001 mg/kg ranges, respectively. In this study, residues of sulfonamides and organophosphorus insecticides that were identified within the 0.01–0.1 mg/kg ranges were noted to exhibit a recovery rate of 90–120% (RSD < 15%) and 80–120% (RSD < 15%), respectively. Moreover, the quantities of spiked analytes used at the lower and higher levels were respectively 5 and 25 ng/g for the 12 sulfonamides (Table S3) and 10 and 50 ng/g for the 18 organophosphorus insecticides (Table S4). Thus, all analyzed sulfonamide and organophosphorus insecticide concentrations were within the acceptable range [22]. Accordingly, our method of extraction proposed herein was robust for the analysis of the investigated sulfonamides and organophosphorus insecticides in the fish samples.

LC-MS/MS demonstrated a positive result for sulfonamide in only 3.85% of the 52 fish samples, and in all positive samples, only sulfamethazine, a TFDA-banned sulfonamide, was observed. Sulfonamide residues were identified in 77.59% of a total of 116 samples of fish in 2016 in China [23], 4.27% of 101 samples of fish over 1994–1998 in Slovenia [24], 17.39% of 138 samples of fish in 2012 in Iran [25], and 1.75% of 171 samples of fish over May September 2008 in South Korea (specifically sulfadiazine and sulfamethoxazole residues) [26]. The TFDA conducted surveys for sulfonamide residues in aquacultured products over 2013–2018 in Taiwan and detected increased illegal use of sulfonamides as antibiotic agents in marine products in order to increase production [27–32]: the detection rate of banned sulfonamide residues in fish samples was 0.50% in 2013 (1/199) [27], 0% in 2014 (0/194) [28], 4.29% in 2015 (3/70) [29], 1.49% in 2016 (1/67) [30], 2.94% in 2017 (2/68) [31], and 2.67% in 2018 (2/75) [32]. However, our current findings are inconsistent with the aforementioned findings from the TFDA surveys. These differences may possibly be due to varied sample sizes. In addition, our study samples were derived from areas of fish production in Taiwan; by contrast, those derived by the TFDA could have been imported fish. Moreover, we included a larger fish sample size and analyzed for more sulfonamides. Therefore, a wider spectrum of banned sulfonamide residues was analyzed in fish in Taiwan.

The predominant sulfonamide residue was sulfamethazine, with its maximum concentration being 0.03 mg/kg (in one tilapia sample). This is consistent with the results observed by Sampaio et al. [1] and Nunes et al. [14]: sulfamethazine is a frequently applied sulfonamide in fish aquaculture. Moreover, sulfamethazine has previously been found to be present at high concentrations (>100 μg/kg) frequently [23]. The aforementioned studies have positively identified sulfathiazole [27], sulfamethoxazole [29,30,32], and sulfadiazine [31] in fish samples, and a TFDA survey positively identified sulfaquinoxaline and sulfathiazole in soft-shell turtle samples [28]. These findings confirm the increased use of sulfonamides as feed additives in aquaculture [11]. In addition, the water sulfamethazine level of 100 ng/L could affect greenhouse gas release, atmospheric ozone depletion, and eutrophication control through denitrification inhibition and N_2O release stimulation [33]. Moreover, high sulfonamide usage appears to be strongly correlated with bacterial resistance to sulfonamides [34], achieved through the development of antibiotic resistance genes, which can pose a potential human health risk [23]. Our results, in combination the findings derived from other surveys described here, demonstrate the exposure of Taiwan's population to trace sulfonamide levels through fish consumption.

Of all 18 analyzed organophosphorus insecticides, only chlorpyrifos was detected in the fish samples. In all fish samples, 5.77% was the detection rate for all studied organophosphorus insecticide residues, with the highest rate being for chlorpyrifos (6.25%) in milk fish and perch samples, followed by tilapia samples (5.0%). During their aquaculture activities, some farmers in Taiwan apply organophosphorus insecticides for ectoparasite treatment [4]. Sun et al. [35] reported an organophosphorus insecticide detection rate of 11.37% in 607 fish samples from traditional markets, regional supermarkets, fish markets, and fish farms in Taiwan over 2001–2003; however, over 2002–2004, the authors reported an organophosphorus insecticide detection rate of 16.83% (only chlorpyrifos) in 814 fish samples from Taiwan markets [36]. Our results concerning organophosphorus insecticide detection rates are much lower than those of Sun et al. [35,36]. These discrepancies are likely due to differences in the sample sizes and collection sources as well as the differences in the organophosphorus insecticide categories analyzed. These could also have occurred because of the Taiwanese government's implementation of the national action plan for reducing veterinary drug use in 2006 [37].

The predominant organophosphorus insecticide in our fish samples was chlorpyrifos—similar to the observations of a similar survey in Egypt [13]. Even though chlorpyrifos rapidly degrades in the environment, the Egyptian study reported that extensive chlorpyrifos use polluted aquatic habitats and caused increased general toxicity to all vertebrates compared with other classes of insecticides. Chlorpyrifos remains a widely used organophosphorus insecticide for controlling pests in agriculture as well as in sanitation industries around the globe [38]. Because of neurotoxicity concerns, Taiwan's government has banned the use of organophosphorus insecticides (except for trichlorfon) in aquaculture and specified strict MRLs for animal husbandry [4]. In the current study, the insecticides detected may have been illegally applied for the direct treatment of ectoparasite infections occurring in fish or for the treatment of infectious diseases caused by the presence of other aquacultured organisms during mixed breeding in aquaculture ponds, as we revealed previously [4,6,7]. Moreover, we detected chlorpyrifos levels of 0.002 mg/kg, which are much lower than the level revealed by Sun et al. (0.463 mg/kg) [36]. The aquacultured fish also exhibited more variation in chlorpyrifos residues. The chlorpyrifos contamination sources could be fish feed and aquaculture environments, including water and sediments. A study reported that chlorpyrifos rapidly degrades in the marine ecosystem [39], but data on chlorpyrifos contamination in fish feed are limited. Thus, we recommend additional studies examining phthalate content in fish feeds. Information on organophosphorus insecticide use for ectoparasite infection treatment in other special fish species, however, remains limited. The detected levels of organophosphorus insecticide varied among all three fish species; this variation is attributable to differences in the sample sizes used. We thus recommend the execution of further research employing large samples of cultured fish.

Several guidelines provide parameters that can aid in assessing human and other organisms' risks of various conditions after exposure to chemical residues through food; these parameters include ADI, target hazard quotient (THQ), and tolerable daily intake (TDI) [2,4,40]. ADI is recommended by JECFA, but apparent discrepancies in consumption rates and eating habits are not considered in ADI [7]. Through the use of TDI, one could evaluate the risks that are related to consuming specific food products tainted with plasticizers and could identify whether exposure to plasticizers such as phthalates at various degrees is harmful to human health [41]. THQ indicates the assessment of health risk of noncarcinogenic harmful effects [40]. The JECFA [42] and US Environmental Protection Agency [43] have proposed EDI, which can be used to estimate chronic dietary intake at a relatively high accuracy level. In the current study, EDIs were not exceeded by the corresponding ADIs. Because few sulfonamide residues were detected, the EDIs revealed that fish consumption led to a substantially lower dietary intake of sulfamethazine, relative to that specified by the corresponding ADIs, in Taiwan's population. In addition, when evaluated against the ADIs, the EDI corresponded to <1% of the ADIs, signifying that the corresponding risk is negligible [2,7,44]. Farmers engaged in aquaculture in Taiwan may also apply organophosphorus insecticides for ectoparasite treatment [4,45]. In the present study, the consumption of aquacultured fish led to exposure to very low levels of the organophosphorus insecticide chlorpyrifos: in the Taiwanese population, its EDI was much lower than its JECFA-recommended ADI (<0.1% of the ADI for organophosphorus insecticide residues). Therefore, these results were similar to those for sulfamethazine for fish consumption. The risk assessment is negligible because the EDI is <1% of the corresponding ADI [2,7,44]. Taken together, sulfonamide and organophosphorus insecticide residue levels in aquacultured fish in Taiwan may not affect human health adversely.

Our results suggest that to ensure commercial food safety, regulatory authorities as well as producers in Taiwan must endeavor to continually monitor aquacultured products and potential contamination sources. Moreover, considering that antibiotics may exert adverse effects on health and aquatic environments, additional studies on the effects exerted by such pollutants are imperative.

4. Materials and Methods

4.1. Samples, Chemicals, and Reagents

In total, 52 fish samples (20 tilapia, 16 milkfish, and 16 perch samples) were collected from Taiwanese aquafarms in major production areas located in Yunlin, Chiayi, Tainan, Kaohsiung, and Pingtung over June 2018 to October 2019 (Figure S3). As surveyed in 2018 by Fisheries Agency of Taiwan, these fish were bred on a large scale in Taiwan [3]. From the obtained fish samples (each weighing ~1200 g and comprising 2 fish), fish muscles were extracted, followed homogenization and storage at −20 °C until analysis. In the analyses executed in this study, three replicates were extracted from each composite sample to determine each tested compound's average concentration.

Sigma-Aldrich (St. Louis, MO, USA) was the source of the analytical compound standards of the 12 sulfonamides used in this study, namely sulfamethazine, sulfathiazole, sulfamethoxazole, sulfamethoxypyridazine, sulfadoxine, sulfamerazine, sulfaethoxypyridazine, sulfameter, sulfamonomethoxine, sulfadimethoxine, sulfapyridine, and sulfadiazine (all >95.0% purity). Moreover, those of organophosphorus insecticide, namely chlorfenvinphos, chlorpyrifos, diazinon, fenitrothion, fenthion, malathion, methacrifos, methamidophos, methidathion, phoxim, profenophos, pyrazophos, and trichlorfon (all >97.0% purity) were from Dr. Ehrenstorfer GmbH (Augsburg, Germany). In addition, fenamiphos (99.5% purity) were obtained from Bayer CropScience AG (Monheim, Germany), formothion (96.0% purity) from Sandoz India (Mumbai, India), iprobenfos (99.6% purity) from Sigma-Aldrich, and prothiofos (95.5% purity) and triazophos (99.5% purity) from ChemService (West Chester, PA, USA). Merck (Darmstadt, Germany) was the source of chromatography-grade acetone, anhydrous sodium sulfate (ASS), methanol (MeOH), formic acid (FA), ethyl acetate, n-pentane, n-hexane, and acetonitrile (ACN). We also purchased, from Agilent Technologies (Wilmington, DE, USA), 15-mL QuEChERS cleanup tubes (Agilent SampliQ QuEChERS EN fatty dispersive-SPE kit, p/n 5982-5156) and a QuEChERS extraction salt packet (Agilent SampliQ QuEChERS EN Extraction kit, p/n 5982-5650; mixture constituents: 1 g sodium citrate, 1 g NaCl, 0.5 g citric acid disodium salt, and 4 g anhydrous magnesium sulfate).

4.2. Instruments and Apparatus

A nitrogen evaporator (N-Evap-111; Organomation Associates, Berlin, MA, USA), centrifuge (Allegra X-22R; Beckman Coulter, Fullerton, CA, USA), nitrogen generator (Model 05B, System Instruments Co., Tokyo, Japan), and vortex mixer (type 37,600 mixer, Barnstead Thermolyne, Dubuque, IA, USA) were employed in this study for sample preparation. For LC-amenable sulfonamides, LC-MS/MS consisted of a mass spectrometer (ABI 4000 QTRAP, Applied Biosystems, Foster City, CA, USA) in electrospray ionization (ESI) mode and an UltiMate 3000 HPLC system (Thermo Fisher Scientific, Waltham, MA, USA). HPLC separation was executed using an Acquity UPLC® HSS T3 C18 column (2.1 mm internal diameter × 1.8 µm film thickness ×100 mm; Waters, Milford, MA, USA) with chromatographic conditions at 40 °C. To determine the levels of residues of LC-amenable organophosphorus insecticides, the liquid chromatography–tandem mass spectrometry (LC-MS/MS) apparatus comprised an UltiMate 3000 HPLC system and a mass spectrometer (TSQ quantiva triple quadrupole; Thermo Fisher Scientific, Austin, TX USA). In addition, LC-amenable organophosphorus insecticides were separated using a CORTECS UPLC C18 column (2.1 mm internal diameter × 1.6 µm film thickness ×100 mm; Waters, Milford, MA, USA). To analyze gas chromatography (GC)-amenable organophosphorus insecticides, GC-tandem mass spectrometry (GC-MS/MS) was executed on a GC system (Thermo Scientific Trace 1310; Thermo Fisher Scientific, Austin, TX USA) as well as a mass spectrometer (TSQ 8000 triple quadrupole, Thermo Fisher Scientific) that was coupled with Rxi®-5Sil MS fused silica capillary column (0.25 mm internal diameter × 0.25 µm film thickness × 30 m; Restek, Bellefonte, PA, USA).

4.3. Standard Solution Preparation

To prepare stock solutions of individual sulfonamides and organophosphorus insecticide standards in this study, a portion of each analyte (100 mg) was accurately weighed and dissolved in volumetric flasks containing 100 mL of acetone, ACN, or MeOH, according to analyte solubility. For the preparation of working standard mixtures, all stock solution types were combined and subsequently diluted to 1 mg/L. For storage, all solutions were maintained at −20 °C; however, before each use, they were brought to room temperature. The prepared working standard solutions were applied to derive a series of calibration standards through serial dilution in the 1–500 ng/mL range.

4.4. Sample Preparation and Extraction

To detect sulfonamide residues, we used TFDA's directions for multiresidue analysis of residues of veterinary drugs in foods, which required the cleaning and homogenization of fish samples first [46]. In brief, 5 g of the homogenate and 25 mL of ACN in 5% MeOH were mixed on a vortex mixer for 3 min. Then, 10 g of ASS was added to the homogenate and mixed for 10 min; subsequently, centrifugation was executed at 3500 g at 4 °C for 10 min, after which only the supernatant layer was retained. The remaining tissue pellets were re-extracted using 25 mL of ACN in 5% MeOH, followed by centrifugation. The first extract was combined with the previously separated ACN layer. The resulting mixture was subjected to liquid–liquid extraction in a separating funnel. The filtrate was then added to 30 mL of ACN-saturated n-hexane and mixed on a vortex mixer for 10 min. The ACN-extracted layer was dried at 40 °C in a nitrogen evaporator. The evaporation residue was dissolved in 1 mL of 50% MeOH and filtered through a 0.2-µm polyvinylidene fluoride filter (Whatman, Maidstone, UK). Next, the derived filtrate was transferred to an autosampler vial before injection into a chromatograph.

For analyzing residues of insecticides in fish samples, we used the European Committee for Standardization-developed QuEChERS extraction procedure [6,7,47]. In brief, 10 g of homogenized fish sample and 10 mL of ACN were mixed vigorously in a 50-mL centrifuge tube for 1 min; then, QuEChERS extraction salt was added, and the mixture was mixed on a vortex for 1 min and then subjected to a 5-min centrifugation process executed at 3000 g. Thereafter, crude ACN extract (~6 mL) was transferred into QuEChERS cleanup tubes. The ACN layer was mixed vigorously for 2 min and centrifuged at 3000 g for 5 min. Next, 1 mL of the extract was filtered through a 0.2-μm filter membrane and transferred into an autosampler vial for LC-MS/MS. Another 1 mL of extracted solution was near-completely dried in the nitrogen evaporator at 40 °C. Finally, the residue was redissolved with 1 mL of a 1:1 (v/v) n-hexane and acetone mixture, filtered through a 0.2-μm filter membrane, and introduced into an autosampler vial for GC-MS/MS.

4.5. LC-MS/MS Parameters

The injection volume used for detecting the sulfonamide and organophosphorus insecticide residues was 10 μL. The mobile phase was binary, comprising eluents A (0.1% FA) and B (0.1% FA in MeOH), and the gradient of the mobile phase was developed as follows: 5% eluent B from 0 to 2 min (flow rate, 0.3 mL/min); followed by a step increase of eluent B to 20% from 2 to 3 min, 25% from 3 to 6 min, 27% from 6 to 8.6 min, and 37% from 8.6 to 14.5 min; then linear increase to 100% eluent B from 14.5 to 14.7 min; and finally decrease to 4% eluent B at 18.7 min, which was maintained from 18.7 to 20 min. MS was determined in positive ESI modes with monitoring of the two most abundant MS/MS (precursor/product) ion transitions by using an MRM program for each analyte. The MS parameter settings are outlined as follows: collision gas argon pressure, 0.12 mL/min; desolvation flow, 1000 L/h; source temperature, 150 °C; desolvation temperature, 500 °C; dwell time for every MRM transition, 5 ms; cone gas flow, 50 L/h; capillary voltage, 3 kV. Table S1 lists the precursor and corresponding product ions with optimum collision energy obtained through the MRM detection for 12 sulfonamides and LC-amenable 6 organophosphorus insecticides.

4.6. GC-MS/MS Parameters

GC-MS/MS analysis was executed in positive and negative electron-impact ionization interface modes. The carrier gas, namely helium, was applied at a constant flow rate of 1 mL/min. The temperatures of injector were 280 °C. In addition, the oven temperature was set at 60 °C—it was initially maintained isothermal for 1 min, next raised to 170 °C at 40 °C/min, and finally maintained at 310 °C for 8 min. The set source and transfer-line temperatures were 300 and 250 °C, respectively. In the splitless mode, the injection volume was determined to be 10.0 μL. In the collision chamber (second quadrupole), these ions were collision-activated with argon at 4.4 mTorr. Table S2 lists the precursor and corresponding product ions with optimum collision energy obtained through the MRM detection for GC-amenable 12 organophosphorus insecticides.

4.7. Quality Assurance and Validation

To validate our described method, recovery, repeatability, linearity, and LOQ were estimated [2,4,6]. For repeatability and recovery estimation, we spiked blank samples (in triplicate) with a standard mixture of the analytes at two concentrations 5 (low) and 25 (high) ng/g for analysis of sulfonamides and 10 (low) and 50 (high) ng/g for analysis of LC- and GC-amenable organophosphorus insecticides. The recovery was then calculated through the comparison of the noted concentrations of samples spiked before extraction with blanks spiked at the same concentration after extraction.

The derived reproducibility is presented herein as RSD (%); moreover, the LOQs were derived as the analyte concentration that generated a peak signal 3–10 times higher than the background noise in the chromatogram. For evaluation of linearity, matrix-matched calibration was executed by using blank sample extracts and then adding the corresponding amount of the working target compound solution at concentrations of 1–500 ng/mL. Linearity of the calibration curves was detected by fitting least-squares regression analysis in a linear mode ($R^2 \geq 0.990$) in the studied concentration range. All sample concentrations lower than their corresponding LOQs were regarded as undetectable [4,7,47].

4.8. Sulfonamide and Organophosphorus Insecticide EDIs Compared with ADIs

To assess the degree of exposure to veterinary drug residues through fish in Taiwanese people, the corresponding EDIs of the sulfonamide and organophosphorus insecticide residues were determined. The JECFA-recommended ADIs for these residues were used for comparison. EDI calculation was executed as follows [2,6,7]:

$$\text{EDI (ng/kg/day)} = [(\text{daily fish consumption in g/day}) \times (\text{mean veterinary drug concentration in ng/g})]/(\text{human body weight in kg})$$

Moreover, data regarding Taiwan residents' daily seafood consumption were extracted from the results of the National Nutrition and Health Survey by Ministry of Health and Welfare, Taiwan: 96.9 g for men and 74.2 g for women [48]. The mean Taiwanese body weight was considered to be 60 kg [48]. The maximal EDIs were determined from maximum residue concentrations.

5. Conclusions

We report an efficient and sensitive LC and GC-MS/MS-based method for detecting sulfonamide and organophosphorus insecticide residues in fish. The most abundant and the only sulfonamide and organophosphorus insecticide residues detected in all fish samples were sulfamethazine and chlorpyrifos, respectively. Nevertheless, all the detected chemicals were present in trace amounts, much below the TFDA-recommended MRLs, in all 52 fish samples. This indicates that the presence of sulfamethazine and chlorpyrifos use during large-scale breeding of fish in Taiwan does not lead to severe contamination. Furthermore, EDIs for these chemicals in Taiwanese adults were considerably lower than the JECFA-defined ADIs—confirming that no immediate health risk is posed by consuming the aquacultured fish. Therefore, the low sulfamethazine and chlorpyrifos intake through the consumption of contaminated fish in Taiwan seems to present a negligible threat to the health of Taiwanese people. Taiwanese regulatory authorities and aquafarmers may use our findings as a reference for improving aquaculture-related food safety regulation.

We, however, are unaware of contaminants other than sulfonamides and organophosphorus insecticides used in fish production that may be consumed during daily meat intake and may exceed their ADIs to a hazardous extent within the general population. Regardless of our findings, global concern regarding veterinary antibiotic and insecticide contamination and adverse effects on the environment and human health is increasing. Thus, a background information system on veterinary antibiotic and insecticide consumption through fish must be established and improved so as to provide an appropriate monitoring and management framework. Moreover, aquatic samples should be continually surveyed for detecting residues of chemicals and ensuring food safety.

Supplementary Materials: The following are available online, Table S1: MS/MS fragmentation conditions for 12 sulfonamides and LC-amenable 6 organophosphorus insecticides. Table S2: MS/MS fragmentation conditions for GC-amenable 12 organophosphorus insecticides. Table S3: Recovery, repeatability, and limit of quantification of veterinary drugs spiked into tilapia samples. Table S4: Recovery, repeatability, and limit of quantification of organophosphorus insecticides spiked into tilapia samples. Figure S1: LC-MS/MS chromatogram of the detected 12 sulfonamides residues at the quantification ion for sulfamethazine in the positive samples. Figure S2: GC–MS/MS chromatograms of the detected 18 organophosphorus insecticide residues at the quantification ion for chlorpyrifos in the positive samples. Figure S3: Location of 52 sampling areas in Taiwan.

Author Contributions: C.-P.C. and P.-H.H. conceived the idea and executed experiments. W.-C.Y., C.-F.W., C.-C.C., M.-Y.T., and H.-P.T. provided assistance in recombinant construction. C.-T.L., Y.-J.X., and J.-H.W. executed data analysis. G.-R.C. wrote, reviewed, and edited the manuscript. All authors have read and agreed to the published version of the manuscript.

Funding: This study was supported by the Ministry of Science and Technology (Taiwan) (MOST 107-2313-B-415-012) and, in part, by Grant 108AS-8.2.4-BQ-B2 and 108AS-8.2.4-BQ-B3 from the Council of Agriculture, Executive Yuan

(Taiwan) and the Taichung Veterans General Hospital (Taiwan) and National Chung-Hsing University (Taiwan) (TCVGH-NCHU-1087608). This manuscript was edited by Wallace Academic Editing.

Conflicts of Interest: No potential conflict of interest was reported by the authors.

References

1. Sampaio, F.G.; Carra, M.L.; Jonsson, C.M.; Gonçalves, V.T.; Dal'Bo, G.; Nunes, K.; Valim, J.H.; Dallago, B.S.; do Nascimento de Queiroz, S.C.; Reyes, F.G. Effects of dietary exposure to sulfamethazine on the hematological parameters and hepatic oxidative stress biomarkers in Nile Tilapia (*Oreochromis niloticus*). *Bull. Environ. Contam. Toxicol.* **2016**, *97*, 528–535. [CrossRef] [PubMed]
2. Tsai, M.Y.; Lin, C.F.; Yang, W.C.; Lin, C.T.; Hung, K.H.; Chang, G.R. Health risk assessment of banned veterinary drugs and quinolone residues in shrimp through liquid chromatography–tandem mass spectrometry. *Appl. Sci.* **2019**, *9*, 2463. [CrossRef]
3. Fisheries Agency, Council of Agriculture, Taiwan. 2018 Fisheries Statistical Yearbook. 2019. Available online: https://www.fa.gov.tw/cht/PublicationsFishYear/ (accessed on 2 January 2020).
4. Chang, H.Y.; Yang, W.C.; Xue, Y.J.; Tsai, M.Y.; Wang, J.H.; Chang, G.R. Phthalates and organophosphorus insecticide residues in shrimp determined by liquid/gas chromatography–tandem mass spectrometry and a health risk assessment. *Mar. Pollut. Bull.* **2019**, *144*, 140–145. [CrossRef] [PubMed]
5. Liao, I.C. Aquaculture practices in Taiwan and its visions. *J. Fish. Soc. Taiwan* **2005**, *32*, 193–206.
6. Chang, G.R.; Chen, H.S.; Lin, F.Y. Analysis of banned veterinary drugs and herbicide residues in shellfish by liquid chromatography-tandem mass spectrometry (LC/MS/MS) and gas chromatography-tandem mass spectrometry (GC/MS/MS). *Mar. Pollut. Bull.* **2016**, *113*, 579–584. [CrossRef]
7. Chang, G.R. Persistent organochlorine pesticides in aquatic environments and fishes in Taiwan and their risk assessment. *Environ. Sci. Pollut. Res. Int.* **2018**, *25*, 7699–7708. [CrossRef]
8. Chang, G.R. Surveys on banned veterinary drugs residues in marine bivalves and gastropods in Taiwan between 2010 and 2015: A mini review. *J. Aquat. Pollut. Toxicol.* **2017**, *1*, 1–5.
9. Li, J.H.; Yu, W.J.; Lai, Y.H.; Ko, Y.C. Major food safety episodes in Taiwan: Implications for the necessity of international collaboration on safety assessment and management. *Kaohsiung J. Med. Sci.* **2012**, *28*, S10–S16. [CrossRef]
10. Peng, G.J.; Chang, M.H.; Fang, M.C.; Liao, C.D.; Tsai, C.F.; Tseng, S.H.; Kao, Y.M.; Chou, H.K.; Cheng, H.F. Incidents of major food adulteration in Taiwan between 2011 and 2015. *Food Control* **2017**, *72*, 145–152. [CrossRef]
11. He, X.T.; Wang, Q.; Nie, X.P.; Yang, Y.T.; Cheng, Z. Residues and health risk assessment of sulfonamides in sediment and fish from typical marine aquaculture regions of Guangdong Provinces, China. *Environ. Sci.* **2013**, *35*, 2728–2734.
12. Saglam, N.; Yonar, M.E. Effects of sulfamerazine on selected haematological and immunological parameters in rainbow trout (*Onchorhynchus mykiss*, Walbaum, 1792). *Aquacul. Res.* **2009**, *40*, 395–404. [CrossRef]
13. Malhat, F.M.; Nasr, I. Organophosphorus pesticides residues in fish samples from the River Nile tributaries in Egypt. *Bull. Environ. Contam. Toxicol.* **2011**, *87*, 689–692. [CrossRef] [PubMed]
14. Nunes, K.S.D.; Vallim, J.H.; Assalin, M.R.; Queiroz, S.C.N.; Paraíba, L.C.; Jonsson, C.M.; Reyes, F.G.R. Depletion study, withdrawal period calculation and bioaccumulation of sulfamethazine in tilapia (*Oreochromis niloticus*) treated with medicated feed. *Chemosphere* **2018**, *197*, 89–95. [CrossRef] [PubMed]
15. Mojsak, P.; Łozowicka, B.; Kaczyński, P. Estimating acute and chronic exposure of children and adults to chlorpyrifos in fruit and vegetables based on the new, lower toxicology data. *Ecotoxicol. Environ. Saf.* **2018**, *159*, 182–189. [CrossRef]
16. Food and Drug Administration, Taiwan (TFDA); Ministry of Health and Welfare, Taiwan. Method of Test for Veterinary Drug Residues In-Method for Multiresidue Analysis. 2019. Available online: https://www.fda.gov.tw/TC/newsContent.aspx?cid=3&id=25664 (accessed on 12 January 2020).
17. Food and Drug Administration, Taiwan (TFDA); Ministry of Health and Welfare, Taiwan. Method of Test for Pesticide Residues in Livestock and Poultry Products-Multiresidue Analysis. 2016. Available online: https://www.fda.gov.tw/TC/newsContent.aspx?cid=3&id=21426 (accessed on 15 January 2020).
18. Sun, F.; Wong, S.S.; Li, G.C.; Chen, S.N. Multiresidue determination of pesticide in fishery products by a tandem solid-phase extraction technique. *J. Food Drug Anal.* **2005**, *2*, 151–158.

19. Sun, X.; Zhu, F.; Xi, J.; Lu, T.; Liu, H.; Tong, Y.; Ouyang, G. Hollow fiber liquid-phase microextraction as clean-up step for the determination of organophosphorus pesticides residues in fish tissue by gas chromatography coupled with mass spectrometry. *Mar. Pollut. Bull.* **2011**, *63*, 102–107. [CrossRef]
20. Li, Y.G.; Mao, Y.N.; Chen, Z.L.; Liu, X.; Luo, P.; Zhou, Y.; Wen, S. Determination of seven organophosphate pesticide residues in salted fish by high performance liquid chromatography-tandem mass spectrometry with gel permeation chromatography. *J. Food Saf. Qual.* **2014**, *5*, 2670–2676.
21. Lazartigues, A.; Wiest, L.; Baudot, R.; Thomas, M.; Feidt, C.; Cren-Olivé, C. Multiresidue method to quantify pesticides in fish muscle by QuEChERS-based extraction and LC-MS/MS. *Anal. Bioanal. Chem.* **2011**, *400*, 2185–2193. [CrossRef]
22. Chen, H.M.; Chao, W.P.; Huang, C.N.; Wu, B.S.; Liao, C.D.; Tseng, S.H.; Kao, Y.M. Development of a liquid chromatography-tandem mass spectrometry method for quantifying amprolium residues in animal food tissues. *Taiwan. J. Agric. Chem. Food Sci.* **2015**, *53*, 160–170.
23. Song, C.; Li, L.; Zhang, C.; Kamira, B.; Qiu, L.; Fan, L.; Wu, W.; Meng, S.; Hu, G.; Chen, J. Occurrence and human dietary assessment of sulfonamide antibiotics in cultured fish around Tai Lake, China. *Environ. Sci. Pollut. Res. Int.* **2017**, *24*, 17493–17499. [CrossRef]
24. Sinigoj-Gacnik, K.; Doganoc, D.Z. Contamination of farm animals and fishes from Slovenia with heavy metals and sulfonamides. *Bull. Environ. Contam. Toxicol.* **2000**, *64*, 235–241. [PubMed]
25. Barani, A.; Fallah, A.A. Occurrence of tetracyclines, sulfonamides, fluoroquinolones and florfenicol in farmed rainbow trout in Iran. *Food Agric. Immunol.* **2015**, *26*, 420–429. [CrossRef]
26. Won, S.Y.; Lee, C.H.; Chang, H.S.; Kim, S.O.; Lee, S.H.; Kim, D.S. Monitoring of 14 sulfonamide antibiotic residues in marine products using HPLC-PDA and LC-MS/MS. *Food Control.* **2011**, *22*, 1101–1107. [CrossRef]
27. Fu, H.P.; Kuo, H.W.; Shih, C.C.; Lin, H.C.; Lin, K.H.; Lin, Y.R.; Chou, H.K.; Shyu, J.F.; Pan, J.Q.; Hsu, C.K.; et al. 2013 Post-market survey on veterinary drug residues in livestock and aquatic products. *Ann. Rept. Food Drug Res.* **2014**, *5*, 81–91.
28. Fu, H.P.; Lin, Y.R.; Su, H.C.; Wang, T.S.; Hsu, C.H.; Liu, F.M.; Feng, R.L.; Hsu, C.K.; Liu, L.W.; Tzeng, G.S.; et al. 2014 Post-market survey on veterinary drugresidues in livestock and aquatic products. *Ann. Rept. Food Drug Res.* **2015**, *5*, 67–75.
29. Chen, C.M.; Fu, H.P.; Lin, Y.Y.; Lin, Y.R.; Chen, T.L.; Chiu, H.Y. 2015 Postmarketing surveillence study on veterinary drug residues in livestock, poultry and aquatic products. *Ann. Rept. Food Drug Res.* **2016**, *7*, 28–36.
30. Chen, C.M.; Hang, S.F.; Lin, Y.Y.; Lin, Y.R.; Chen, T.L. Post-market surveillance study on veterinary drug residues in poultry, livestock and aquatic products in 2016. *Ann. Rept. Food Drug Res.* **2017**, *8*, 54–62.
31. Fu, H.P.; Chen, C.M.; Hang, S.F.; Lin, Y.Y.; Lin, Y.R.; Chen, T.L. Post-market surveillance study on veterinary drug residues in poultry, livestock and aquatic products in 2017. *Ann. Rept. Food Drug Res.* **2018**, *9*, 115–124.
32. Fu, H.P.; Hang, Z.J.; Lin, Y.R.; Chen, T.L. Post-market surveillance study on veterinary drug residues in poultry, livestock and aquatic products in 2018. *Ann. Rept. Food Drug Res.* **2019**, *10*, 50–57.
33. Hou, L.; Yin, G.; Liu, M.; Zhou, J.; Zheng, Y.; Gao, J.; Zong, H.; Yang, Y.; Gao, L.; Tong, C. Effects of sulfamethazine on denitrification and the associated N_2O release in estuarine and coastal sediments. *Environ. Sci. Technol.* **2015**, *49*, 326–333. [CrossRef]
34. Zhang, Q.; Jia, A.; Wan, Y.; Liu, H.; Wang, K.; Peng, H.; Dong, Z.; Hu, J. Occurrences of three classes of antibiotics in a Natural River basin: Association with antibiotic-resistant *Escherichia coli*. *Environ. Sci. Technol.* **2014**, *48*, 14317–14325. [CrossRef] [PubMed]
35. Sun, F.; Wong, S.S.; Li, G.C.; Chen, S.N. A preliminary assessment of consumer's exposure to pesticide residues in fisheries products. *Chemosphere* **2006**, *62*, 674–680. [CrossRef] [PubMed]
36. Sun, F.; Chen, H.S. Monitoring of pesticide chlorpyrifos residue in farmed fish: Investigation of possible sources. *Chemosphere* **2008**, *71*, 1866–1869. [CrossRef] [PubMed]
37. Fu, H.P.; Chen, H.C.; Chou, H.K.; Chiou, H.S.; Chou, K.L.; Lai, K.M.; Ku, C.C.; Hsu, C.T.; Hsu, Y.H.; Chen, S.O.; et al. Survey on veterinary drug residues in food of animal origin. *Ann. Rept. Food Drug Res.* **2011**, *2*, 145–153.
38. Liu, B.; McConnell, L.L.; Torrents, A. Hydrolysis of chlorpyrifos in natural waters of the Chesapeake Bay. *Chemosphere* **2001**, *44*, 1315–1323. [CrossRef]
39. Kales, S.P.; Carvalho, F.P.; Raghu, K.; Sherkhane, P.D.; Pandit, G.G.; Rao, A.M.; Mukherjee, P.K.; Murthy, M. Studies on degradation of 14C-chlorpyrifos in the marine environment. *Chemosphere* **1999**, *39*, 969–976. [CrossRef]

40. Gu, Y.G.; Lin, Q.; Wang, X.H.; Du, F.Y.; Yu, Z.L.; Huang, H.H. Heavy metal concentrations in wild fishes captured from the South China Sea and associated health risks. *Mar. Pollut. Bull.* **2015**, *96*, 508–512. [CrossRef]
41. Tsai, M.Y.; Ho, C.H.; Chang, H.Y.; Yang, W.C.; Lin, C.F.; Lin, C.T.; Xue, Y.J.; Lai, J.M.; Wang, J.H.; Chang, G.R. Analysis of pollution of phthalates in pork and chicken in Taiwan using liquid chromatography-tandem mass spectrometry and assessment of health risk. *Molecules* **2019**, *24*, 3817. [CrossRef]
42. Prado, C.K.; Ferreira, F.D.; Bando, E.; Machinski, M., Jr. Oxytetracycline, tetracycline, chlortetracycline and doxycycline in pasteurised cow's milk commercialised in Brazil. *Food Addit. Contam. Part B Surveill.* **2015**, *8*, 81–84. [CrossRef]
43. Zhang, G.; Pan, Z.; Bai, A.; Li, J.; Li, X. Distribution and bioaccumulation of organochlorine pesticides (OCPs) in food web of Nansi Lake, China. *Environ. Monit. Assess.* **2014**, *186*, 2039–2051. [CrossRef]
44. Vragović, N.; Bazulić, D.; Njari, B. Risk assessment of streptomycin and tetracycline residues in meat and milk on Croatian market. *Food Chem. Toxicol.* **2011**, *49*, 352–355. [CrossRef] [PubMed]
45. Hsueh, Y.S. The Transformation of Taiwan into the empire of the giant tiger prawn (1968–1988): The roles of the government and the people. *Acad. Hist.* **2010**, *24*, 139–176.
46. Lee, H.C.; Chen, C.M.; Wei, J.T.; Chiu, H.Y. Analysis of veterinary drug residue monitoring results for commercial livestock products in Taiwan between 2011 and 2015. *J. Food Drug Anal.* **2018**, *26*, 565–571. [CrossRef] [PubMed]
47. Shen, Y.R.; Cheng, M.W.; Wu, B.S.; Yang, K.C.; Chang, Y.H.; Tseng, S.H.; Kao, Y.M.; Chiueh, L.C.; Shih, D.Y.C. Developement of the method for analysis of multiple pesticide residues in animal matrices by QuEChERS method. *Taiwan. J. Agric. Chem. Food Sci.* **2013**, *51*, 148–160.
48. Wu, S.J.; Chang, Y.H.; Fang, C.W.; Pan, W.H. Food sources of weight, calories, and three macro-nutrients-NAHSIT 1993–1996. *Nutr. Sci. J.* **1999**, *24*, 41–58.

Sample Availability: Samples of the compounds are available from the authors.

molecules

MDPI

Article

Briarenols I—K, New Anti-inflammatory 8,17-Epoxybriaranes from the Octocoral *Briareum excavatum* (Briareidae)

Thanh-Hao Huynh [1,2], Lee-Shing Fang [3,4], Yu-Hsin Chen [2], Bo-Rong Peng [2], You-Ying Chen [2], Li-Guo Zheng [2], Yu-Jen Wu [5], Zhi-Hong Wen [6], Jih-Jung Chen [7], Tzu-Chi Lin [8,*] and Ping-Jyun Sung [1,2,6,9,10,*]

1 Graduate Institute of Marine Biology, National Dong Hwa University, Pingtung 94450, Taiwan; haohuynh0108@gmail.com

2 National Museum of Marine Biology and Aquarium, Pingtung 94450, Taiwan; kb5634@yahoo.com.tw (Y.-H.C.); pengpojung@gmail.com (B.-R.P.); zoeblack0108@gmail.com (Y.-Y.C.); t0919928409@gmail.com (L.-G.Z.)

3 Center for Environmental Toxin and Emerging-Contaminant Research, Cheng Shiu University, Kaohsiung 83347, Taiwan; lsfang@gcloud.csu.edu.tw

4 Super Micro Mass Research and Technology Center, Cheng Shiu University, Kaohsiung 83347, Taiwan

5 Department of Nursing, Meiho University, Pingtung 91202, Taiwan; x00002180@meiho.edu.tw

6 Department of Marine Biotechnology and Resources, National Sun Yat-sen University, Kaohsiung 80424, Taiwan; wzh@mail.nsysu.edu.tw

7 Faculty of Pharmacy, School of Pharmaceutical Sciences, National Yang-Ming University, Taipei 11221, Taiwan; chenjj@ym.edu.tw

8 Department of Emergency Medicine, Antai Medical Care Corporation Antai Tian-Sheng Memorial Hospital, Pingtung 92842, Taiwan

9 Chinese Medicine Research and Development Center, China Medical University Hospital, Taichung 40447, Taiwan

10 Graduate Institute of Natural Products, Kaohsiung Medical University, Kaohsiung 80708, Taiwan

* Correspondence: Bryan1110@hotmail.com (T.-Z.L.); pjsung@nmmba.gov.tw (P.-J.S.); Tel.: +886-8-882-5037 (T.-Z.L.); Fax: +886-8-882-5087 (P.-J.S.)

Received: 7 February 2020; Accepted: 16 March 2020; Published: 19 March 2020

Abstract: Five 8,17-epoxybriaranes, including three new compounds—briarenols I–K (**1–3**), along with two known analogues, briaexcavatolide P (**4**) and briaexcavatin P (**5**), were isolated from the octocoral *Briareum excavatum*. The structures of briaranes **1–3** were elucidated by spectroscopic methods, including 1D and 2D NMR studies and (+)-HRESIMS. Briarane **4** exerted inhibition effects on inducible nitric oxide synthase (iNOS) and cyclooxygenase-2 (COX-2) release from RAW 264.7.

Keywords: *Briareum excavatum*; briarenol; briarane; anti-inflammatory; iNOS; COX-2

1. Introduction

Octocorals of the genus *Briareum* (family Briareidae) [1–4] are proven to be the most important source to produce briarane-type diterpenoids [5]. The compounds of this type are only found in marine invertebrates, particularly in octocorals and demonstrated a wide spectrum of bioactivities, such as anti-inflammatory activity [6] and cytotoxicity [7]. In our continuing research into the chemical constituents of an octocoral *B. excavatum* (Nutting 1911), which was distributed extensively in the coral reefs of Taiwan, have resulted in isolation of three previously unreported 8,17-epoxybriaranes–briarenols I–K (**1–3**) along with two known analogues, briaexcavatolide P (**4**) [8] and briaexcavatin P (**5**) [9],

(Figure 1). In the current study, the comprehensive workflow of isolation, structure determination, and anti-inflammatory activity evaluation, was implemented on briaranes **1–5**.

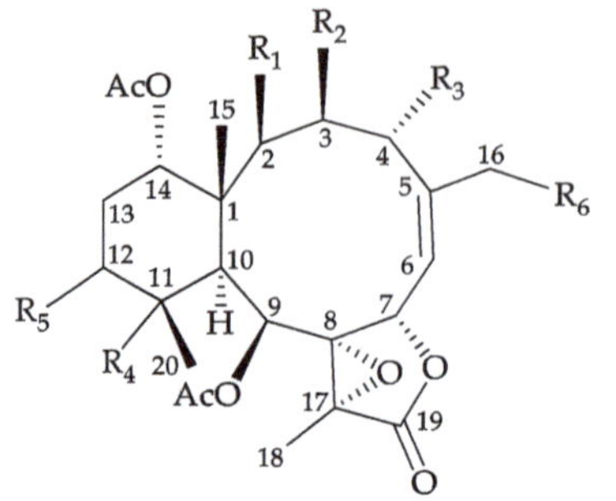

1: R_1 = OAc, R_2 = R_6 = OH, R_3 = *n*-OC(O)Pr, R_4 = H, R_5 = β-OH
2: R_1 = R_2 = OH, R_3 = OAc, R_4 = R_6 = H, R_5 = β-OH
3: R_1 = OAc, R_2 = R_3 = R_6 = H, R_4 = OH, R_5 = β-OH
4: R_1 = OAc, R_2 = OH, R_3 = *n*-OC(O)Pr, R_4 = R_6 = H, R_5 = β-OH
5: R_1 = R_3 = OAc, R_2 = OH, R_4 = R_6 = H, R_5 = β-OH
6: R_1 = OAc, R_2 = *n*-OC(O)Pr, R_3 = R_4 = R_6 = H, R_5 = β-OH
7: R_1 = OAc, R_2 = R_3 = R_6 = H, R_4 = OH, R_5 = α-OH

Figure 1. Structures of briarenols I–K (**1–3**), briaexcavatolide P (**4**), briaexcavatin P (**5**), excavatolide B (**6**), and briareolide B (**7**).

2. Results and Discussion

Briarenol I (**1**) was isolated as an amorphous powder and displayed a sodiated adduct ion at *m*/*z* 649.24677 in the (+)-HRESIMS, which indicated its molecular formula was $C_{30}H_{42}O_{14}$ (calculated for $C_{30}H_{42}O_{14}$ + Na, 649.24668; unsaturation degrees = 10). The IR spectrum revealed absorptions for hydroxy (ν_{max} 3524 cm^{-1}), γ-lactone (ν_{max} 1783 cm^{-1}), and ester carbonyl (ν_{max} 1736 cm^{-1}) moieties. Resonances in the ^{13}C NMR of **1** at δ_C 172.9, 172.3, 170.5, 170.0, and 170.0 (5 × C) supported the presence of a γ-lactone and four esters (Table 1). Three of the esters were identified as acetates by the presence of three methyl singlet resonances in the ^1H NMR spectrum at δ_H 2.34, 2.15, and 2.08 (Table 2) and the remaining ester was found to be an *n*-butyroxy group based on ^1H NMR studies, including a correlation spectroscopy (COSY) experiment, which revealed seven contiguous protons (δ_H 2.30, 2H, t, *J* = 7.2 Hz; 1.63, 2H, tq, *J* = 7.2 Hz; 0.95, 3H, t, *J* = 7.2 Hz). From the COSY spectrum (Figure 2), the proton sequences from H-6/H-7, H-9/H-10/H-11/H-12/H_2-13/H-14, and H-11/H_3-20 were established. The hydroxy proton signals at δ_H 4.30 (1H, d, *J* = 12.0 Hz), 1.49 (1H, d, *J* = 4.0 Hz), and 3.49 (1H, dd, *J* = 9.6, 4.4 Hz) were found to correlate with H-3 (δ_H 4.59, d, *J* = 12.0 Hz), H-12 (δ_H 4.10, m), and H_2-16 (δ_H 4.35, 1H, dd, *J* = 13.6, 4.4 Hz; 4.04, 1H, dd, *J* = 13.6, 9.6 Hz), respectively. Thus, the hydroxy groups should be positioned at C-3, C-12, and C-16, respectively. Olefinic resonances in the ^{13}C NMR at δ_C 125.5 (CH-6) and 142.0 (C-5) indicated the presence of a trisubstituted carbon–carbon double bond. On the basis of these data and the heteronuclear multiple bond correlation (HMBC) experiment (Figure 2), the connectivity from C-1 to C-14 was established. A hydroxymethyl group at C-5 was revealed by the HMBC between C-16 oxymethylene protons to C-4, C-5, and C-6. The C-15 methyl group at C-1 was confirmed by the HMBC between H_3-15/C-1, C-2, C-10, C-14, and H-10/C-15. The *n*-butyrate positioned at C-4 was confirmed from the connectivity between H-4 (δ_H 6.14) and the carbonyl carbon of *n*-butyrate group (δ_C 172.3). HMBC from the oxymethine protons at δ_H 4.53 (H-2), 5.32 (H-9), and 4.88 (H-14) to the acetate carbonyls at δ_C 172.9, 170.0, and 170.0, placed the acetoxy groups on C-2, C-9, and C-14, respectively. Thirteen of the fourteen oxygen atoms in the molecular formula of 1 could be accounted for from the presence of a γ-lactone, four esters, and three hydroxy groups. The remaining oxygen atom had to be placed between C-8 and C-17 to form a tetrasubstituted epoxide based on the ^{13}C NMR evidences at δ_C 70.8 (C-8) and 62.5 (C-17) and the ^1H NMR chemical shift of a tertiary methyl at δ_H 1.66 (3H, s, H_3-18).

Table 1. ^{13}C NMR (δ_C 100 MHz, $CDCl_3$) data for briaranes **1–3**.

Position	1	2	3
1	43.3, C [a]	43.7, C	47.7, C
2	87.4, CH	85.6, CH	74.9, CH
3	73.1, CH	73.6, CH	31.6, CH_2
4	66.0, CH	65.9, CH	28.4, CH_2
5	142.0, C	139.3, C	144.8, C
6	125.5, CH	124.3, CH	118.4, CH
7	74.1, CH	74.0, CH	74.9, CH
8	70.8, C	69.9, C	70.8, C
9	66.2, CH	67.1, CH	67.4, CH
10	40.5, CH	41.4, CH	49.0, CH
11	37.2, CH	36.4, CH	78.2, C
12	66.6, CH	67.0, CH	73.4, CH
13	30.2, CH_2	30.4, CH_2	30.2, CH_2
14	80.5, CH	80.1, CH	74.8, CH
15	18.6, CH_3	19.1, CH_3	14.3, CH_3
16	62.5, CH_2	16.8, CH_3	27.2, CH_3
17	62.5, C	61.8, C	66.5, C
18	10.3, CH_3	10.3, CH_3	10.4, CH_3
19	170.5, C	170.9, C	170.4, C
20	8.9, CH_3	8.7, CH_3	16.9, CH_3
OAc-2	172.9, C		170.2, C
	21.2, CH_3		21.2, CH_3
OAc-4		169.5, C	
		21.0, CH_3	
OAc-9	170.0, C	169.2, C	168.1, C
	21.5, CH_3	21.1, CH_3	21.5, CH_3
OAc-14	170.0, C	170.0, C	170.4, C
	21.2, CH_3	21.0, CH_3	21.3, CH_3
n-OC(O)Pr-4	172.3, C		
	35.9, CH_2		
	18.2, CH_2		
	13.7, CH_3		

[a] Multiplicity deduced by DEPT and HSQC spectra.

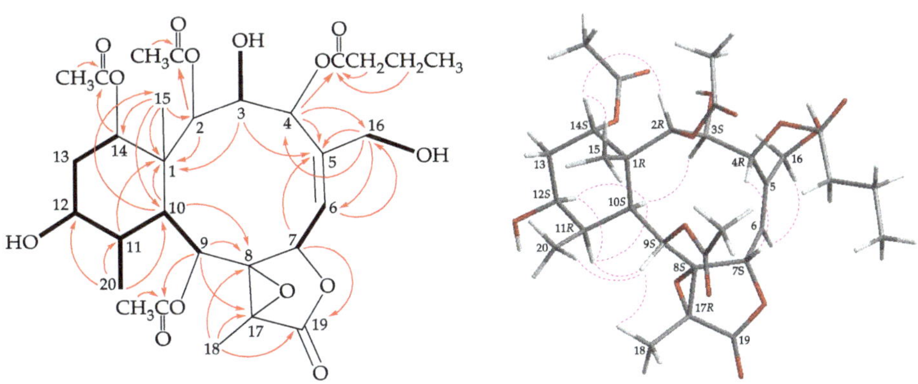

Figure 2. The COSY (▬) correlations, selective HMBC (⌒), and protons with key NOESY correlations (⌒) of **1**.

Table 2. ^{1}H NMR (δ_H, 400 MHz in $CDCl_3$) data (J in Hz) for briaranes **1–3**.

Position	1	2	3
2	4.53 s	3.45 d (10.4)	5.13 d (8.4)
3α/β	4.59 d (12.0)	4.27 d (10.4)	1.67 m; 2.60 ddd (16.0, 14.8, 6.0)
4/4′	6.14 s	6.05 d (1.2)	2.48 br d (16.0); 1.90 m
6	5.53 d (6.0)	5.29 dq (6.4, 1.6)	5.19 s
7	5.62 d (6.0)	5.71 d (6.4)	5.19 s
9	5.32 d (8.8)	5.26 d (9.2)	5.78 d (1.2)
10	2.64 dd (8.8, 4.8)	2.55 dd (9.2, 5.2)	2.14 br s
11	2.41 m	2.47 m	
12	4.10 m	4.05 m	3.72 dd (12.4, 4.8)
13α/β	1.75 m; 2.01 m	1.69 m; 2.00 m	1.67 m; 2.04 m
14	4.88 dd (2.8, 2.8)	4.92 dd (2.8, 2.8)	4.79 dd (2.4, 2.0)
15	0.99 s	0.99 br s	1.21 s
16a/b	4.35 dd (13.6, 4.4); 4.04 dd (13.6, 9.6)	1.89 br s	1.99 s
18	1.66 s	1.66 s	1.77 s
20	1.12 d (6.8)	1.07 d (7.2)	1.15 s
OH-2		2.79 d (10.4)	
OH-3	4.30 d (12.0)	2.87 d (10.4)	
OH-12	1.49 d (4.0)	1.43 d (4.0)	-
OH-16	3.49 dd (9.6, 4.4)		
OAc-2	2.08 s		2.00 s
OAc-4		2.14 s	
OAc-9	2.34 s	2.32 s	2.22 s
OAc-14	2.15 s	2.16 s	2.03 s
n-OC(O)Pr-4	0.95 t (7.2) 1.63 tq (7.2) 2.30 t (7.2)		

The stereochemistry of **1** was deduced from an NOESY experiment (Figure 2) and biogenetic considerations. The NOE correlations of H-10/H-11, H-10/H-12, and H-11/H-12 indicated that these protons were situated on the same face of the structure and were assigned as the α protons since the C-15 methyl is the β-substituent at C-1. The NOE correlation between H_3-15 and H-14 implied that H-14 had a β-orientation. H-3 exhibited a correlation with H-10, and, as well as a lack of coupling constants were detected between H-2/H-3 and H-3/H-4, indicating the dihedral angles between H-2/H-3 and H-3/H-4 were approximately 90° and the 2-acetoxy, 3-hydroxy, and 4-*n*-butyroxy groups were β-, β-, and α-oriented, respectively. A correlation from H-4 to H-7, suggested that H-7 was β-oriented. The *Z*-configuration of C-5/6 double bond was confirmed based on the fact that the C-6 olefinic proton (δ_H 5.53) correlated to one of the C-16 hydroxymethyl protons (δ_H 4.04). H-9 was found to correlate with H-11, H_3-18, and H_3-20. From a consideration of molecular model, H-9 was found to be reasonably close to H-11, H_3-18, and H_3-20, thus, H-9 should be placed on the α face, and Me-18 was β-oriented in the γ-lactone moiety, and the 8,17-epoxy group should be α-oriented. It was found that the NMR signals of **1** were similar to those of a known briarane, briaexcavatolide P (**4**) (Figure 1) [8], except that the signals corresponding to the Me-16 vinyl methyl in **4** were replaced by signals for a hydroxymethyl group in **1**. Additionally, as briaranes **1–5** were isolated along with a known briarane, excavatolide B (**6**) [6,10,11] from the same target organism, *B. excavatum*, and the absolute configuration of **6** was determined by a single-crystal X-ray diffraction analysis [6,11]. Therefore, on biogenetic grounds to assume that briaranes **1–5** had the same absolute stereochemistry as that of **6**, tentatively, and the configurations of stereogenic carbons of **1** were determined as 1*R*,2*R*,3*S*,4*R*,7*S*,8*S*, 9*S*,10*S*,11*R*,12*S*,14*S*, and 17*R* (Supplementary Materials, Figures S1–S10).

Briarenol J (**2**) had a molecular formula $C_{26}H_{36}O_{12}$ by its (+)-HRESIMS at *m/z* 563.21007 (calculated for $C_{26}H_{36}O_{12}$ + Na, 563.20990). The IR spectrum showed bands at 3483, 1779, and 1727 cm^{-1}, consistent with the presence of hydroxy, γ-lactone, and ester groups, respectively, in **2**. From the ^{13}C and DEPT data (Table 2), one trisubstituted double bond was deduced from the signals of two carbons at δ_C 139.3 (C-5) and 124.3 (CH-6). A methyl-containing tetrasubstituted epoxy group was confirmed from the signals of two oxygenated quaternary carbons at δ_C 69.9 (C-8) and 61.8 (C-17), and from the chemical shift of a tertiary methyl (δ_H 1.66, 3H, s; δ_C 10.3, CH_3-18; Tables 1 and 2). Four carbonyl resonances at δ_C 170.9, 170.0, 169.5, and 169.2 in the ^{13}C spectrum confirmed the presence of a γ-lactone and three esters. All the esters were identified as acetates by the presence of three methyl singlet resonances in the ^1H NMR spectrum at δ_H 2.32, 2.16, and 2.14, respectively.

Coupling constants information in the COSY spectrum of **2** enabled identification of H-6/H-7, H-9/H-10/H-11/H-12/H_2-13/H-14, H-11/H_3-20, and H-6/H_3-16 (by allylic coupling; Figure 3), these data, together with the HMBC experiment (Figure 3), the molecular framework of **2** could be established. The HMBC also indicated that the acetoxy groups should be attached at C-4, C-9, and C-14, respectively. Thus, the remaining hydroxy groups have to be positioned at C-2, C-3, and C-12, as indicated by the COSY correlations between H-2/OH-2, H-3/OH-3, and H-12/OH-12.

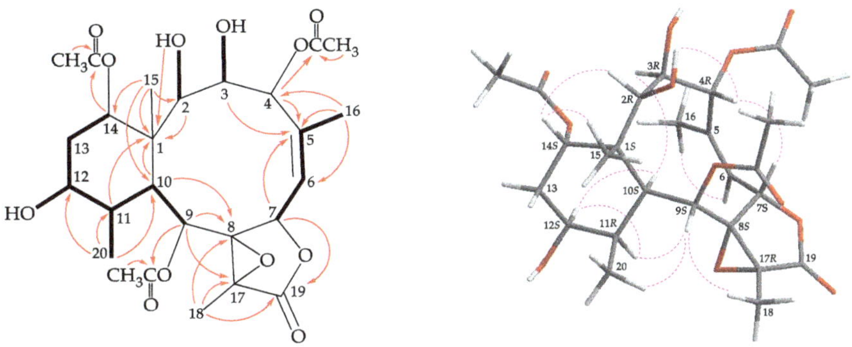

Figure 3. The COSY (▬) correlations, selective HMBC (⌒), and protons with key NOESY correlations (⌒) of **2**.

The stereochemistry of **2** was elucidated from the NOE interactions observed in a NOESY experiment (Figure 3) and by the vicinal ^1H–^1H coupling constant analysis. In the NOESY spectrum, correlations were observed between H-10 with H-3 and H-12; and H-12 correlated with H-11, indicating that these protons should be α-oriented. H-14 gave a correlation with H_3-15, confirming the β-orientation for this proton. H-2 showed a correlation with H-14, and a lack of coupling constant was detected between H-2/H-3, indicating the dihedral angle between H-2/H-3 is approximately 90° and the 2-hydroxy group was β-oriented. H-4 exhibited correlations with H-7 and 2-hydroxy proton, confirming the β-orientations for H-4 and H-7. H-9 was found to show correlations with H-11, H_3-18, and H_3-20, and from molecular models, H-9 and H_3-18 should be placed on the α- and β-face, respectively. The *Z*-configuration of C-5/C-6 double bond was elucidated by a correlation between H-6 and H_3-16. The NMR data of **2** were found to be similar to those of a known briarane, briaexcavatin P (**5**) [9]. It was found that the 2-acetoxy substituent in **5** was replaced by a hydroxy group in **2**. By comparison of the proton and carbon chemical shifts, coupling constants, NOESY correlations, and rotation value of **2** with those of **5**, the stereochemistry of **2** was confirmed to be the same as that of **5**, and the configurations of the stereogenic centers of **2** were assigned as 1*S*,2*R*,3*R*,4*R*, 7*S*,8*S*,9*S*,10*S*,11*R*,12*S*,14*S*, and 17*R* (Supplementary Materials, Figures S11–S20).

Briarane **3** (briarenol K) was found to have a molecular formula of $C_{26}H_{36}O_{11}$ based on its (+)-HRESIMS peak at *m/z* 547.21514 (calculated for $C_{26}H_{36}O_{11}$ + Na, 547.21498). Its absorption peaks in the IR spectrum showed ester carbonyl, γ-lactone, and broad OH stretching at 1739, 1780, and 3468 cm^{-1}, respectively. The ^{13}C NMR spectrum indicated that three esters and a γ-lactone were present, as carbonyl resonances were observed at δ_C 168.1, 170.2, 170.4, and 170.4, respectively (Table 1). The ^1H NMR data also indicated that presence of three acetate methyls at δ_H 2.22, 2.03, and 2.00 (each 3H × s; Table 2). It was found that the spectroscopic data of **3** were similar to those of a known briarane, briareolide B (**7**) [12]; however, by comparison of the ^1H and ^{13}C NMR chemical shifts of CH-12 oxymethine (δ_H 3.72, 1H, dd, *J* = 12.4, 4.8 Hz; δ_C 73.4), CH_2-13 sp^3 methylene (δ_H 1.67, 1H, m; 2.04, 1H, m; δ_C 30.2), C-11 oxygenated quaternary carbon (δ_C 78.2), and Me-20 tertiary methyl (δ_H 1.15, 3H, s; δ_C 16.9) of **3** with those of **7** (δ_H 3.56, 1H, m; δ_C 73.9, CH-12; δ_H 2.03, 1H, m; 2.12, 1H, m; δ_C 27.6, CH_2-13; δ_C 74.7, C-11; δ_H 1.16, 3H, s; δ_C 22.5, Me-20) [12] showed that the hydroxy group at C-12 in **3** was β-oriented. The locations of the functional groups were further confirmed by other HMBC and COSY correlations (Figure 4), and hence briarenol K was assigned the structure of **3**. The NOESY spectrum exhibited a correlation from H-10 to H-12, further supporting that H-12 was α-oriented and the stereogenic centers of **3** were assigned as 1*S*,2*S*,7*S*,8*S*,9*S*,10*S*,11*S*,12*S*,14*S*, and 17*R*, by the correlations observed in a NOESY spectrum (Figure 4) and this compound was found to be the 12-epimer of briareolide B (**7**) [12] (Supplementary Materials, Figures S21–S30).

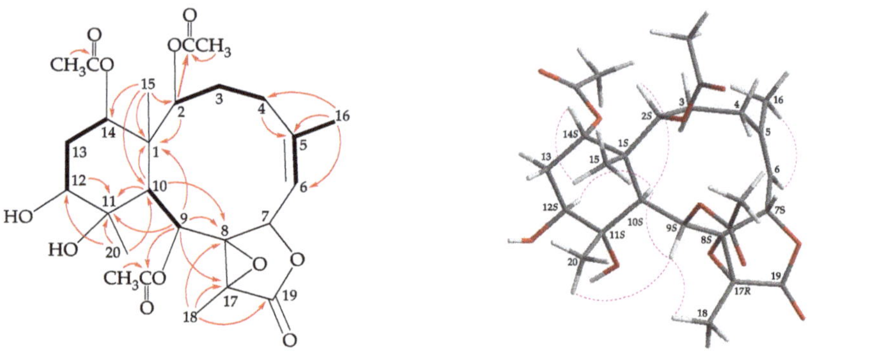

Figure 4. The COSY (▬) correlations, selective HMBC (⌒), and protons with key NOESY correlations (⌒) of **3**.

The inhibition effects of briaranes **1–5** on the release of inducible nitric oxide synthase (iNOS) and cyclooxygenase-2 (COX-2) protein from lipopolysaccharides (LPS)-stimulated RAW 264.7 were assessed. The results showed that briarane **4** reduced the release of iNOS and COX-2 to 35.37% and 54.61% at a concentration of 10 μM, respectively (Figure 5 and Table 3). Briarane **1** was found to be weaker than those of **4** in term of reducing the expression of iNOS and COX-2, indicating that the hydroxy group at C-16 in **1** reduced the activity.

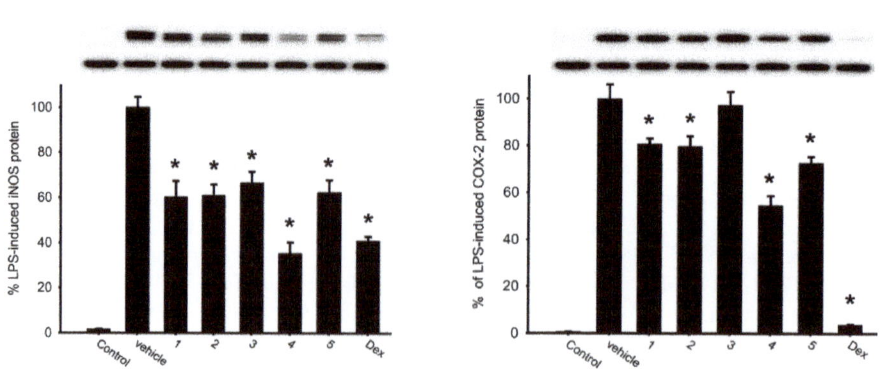

Figure 5. Western blotting showed that briarane **4** downregulated the expression of iNOS and COX-2. Data were normalized to the cells treated with LPS only, and cells treated with dexamethasone (Dex; 10 μM) were used as a positive control. Data are expressed as the mean ± SEM (n = 2~4). * Significantly different from cells treated with LPS ($p < 0.05$).

Table 3. Effects of briaranes **1–5** on LPS-induced pro-inflammatory iNOS and COX-2 protein expression in macrophages.

	iNOS	COX-2	β-Actin	
	Expression (% of LPS)	Expression (% of LPS)	Expression (% of LPS)	*n*
Negative Control	1.71 ± 0.13	0.62 ± 0.09	120.48 ± 1.28	2
LPS	100.00 ± 4.53	100.00 ± 6.05	100.00 ± 3.09	4
1	60.27 ± 7.05	80.63 ± 2.32	100.29 ± 2.46	4
2	60.94 ± 4.89	79.65 ± 4.27	98.29 ± 3.35	4
3	66.64 ± 4.79	97.28 ± 5.66	100.49 ± 6.44	4
4	35.37 ± 4.94	54.61 ± 4.03	104.56 ± 2.83	4
5	62.36 ± 5.42	72.63 ± 2.6	104.79 ± 2.76	4
Dexamethasone	41.00 ± 2.63	3.73 ± 0.35	104.24 ± 5.82	2

Data were normalized to those of cells treated with LPS alone, and cells treated with dexamethasone were used as a positive control. Data are expressed as the mean ± SEM (n = 2–4).

3. Materials and Methods

3.1. General Experimental Procedures

Optical rotation values were measured using a Jasco P-1010 digital polarimeter (Japan Spectroscopic, Tokyo, Japan). IR spectra were measured on a Thermo Scientific Nicolet iS5 FT-IR spectrophotometer (Waltham, MA, USA). NMR spectra were taken on a Jeol Resonance ECZ 400 S NMR spectrometer (Tokyo, Japan), using the residual $CHCl_3$ signal (δ_H 7.26 ppm) and $CDCl_3$ (δ_C 77.1 ppm) as the internal standard for ^1H and ^{13}C NMR, respectively; coupling constants (J) are presented in Hz. ESIMS and HRESIMS were recorded using a Bruker 7 Tesla solariX FTMS system. Column chromatography was carried out with silica gel (230–400 mesh, Merck, Darmstadt, Germany). TLC was performed on plates precoated with Kieselgel 60 F_{254} (0.25-mm-thick, Merck, Darmstadt, Germany), then sprayed with 10% H_2SO_4 solution followed by heating to visualize the spots. Normal-phase HPLC (NP-HPLC) was performed using a system comprised of a Hitachi L-7100 pump (Tokyo, Japan) and a Rheodyne 7725i injection port (Rohnert Park, CA, USA). Reverse-phase HPLC (RP-HPLC) was performed using a system comprised of a Hitachi L-2130 pump (Tokyo, Japan), a Hitachi L-2455 photodiode array detector (Tokyo, Japan), and a Rheodyne 7725i injection port (Rohnert Park, CA, USA). A semipreparative normal-phase column (YMC-Pack SIL, S-5 μm, 250 mm × 20 mm, Sigma-Aldrich, St. Louis, MO, USA) was used for NP-HPLC. A semipreparative reverse-phase column (Luna, 5 μm, C18(2) 100 Å, AXIA Packed, 250 mm × 21.2 mm; Phenomenex, Torrance, CA, USA) was used for RP-HPLC.

3.2. Animal Material

Specimens of *B. excavatum* were collected in June 2017 by hand with self-contained underwater breathing apparatus (SCUBA) divers off the coast of Lanyu Island (Orchid Island), Taiwan. The samples were then stored in a –20 °C freezer until extraction. A voucher specimen was deposited in the National Museum of Marine Biology and Aquarium, Taiwan (NMMBA-TW-SC-2017-418). Identification of the species of this organism was performed by comparison as described in previous publications [1–4].

3.3. Extraction and Isolation

The freeze-dried and sliced bodies (wet/dry weight = 1344/568 g) of the specimen were extracted with supercritical CO_2 to give 58.9 g of extract. Partial extract (36.4 g) was then applied on silica gel column and eluted with gradients of n-hexane/EtOAc to furnish fractions A–K. Fraction F was purified by NP-HPLC using a mixture of n-hexane/acetone (4:1) to yield fractions F1–F13. Fraction F6 was repurified by RP-HPLC, using a mixture of MeOH/H_2O (60:40; at a flow rate = 4 mL/min) to afford **4** (6.7 mg). Fraction G was separated by NP-HPLC, using a mixture of n-hexane/acetone (3:1) to yield fractions G1–G12. Fractions G6 and G7 were repurified by RP-HPLC using a mixture of MeOH/H_2O

(60:40; at a flow rate = 4.0 mL/min) to afford **5** (1.3 mg) and **3** (1.0 mg), respectively. Fraction H was separated by NP-HPLC using a mixture of *n*-hexane and acetone (3:1) to yield fractions H1−H18. Fractions H12 and H15 were repurified by RP-HPLC, using a mixture of MeOH/ H_2O (60:40; at a flow rate = 4.0 mL/min) to afford **2** (2.1 mg) and **1** (0.6 mg), respectively.

Briarenol I (**1**): Amorphous powder; $[\alpha]^{22}_{D}$ + 207 (*c* 0.03, $CHCl_3$), IR (ATR) ν_{max} 3524, 1783, 1736, 1222, 891 cm^{-1}; ^{13}C (100 MHz, $CDCl_3$) and ^1H (400 MHz, $CDCl_3$) NMR data (see Tables 1 and 2); ESIMS: *m/z* 649 $[M + Na]^+$; HRESIMS *m/z* 649.24677 (calculated for $C_{30}H_{42}O_{14}$ + Na, 649.24668).

Briarenol J (**2**): Amorphous powder; $[\alpha]^{26}_{D}$ + 140 (*c* 0.08, $CHCl_3$), IR (ATR) ν_{max} 3483, 1779, 1727, 1220, 890 cm^{-1}; ^{13}C (100 MHz, $CDCl_3$) and ^1H (400 MHz, $CDCl_3$) NMR data (see Tables 1 and 2); ESIMS: *m/z* 563 $[M + Na]^+$; HRESIMS *m/z* 563.21007 (calculated for $C_{26}H_{36}O_{12}$ + Na, 563.20990).

Briarenol K (**3**): Amorphous powder; $[\alpha]^{23}_{D}$ + 37 (*c* 0.06, $CHCl_3$), IR (ATR) ν_{max} 3468, 1780, 1739, 1255, 892 cm^{-1}; ^{13}C (100 MHz, $CDCl_3$) and ^1H (400 MHz, $CDCl_3$) NMR data (see Tables 1 and 2); ESIMS: *m/z* 547 $[M + Na]^+$; HRESIMS *m/z* 547.21514 (calculated for $C_{26}H_{36}O_{11}$ + Na, 547.21498).

Briaexcavatolide P (**4**): Amorphous powder; $[\alpha]^{24}_{D}$ + 182 (*c* 0.3, $CHCl_3$) (ref. [8], [α]27D + 167 (*c* 1.0, $CHCl_3$)), IR (ATR) ν_{max} 3513, 1783, 1731, 1218, 889 cm^{-1}; ^1H and ^{13}C NMR data were found to be in agreement with previous study [8]; ESIMS: *m/z* 633 $[M + Na]^+$.

Briaexcavatin P (**5**): Amorphous powder; $[\alpha]^{23}_{D}$ + 134 (*c* 0.05, $CHCl_3$) (ref. [9], $[\alpha]^{25}_{D}$ + 198 (*c* 0.08, $CHCl_3$)), IR (ATR) ν_{max} 3503, 1785, 1735, 1240, 889 cm^{-1}; ^1H and ^{13}C NMR data were found to be in agreement with previous study [9]; ESIMS: *m/z* 605 $[M + Na]^+$.

3.4. *In Vitro Anti-inflammatory Assay*

The proinflammatory suppression assay was employed to assess the activities of the isolated compounds **1–5** against the release of iNOS and COX-2 from macrophage cells as the literature reported [13–15].

4. Conclusions

B. excavatum was demonstrated to have a wide structural diversity of briarane-type diterpenoids that possessed various pharmacological properties, especially in anti-inflammatory activity. In our continued study on *B. excavatum*, three previously unreported briaranes, briarenols I–K (**1–3**), along with the known analogues, briaexcavatolide P (**4**) and briaexcavatin P (**5**), were isolated. In the present study, the anti-inflammatory activity of **1–5** was assessed using inhibition of pro-inflammatory iNOS and COX-2 release from macrophages. The results indicated that briaexcavatolide P (**4**) showed the most potent suppressive effect on iNOS release.

Supplementary Materials: The Supplementary Materials are available online. ESIMS, HRESIMS, IR, 1D and 2D NMR spectra of new compounds **1–3**.

Author Contributions: Conceptualization, L.-S.F., Y.-J.W., Z.-H.W. and P.-J.S; investigation, T.-H.H., Y.-H.C., B.-R.P., Y.-Y.C., L.-G.Z. and J.-J.C.; writing—original draft preparation, T.-H.H. and P.-J.S.; writing—review and editing, T.-C.L. and P.-J.S. All authors have read and agreed to the published version of the manuscript.

Funding: This research was supported by grants from the National Museum of Marine Biology and Aquarium; the National Dong Hwa University; and the Ministry of Science and Technology, Taiwan (Grant Nos: MOST 106-2320-B-291-001-MY3 and 107-2320-B-291-001-MY3) awarded to Ping-Jyun Sung.

Conflicts of Interest: The authors declare no conflicts of interest.

References

1. Bayer, F.M. Key to the genera of octocorallia exclusive of Pennatulacea (Coelenterata: Anthozoa), with diagnoses of new taxa. *Proc. Biol. Soc. Wash.* **1981**, *94*, 902–947.
2. Benayahu, Y.; Jeng, M.-S.; Perkol-Finkel, S.; Dai, C.-F. Soft corals (Octocorallia: Alcyonacea) from Southern Taiwan. II. Species diversity and distributional patterns. *Zool. Stud.* **2004**, *43*, 548–560.
3. Miyazaki, Y.; Reimer, J.D. Morphological and genetic diversity of *Briareum* (Anthozoa: Octocorallia) from the Ryukyu Archipelago, Japan. *Zool. Sci.* **2014**, *31*, 692–702. [CrossRef] [PubMed]

4. Samimi-Namin, K.; van Ofwegen, L.P. Overview of the genus *Briareum* (Cnidaria, Octocorallia, Briareidae) in the Indo-Pacific, with the description of a new species. *Zookeys* **2016**, *557*, 1–44. [CrossRef] [PubMed]
5. Su, Y.-D.; Su, J.-H.; Hwang, T.-L.; Wen, Z.-H.; Sheu, J.-H.; Wu, Y.-C.; Sung, P.-J. Briarane diterpenoids isolated from octocorals between 2014 and 2016. *Mar. Drugs* **2017**, *15*, 44. [CrossRef] [PubMed]
6. Hu, C.-C.; Lin, N.-C.; Peng, B.-R.; Chen, Y.-Y.; Wen, Z.-H.; Chang, Y.-C.; Lee, G.-H.; Wu, Y.-C.; Sung, P.-J. New trihydroxybriarane diterpenoids from an octocoral *Briareum* sp. *Phytochem. Lett.* **2020**, *35*, 23–27. [CrossRef]
7. Molina, S.L.; Forero, A.M.; Ayala, F.I.; Puyana, M.; Zea, S.; Castellanos, L.; Muñoz, D.; Arboleda, G.; Sandoval-Hernández, A.G.; Ramos, F.A. Metabolic profiling of the soft coral *Erythropodium caribaeorum* (Alcyonacea: Anthothelidae) from the Colombian Caribbean reveals different chemotypes. *Mar. Drugs* **2020**, *18*, 4. [CrossRef] [PubMed]
8. Wu, S.-L.; Sung, P.-J.; Chiang, M.Y.; Wu, J.-Y.; Sheu, J.-H. New polyoxygenated briarane diterpenoids, briaexcavatolides O–R, from the gorgonian *Briareum excavatum*. *J. Nat. Prod.* **2001**, *64*, 1415–1420. [CrossRef] [PubMed]
9. Sung, P.-J.; Lin, M.-R.; Hwang, T.-L.; Fan, T.-Y.; Su, W.-C.; Ho, C.-C.; Fang, L.-S.; Wang, W.-H. Briaexcavatins M–P, four new briarane-related diterpenoids from cultured octocoral *Briareum excavatum* (Briareidae). *Chem. Pharm. Bull.* **2008**, *56*, 930–935. [CrossRef] [PubMed]
10. Sheu, J.-H.; Sung, P.-J.; Cheng, M.-C.; Liu, H.-Y.; Fang, L.-S.; Duh, C.-Y.; Chiang, M.Y. Novel cytotoxic diterpenes, excavatolides A–E, isolated from the Formosan gorgonian *Briareum excavatum*. *J. Nat. Prod.* **1998**, *61*, 602–608. [CrossRef] [PubMed]
11. Yao, J.-W.; Chi, W.-C.; Lee, G.-H.; Su, J.-H.; Hwang, T.-L.; Wu, Y.-J.; Su, T.-R.; Sheu, J.-H.; Sung, P.-J. 2-Acetoxybriaranes from *Briareum violaceum*. *Tetrahedron* **2019**, *75*, 3751–3757. [CrossRef]
12. Pordesimo, E.O.; Schmitz, F.J.; Ciereszko, L.S.; Hossain, M.B.; van der Helm, D. New briarein diterpenes from the Caribbean gorgonians *Erythropodium caribaeorum* and *Briareum* sp. *J. Org. Chem.* **1991**, *56*, 2344–2357. [CrossRef]
13. Jean, Y.-H.; Chen, W.-F.; Sung, C.-S.; Duh, C.-Y.; Huang, S.-Y.; Lin, C.-S.; Tai, M.-H.; Tzeng, S.-F.; Wen, Z.-H. Capnellene, a natural marine compound derived from soft coral, attenuates chronic constriction injury-induced neuropathic in rats. *Br. J. Pharmacol.* **2009**, *158*, 713–725. [CrossRef] [PubMed]
14. Jean, Y.-H.; Chen, W.-F.; Duh, C.-Y.; Huang, S.-Y.; Hsu, C.-H.; Lin, C.-S.; Sung, C.-S.; Chen, I.-M.; Wen, Z.-H. Inducible nitric oxide synthase and cyclooxygenase-2 participate in anti-inflammatory and analgesic effects of the natural marine compound lemnalol from Formosan soft coral *Lemnalia cervicorni*. *Eur. J. Pharmacol.* **2008**, *578*, 323–331. [CrossRef] [PubMed]
15. Chen, L.-C.; Lin, Y.-Y.; Jean, Y.-H.; Lu, Y.; Chen, W.-F.; Yang, S.-N.; Wang, H.-M.D.; Jang, I.-Y.; Chen, I.-M.; Su, J.-H.; et al. Anti-inflammatory and analgesic effects of the marine-derived compound comaparvin isolated from the crinoid *Comanthus bennetti*. *Molecules* **2014**, *19*, 14667–14686. [CrossRef] [PubMed]

molecules

MDPI

Article

Identification of 1-Butyl-Lysergic Acid Diethylamide (1B-LSD) in Seized Blotter Paper Using an Integrated Workflow of Analytical Techniques and Chemo-Informatics

Emmanouil Tsochatzis [1], Joao Alberto Lopes [1,*], Fabiano Reniero [1], Margaret Holland [1], Jenny Åberg [2] and Claude Guillou [1,*]

1 European Commission, Joint Research Centre (JRC), I-20127 Ispra, Italy; Emmanouil.Tsochatzis@ec.europa.eu (E.T.); Fabiano.RENIERO@ec.europa.eu (F.R.); Margaret.HOLLAND@ec.europa.eu (M.H.)

2 Swedish Customs Laboratory, Box 6055, SE-171 06 Solna, Sweden; Jenny.ABERG@tullverket.se

* Correspondence: Joao-Filipe.ALBERTO-LOPES@ec.europa.eu (J.A.L.); Claude.GUILLOU@ec.europa.eu (C.G.); Tel.: +32-1457-3032 (J.A.L.); +39-0332785678 (C.G.)

Received: 24 January 2020; Accepted: 5 February 2020; Published: 7 February 2020

check for updates

Abstract: The rapid dispersion of new psychoactive substances (NPS) presents challenges to customs services and analytical laboratories, which are involved in their detection and characterization. When the seized material is limited in quantity or of a complex nature, or when the target substance is present in very small amounts, the need to use advanced analytical techniques, efficient workflows and chemo-informatics tools is essential for the complete identification and elucidation of these substances. The current work describes the application of such a workflow in the analysis of a single blotter paper, seized by Swedish customs, that led to the identification of a lysergic acid diethylamide (LSD) derivative, 1-butyl-lysergic acid diethylamide (1B-LSD). Such blotter paper generally contains an amount in the range of 30–100 ug. This substance, which is closely related to 1-propionyl-lysergic acid diethylamide (1P-LSD), seems to have only recently reached the drug street market. Its identification was made possible by comprehensively combining gas chromatography with mass spectrometry detection (GC–MS), liquid chromatography coupled with high-resolution tandem MS (LC–HR-MS/MS), Orbitrap-MS and both 1D and 2D nuclear-magnetic-resonance (NMR) spectroscopy. All the obtained data have been managed, assessed, processed and evaluated using a chemo-informatics platform to produce the effective chemical and structural identification of 1B-LSD in the seized material.

Keywords: blotter paper sample; 1-butyl-lysergic acid diethylamide (1B-LSD); HR–MS/MS; NMR; GC–MS

1. Introduction

During the past decade, there has been a significant increase in the circulation of new psychoactive substances (NPS) within the EU. This evolution in the drugs market, and the speed at which new substances are being created, is a cause of great concern for regulatory bodies. In 2016, NPS were reported to the EU Early Warning System at a rate of one per week, and it is estimated that a similar trend continued during the following years [1]. At the EU level, two agencies have a particular interest in this field—the European Monitoring Center for Drugs and Drug Addiction (EMCDDA) and EUROPOL [1–3].

Customs authorities are responsible for controlling the flow of goods into the EU, acting as the first control and contact point for NPS. However, the majority of customs laboratories are not equipped for the analysis of these substances, as they normally lack the advanced analytical and chemo-informatics tools that enable the complete identification and characterization of any new and/or relatively unknown NPS. These tools and expertise are available at the European Commission's Joint Research Centre (JRC), and, therefore, a collaborative project with the Customs Laboratories European Network (CLEN) has been established. Whenever an EU customs laboratory cannot completely identify an unknown seized substance/mixture, it is sent to the JRC for additional analysis. Since 2014, dozens of substances have been identified in the frame of this collaboration, including some which were previously unreported [3].

One of the most well-known and characterized psychoactive substances is d-lysergic acid diethylamide (LSD), which is most commonly circulated through small pieces of paper called "blotter paper" and is widespread all over the world because of its strong hallucinogenic effect [4]. In recent years, structurally modified LSD-type NPS have been found on the market, as is reported in several scientific papers [5–9]. One of them, 1-propionyl-d-lysergic acid diethylamide (1P-LSD), was identified and characterized after extraction from a seized sample by applying ^{1}H and ^{13}C-NMR along with GC–MS and UHPLC-qTOF-MS analysis [5]. Another closely related LSD derivative, 1-butanoyl-d-lysergic acid diethylamide (1B-LSD), appears to have already reached the street market, although no seizure of this drug has yet been reported (Figure 1). The only articles published present the analysis of commercial standards, using some analytical techniques (GC–MS, NMR, LC-MS) [9], or the screening procedures for the detection of nine LSD derivatives in rat urine, including 1B-LSD, using LC–HR-MS/MS [10].

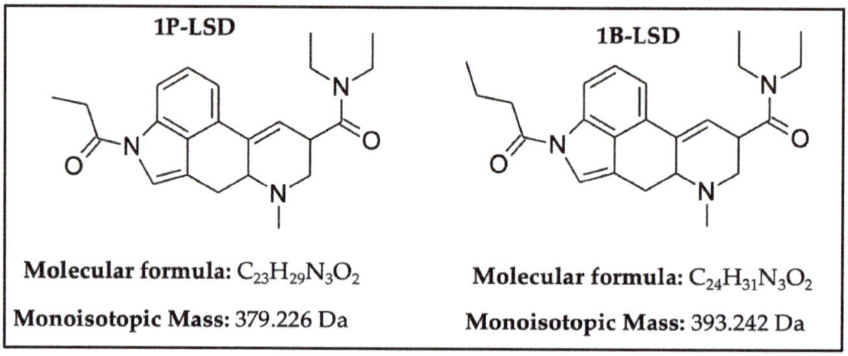

Figure 1. Chemical structures, molecular formulas and molecular masses (Da) of 1-butyl-lysergic acid diethylamide (1B-LSD) and 1-propionyl-lysergic acid diethylamide (1P-LSD).

The current study reports the integrated approach used for the extraction and identification of 1B-LSD, from a blotter paper sample (labeled "1B-LSD Blotters (25 MCG)") found in a package seized by the Swedish customs on 05/11/2018 at Arlanda airport. The chemical identification was performed using GC–MS, HR–MS, NMR and chemo-informatics tools.

2. Results

2.1. GC-MS Analysis

From the chromatogram of the GC–MS analysis (Figure S1), it could be concluded that the matrix was quite complex. The main peak was among the last to be eluted, with a retention time of 23.5 min, and was identified as potentially 1B-LSD (see Figure 2), in agreement with the fragmentation patterns of 1P-LSD and the 1B-LSD, as reported by Brandt et al. [5,9]. The complete identification has been verified and confirmed by the results obtained from NMR and HR-MS.

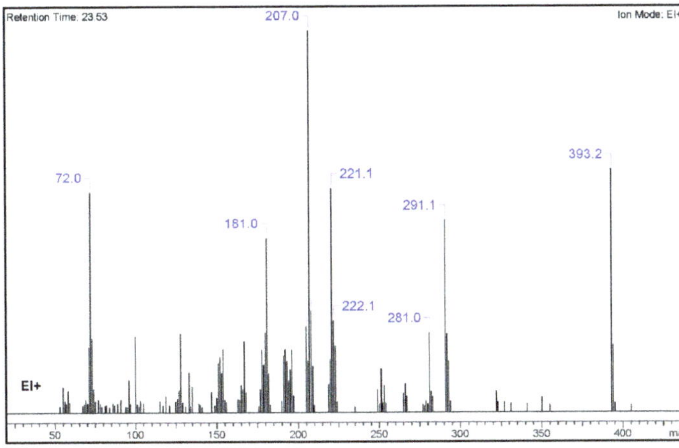

Figure 2. MS spectrum resulting from the GC–MS peak, identified as 1B-LSD.

The resulting identification of the most relevant fragments, as received from the chemo-informatics tool for the GC–MS analysis, are presented in Table 1. Nevertheless, it shall be highlighted that all GC–MS fragmentation patterns were consistent with those reported by Brandt et al. [9].

Table 1. GC–MS-identified fragments for 1B-LSD (ACD/Spectrus Processor 2017.2.1).

No.	Fragment New Structure	Formula	Label	m/z Exp. [1]	RI Exp. (%) [2]
1	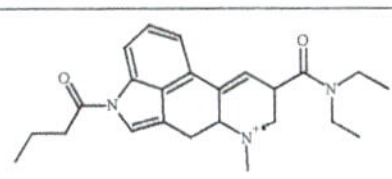	C24H31N3O2(+)	M	393.2	100.0
2	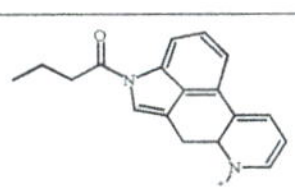	C18H15N2O2(+)	M-C6H16N	291.1	79.3
3	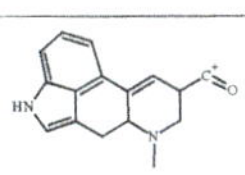	C16H15N2O(+)	M-C8H16NO	251.1	15.5
4	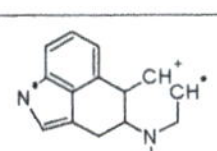	C15H13N2(+)	M-C9H18NO2	221.1	58.665
5	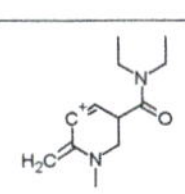	C12H19N2O(+)	M-C12H12NO	207.0	69.3
6	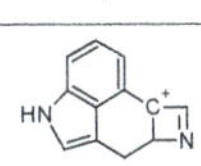	C12H9N2(+)	M-C12H11NO	181.0	45.6
7	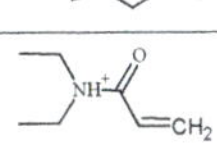	C7H14NO(+)	M-C17H20N2O	128.1	20.6
8	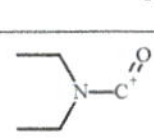	C5H10NO(+)	M-C19H21N2O	100.0	31.4
9	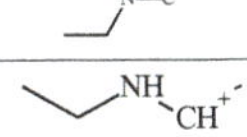	C4H10N(+)	M-C20H21N2O2	72.0	90.8

[1] m/z Exp.: experimental m/z; [2] RI Exp. (%): Relative Intensity (%) of experimental m/z.

2.2. UHPLC–HR-MS Experiment

The full-scan UHPLC–qTOF-MS analysis (positive mode) of the blotter paper methanol extract revealed the TIC of a relatively complex mixture, in accordance with what had been observed in the GC–MS analysis (Figure S2). The extraction of the ion chromatogram (XIC) with the $[M + H]^+$ of 1B-LSD (*m/z* 394.249), as received from the LC–qTOF-MS analysis, revealed a clear and significant peak. The analytical system allowed for a chromatographic separation to be performed whilst producing untargeted (all-ion mode) MS^2 spectra (Figure 3).

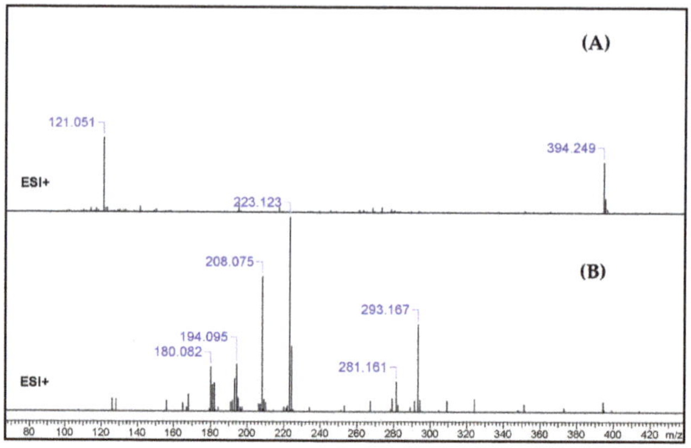

Figure 3. U(H)PLC–qTOF-MS (**A**) full-scan MS (**B**) untargeted MS^2 spectrum of *m/z* 394.249 ($[M + H]^+$) of the extract.

For the Orbitrap-MS experiments, the sample extract was directly infused into the system for analysis. Full-scan MS was performed initially, with a multitude of ions being found, as the blotter paper methanol extract, containing several components, was injected directly. In fact, the lack of a chromatographic technique coupled to the Orbitrap, increased the complexity of the results. However, the previous results, indicating the possible presence of 1B-LSD, allowed an oriented extraction of its protonated molecular ion $[M + H]^+$, which was found and confirmed as m/z 394.25004. After this analysis, the ion was isolated and fragmented successively (MS^2, MS^3 and MS^4), and Figure 4 shows the fragmentation spectrum of this ion (MS^2).

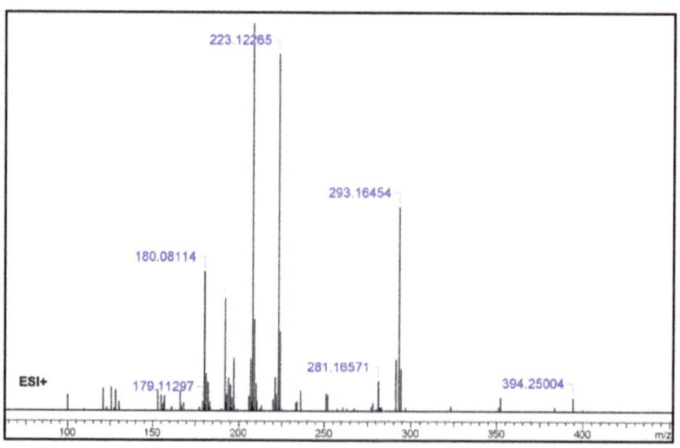

Figure 4. MS^2 spectrum from Infusion to Orbitrap MS of the extract [selected molecular ion *m/z* was 394.25004 assigned to the protonated molecular mass $[M + H]^+$.

Some of the fragments identified with the Orbitrap-MS analysis are also consistent with those of the GC–MS, even if different ionization modes were used (ESI+ vs. EI+). All the identification was based on spectral data (Agilent MassHunter), the literature and the application of chemo-informatics tools (ACD/Labs Spectrus Processor and MS Fragmenter), a software platform that allows the prediction of fragmentation and the comparison of experimental with theoretical data. The resulting identification of the most relevant fragments, as received from the chemo-informatics tool for the Orbitrap-MS, are presented in Table 2. All results were consistent with both the qTOF-MS fragmentation and the fragmentation patterns of 1B-LSD, as reported by Brandt et al. [9].

Table 2. Orbitrap-MS-identified fragments for 1B-LSD (ACD/Spectrus Processor 2017.2.1).

No.	Fragment New Structure	Formula	Label	*m/z* Exp. [1]	RI Exp. (%) [2]
1	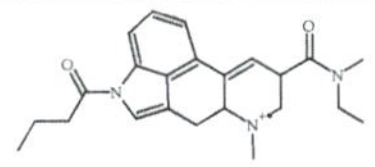	C24H31N3O2(+)	M	394.24890	3.5
2	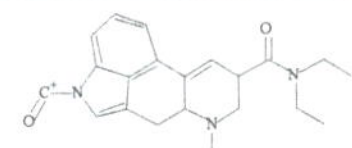	C21H26N3O2(+)	M + H-C3H6	352.20195	3.6
3	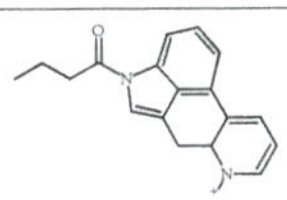	C19H21N2O(+)	M + H-C5H11NO	293.16454	52.7
4	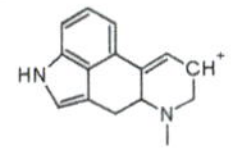	C15H15N2(+)	M-C9H16NO2	223.12265	27.5
5	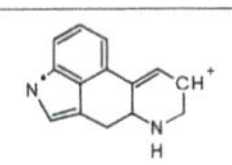	C12H20N2O(+)	M-C12H12NO	208.07528	69.3
6	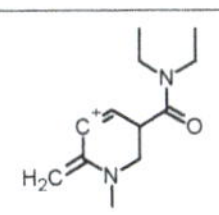	C12H19N2O(+)	M + H-C12H13NO	207.14919	13.5
7	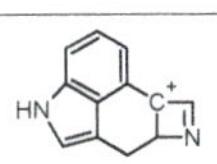	C12H8N2(+)	M-C12H11NO	180.06820	36.0
8	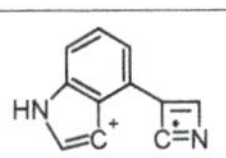	C11H6N2(+)	M-C11H9NO	166.05255	5.1
9	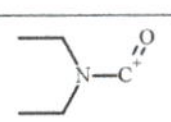	C5H10NO(+)	M-C19H22N2O	100.07569	4.5

[1] *m/z* Exp.: experimental *m/z*; [2] RI Exp. (%): Relative Intensity (%) of experimental *m/z*.

2.3. NMR

Despite the fact that the sample aliquot was so small (the amount of LSD in a blotter is usually in the range 30–100 μg), the ^{1}H-NMR spectrum revealed a complex mixture (Figure 5) in which many signals could be identified as typical of LSD-like substances, and which share the same backbone structure. Nevertheless, the most noteworthy signals were those which allowed the identification of 1B-LSD in this mixture, i.e., those assigned to the butyl (CH_3 and CH_2) part of the molecule. On the other hand, a direct 1D ^{13}C NMR measurement was not possible due to the reduced quantity of the sample and the presence of relatively large quantities of other interfering substances extracted from the original matrix. The bond connection between the butyl proton resonances was confirmed with the 2D DQF-COSY experiment (Figure S4). The 2D HSQC confirmed the presence of the methyl and methylene groups and HMBC showed the connection of the methylene groups with a signal of a quaternary carbon at 171.6 ppm in line with the C=O reported by Brandt et al. [9]. All our NMR results are consistent with those for 1B-LSD as reported by Brandt et al. [9].

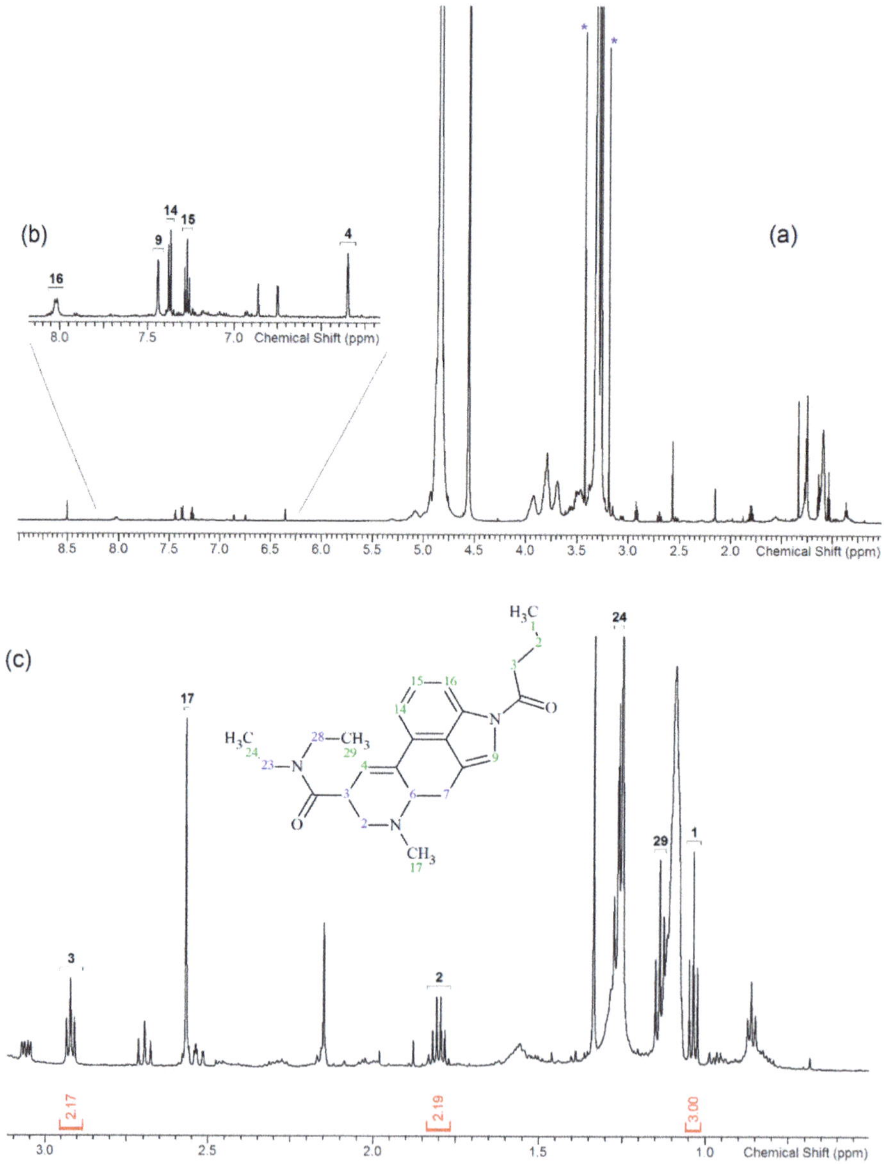

Figure 5. ^{1}H NMR spectrum 1B-LSD (**a**) full spectrum, (**b**) expansion of the aromatic region (**c**) expansion of the aliphatic region showing the region of interest for the identification of the 1-butyl group. The peaks labelled with * at 3.1 and 3.4 ppm correspond to the ^{13}C satellite signals of the residual protonated NMR solvent (methanol-d4 ≥ 99.8 atom % D).

3. Discussion

3.1. GC–MS and HR-MS

From the GC–MS spectrum (Figure 1), the 1B-LSD molecular ion (*m*/*z* 393) could be identified, as well as some characteristic clusters of these types of substances. More precisely, the clusters at *m*/*z* 293, *m*/*z* 292 and *m*/*z* 291, were consistent with the additional CH_2 substituent of 1B-LSD compared to that reported by Brandt et al. [9], when compared with 1P-LSD, which presents a fragment cluster at *m*/*z* 277, *m*/*z* 278 and *m*/*z* 279, as also previously reported [5]. In the case of LSD, with no substituent at the indole nitrogen, this fragment cluster of ions is shifted towards the cluster at *m*/*z* 221, *m*/*z* 222 and *m*/*z* 223 [5,9]. With the exception of *m*/*z* 322, which represents the loss of the acyl radical from 1B-LSD, the remaining three species might have represented the butyl-substituted counterparts of those species, previously described as LSD, detected at *m*/*z* 280 (retro-Diels Alder), *m*/*z* 223 and *m*/*z* 196 (loss of *N*,*N*-diethylacrylamide). A neutral loss of *N*,*N*-diethylformamide from the $M\bullet^+$ might have led to the formation of the *m*/*z* 291 ion [5]. The EI+ characterization of 1B-LSD, 1P-LSD and LSD, presented by Brandt and other authors, was perfectly consistent with the results of the present study, and allows the unequivocal identification of 1B-LSD [5–9].

In the case of UHPLC–qTOF-MS analysis, the extraction of the ion chromatogram (XIC) with the $[M + H]^+$ of 1B-LSD (*m*/*z* 394.249) revealed a clear and abundant peak. From the respective MS^2 fragmentation pattern, five abundant ions were identified which related to the suspected molecular structure, namely the *m*/*z* 293.165 (loss of [(diethylamino)methylidyne]oxidanium ion), 223.124 (loss of but-1-en-1-one), 208.075 (loss of a methyl), 192.091 (loss of methylamine) and 180.081 (internal rearrangement of tetrahydropyridine with loss of a methyl). The obtained results were in accordance with previously reported works on the ESI+ analysis of 1B-LSD itself, and also 1P-LSD [5,9,11–14].

Regarding the Orbitrap-MS, the MS^2 fragmentation pattern was in full accordance with the qTOF-MS^2 results, with the four most abundant ions being *m*/*z* 293.16454 (loss of [(diethylamino)methylidyne]oxidanium ion), 223.12265 (loss of but-1-en-1-one), 208.07528 (loss of a methyl) and fragments 180.08114 and 181.08117 (internal rearrangement of tetrahydropyridine with loss of a methyl). Once again, these fragments are typical of 1B-LSD and 1-P-LSD [5,9,11–14].

For both the qTOF and Orbitrap-MS analysis, some of the fragments identified with the Orbitrap-MS analysis were also consistent with those of the GC–MS analysis, even though different ionization modes were used (ESI+ vs. EI+).

3.2. NMR

Despite the fact that the sample aliquot was so small, the ^{1}H-NMR spectrum still revealed the identity of a complex mixture (Figure 4) in which many signals could be identified (like protons 14 to 16 of the aromatic ring) as typical of LSD-like substances which share the same backbone structure. However, the signals that are most noteworthy are those which allow the identification of 1B-LSD in this mixture, i.e., those assigned to the butyl (CH_3 and CH_2) part of the molecule with δH (MeOD-d_4): 1.09 ppm [4] (3H, t, J = 7.4 Hz), 1.86 ppm [3] (2H, m, J = 7.4 Hz) and 2.98 ppm [2] (2H, t, J = 7.4 Hz). These results are consistent with a recent study by Brandt et al. on this substance [9].

On the other hand, the direct 1D ^{13}C NMR measurement was not possible due to the reduced quantity of the sample and the presence of relatively large quantities of other substances extracted from the original matrix. However, with 2D HSQC, it was possible to confirm the presence of the methyl and methylene groups. The bond connections between the protons in this butyl group were confirmed by the 2D DQF-COSY. Furthermore, the long-range ^{1}H-^{13}C HMBC showed the correlation of these two methylene groups with a C=O at 171.6 ppm. These results are consistent with a recent work of Brandt et al. on this substance [9].

4. Materials and Methods

4.1. Chemical and Reagents

All solvents for the NMR, HR-MS/MS and GC–MS, were obtained from Sigma-Aldrich (Steinheim, Germany) and all LC-MS solvents were ChromaSolv grade, obtained from Fluka Analytical. Ultrapure water (18.2 MΩ) was obtained from a Milli-Q system (Millipore, Bedford, MA, USA).

4.2. Seized Blotter Sample

A plastic bag containing one blotter paper labeled as "1B-LSD Blotters 25 MCG" was seized at Stockholm Arlanda Airport, Sweden, together with one plastic bag containing 10 flualprazolam tablets and one plastic bag containing 0.5 g of 2-fluorodeschloroketamine (both bags were correctly labelled). No sub-sample was available for any of the seized materials.

4.3. Sample Preparation

The material was extracted at the Swedish Customs Laboratory with 0.5 mL of MeOH, followed by filtration. After GC–MS analysis, aliquots of the samples were shipped to the JRC for further analysis. A direct injection of the methanol sample was carried out for the UHPLC–qTOF-MS analysis, while for the HR-Orbitrap-MS, the extract was diluted 1:50 with acetonitrile.

4.4. Instrumental Analysis

4.4.1. GC–MS

An Agilent 7890 Gas Chromatograph equipped with a 5977 MS (Agilent, Santa Clara, CA, USA) was used for the analysis of the sample. The GC-column was a HP5-MS UI (Agilent Technologies, 30 m, 0.250 mm, 0.25 mm) with helium as the carrier gas, at a flow of 0.8 mL/min. An initial oven temperature of 100 °C was set with a 5 min isothermal period followed by heating up to 300 °C at a rate of 30 °C/min and held for 13 min. The total run time was 25 min. The injection volume was 1 μL in split mode (20:1), and the injector temperature was set at 250 °C. The GC–MS operated at scan mode m/z ranging from 30–550. GC–MS analysis was performed by the Swedish Customs Laboratory, as well as at the JRC. Data have been processed with an ACD/labs spectrus processor.

4.4.2. NMR

For the acquisition of a ^{1}H NMR spectrum, the remaining methanol extract was mixed with deuterated methanol MeOD-d_4 (up to a final volume of 600 μL), which was used as an internal lock and chemical shift reference at δH = 3.3 ppm. The ^{1}H NMR experiments were performed at 300 K on a Bruker (Rheinstetten, Germany) spectrometer Avance III HD 600 (nominal proton frequency 600.13 MHz) equipped with a 5 mm QCI cryo-probe (^{1}H, ^{13}C, ^{15}N and ^{19}F). The ^{1}H and ^{13}C chemical shifts are expressed in ppm, referenced to the proton signal of the MeOD-d_4 (3.3 ppm for ^{1}H and 47.6 for ^{13}C). Compounds were characterized by one-dimensional ^{1}H, as well as two-dimensional ^{1}H/^{1}H COSY, ^{1}H/^{1}H TOCSY, ^{1}H/^{13}C HSQC, ^{1}H/^{13}C HMBC and ^{1}H/^{15}N HMBC experiments.

4.4.3. HR-MS/MS

For UHPLC–qTOF-MS analysis, a UHPLC system (Agilent 1290) with a quadrupole Time-Of-Flight (TOF) mass spectrometer (Agilent 6540 UHD Accurate-Mass, Agilent, Waldbronn, Germany), using an ESI interface, operating in positive ionization mode with a 4 kV capillary voltage, was used. The source operated at 325 °C and nitrogen was used as both the drying (40 psi) and nebulizing gas (10 L min^{-1}). The injection volume was 5 μL. The TOF-MS detector was set to acquire MS data over an m/z range of 100–1600. All-Ions MS/MS experiments were performed to screen and quantify constituents of the sample in a single analysis. Under the aforementioned conditions, fragmentation collision energies ranged from 5–60 eV.

The analytical column was a Waters BEH C18 100 × 2.1 mm, 1.7 μm particle size (Waters, Milford, MA, USA), temperature-controlled at 40 °C. The mobile phase consisted of water with 0.1% formic acid (A) and methanol with 0.1% formic acid (B) at a flow rate of 200 μL min^{-1}. The gradient program changed linearly from 50% to 95% (B) in 25 min, followed by an isocratic elution for 4 min. An equilibration time of 1 min was set for the mobile phase to reach initial conditions again.

For Direct infusion Orbitrap-MS, and in correlation with the UHPLC–qTOF-MS analysis, additional measurements were performed using an Orbitrap MS system, to confirm the molar mass (M) and the mass fragmentation patterns of the suspected NPS. This analysis was performed in positive-ion mode with Electro Spray Ionization (ESI) using a Thermo LTQ Orbitrap MS (Thermo Scientific, Bremen, Germany), and operated with mass resolution of 140,000 at *m*/*z* 200. The sample was infused at a flow rate of 5 μL/min on the system.

4.5. Subsection Chemoinformatics Tools

The ACD/Labs platform (ACD/Labs, Toronto, Canada) was used in combination with Agilent's MassHunter (Agilent Technologies) and XCalibur (Thermo Scientific) for the assessment and evaluation of the obtained data, including all the obtained chromatographic and MS data. This platform allows the confirmation and checking of the consistency of any suggested chemical structures with NMR experimental data, as well as assigning experimental spectra to structures, enabling the creation of a central, fully searchable repository of the assigned NMR spectra. The software was also used to project fragmentation paths, by comparing experimental MS and MS/MS data with theoretical data.

5. Conclusions

The identification of NPS in seized samples still remains a challenge for many laboratories. In cases where a novel unreported substance is found, or when a sample is seized, either in very small amounts or in a complex mixture, routine analytical techniques are often not sufficient. In these cases, an analytical workflow that combines hyphenated techniques with HR-MS, NMR and chemo-informatics tools is the most effective approach to identify and/or confirm the presence of an NPS with sufficient precision.

By following such an analytical workflow, it was possible, in this study, to identify 1B-LSD in the methanol extract of a single blotter paper by using four different techniques and chemo-informatics tools. Although previously characterized at analytical-standard levels, it is the first time that this substance was found and identified in samples circulating in the street market. This work can also be seen as proof that scientific cooperation between modern forensic laboratories can lead to the correct and reliable identification of NPS, even if present in trace amounts.

Supplementary Materials: The following are available online, Figure S1: GC chromatogram analysis (Full scan mode) of the blotter paper methanol extract, Figure S2: UHPLC-qTOF-MS analysis of the blotter paper methanol extract (A) and the extracted ion chromatogram (XIC) of the 1B-LSD $[M + H]^+$, Figure S3: 2D COSY NMR spectra of the blotter paper methanol extract; Figure S4: HMBC NMR spectra of the blotter paper methanol extract.

Author Contributions: Investigation and experimental analysis, E.T., J.A.L., M.H., F.R., J.Å. and C.G.; Writing—original draft, E.T. and J.A.L.; Revision and experimental follow-ups: E.T., J.A.L., M.H., F.R., J.Å. and C.G.; Visualization and final revision: E.T., J.A.L., M.H., F.R. and C.G. All authors have read and agreed to the published version of the manuscript.

Funding: This research was funded by EUROPEAN COMMISSION DIRECTORATE GENERAL FOR TAXATION AND CUSTOMS UNIONS (DG TAXUD) for the administrative and financial support provided to the JRC through the Administrative Arrangement CLEN2SAND II (JRC-Nr 34606-DG TAXUD-Nr TAXUD/2016/DE/331).

Acknowledgments: Hendrik Emons (JRC) is acknowledged for reviewing this manuscript.

Conflicts of Interest: The authors declare no conflict of interest.

References

1. EMCDDA. European Drug Report 2017: Trends and Developments. Publications of of the European Union 2017. Available online: http://www.emcdda.europa.eu/system/files/publications/4541/TDAT17001ENN.pdf (accessed on 27 December 2019).
2. Gaspar, H.; Bronze, S.; Oliveira, C.; Victor, B.L.; Machuqueiro, M.; Pacheco, R.; Caldeira, M.J.; Santos, S. Proactive response to tackle the threat of emerging drugs: Synthesis and toxicity evaluation of new cathinones. *Forens. Sci. Int.* **2018**, *290*, 146–156. [CrossRef] [PubMed]
3. Lobo Vicente, J.; Chassaigne, H.; Holland, M.V.; Reniero, F.; Kolář, K.; Tirendi, S.; Vandecasteele, I.; Vinckier, I.; Guillou, C. Systematic analytical characterization of new psychoactive substances: A case study. *Forens. Sci. Int.* **2016**, *265*, 107–115. [CrossRef] [PubMed]
4. Hida, M.; Mitsui, T. Rapid identification of lysergic acid diethylamide in blotter paper by microscope FT-IR. *Anal. Sci.* **1999**, *15*, 289–291. [CrossRef]
5. Brandt, S.D.; Kavanagh, P.; Westphal, F.; Stratford, A.; Elliott, S.P.; Hoang, K.; Wallach, J.; Halberstadt, A.L. Return of the lysergamides. Part I: Analytical and behavioural characterization of 1-propionyl-D-lysergic acid diethylamide (1P-LSD): Analytical and behavioural characterization of 1-propionyl-LSD. *Drug Test. Anal.* **2016**, *8*, 891–902. [CrossRef] [PubMed]
6. Brandt, S.D.; Kavanagh, P.V.; Westphal, F.C.; Elliott, S.P.D.; Wallach, J.E.; Stratford, A.F.; Nichols, D.E.; Halberstadt, A.L. Return of the lysergamides. Part III: Analytical characterization of N6-ethyl-6-norlysergic acid diethylamide (ETH-LAD) and 1-propionyl ETH-LAD (1P–ETH-LAD). *Drug Test. Anal.* **2017**, *9*, 1641–1649. [CrossRef] [PubMed]
7. Brandt, S.D.; Kavanagh, P.V.; Westphal, F.; Elliott, S.P.; Wallach, J.; Colestock, T.; Burrow, T.E.; Chapman, S.J.; Stratford, A.; Nichols, D.E.; et al. Return of the lysergamides. Part II: Analytical and behavioural characterization of N6-allyl-6-norlysergic acid diethylamide (AL-LAD) and (2'S,4'S)-lysergic acid 2,4-dimethylazetidide (LSZ). *Drug Test. Anal.* **2017**, *9*, 38–50. [CrossRef] [PubMed]
8. Brandt, S.D.; Kavanagh, P.V.; Twamley, B.; Westphal, F.; Elliott, S.P.; Wallach, J.; Stratford, A.; Klein, L.M.; McCorvy, J.D.; Nichols, D.E.; et al. Return of the lysergamides. Part IV: Analytical and pharmacological characterization of lysergic acid morpholide (LSM-775). *Drug Test. Anal.* **2018**, *10*, 310–322. [CrossRef] [PubMed]
9. Brandt, S.D.; Kavanagh, P.V.B.; Westphal, F.C.; Stratford, A.D.; Elliott, S.P.E.; Dowling, G.B.F.; Wallach, J.G.; Halberstadt, A.L. Return of the lysergamides. Part V: Analytical and behavioural characterization of 1-butanoyl-d-lysergic acid diethylamide (1B-LSD). *Drug Test. Anal.* **2019**, *11*, 1122–1133. [CrossRef] [PubMed]
10. Wagmann, L.; Richter, L.H.J.; Kehl, T.; Wack, F.; Bergstrand, M.P.; Brandt, S.D.; Stratford, A.; Maurer, H.H.; Meyer, M.R. In Vitro metabolic fate of nine LSD-based new psychoactive substances and their analytical detectability in different urinary screening procedures. *Anal. Bioanal. Chem.* **2019**, *411*, 4751–4763. [CrossRef] [PubMed]
11. Romão, W.; Sabino, B.D.; Bueno, M.I.M.; Vaz, B.G.; Júnior, A.C.; Maldaner, A.O.; de Castro, E.V.; Lordeiro, R.A.; Nascentes, C.C.; Eberlin, M.N.; et al. LSD and 9,10-dihydro-LSD Analyses in Street Drug Blotter Samples via Easy Ambient Sonic-Spray Ionization MassSpectrometry (EASI-MS). *J. Forensic Sci.* **2012**, *57*, 1307–1312. [CrossRef] [PubMed]
12. Clark, C.C. Differentiation of lysergic acid diethylamide (LSD) from *N*-methyl-*N*-propyl and *N*-butyl amides of lysergic acid. *J. Forensic Sci.* **1989**, *34*, 532–546. [CrossRef]
13. Laing, R.R. (Ed.) Hallucinogens. In *A Forensic Drug Handbook*; Elsevier Science Ltd.: London, UK, 2003.
14. Siegel, J.A. Part II: Mass Spectrometry of Lysergic Acid Diethylamide (LSD). In *Handbook of Forensic Drug Analysis*; Smith, F.P., Ed.; Elsevier Academic Press: San Diego, CA, USA, 2004; pp. 186–187.

Sample Availability: No sample is available.

molecules

MDPI

Article

Phytochemistry and Comprehensive Chemical Profiling Study of Flavonoids and Phenolic Acids in the Aerial Parts of *Allium Mongolicum* Regel and Their Intestinal Motility Evaluation

Yongzhe Dong [1,2,†], Jingya Ruan [1,2,†], Zhijuan Ding [2], Wei Zhao [1], Mimi Hao [2], Ying Zhang [2], Hongyu Jiang [2], Yi Zhang [1,2,*] and Tao Wang [1,2,*]

1 Tianjin Key Laboratory of TCM Chemistry and Analysis, Tianjin University of Traditional Chinese Medicine, 10 Poyanghu Road, West Area, Tuanbo New Town, Jinghai District, 301617 Tianjin, China; dongyongzhe44@hotmail.com (Y.D.); Ruanjy19930919@163.com (J.R.); zhaowei126123@126.com (W.Z.)

2 Institute of TCM, Tianjin University of Traditional Chinese Medicine, 10 Poyanghu Road, West Area, Tuanbo New Town, Jinghai District, 301617 Tianjin, China; 15222792071@163.com (Z.D.); haomimi126@126.com (M.H.); zyingtzy@163.com (Y.Z.); jhy15731602454@163.com (H.J.)

* Correspondence: zhwwxzh@tjutcm.edu.cn (Y.Z.); wangtao@tjutcm.edu.cn (T.W.); Tel./Fax: +86-22-5959-6168 (T.W.)

† These authors contributed equally to this manuscript.

Academic Editor: Ping-Chung Kuo
Received: 15 December 2019; Accepted: 24 January 2020; Published: 29 January 2020

check for updates

Abstract: To clarify whether flavonoids and phenols in *Allium mongolicum* Regel have the effect of improving gastrointestinal function and analyze its quality, this study was designed to isolate and identify them from the aerial parts of *A. mongolicum* by using various chromatographic and spectrophotometric methods, a bioassay on motility of mouse isolated intestine tissue, as well as qualitative analysis using liquid chromatography/mass spectrometry (LC-MS) analysis. As a result, 31 flavonoids and phenolic acids were obtained and identified, including six new flavonoid glycosides, mongoflavonosides A_1 (**1**), A_2 (**2**), A_3 (**3**), A_4 (**4**), B_1 (**5**), B_2 (**6**), and four new phenolic acid glycosides, mongophenosides A_1 (**7**), A_2 (**8**), A_3 (**9**), B (**10**). Among them, eleven flavonoids and three phenolic acids showed significant increase in the height of mouse small intestinal muscle. It was a first systematic bioactive constituents' study for *A. mongolicum* on gastrointestinal tract. Furthermore, according to the retention time (t_R) and the exact mass-to-charge ratio (*m/z*), thirty-one compounds were unambiguously identified by comparing to the standard references by using LC-MS. Then, on the basis of generalized rules of MS/MS fragmentation pattern, chromatographic behaviors, as well as biosynthetic laws of the 31 isolates, five flavonoid glycosides and one phenolic acid glycoside were tentatively speculated. On the basis of the study, a fast analysis method for flavonoids and phenolic acids in *A. mongolicum* was established.

Keywords: *Allium mongolicum* Regel; flavonoids; phenolic acids; mouse isolated intestine tissue; qualitative analysis

1. Introduction

Flavonoids and phenolic acids are secondary metabolites found in most *Allium* vegetables, such as onion (*Allium cepa* L.) [1], scallion (*Allium fistulosum* L.) [2], garlic (*Allium sativum* L.) [3], fruits, and traditional medicine. They also exert multiple biological properties, such as antitumor [4,5] antioxidant [6,7], anti-inflammation and gastrointestinal motility effect improvement [8–11], which make them show high correlations in the inhibition or management of many chronic diseases, such as cardiovascular and cerebrovascular diseases, diabetes, cancer, digestive system diseases, and so on.

As a traditional Mongolian medicinal herb, *Allium mongolicum* Regel (Liliaceae family) is mainly found in the high altitude desert of the Inner Mongolia Autonomous Region, Ningxia Hui Autonomous Region, Qinghai, Gansu, and Shanxi provinces [12], and has various special properties, such as stimulating the appetite, lowering blood pressure, hypolipidemic, replenishing the kidneys, and acting as an aphrodisiac [7]. Though it has been used to stimulate the appetite, studies were mainly focused on animal cultivation and improvement on meat products [12–14] up until now. Experiments related to the gastrointestinal tract have not been reported for the plant. On the other hand, pharmacological investigations showed the biological activities of *A. mongolicum*, including anti-inflammatory, antimicrobial, and antioxidant, which were attributed to the presence of different phytochemical groups like flavonoids and phenols [7,15,16]. However, it is still limited to the activity of total flavonoids and phenolic acids. Until now, the phytochemistry investigation of flavonoids is only reported by Dr. Zhao [16] and our laboratory [17,18]—there is no report for phenolic acids investigation. Moreover, there is no comprehensive chemical profiling study for the plant.

The present study was designed to isolate and identify flavonoids and phenolic acids from the aerial parts of *A. mongolicum* followed by bioactivity study on the motility of mouse isolated intestine tissue and qualitative analysis using liquid chromatography/mass spectrometry (LC-MS) analysis.

2. Results and Discussion

The fresh, aerial parts of *A. mongolicum* (17.8 kg) was successively heated under reflux with 95% EtOH for 3 h and 50% EtOH for 2 h one time each to obtain dry extract of *A. mongolicum* aerial parts (AM, 515.0 g). Then 470.1 g of it was partitioned with EtOAc/H_2O (1:1, 8L/8L) to yield EtOAc layer dry extract (AME, 64.9 g) and H_2O layer dry extract (AMH, 381.0 g).

Then, AM, AMH, and AME were tested for frequency and height by using a tissue perfusion method. As results, AM and AMH showed significant increase in the contraction amplitude of mouse small intestinal muscle at 200 μg/mL (relative height for AM: 137.4 ± 11.8%* and AMH: 121.8 ± 1.0%**, respectively), but had no significant effect on frequency (relative frequency for AM: 95.2 ± 2.8% and AMH: 100.1 ± 1.9%, respectively). While AME displayed no significant effect on both of them (relative height: 127.9 ± 20.8%; relative frequency: 100.0 ± 9.18%).

Therefore, AMH was further fractionated by column chromatography (CC) and purified by HPLC to afford six new flavonoid glycosides, named as mongoflavonosides A_1–A_4 (**1–4**), B_1 (**5**), B_2 (**6**), four new phenolic acid glycosides, named as mongophenosides A_1–A_3 (**7–9**), B (**10**) (Figure 1), along with known compounds, kaempferol-3,7,4′-tri-*O*-β-glucoside (**18**) [19], quercetin-3-*O*-β-D-rutinoside-7-*O*-β-D-glucuronide (**25**) [20], quercetin-3,7,4′-tri-*O*-glucoside (**26**) [21], isorhamnetin 3-*O*-β-D-glucopyranoside (**27**) [22], as well as three phenols, *trans-p*-hydroxycinnamate sophorose (**28**) [23], tuberonoid A (**29**) [24], trans-caffeic acid (**30**) [25], and benzyl-*O*-β-D-glucopyranoside (**31**) [26] (Figure 2). Among the known isolates, **28** and **30** were obtained from the *Allium* genus for the first time; **18**, **25–27**, **29**, and **31** were firstly found from the plant.

Furthermore, the improved effects on the motility of the mouse isolated intestine tissue of the above-mentioned compounds as well as our previously reported flavonoids, kaempferol-3-*O*-β-D-glucopyranoside (**11**), kaempferol-3-*O*-β-D-glucopyranosyl(1→4)-β-D-glucopyranoside (**12**), kaempferol-3-*O*-β-D- rutinoside (**13**), kaemperol-3-*O*-β-D-glucopyranosyl(1→4)[α-L-rhamanopyranosyl(1→6)]-β-D-glucopyranoside (**14**), kaempferol-3-*O*-rutinoside-7-*O*-glucuronide (**15**), kaempferol-3-rutinoside-4′-glucopyranoside (**16**), kaempferol-3-*O*-gentiobioside-4′-*O*-glucopyranoside (**17**) [17], isoquercetin (**19**), quercetin-3-*O*-(6′′-*O*-acetyl)-β-D-glucopyranoside (**20**), quercetin-3-*O*-β-D-glucopyranosyl(1→4)-β-D-glucopyranoside (**21**), rutin (**22**), quercetin-3,4′-di-*O*-β-D-glucopyranoside (**23**), quercetin 3-*O*-(6′′-*O*-α-L-rhamnopyranosyl)-β-D-glucopyranoside-7-*O*-β-D-glucopyranoside (**24**) [18] (Figure 2) were reported here. Then, qualitative analysis for the aerial parts of *A. mongolicum* by using LC-MS spectrometry technology was developed.

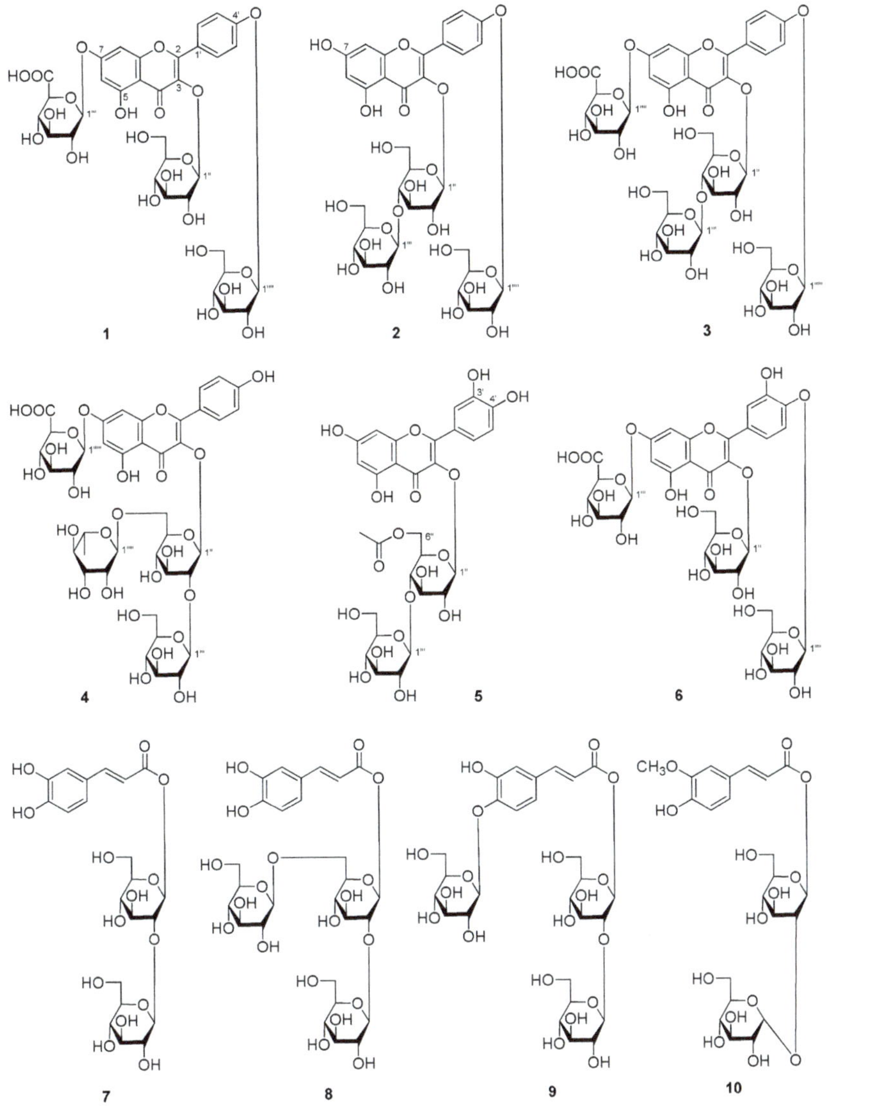

Figure 1. The new compounds obtained from the aerial parts of *A. mongolicum* (**1–10**).

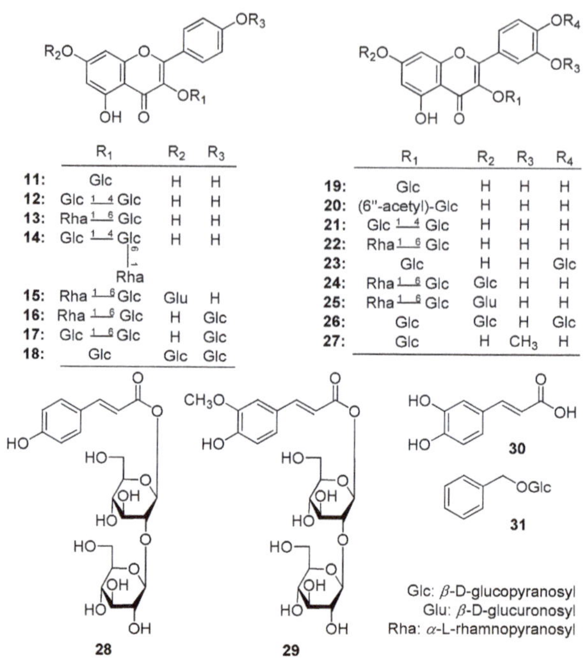

Figure 2. The known compounds obtained from the aerial parts of *A. mongolicum* (**11**–**31**).

2.1. Identification of Compounds

Mongoflavonoside A_1 (**1**) was isolated as a yellow powder with negative optical rotation ($[\alpha]_D^{25}$ −54.0, H_2O). Its molecular formula was deduced to be $C_{33}H_{38}O_{22}$ by the negative-ion Electron Spray Ionization-Quadrupole-Orbitrap-Mass Spectrometry (ESI-Q-Orbitrap MS) analysis (*m/z* 785.17883 $[M-H]^-$, calculated for $C_{33}H_{37}O_{22}$, 785.17710). The IR spectrum displayed the absorption bands assignable to hydroxyl (3354 cm^{-1}), carbonyl (1716 cm^{-1}), α,β-unsaturated ketone carbonyl (1652 cm^{-1}), aromatic ring (1601, 1507, 1457 cm^{-1}), and ether functions (1072 cm^{-1}), respectively. The ^1H and ^{13}C-NMR (Table 1) spectra suggested that **1** was a flavonoid glycoside with a kaempferol aglycone [δ 6.48 (1H, br. s, H-6), 6.88 (1H, br. s, H-8), 7.19 (2H, d, *J* = 9.0 Hz, H-3′,5′), 8.16 (2H, d, *J* = 9.0 Hz, H-2′,6′)] and three glycosyl groups [δ 5.05 (1H, d, *J* = 7.0 Hz, H-1′′′′), 5.26 (1H, d, *J* = 6.5 Hz, H-1′′′), 5.50 (1H, d, *J* = 7.0 Hz, H-1′′)]. Acid hydrolysis of **1** with 5% aqueous H_2SO_4 solution–1,4-dioxane (1:1, *v/v*) under 110 °C for 2 h to afford D-glucuronic acid and D-glucose, whose absolute configurations were determined by GC-MS analysis of their trimethysilyl thiazolidine derivatives [27]. Meanwhile, correlations were observed between the following proton and carbon pairs in its HSQC-TOCSY spectrum: H-1′′ and C-1′′–C-5′′; δ_H 3.35, 3.58 (H_2-6′′) and C-4′′–C-6′′; H-1′′′ and C-1′′′–C-4′′′; δ_H 3.98 (H-5′′′) and C-2′′′–C-5′′′; H-1′′′′ and C-2′′′′–C-4′′′′; δ_H 3.50 (H_2-6′′′′) and C-4′′′′–C-6′′′′. Combining with the correlations displayed in its ^1H ^1H COSY and HSQC spectrum, the NMR data of three glycosyls were assigned in detail. Finally, according to the long-range correlations from H-1′′ to C-3; H-1′′′ to C-7; H-1′′′′ to C-4′ showed in its HMBC experiment (Figure 3), the connections between glycosyl groups and aglycone were determined. On the basis of above-mentioned evidence, the structure of mongoflavonoside A_1 (**1**) was identified.

Table 1. ^{13}C-NMR data for **1–6** in DMSO-d_6.

No.	1	2	3	4	5	6	No.	1	2	3	4	5	6
2	156.0	155.5	156.1	157.9	156.3	156.0	<u>C</u>OCH$_3$					169.6	
3	133.9	133.5	133.9	133.1	132.8	134.0	CO<u>C</u>H$_3$					19.9	
4	177.7	177.3	177.6	177.3	177.0	177.7	1‴	99.0	103.0	103.0	103.4	103.2	99.2
5	160.8	161.1	160.7	160.4	161.1	160.7	2‴	72.6	73.2	73.1	73.0	73.0	72.7
6	99.2	98.9	99.4	99.3	98.8	99.2	3‴	75.7	76.3	76.3	76.3	76.3	76.1
7	162.5	164.8	162.9	162.3	164.9	162.7	4‴	71.3	69.9	69.9	69.9	69.9	71.6
8	94.3	93.8	94.4	94.6	93.6	94.3	5‴	75.0	76.7	76.7	76.7	76.7	74.3
9	156.2	156.4	156.0	156.2	156.5	156.0	6‴	170.5	60.9	60.9	60.9	61.0	171.4
10	105.8	103.8	105.6	105.6	103.4	105.6	1⁗	99.8	99.9	99.3	101.0		101.3
1′	123.5	123.6	123.4	120.7	120.8	124.2	2⁗	73.1	73.1	72.8	70.4		73.2
2′	130.6	130.5	130.6	130.9	116.0	116.5	3⁗	76.4	76.4	76.2	70.9		75.7
3′	115.7	115.7	115.7	115.0	144.8	146.0	4⁗	69.5	69.5	71.7	71.6		69.6
4′	159.2	159.1	159.3	159.9	148.5	147.5	5⁗	77.0	77.0	73.9	68.1		77.1
5′	115.7	115.7	115.7	115.0	115.0	115.3	6⁗	60.5	60.5	171.4	17.5		60.6
6′	130.6	130.5	130.6	130.9	121.3	121.0	1′′′′′			99.8	98.7		
1″	100.6	100.5	100.4	100.8	100.5	100.5	2′′′′′			73.1	72.8		
2″	74.1	73.8	73.1	73.7	73.5	74.0	3′′′′′			76.4	76.2		
3″	76.3	74.6	74.6	74.6	74.5	76.4	4′′′′′			69.5	71.6		
4″	69.8	80.0	80.0	80.8	80.0	69.9	5′′′′′			77.0	73.7		
5″	77.5	75.3	75.4	73.7	71.9	77.6	6′′′′′			60.5	171.7		
6″	60.8	60.1	60.0	67.0	62.1	60.8							

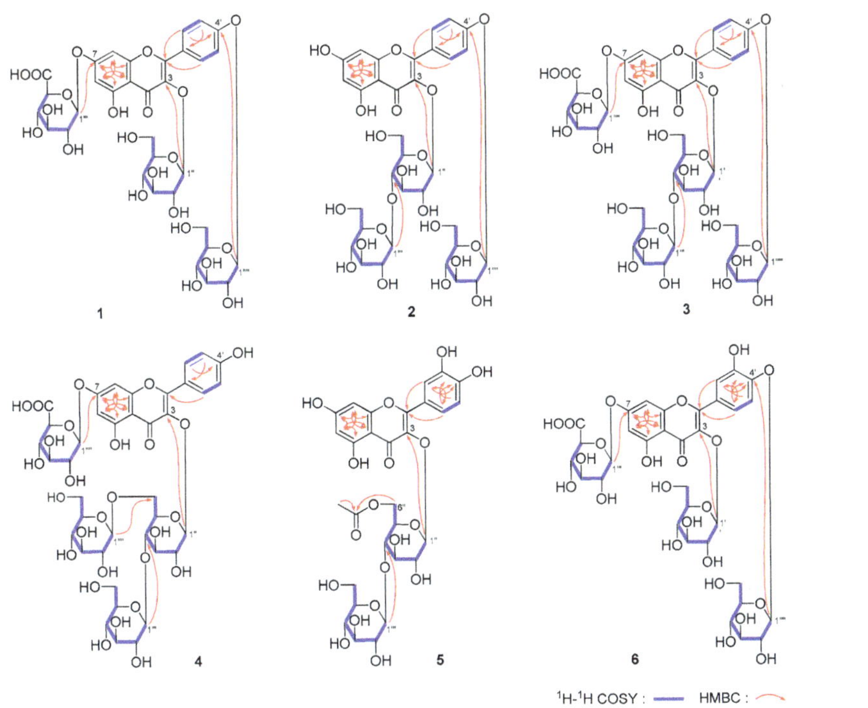

Figure 3. Key ^{1}H ^{1}H COSY and HMBC correlations of **1–6**.

Mongoflavonoside A$_2$ (**2**) is a yellow powder with negative optical rotation ($[\alpha]_D^{25}$ −26.0, MeOH). Its ESI-Q-Orbitrap MS spectrum showed a peak at *m/z* 771.19971 $[M - H]^-$ (calculated for $C_{33}H_{39}O_{21}$, 771.19783), and its molecular formula was deduced to be $C_{33}H_{40}O_{21}$. After hydrolyzing with 1 M HCl, the product was analyzed by using HPLC with an optical rotation detector. As a result, D-glucose was detected [28]. The ^{1}H and ^{13}C-NMR (Table 1) spectra suggested that compound **2** had the same aglycone, kaempferol [δ 6.21 (1H, br. s, H-6), 6.44 (1H, br. s, H-8), 7.17 (2H, d, *J* = 9.0 Hz, H-3′,5′), 8.11 (2H, d, *J* = 9.0 Hz, H-2′,6′)] as that of **1**. In addition, there were three β-D-glucopyranosyl moieties [δ 4.27 (1H, d, *J* = 8.0 Hz, H-1‴), 5.03 (1H, d, *J* = 7.5 Hz, H-1‴‴), 5.51 (1H, d, *J* = 8.0 Hz, H-1″)]. To solve the overlapping problem of three β-D-glucopyranosyl groups, the HSQC-TOCSY experiment was performed. The correlations between C-1″ and δ_H 3.26 (H-2″), 3.39 (H-4″), 3.41 (H-3″), 5.51 (H-1″); δ_H 3.51, 3.63 (H_2-6″) and C-4″–C-6″; H-1‴ and C-1‴–C-4‴; δ_H 3.42, 3.71 (H_2-6‴) and C-4‴–C-6‴; H-1‴‴ and C-1‴‴–C-4‴‴; δ_H 3.50, 3.71 (H_2-6‴‴) and C-4‴‴–C-6‴‴ were found in it. Moreover, the HMBC displayed the long-range correlations from H-1″ to C-3; H-1‴ to C-4″; H-1‴‴ to C-4′ (Figure 3). Consequently, the structure of mongoflavonoside A_2 (**2**) was elucidated to be kaempferol 3-*O*-β-D-glucopyranosyl(1→4)-β-D-glucopyranosyl-4′-*O*-β-D-glucopyranoside.

Mongoflavonoside A_3 (**3**) exhibited negative optical rotation ($[\alpha]_D^{25}$ −64.7, H_2O). Its molecular formula was revealed to be $C_{39}H_{48}O_{27}$ by negative ESI-Q-Orbitrap MS analysis (*m/z* 947.23242 [M − H]$^-$, calculated for $C_{39}H_{47}O_{27}$, 947.22992). The ^{1}H, ^{13}C-NMR (Table 1) along with various 2D NMR (^{1}H ^{1}H COSY, HSQC, HMBC, and HSQC-TOCSY) spectra denoted that **3** had the same moiety, kaempferol 3-*O*-β-D-glucopyranosyl(1→4)-β-D-glucopyranosyl-4′-*O*-β-D-glucopyranosyl [δ 4.26 (1H, d, *J* = 7.0 Hz, H-1‴), 5.04 (1H, d, *J* = 7.0 Hz, H-1‴‴), 5.53 (1H, d, *J* = 7.5 Hz, H-1″), 6.46 (1H, br. s, H-6), 6.85 (1H, br. s, H-8), 7.18 (2H, d, *J* = 8.5 Hz, H-3′,5′), 8.14 (2H, d, *J* = 8.5 Hz, H-2′,6′)] as that of **2**. Meanwhile, one more β-D-glucuropyranosyl [δ 5.12 (1H, d, *J* = 7.0 Hz, H-1⁗)] appeared in **3**. On the other hand, the proton signals at H-6 and H-8 and the carbon signal at C-7 shifted to the lower field in comparison with those of **2**, which suggested that the β-D-glucuropyranosyl linked with 7-position of kaempferol. It was clarified by the long-range correlation from H-1⁗ to C-7 (Figure 3). Then, the structure of mongoflavonoside A_3 (**3**) was determined.

The molecular formula of mongoflavonoside A_4 (**4**) was measured to be $C_{39}H_{48}O_{26}$ by negative ESI-Q-Orbitrap MS analysis (*m/z* 931.23785 [M − H]$^-$, (calculated for $C_{39}H_{47}O_{26}$, 931.23501). Its acid hydrolysis product was derived to obtain trimethylsilane thiazolidine derivatives, then the existence of D-glucuronic acid, D-glucose, and L-rhamnose were clarified by GC analysis [27]. Its ^{1}H, ^{13}C-NMR (Table 1) and 2D NMR spectra indicated that **4** had the same moiety, kaempferol 3-*O*-β-D-glucopyranosyl(1→4)[α-L-rhamanopyranosyl(1→6)]-β-D-glucopyranosyl [δ 4.14 (1H, d, *J* = 7.5 Hz, H-1‴), 4.40 (1H, br. s, H-1⁗), 5.23 (1H, d, *J* = 7.5 Hz, H-1″), 6.37 (1H, br. s, H-6), 6.73 (1H, br. s, H-8), 6.88 (2H, d, *J* = 9.0 Hz, H-3′,5′), 8.00 (2H, d, *J* = 9.0 Hz, H-2′,6′)] as that of kaempferol 3-*O*-β-D-glucopyranosyl(1→4)[α-L-rhamanopyranosyl(1→6)]-β-D-glucopyranoside [29]. In addition, one β-D-glucuropyranosyl [δ_H 5.16 (1H, d, *J* = 7.0 Hz, H-1‴‴); δ_C 71.6 (C-4‴‴), 72.8 (C-2‴‴), 73.7 (C-5‴‴), 76.2 (C-3‴‴), 98.7 (C-1‴‴), 171.7 (C-6‴‴)] appeared in **4**. The long-range correlation observation from H-1‴‴ to C-7 (Figure 3) in its HMBC spectrum suggested the β-D-glucuropyranosyl connected with C-7 of kaempferol 3-*O*-β-D-glucopyranosyl(1→4)[α-L-rhamanopyranosyl(1→6)]-β-D-glucopyranosyl. Then, the structure of mongoflavonoside A_4 (**4**) was constructed.

Mongoflavonoside B_1 (**5**) was isolated as a yellow powder and showed negative optical rotation ($[\alpha]_D^{25}$ −12.0, MeOH). The molecular formula, $C_{29}H_{32}O_{18}$ of **5** was determined from ESI-Q-Orbitrap MS (*m/z* 667.15228 [M − H]$^-$; calculated for $C_{29}H_{31}O_{18}$, 667.15049) analysis. Its IR spectrum exhibited characteristic absorptions of hydroxyl (3362 cm^{-1}), ester carbonyl (1721 cm^{-1}), α,β-unsaturated ketone carbonyl (1654 cm^{-1}), aromatic ring (1605, 1507, 1448 cm^{-1}), and ether functions (1070 cm^{-1}). The ^{1}H and ^{13}C-NMR spectra displayed signals of a quercetin moiety [δ 6.18 (1H, br. s, H-6), 6.39 (1H, br. s, H-8), 6.83 (1H, d, *J* = 8.5 Hz, H-5′), 7.50 (1H, dd, *J* = 2.0, 8.5 Hz, H-6′), 7.51 (1H, d, *J* = 2.0 Hz, H-2′)], two β-D-glucopyranosyl groups [δ 4.21 (1H, d, *J* = 8.0 Hz, H-1‴), 5.40 (1H, d, *J* = 8.0 Hz, H-1″)], along with an acetyl [δ_H 1.71 (3H, s, 6″-$COCH_3$); δ_C 19.9 (6″-$COCH_3$), 169.6 (6″-$COCH_3$)]. As shown in Figure 3, the ^{1}H ^{1}H COSY experiment on **5** indicated the presence of partial structures written in bold lines. Moreover, in the HMBC spectrum, the long-range correlations from H-1″ to C-3; H-1‴ to C-4″;

6″-COCH_3 to 6″-$COCH_3$; H_2-6″ to 6″-$COCH_3$ were observed. Finally, after treating **5** with 1 M HCl, D-glucose was detected from its acid hydrolysis product [28]. Consequently, the structure of **5** was identified, and named as mongoflavonoside B_1.

The molecular formula, $C_{33}H_{38}O_{23}$ of **6** was measured on ESI-Q-Orbitrap MS (*m/z* 801.17407 [M − H]$^-$, calculated for $C_{33}H_{37}O_{23}$, 801.17201) analysis. The ^{1}H, ^{13}C NMR (Table 1) and 2D NMR (^{1}H ^{1}H COSY, HSQC, HMBC, HSQC-TOCSY) spectra suggested **6** had the same glycosyl moieties with **1**: two β-D-glucopyranosyls [δ 4.88 (1H, d, *J* = 7.0 Hz, H-1⁗), 5.52 (1H, d, *J* = 7.0 Hz, H-1″)], and one β-D-glucuropyranosyl [5.16 (1H, d, *J* = 7.0 Hz, H-1‴)]. Meanwhile, **6** possessed the same aglycone, quercetin [δ 6.45 (1H, br. s, H-6), 6.85 (1H, br. s, H-8), 7.23 (1H, d, *J* = 8.5 Hz, H-5′), 7.64 (1H, d, *J* = 1.5, 8.5 Hz, H-6′), 7.69 (1H, d, *J* = 1.5 Hz, H-2′)] as that of **5**. Finally, the connectivities of glycosyl moieties with aglycone were determined by the correlations from H-1″ to C-3; H-1‴ to C-7; H-1⁗ to C-4′ (Figure 3) showed in its HMBC spectrum.

Mongophenoside A_1 (**7**) was obtained as a white powder with negative optical rotation ($[\alpha]_D^{25}$ −21.0, MeOH). ESI-Q-Orbitrap MS of **7** exhibited quasimolecular ion peak at *m/z* 503.14151 [M − H]$^-$ (calculated for $C_{21}H_{27}O_{14}$, 503.13953), and its molecular formula was deduced to be $C_{21}H_{28}O_{14}$. The IR spectrum of it showed absorption bands ascribable to hydroxyl (3362 cm^{-1}), α,β-unsaturated ester carbonyl (1709 cm^{-1}), aromatic ring (1601, 1521, 1447 cm^{-1}), and ether function (1074 cm^{-1}). Acid hydrolysis of **7** liberated D-glucose, which was identified by HPLC analysis [28]. Its ^{1}H, ^{13}C NMR (Table 2) spectra indicated the existence of one *trans*-caffeoyl [δ_H 6.27 (1H, d, *J* = 16.0 Hz, H-8), 6.75 (1H, d, *J* = 7.5 Hz, H-5), 7.01 (1H, br. d, ca. *J* = 8 Hz, H-6), 7.06 (1H, br. s, H-2), 7.55 (1H, d, *J* = 16.0 Hz, H-7); δ_C 113.3 (C-8), 146.2 (C-7), 164.9 (C-9)], along with two β-D-glucopyranosyl groups [δ 4.42 (1H, d, *J* = 8.0 Hz, H-1″), 5.56 (1H, d, *J* = 8.0 Hz, H-1′)]. Meanwhile, the partial structures written in bold lines shown in Figure 4 were determined by proton and proton correlations observed in its ^{1}H ^{1}H COSY experiment. The planar structure of **5** was finally elucidated according to the long-range correlations from H-1′ to C-9; H-1″ to C-2′ (Figure 4) found in HMBC experiment, and the structure of **7** was named as mongophenoside A_1.

Table 2. ^{13}C-NMR data for **7–10**.

No.	7 [a]	8 [a]	9 [a]	10 [b]	No.	7 [a]	8 [a]	9 [a]	10 [b]
1	125.4	125.2	128.5	127.5	5′	77.4	75.9	77.1	78.0
2	114.6	114.6	115.0	111.7	6′	60.3	67.7	60.6	62.8
3	145.5	145.6	146.7	149.5	1″	104.4	104.4	92.4	94.0
4	148.6	149.0	147.5	151.0	2″	74.4	74.4	81.6	73.9
5	115.6	115.6	115.9	116.6	3″	76.0	76.1	75.6	74.9
6	121.6	121.7	120.9	124.2	4″	69.3	69.3	69.0	71.8
7	146.2	146.3	145.5	146.9	5″	76.6	76.7	77.5	73.0
8	113.3	113.1	115.5	115.1	6″	60.3	60.3	60.3	62.9
9	164.9	164.9	164.7	169.8	1‴		103.0	104.5	
3-OCH_3				56.5	2‴		73.4	74.5	
1′	92.3	92.3	101.4	98.2	3‴		76.6	76.0	
2′	81.5	81.4	73.1	76.3	4‴		69.9	69.3	
3′	75.6	75.5	75.7	78.1	5‴		76.8	76.6	
4′	69.0	68.8	69.7	71.9	6‴		60.9	60.3	

Determined in [a] DMSO-d_6, [b] CD_3OD.

Mongophenoside A_2 (**8**), a white powder, showed negative optical rotation ($[\alpha]_D^{25}$ −14.5, MeOH). ESI-Q-Orbitrap MS analysis suggested its molecular formula was $C_{27}H_{38}O_{19}$ (665.19427 [M − H]$^-$; calculated for $C_{27}H_{37}O_{19}$, 665.19236). The ^{1}H and ^{13}C-NMR (Table 2) spectra indicated **8** possessed the same moiety, trans-caffeic acid-9-*O*-β-D-glucopyranosyl(1→2)-β-D-glucopyranosyl [δ 4.42 (1H, d, *J* = 7.5 Hz, H-1″), 5.56 (1H, d, *J* = 7.0 Hz, H-1′), 6.27 (1H, d, *J* = 16.0 Hz, H-8), 6.76 (1H, d, *J* = 7.5 Hz, H-5), 7.02 (1H, br. d, ca. *J* = 8 Hz, H-6), 7.06 (1H, br. s, H-2), 7.55 (1H, d, *J* = 16.0 Hz, H-7)] as that of **7**. Except for that, one more β-D-glucopyranosyl [δ 4.17 (1H, d, *J* = 7.5 Hz, H-1‴)] appeared in **8**. Meanwhile, C-6′ of it was found to significantly shift to lower field (δ_C 67.7 for **8**; 60.3 for **7**) comparing with **7**, which suggested C-6′ was substituted by the β-D-glucopyranosyl. In the HMBC spectrum, the long-range correlations from H-1′′′ to C-6′; H-1′′ to C-2′; H-1′ to C-9 (Figure 4) were observed. Moreover, treated **8** with 1 M HCl, D-glucose was yielded [28]. Consequently, the structure of mongophenoside A_2 (**8**) was elucidated.

The ESI-Q-Orbitrap MS spectrum of mongophenoside A_3 (**9**) displayed the same molecular formula, $C_{27}H_{38}O_{19}$ (*m/z* 665.19452 $[M - H]^-$; calculated for $C_{27}H_{37}O_{19}$, 665.19236) as that of **8**. Meanwhile, the ^1H, ^{13}C NMR (Table 2) and 2D NMR (^1H ^1H COSY, HSQC, HMBC, HSQC-TOCSY) spectra suggested they had same functional groups as following: trans-caffeic acid aglycone [δ 6.45 (1H, d, J = 16.0 Hz, H-8), 7.19 (1H, br. s, H-2), 7.12 (2H, m, H-5 and H-6), 7.61 (1H, d, J = 16.0 Hz, H-7)] and three β-D-glucopyranosyl groups [δ 4.43 (1H, d, J = 7.5 Hz, H-1′′′), 4.80 (1H, d, J = 7.5 Hz, H-1′), 5.57 (1H, d, J = 7.0 Hz, H-1′′)]. Finally, the connectivities of the above-mentioned groups were clarified by the long-range correlations from H-1′ to C-4; H-1′′ to C-9; H-1′′′ to C-2′′ (Figure 4), as shown in its HMBC experiment.

Mongophenoside B (**10**) was obtained as a white powder with positive optical rotation ($[\alpha]_D^{25}$ +8.0, MeOH). Its ESI-Q-Orbitrap MS spectrum showed the negative ion peak at *m/z* 517.15668 $[M - H]^-$ (calculated for $C_{22}H_{29}O_{14}$, 517.15518), which indicated the molecular formula of it was $C_{22}H_{30}O_{14}$. Acid hydrolysis **10** with 1 M HCl, D-glucose was liberated [28]. The ^1H, ^{13}C NMR (Table 2) and 2D NMR spectra of it suggested the existence of one trans-feruloyl [δ 6.34 (1H, d, J = 16.0 Hz, H-8), 6.80 (1H, d, J = 8.0 Hz, H-5), 7.06 (1H, br. d, ca. J = 8 Hz, H-6), 7.17 (1H, br. s, H-2), 7.60 (1H, d, J = 16.0 Hz, H-7), 3.88 (3H, s, 3-OCH_3)], one β-D-glucopyranosyl [δ 4.48 (1H, d, J = 8.0 Hz, H-1′)], together with one α-D-glucopyranosyl [δ 5.10 (1H, J = 3.5 Hz, H-1′′)]. Moreover, the long-range correlations from H-1′ to C-9; H-1′′ to C-2′ were observed in its HMBC experiment. On the basis of above-mentioned evidence, the structure of mongophenoside B (**10**) was identified as trans-ferulic acid-9-*O*-α-D-glucopyranosyl(1→2)-β-D-glucopyranoside.

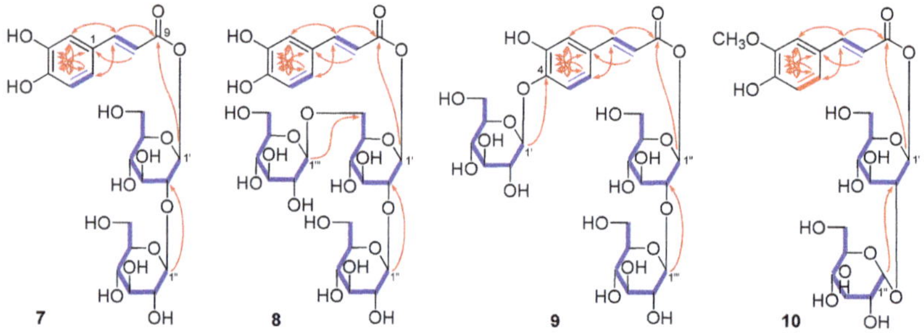

Figure 4. Key ^1H ^1H COSY and HMBC correlations of **7**–**10**.

2.2. *Inhibitory Effects of Obtained Compounds* **1**–**31** *on the Motility of Mouse Isolated Intestine Tissue*

Moreover, the obtained constituents of **1**–**31** were tested for frequency and height by using a tissue perfusion method [30]. Through tissue perfusion experiments, it was found that all compounds displayed no effect on isolated intestinal tissue contraction frequency (Table 3). While almost all isolates exhibited the tendency of increasing the contraction amplitude of mouse small intestinal muscle though only flavonoids **3**, **4**, **11**–**15**, **21**–**23**, and **26**, as well as phenolic acids **7**, **29**, and **30** showed significant difference comparing with normal group.

Table 3. Inhibitory effects of compounds **1–31** on motility of mouse isolated intestine tissue.

Compd.	Intestine Motility (%)		Compd.	Intestine Motility (%)	
	Relative Height	Relative Frequency		Relative Height	Relative Frequency
N	100.0 ± 4.9	100.0 ± 1.2	**16**	116.1 ± 10.4	81.3 ± 11.1
P	190.8 ± 19.2 **	85.6 ± 2.6	**17**	108.1 ± 6.1	97.3 ± 2.3
1	112.3 ± 2.3	99.4 ± 1.2	**18**	116.6 ± 4.4	106.6 ± 4.5
2	105.1 ± 19.5	98.8 ± 2.9	**19**	107.7 ± 3.1	99.3 ± 3.5
3	148.9 ± 4.5 **	100.8 ± 0.9	**20**	107.8 ± 32.7	99.4 ± 0.6
4	170.0 ± 6.4 *	98.6 ± 2.1	**21**	121.9 ± 6.6 *	95.2 ± 4.4
5	99.7 ± 25.0	101.6 ± 7.7	**22**	125.2 ± 8.1 *	97.4 ± 4.0
6	107.9 ± 18.9	96.7 ± 3.1	**23**	142.2 ± 11.2 *	101.1 ± 3.7
7	157.4 ± 20.8 *	94.9 ± 2.7	**24**	117.9 ± 12.7	98.3 ± 2.8
8	150.5 ± 25.9	95.6 ± 2.8	**25**	104.9 ± 7.7	96.3 ± 1.9
9	149.1 ± 36.2	94.6 ± 3.8	**26**	137.4 ± 2.4 **	96.2 ± 2.8
10	121.1 ± 21.6	98.7 ± 3.0	**27**	105.6 ± 32.3	91.8 ± 3.8
11	123.2 ± 6.8 *	99.1 ± 7.1	**28**	136.8 ± 12.4	97.1 ± 2.0
12	144.2 ± 14.3 *	95.7 ± 1.3	**29**	148.1 ± 6.8 *	105.7 ± 4.1
13	151.5 ± 17.1 *	98.8 ± 4.4	**30**	125.0 ± 1.6 **	98.8 ± 2.3
14	120.4 ± 9.2 *	98.4 ± 1.7	**31**	97.8 ± 1.9	98.8 ± 1.2
15	143.7 ± 1.3 **	100.6 ± 3.1			

Values are the means ± standard error of measurement, significantly different from the control group, * $p < 0.05$, ** $p < 0.01$, n = 6. Normal (N): isolated intestine tissue; Positive control (P): Mosapride citrate dihydrate, final concentration was 200 μg/mL. Compounds **1–31**: final concentration was 50 μM. Frequency, and height of normal group was set as 100%, relative values of them were calculated as: (sample/normal) × 100%.

2.3. *Qualitative Analysis*

As an important edible medicinal plant for Mongolian people, *A. mongolicum* has made a great contribution to the development of the local economy, yet there is a lack of analysis of its quality until now.

Our systematic phytochemistry isolation results indicated the main constituents of AM were flavonoids and phenolic acids. The aglycones in the plant mainly included quercetin, kaempferol, as well as isorhamnetin for flavonoid glycosides; while coumaric acid, caffeic acid, and ferulic acid for phenolic acid glycosides. The sugars consisted of β-D-glucopyranoside (Glc), α-D-glucopyranoside (α-Glc), β-D-glucuronic acid (Glu), and α-L-rhamnopyranoside (Rha). While α-Glc was only found in phenolic acid glycosides, Glu and Rha substituted only for flavonoid glycosides.

As for flavonoids, 3-, 7-, and 4′-OH of quercetin, kaempferol, and isorhamnetin were easily substituted by various glycosyls to format *O*-glycosides. Among them, 7- and 4′-OH was substituted by monosaccharose such as Glc and Glu, while Glu was found to only link with their 7-position. Meanwhile, 3-OH was with a high degree of glycosylation, having one to three sugar moieties, and all of the glycosyl groups directly linked to flavonoid was Glc group, then its 2-, 4-, or 6-position was substituted by another Glc continuously; moreover, its 6-position could also be replaced by Rha [to form rutinosyl (Rut)] or acetyl group (Figure 5).

On the other hand, the carboxyl of obtained phenolic acids from AM was easily substituted by sugar moiety such as Glc(1→2)Glc–, α-Glc(1→2)Glc–, or Glc(1→6)Glc(1→2)Glc– on their 9-position, while 4-OH of them was only substituted by monosaccharose, Glc (Figure 6).

Herein, on the basis of above-mentioned phytochemistry study, a fast analysis method for flavonoids and phenolic acids in AM was established by LC-MS on an ESI-Q-Orbitrap MS in negative ion mode (Figure 7). According to the chromatographic retention time (t_R) and the exact mass-to-charge ratio (*m/z*), 31 compounds (**1–31**) were unambiguously identified by comparing to the standard references. Meanwhile, the rules of the MS/MS fragmentation pattern and chromatographic elution order have been generalized. Then, five flavonoid glycosides (**32–36**) and one phenolic acid glycoside (**37**) were tentatively speculated. Among them, **36** was a potential new compound (Table S1, Figure 8).

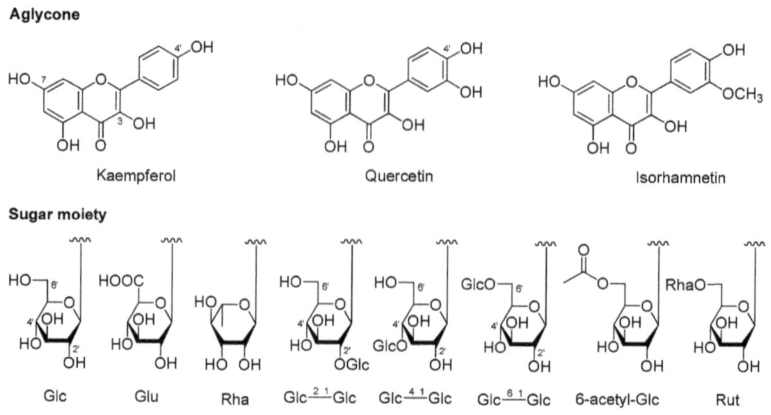

Figure 5. The structure of aglycones and glycosyls of flavonoids from the aerial parts of *A. mongolicum*.

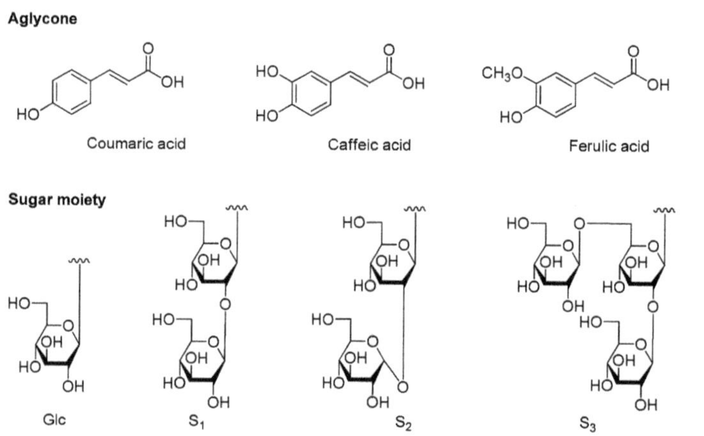

Figure 6. The structure of aglycones and glycosyls of phenolic acids from the aerial parts of *A. mongolicum*.

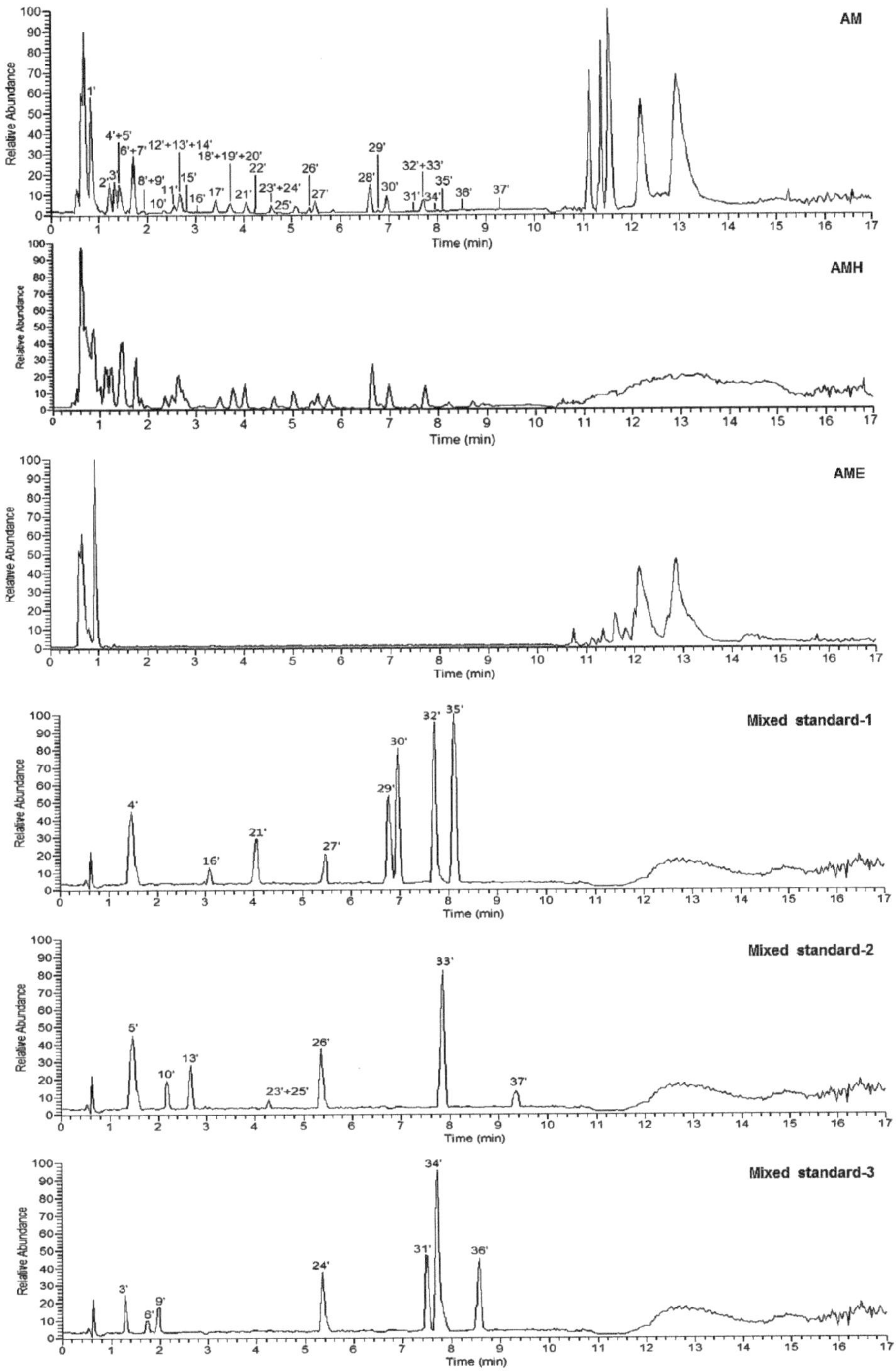

Figure 7. *Cont.*

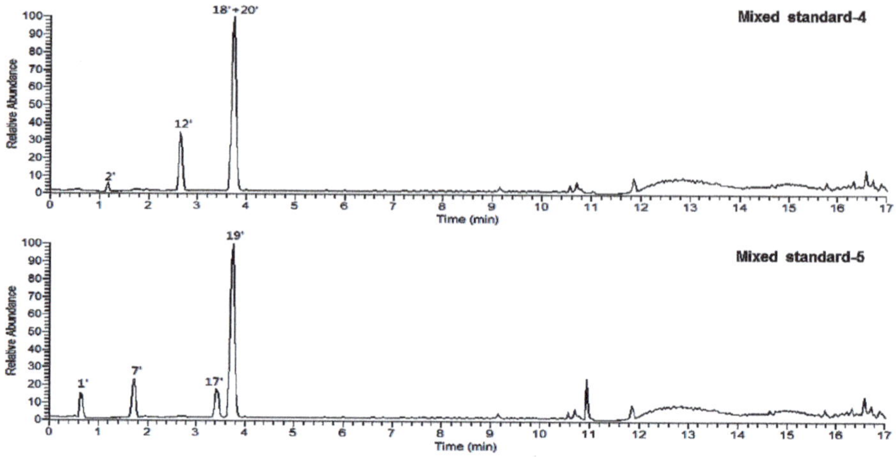

Figure 7. Base peak chromatograms of AM (the extract obtained from fresh the aerial parts of *A. mongolicum* heated reflux with 95% EtOH and 50% EtOH one time each, successively), AMH (H_2O layer extract), AME (EtOAc layer extract), and mixed standard references.

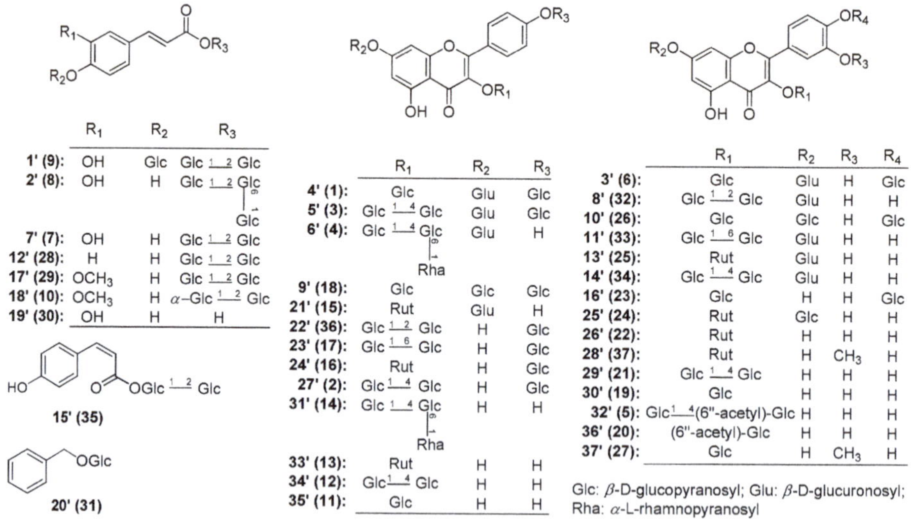

Figure 8. The structures of tentatively presumed compounds from the aerial parts of *A. mongolicum*.

2.3.1. Structural Elucidation of Flavonoids

Peaks 3′–6′, 9′, 10′, 13′, 16′, 21′, 23′–27′ and 29′–37′ were identified by comparison with reference standards (Table S1, Figure 7).

Figure 9 and Figure S74 showed the MS/MS fragmentation pattern of flavonoid glycosides with kaempferol and quercetin aglycones, which suggested both of two kinds of flavonoid glycosides could be ionized to generate heterolytic cleavage with fragments ion peak at *m/z* 285.03936 (Y_{K0}^-) for kaempferol and *m/z* 301.03428 (Y_{Q0}^-) for quercetin, as well as hemolytic cleavage with fragments ion peak at *m/z* 284.03154 $[Y_{K0}^- - H]^-$ for kaempferol and *m/z* 300.02645 $[Y_{Q0}^- - H]^-$ for quercetin, respectively. Then, kaempferol aglycone could be further cleavage to generate fragment ion peaks at *m/z* 255.02880, 179.02371, and 151.00259. The fragment ion peaks at *m/z* 271.02371, 255.02880, 243.02880, 179.02371, as well as 151.00259, were yielded from quercetin by a series of reactions including decarbonylation, dehydrogenation, retro Diels–Alder reaction, and the reaction to remove the B ring. The above-mentioned characteristic fragment ions could be used to distinguish the type of aglycone.

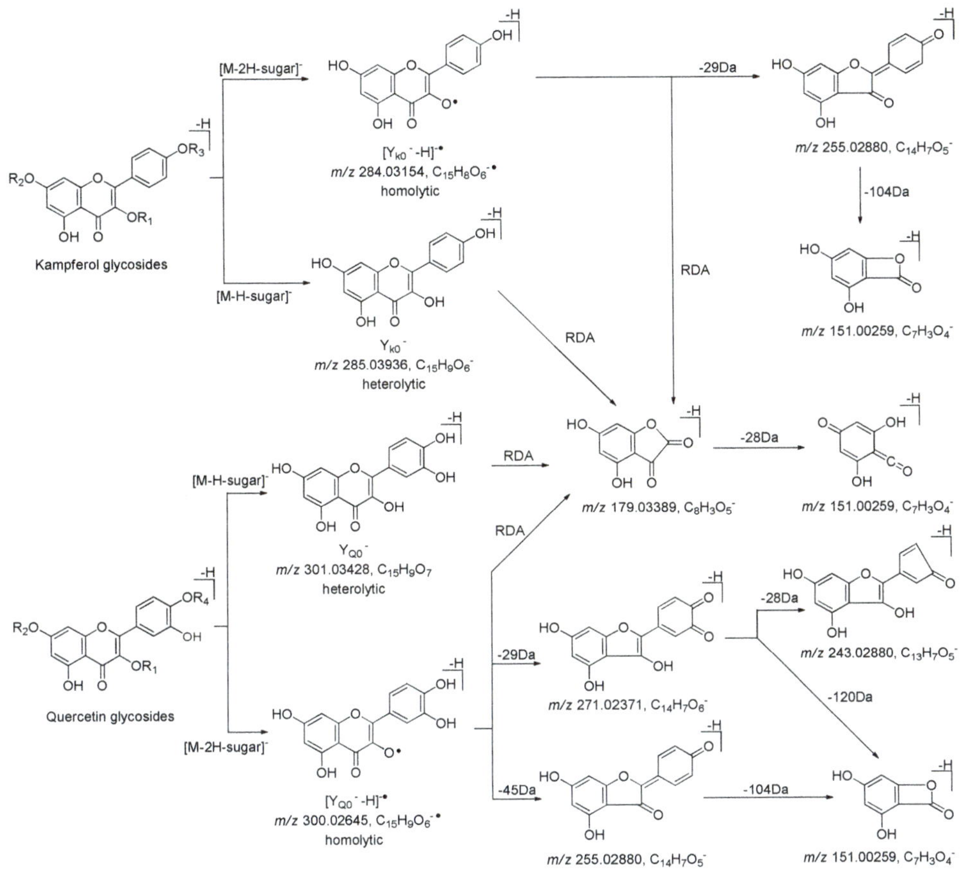

Figure 9. The proposed fragmentation pathways of kaempferol and quercetin glycosides.

Meanwhile, when the 4′-position of flavanol aglycone was glycosylated to format *O*-glycoside, the debris ions peaks ($[Y_{K0}{}^{-}2H]^{-}$) at *m/z* 283.02371 for kaempferol and *m/z* 299.01863 ($[Y_{Q0}{}^{-}2H]^{-}$) for quercetin glycosides were stronger than those of *m/z* 284.03154 and 300.02645, respectively (Table S1). Therefore, their ionic strength could be used to quickly determine whether the C-4′ position of the aglycone was replaced by sugar.

Peaks 9′, 22′, 23′, and 27′ were obtained by extracting ion of *m/z* 771.19783 from the total ion chromatogram of AM (Figure 10), among them, 9′, 23′, and 27′ were clarified to be kaempferol-3,7,4′-tri-*O*-β-glucoside (**18**), kaempferol-3-*O*-gentiobioside-4′-*O*-glucopyranoside (**17**), and mongoflavonoside A_2 (**2**) by comparing with reference standards. Then, according to the above-mentioned biosynthetic pathway of substituted sugar, peak 22′ was tentatively presumed to be kaempferol-3-*O*-β-D-glucopyranosyl(1→2)-*O*-β-D-glucopyranosyl-4′-*O*-β-D-glucopyranoside (**36**), which was one new compound.

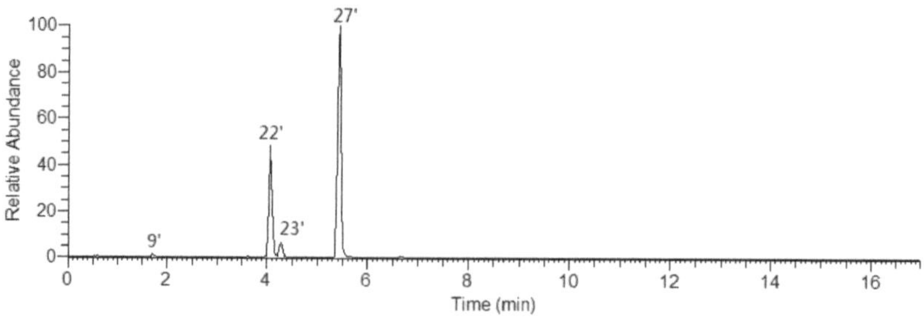

Figure 10. The EIC of the *m/z* 771.19783.

Moreover, during the comparison of the chromatographic retention behavior of peaks 9′, 22′, 23′, and 27′, we discovered the effect of sugar substitution position on t_R was 3,7,4′-tri-*O*-Glc < 3-*O*-Glc(1→2)-Glc-4′-*O*-Glc < 3-*O*-Glc(1→6)-Glc-4′-*O*-Glc < 3-*O*-Glc(1→4)-Glc-4′-*O*-Glc.

The molecular formula of peaks 3′ (*m/z* 801.17407), 8′ (*m/z* 801.17462), 11′ (*m/z* 801.17389), and 14′ (*m/z* 801.17200) were all $C_{33}H_{38}O_{23}$ (Figure 11). Peak 3′ was unambiguously identified as mongoflavonoside B_2 (**6**) by comparison with reference standard. According to the MS/MS fragment ion peaks at *m/z* 301.03428, 300.02645, 299.01863, 271.02371, 255.02880, and 151.00259, peaks 8′, 11′, and 14′ were deduced to be with quercetin aglycone. On the other hand, the fragment ion peaks at *m/z* 625.13993 $[M - H - 176]^-$, 301.03428 $[M - H - 176 - 162 - 162]^-$ suggested the presences of one β-D-Glu and two β-D-Glc in them. Since the strength of fragment ion peak at *m/z* 299.01863 was weaker than that of *m/z* 300.02645, we could propose that 4′-OH of quercetin was not be glycosidated. According to the above-mentioned biosynthetic pathway of substituted sugar and effect of sugar substitution position on t_R, peaks 8′, 11′, and 14′ were tentatively presumed to be quercetin-3-*O*-β-D-glucopyranosyl(1→2)-β-D-glucopyranosyl-7-*O*-β-D-glucuronide (**32**), quercetin-3-*O*-β-D-glucopyranosyl(1→6)-β-D-glucopyranosyl-7-*O*-β-D-glucuronide (**33**), and quercetin-3-*O*-β-D-glucopyranosyl(1→4)-β-D-glucopyranosyl-7-*O*-β-D-glucuronide (**34**), respectively.

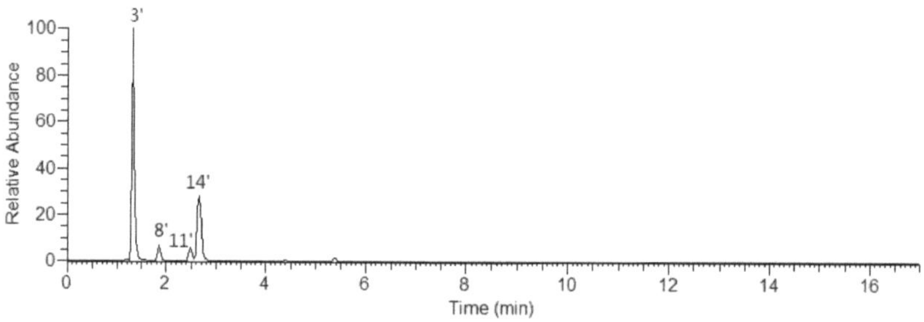

Figure 11. The EIC of the *m/z* 801.17201.

The molecular formula of peak 28′ (*m/z* 799.19391) was $C_{33}H_{38}O_{23}$. Its MS/MS fragment ion peaks displayed at *m/z* 623.15869 [M − H − 176], 315.05048 $[M - H - 176 - 162 - 146]^-$, 300.02713, 271.02469, and 243.02880 suggested the aglycone of it was isorhamnetin and the substituted sugar moieties included one Glu, one Glc, and one Rha. According to the biosynthesis laws summarized above, peak 28′ was deduced to be isorhamnetin-3-*O*-rutinosyl-7-*O*-β-D-glucuronide (**37**) (Table S1, Figure S75).

2.3.2. Structural Elucidation of Phenolic Acids

Peaks 1′, 2′, 7′, 12′, and 17′–20′ were identified unequivocally by comparing with reference standards (Table S1, Figure 7). As what have been mentioned above, the aglycones of phenolic acid glycosides included coumaric acid, caffeic acid, and ferulic acid. It was well known that the characteristic ions of coumaroyl, caffeoyl, and feruloyl were at *m/z* 163.03897 ([coumaroyl − H]$^-$),

179.03389 ([caffeoyl − H]$^-$), and 193.04954 ([feruloyl − H]$^-$), respectively [31]. Then, all of the ions would further generate fragment ion peaks (as shown in Figure 12) by removing 44 Da (–CO_2), 28 Da (–CO), and 18 Da (–H_2O), respectively.

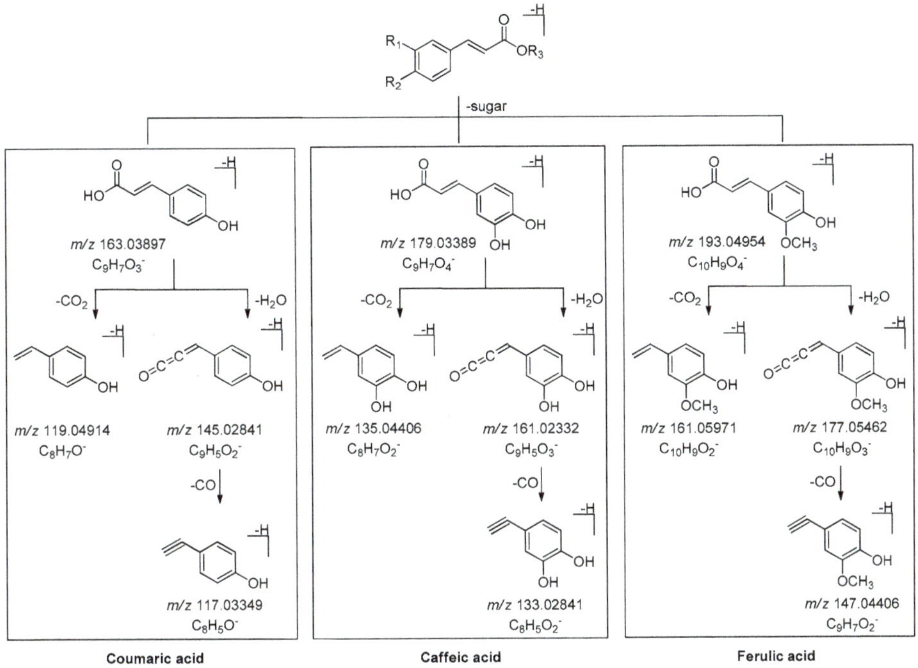

Figure 12. The proposed fragmentation pathways of coumaric acid, caffeic acid, and ferulic acid glycosides.

Meanwhile, the phenomenon of the neutral loss 120 Da on the basis of [M − H]$^-$ were only found in β-D-glucopyranosyl(1→2)-β-D-glucopyranosyl substituted phenolic acid glycosides **7** (peak 7′), **28** (peak 12′), and **29** (peak 17′) (Figure 13 and Figure S76), which could be used to distinguish the type of substituted sugar moieties.

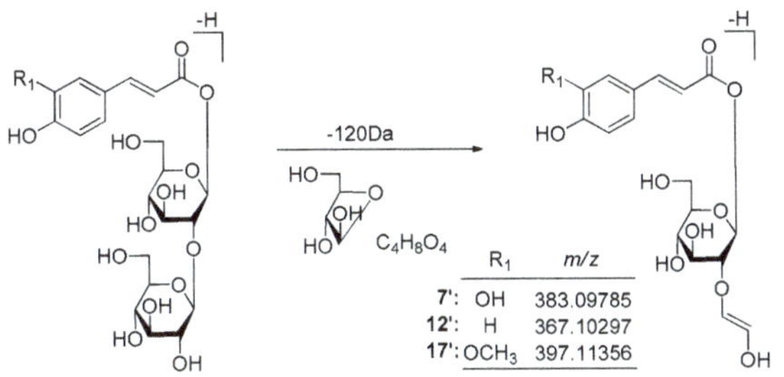

Figure 13. The proposed fragmentation pathways of β-D-glucopyranosyl(1→2)-β-D-glucopyranosyl substituted phenolic acid glycosides.

Moreover, comparing the t_R of compounds **10** (peak 18′) and **29** (peak 17′), α-D-glucopyranosyl-substituted phenolic acid glycoside was found to have the shorter t_R than that of β-D-glucopyranosyl-substituted ones.

The molecular formula of peak 15′ (*m/z* 487.14313) was $C_{21}H_{28}O_{13}$. Its MS/MS fragment ion peaks displayed at *m/z* 367.10297, 163.03888, and 145.02829, which was similar to those of peak 12′ (Table S1, Figure 14). According to the above-mentioned chromatographic retention behavior, we could deduce that peak 15′ was not *p*-hydroxycinnamic acid-9-*O*-α-D-glucopyranosyl(1→2)-β-D-glucopyranoside. As Han et al. reported, the t_R of *cis*-phenylpropane glycoside was longer than that of *trans* one when they were analysed by HPLC with the acetonitrile-water system [24]. Consequently, peak 15′ was tentatively presumed to be *cis*-*p*-hydroxycinnamate sophorose (**35**).

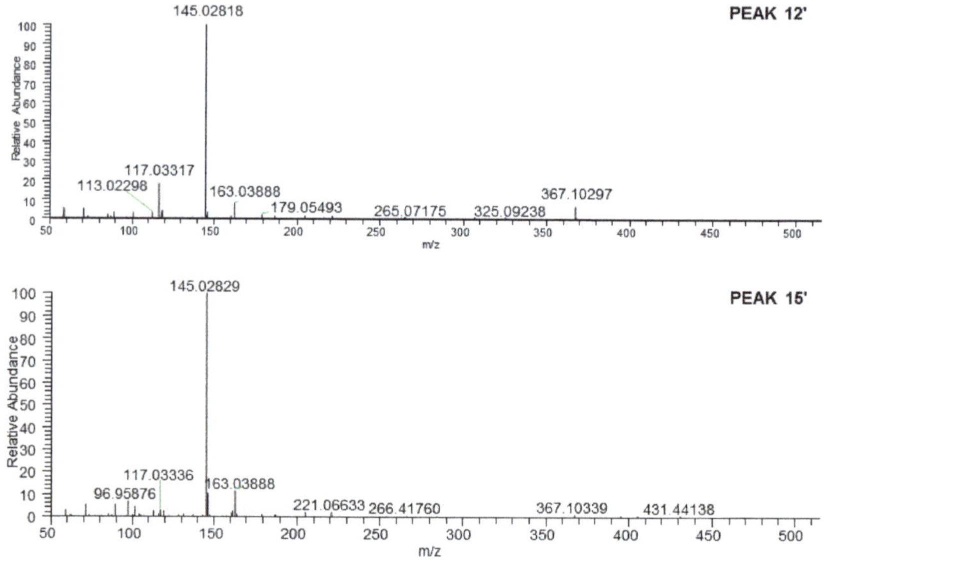

Figure 14. The tandem MS of the [M − H]$^-$ ions for peaks 12′ and 15′.

3. Materials and Methods

3.1. Materials and Methods for Phytochemistry Research

3.1.1. General Experimental Procedures

UV and IR spectra were recorded on a Varian Cary 50 UV-Vis and Varian 640-IR FT-IR spectrophotometer, respectively. Optical rotations were measured on a Rudolph Autopol® IV automatic polarimeter. NMR spectra were determined on a Bruker 500 MHz NMR spectrometer at 500 MHz for ^{1}H and 125 MHz for ^{13}C-NMR (internal standard: TMS). Negative-ion mode ESI-Q-Orbitrap MS were obtained on a Thermo UltiMate 3000 UHPLC instrument (Thermo, Waltham, MA, USA).

Column chromatographies (CC) were performed on macroporous resin D101 (Haiguang Chemical Co., Ltd., Tianjin, China), silica gel (48–75 μm, Qingdao Haiyang Chemical Co., Ltd., Qingdao, China), ODS (40–63 μm, YMC Co., Ltd., Tokyo, Japan), and Sephadex LH-20 (Ge Healthcare Bio-Sciences, Uppsala, Sweden). Preparative high performance liquid chromatography (pHPLC) column, Cosmosil 5C$_{18}$-MS-II (20 mm i.d. × 250 mm, Nakalai Tesque, Inc., Tokyo, Japan) were used to separate the constituents.

3.1.2. Plant Material

The fresh aerial parts of *Allium mongolicum* Regel were collected from Alxa League, Inner Mongolia Autonomous Region, China, and identified by Dr. Li Tianxiang (The Hall of TCM Specimens, Tianjin

University of TCM, China). The voucher specimen was deposited at the Academy of Traditional Chinese Medicine of Tianjin University of TCM.

3.1.3. Extraction and Isolation

See supporting information.

3.1.4. Acid Hydrolysis of 1, 3, 4 and 6

Solution of **1**, **3**, **4** and **6** (each 2.0 mg) in 5% aqueous H_2SO_4-1,4-dioxane were heated under reflux for 1 h, respectively. After cooling, the reaction mixture was neutralized with Amberlite IRA-400 (OH^- form), removed by filtration, subjected to ODS CC (H_2O), and the H_2O eluate was reacted with L-cysteine methyl ester hydrochloride in pyridine and *N*,*O*-bis(trimethylsilyl)trifluoroacetamide (BSTFA), successively. Finally, the reaction product was elucidated by GC analysis (GC conditions, column: Agilent Technologies INC Catalog 19,091 J-413 HP-5, 30 m × 0.320 mm (i.d.) capillary column; column temperature: 230 °C; carrier gas: N_2), and D-glucuronic acid and D-glucose hydrolysates were identified from **1**, **3**, and **6**; D-glucuronic acid, D-glucose, as well as L-rhamnose hydrolysates were detected from **4** by comparing it retention times (t_R: D-glucuronic acid, 23.3 min; D-glucose, 19.6 min; L-rhamnose, 11.4 min) with those of their authentic samples treated in the same way.

3.1.5. Acid Hydrolysis of **2**, **5** and **7–10**

The solution of compounds **2**, **5** and **7–10** (each 1.5 mg) in 1 M HCl (1.0 mL) was heated under reflux for 3 h. After cooling, the reaction mixture was neutralized with Amberlite IRA-400 (OH^- form), then analyzed by HLPC [column, Kaseisorb LC NH_2-60-5, 4.6 mm i.d. × 250 mm (Tokyo Kasei Co., Ltd., Tokyo, Japan); mobile phase, CH_3CN-H_2O (75:25, *v/v*; flow rate, 1.0 mL/min)]. As a result, D-glucose was detected from the aqueous phase of **2**, **5** and **7–10** by comparison of its retention time and optical rotation with those of the authentic sample, D-glucose (t_R 12.5 min, positive), respectively.

3.2. Materials and Methods for Bioassay

The activities of compounds **1–31** were tested for frequency and height by using tissue perfusion method reported before [30]. Samples in DMSO solution were added after 15 min equilibrate incubation; the final DMSO concentration was 0.1% and final concentration of samples were 50 μM. Mosapride citrate dihydrate (Xi'an Janssen Pharmaceutical Ltd., Xi'an, China), final concentration was 200 μg/mL.

Data were analyzed by SPSS 22.0 software. All values were expressed as mean ± S.D. A *p*-value of 0.05 was considered to indicate statistical significance. One-way analysis of variance (ANOVA) and Tukey's studentized range test were used for the evaluation of the significant differences between means and post hoc, respectively.

3.3. Materials and Methods for Qualitative Analysis

3.3.1. Materials

The isolated 31 compounds including 24 flavonoids, and 7 phenolic acids were used for reference standards. Their purities were > 98%.

HPLC grade Acetonitrile (Thermo-fisher, Waltham, MA, USA), formic acid (Roe Scientific Inc., Newark, NJ, USA), and ultra-pure water prepared with a Milli-Q purification system (Millipore, MA, USA) were used for LC-MS analysis.

3.3.2. Sample Preparation

Preparation of Standard Solutions

Standard test solutions of the above-mentioned standard references were prepared in MeOH at a final concentration of 1 μg/mL approximately. All stock solutions were stored at 4 °C in darkness and brought to room temperature before use.

Preparation of the Aerial Parts of *A. mongolicum* Extract Test Solutions

A. mongolicum extract (AM) was prepared by using the same method as described in "Extraction and Isolation" section. The AM was dissolved with MeOH and filtered with 0.22 μm microporous membrane to get test stock solution at a final concentration of 30 mg/mL, which was stored at 4 °C in darkness and brought to room temperature before use.

3.3.3. UHPLC

A Thermo UltiMate 3000 UHPLC instrument (Thermo, Waltham, MA) equipped with a quaternary pump, an autosampler was used to accomplish the analysis. Samples were separated on a Waters ACQUITY UPLC®HSS C18 (2.1 × 100 mm, 1.8 μm) using a mobile phase composed of H_2O with 0.1% formic acid (A) and CH_3CN with 0.1% formic acid (B) in the gradient program: 0–2 min, 9–10% B; 2–5 min, 10–17% B; 5–7 min, 17–20% B; 7–9 min, 20% B; 9–10 min, 20–86% B; 10–14 min, 86–100% B; 14–17 min, 100% B; An equilibration of 3 min was used between successive injections. The flow rate was 0.4 mL/min, and column temperature was 35 °C. An aliquot of 1 μL of each sample was injected for analysis.

3.3.4. ESI-Q-Orbitrap MS and Automatic Components Extraction

For tandem mass spectrometry analysis, a Thermo ESI-Q-Orbitrap MS mass spectrometer was connected to the UltiMate 3000 UHPLC instrument via ESI interface. Ultra-high purity nitrogen (N_2) was used as the collision gas and the sheath/auxiliary gas. The ESI source parameters were set as follows: ion spray voltage 3.2 kV, capillary temperature 350 °C, ion source heater temperature 300 °C, sheath gas (N_2) 40 L/h, auxiliary gas (N_2) 10 L/min, and a normalized collision energy (NCE) of −35 V was used. The Orbitrap analyzer scanned the mass range from *m/z* 150 to 1500 in negative ion mode. Monitoring time was 0–17 min. Detection was obtained by full mass-dd mass mode. The MS data were recorded in both profile and centroid formats. Data recording and processing were performed using the Xcalibur 4.0 software (Thermo Fisher Scientific, Inc., Waltham, MA, USA). The accuracy error threshold was fixed at 5 ppm.

Software-aided, automatic background subtraction and components extraction technique was used to generate a peak list containing all the components profiled from the aerial part of *A. mongolicum*. Sieve v2.2 SP2 (Thermo Fisher Scientific) was used for the automatic components extraction: time range, 1–17 min; BP minimum count, 10,000; BP minimum scans, 5; Background SN, 3; MZ Step, 10; and Frame, >1.

4. Conclusions

This paper displayed a study—the first of its kind—focused on the systematic bioactive constituents of the aerial parts of *A. mongolicum* in the gastrointestinal tract. As a result, AM and AMH showed a significant increase in the contraction amplitude of mouse small intestinal muscles, which indicated they might have therapeutic effects on constipation. During this process, we made several achievements:

The first comprehensive phytochemistry investigation was developed for AM by using various spectral and chromatographic methods: six new flavonoid glycosides, mongoflavonosides A_1 (**1**), A_2 (**2**), A_3 (**3**), A_4 (**4**), B_1 (**5**), B_2 (**6**), four new phenolic acid glycosides, mongophenosides A_1 (**7**), A_2 (**8**), A_3 (**9**), B (**10**), as well as 21 known compounds were yielded. They were mainly flavonoids and phenolic acids.

The flavonoids and phenolic acids were clarified for the first time to be the main bioactive constituents of *A. mongolicum* on gastrointestinal tract: flavonoids **3**, **4**, **11–15**, **21–23**, and **26**, as well as phenolic acids **7**, **29**, and **30** showed significant increase in the height of mouse small intestinal muscle.

Furthermore, a fast analysis method for flavonoids and phenolic acids in the aerial parts of *A. mongolicum* was established for the first time by using LC-MS. According to t_R and *m/z*, 31 compounds (**1–31**) were unambiguously identified by comparing to the standard references. Then, on the basis of generalized rules of MS/MS fragmentation pattern, chromatographic behaviors, as well as their biosynthetic laws, five flavonoid glycosides (**32–36**) and one phenolic acid glycoside (**37**) were tentatively speculated. Among them, peak **36** was a potential new one. Thus, the first evidence for quality control of *A. mongolicum* has been duly provided.

Supplementary Materials: Supplementary data (The NMR and HRESIMS spectra of compounds **1–10**, as well as extraction and isolation process) associated with this article can be found in the online version.

Author Contributions: Y.Z. (Yi Zhang) and T.W. designed the research and wrote the manuscript; Y.D., J.R., and Z.D. performed the experimental work; W.Z. and H.J. corrected the data and reviewed the literatures; M.H. and Y.Z. (Ying Zhang) perfected the language. All authors discussed, edited, and approved the final version. All authors have read and agreed to the published version of the manuscript.

Funding: This research was supported by Program for National Natural Science Foundation of China (81673688), and the Important Drug Development Fund, Ministry of Science and Technology of China (2018ZX09711001-009-010, 2018ZX09735-002).

Conflicts of Interest: The authors declare no conflict of interest.

References

1. Gulzar, A.; Siddiqui, M.B.; Bi, S. Phenolic acid allelochemicals induced morphological, ultrastructural, and cytological modification on Cassia sophera L. and Allium cepa L. *Protoplasma* **2015**, *253*, 1211–1221. [CrossRef] [PubMed]
2. Chang, T.-C.; Jang, H.-D.; Lin, W.-D.; Duan, P.-F. Antioxidant and antimicrobial activities of commercial rice wine extracts of Taiwanese Allium fistulosum. *Food Chem.* **2016**, *190*, 724–729. [CrossRef] [PubMed]
3. Onyeoziri, U.P.; Romanus, E.N.; Onyekachukwu, U.I. Assessment of antioxidant capacities and phenolic contents of nigerian cultivars of onions (Allium cepa L) and garlic (Allium sativum L). *Pak. J. Pharm. Sci.* **2016**, *29*, 1183–1188. [PubMed]
4. Bondonno, N.P.; Dalgaard, F.; Kyrø, C.; Murray, K.; Bondonno, C.P.; Lewis, J.R.; Croft, K.D.; Gislason, G.; Scalbert, A.; Cassidy, A.; et al. Flavonoid intake is associated with lower mortality in the Danish Diet Cancer and Health Cohort. *Nat. Commun.* **2019**, *10*, 3651–3661. [CrossRef]
5. Ivanovna, L.; Mikhailovna, T.; Vladimirovich, D.; Valentinovich, M.; Aleksandrovna, E.; Babaskina, L.I.; Litvinova, T.M.; Babaskin, D.V.; Kiselevsky, M.V.; Savinova, O.V.; et al. Influence of Flavonoids on the Cytotoxic Activity of Mononuclear Blood Cells in Model Tests. *Open Access Maced. J. Med. Sci.* **2019**, *7*, 1900–1904. [CrossRef]
6. Li, Y.-D.; Guan, J.-P.; Tang, R.-C.; Qiao, Y.-F. Application of Natural Flavonoids to Impart Antioxidant and Antibacterial Activities to Polyamide Fiber for Health Care Applications. *Antioxidants* **2019**, *8*, 301. [CrossRef]
7. Wang, W.; Li, J.; Zhang, H.; Wang, X.; Fan, J.; Zhang, X. Phenolic compounds and bioactivity evaluation of aqueous and methanol extracts of Allium mongolicum Regel. *Food Sci. Nutr.* **2019**, *7*, 779–787. [CrossRef]
8. Chen, Y.-A.; Tsai, J.-C.; Cheng, K.-C.; Liu, K.-F.; Chang, C.-K.; Hsieh, C.-W. Extracts of black garlic exhibits gastrointestinal motility effect. *Food Res. Int.* **2018**, *107*, 102–109. [CrossRef]
9. Zhang, Y.; Huang, P.; He, W.; Sakah, K.J.; Ruan, J.; Li, Z.; Wang, T. Bioactive constituents obtained from the fruits of Citrus aurantium. *J. Nat. Med.* **2019**, *73*, 146–153.
10. He, W.; Liu, M.; Li, Y.; Yu, H.; Wang, D.; Chen, Q.; Chen, Y.; Zhang, Y.; Wang, T. Flavonoids from Citrus aurantium ameliorate TNBS-induced ulcerative colitis through protecting colonic mucus layer integrity. *Eur. J. Pharmacol.* **2019**, *857*, 172456. [CrossRef]
11. He, W.; Li, Y.; Liu, M.; Yu, H.; Chen, Q.; Chen, Y.; Ruan, J.; Ding, Z.; Zhang, Y.; Wang, T. Citrus aurantium L. and Its Flavonoids Regulate TNBS-Induced Inflammatory Bowel Disease through Anti-Inflammation and Suppressing Isolated Jejunum Contraction. *Int. J. Mol. Sci.* **2018**, *19*, 3057. [CrossRef] [PubMed]

12. Qi, S.; Wang, T.; Chen, R.; Wang, C.; Ao, C. Effects of flavonoids from Allium mongolicum Regel on growth performance and growth-related hormones in meat sheep. *Anim. Nutr.* **2017**, *3*, 33–38. [CrossRef] [PubMed]
13. Du, H.; Erdene, K.; Chen, S.; Qi, S.; Bao, Z.; Zhao, Y.; Wang, C.; Zhao, G.; Ao, C.; Khas, E.; et al. Correlation of the rumen fluid microbiome and the average daily gain with a dietary supplementation of Allium mongolicum Regel extracts in sheep1. *J. Anim. Sci.* **2019**, *97*, 2865–2877. [CrossRef]
14. Li, M.-Y.; Guo, W.-Q.; Guo, G.-L.; Zhu, X.-M.; Niu, X.-T.; Shan, X.-F.; Tian, J.-X.; Wang, G.-Q.; Zhang, D.-M. Effect of sub-chronic exposure to selenium and Allium mongolicum Regel flavonoids on Channa argus: Bioaccumulation, oxidative stress, immune responses and immune-related signaling molecules. *Fish Shellfish. Immunol.* **2019**, *91*, 122–129. [CrossRef]
15. Wang, T.; Wang, C.; Dan, N.; Sa, R.; Du, H.; Guo, C.; Khas-Erdene; Cao, Q.; Ao, C. Effects of total flavonoids from Allium mongolicum Regel on inflammatory mediators induced by lipopolysaccharide of mouse peritoneal macrophages. *Dongwu Yingyang Xuebao* **2018**, *30*, 3702–3709.
16. Zhao, C. *Studies on Separation, Purification and Structural Characterizations of Flavonoids from Allium mongolium Regel and Its Effects on Immunity and Antioxidant Function in Mice*; Inn. Mongolia Agricultural University: Hohhot, China, 2008.
17. Dong, Y.; Qu, L.; Li, X.; Han, L.; Wang, T.; Zhang, Y. Isolation and identification of constituents from Allium mongolicum Regel I. *Zhongguo Yaowu Huaxue Zazhi* **2015**, *25*, 298–302.
18. Dong, Y.; Shi, W.; Yang, S.; Li, X.; Zhang, Y.; Wang, T. Isolation and structural identification of constituents from Allium mongolicum Regel II. *Tianjin Zhongyiyao Daxue Xuebao* **2016**, *35*, 404–408.
19. Manguro, L.O.A.; Ugi, I.; Lemmen, P.; Hermann, R. Flavonol glycosides of Warburgia ugandensis leaves. *Phytochem.* **2003**, *64*, 891–896. [CrossRef]
20. Budzianowski, J. Six flavonol glucuronides from Tulipa gesneriana. *Phytochemistry* **1991**, *30*, 1679–1682. [CrossRef]
21. Kokubo, T.; Nakamura, M.; Yamakawa, T.; Noguchi, H.; Kodama, T. Quercetin 3,7,4′-triglucoside formation from quercetin by Vitis hybrid cell cultures. *Phytochemistry* **1991**, *30*, 829–831. [CrossRef]
22. Yi, Y.; Wu, X.; Wang, Y.; Ye, W.-C.; Zhang, Q.-W. [Studies on the flavonoids from the flowers of Hylocereus undatus]. *Zhong Yao Cai = Zhongyaocai = J. Chin. Med. Mater.* **2011**, *34*, 712–716.
23. Hardorne, J.B. Plant polyphenols–XI. The structure of acylated anthocyanins. *Phytochemistry* **1964**, *3*, 151–160. [CrossRef]
24. Han, S.H.; Suh, W.S.; Park, K.J.; Kim, K.H.; Lee, K.R. Two new phenylpropane glycosides from Allium tuberosum Rottler. *Arch. Pharmacal. Res.* **2015**, *38*, 1312–1316. [CrossRef] [PubMed]
25. Fei, Y.; Chen, Z.; Li, X.; Xu, Q.; Yang, S. Chemical constituents from seeds of Helianthus annuus. *Zhongcaoyao* **2014**, *45*, 631–634.
26. Chan, J.; Yang, L.; Rena, K. Studies on chemical constituents from EtOAc fraction of Sorbus tiansschanica. *Zhongguo Zhongyao Zazhi* **2009**, *34*, 175–176.
27. Zhang, Y.; Li, X.; Ruan, J.; Wang, T.; Dong, Y.; Hao, J.; Liu, E.; Han, L.; Gao, X.; Wang, T. Oleanane type saponins from the stems of Astragalus membranaceus (Fisch.) Bge. var. mongholicus (Bge.) Hsiao. *Fitoterapia* **2016**, *109*, 99–105. [CrossRef]
28. Yang, S.; Sun, F.; Ruan, J.; Yan, J.; Huang, P.; Wang, J.; Han, L.; Zhang, Y.; Wang, T. Anti-inflammatory constituents from Cortex Dictamni. *Fitoterapia* **2019**, *134*, 465–473. [CrossRef]
29. Lee, K.R.; Jeong, E.-K.; Choi, S.U.; Hong, J.; Lee, I.K. New Flavonol Glycosides from Cardamine komarovii. *Heterocycles* **2011**, *83*, 2615. [CrossRef]
30. Shi, W.; Ruan, J.; Guo, Y.; Ding, Z.; Yan, J.; Qu, L.; Zheng, C.; Zhang, Y.; Wang, T. Bioactive constituents study of Pugionium cornutum L. Gaertn on intestinal motility. *Fitoterapia* **2019**, *138*, 104291. [CrossRef]
31. Ruan, J.; Yan, J.; Zheng, D.; Sun, F.; Wang, J.; Han, L.; Zhang, Y.; Wang, T. Comprehensive Chemical Profiling in the Ethanol Extract of Pluchea indica Aerial Parts by Liquid Chromatography/Mass Spectrometry Analysis of Its Silica Gel Column Chromatography Fractions. *Molecules* **2019**, *24*, 2784. [CrossRef]

Sample Availability: Samples of all compounds are available from the authors.

molecules

MDPI

Article

Synthesis, Anticancer Activity, and Preliminary Pharmacokinetic Evaluation of 4,4-Disubstituted Curcuminoid 2,2-bis(Hydroxymethyl)Propionate Derivatives

Der-Yen Lee [1], Yu-Chi Hou [2], Jai-Sing Yang [3], Hui-Yi Lin [4], Tsu-Yuan Chang [2], Kuo-Hsiung Lee [5,6], Sheng-Chu Kuo [2,4,*] and Min-Tsang Hsieh [2,4,6,*]

1 Graduate Institute of Integrated Medicine, China Medical University, No. 91, Hsueh-Shih Road, Taichung 40402, Taiwan; deryen.lee@mail.cmu.edu.tw

2 School of Pharmacy, China Medical University, Taichung 40402, Taiwan; houyc@mail.cmu.edu.tw (Y.-C.H.); u103003309@cmu.edu.tw (T.-Y.C.)

3 Department of Medical Research, China Medical University Hospital, China Medical University, Taichung 40447, Taiwan; jaisingyang@gmail.com

4 Research Center for Chinese Herbal Medicine, China Medical University, Taichung 404, Taiwan; pingababy@yahoo.com.tw

5 Natural Products Research Laboratories, UNC Eshelman School of Pharmacy, University of North Carolina, Chapel Hill, NC 27599, USA; khlee@unc.edu

6 Chinese Medicinal Research and Development Center, China Medical University Hospital, Taichung 40447, Taiwan

* Correspondence: sckuo@mail.cmu.edu.tw (S.-C.K.); t21917@mail.cmu.edu.tw (M.-T.H.); Tel.: +886-4-22030760 (S.-C.K.); +886-4-22053366-5605 (M.-T.H.)

Academic Editor: Ping-Chung Kuo

Received: 13 January 2020; Accepted: 22 January 2020; Published: 22 January 2020

check for updates

Abstract: Compound **1** is a curcumin di-*O*-2,2-bis(hydroxymethyl)propionate that shows significant in vitro and in vivo inhibitory activity against MDA-MB-231 cells with eight to ten-fold higher potency than curcumin. Here, we modified the α-position (C-4 position) of the central 1,3-diketone moiety of **1** with polar or nonpolar functional groups to afford a series of 4,4-disubstituted curcuminoid 2,2-bis(hydroxymethyl)propionate derivatives and evaluated their anticancer activities. A clear structure–activity relationship of compound **1** derivatives focusing on the functional groups at the C-4 position was established based on their anti-proliferative effects against the MDA-MB-231 and HCT-116 cell lines. Compounds **2–6** are 4,4-dimethylated, 4,4-diethylated, 4,4-dibenzylated, 4,4-dipropargylated and 4,4-diallylated compound **1**, respectively. Compounds **2m–6m**, the ester hydrolysis products of compounds **2–6**, respectively, were synthesized and assessed for anticancer activity. Among all compound **1** derivatives, compound **2** emerged as a potential chemotherapeutic agent for colon cancer due to the promising in vivo anti-proliferative activities of **2** (IC_{50} = 3.10 ± 0.29 μM) and its ester hydrolysis product **2m** (IC_{50} = 2.17 ± 0.16 μM) against HCT-116. The preliminary pharmacokinetic evaluation of **2** implied that **2** and **2m** are main contributors to the in vivo efficacy. Compound **2** was further evaluated in an animal study using HCT-116 colon tumor xenograft bearing nude mice. The results revealed a dose-dependent efficacy that led to tumor volume reductions of 27%, 45%, and 60% at 50, 100, and 150 mg/kg doses, respectively. The established structure–activity relationship and pharmacokinetic outcomes of **2** is the guidance for future development of 4,4-disubstituted curcuminoid 2,2-bis(hydroxymethyl)- propionate derivatives as anticancer drug candidates.

Keywords: curcuminoid derivatives; prodrug; colon cancer; breast cancer; active metabolites

1. Introduction

Curcumin [(1*E*,6*E*)-1,7-bis(4-hydroxy-3-methoxyphenyl)hepta-1,6-diene-3,5-dione, the structure of which is shown in Scheme 1], the naturally occurring phytochemical from the rhizome of *Curcuma longa* L., is a polyphenol with a symmetrical structure composed of two *ortho*-methoxyphenol rings connected to each other through a flexible conjugated hydrocarbon chain. It is a versatile therapeutic agent against cancer [1] and exhibits diverse pharmacological effects, including antidiabetic [2], antiviral [3], analgesic [4], nephroprotective [5], and cardioprotective effects [6], via regulating the gene expression [7] and protein binding [8]. Similar to some phenolic natural products, curcumin is a multi-target anticancer agent [9] with the ability to interfere with several signaling pathways associated with tumor progression, metastasis, apoptosis, and angiogenesis [10].

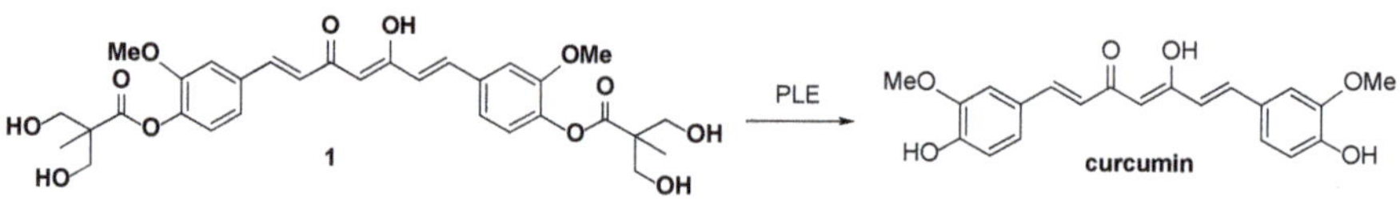

Scheme 1. PLE mediated hydrolysis of **1**.

While curcumin shows promising therapeutic properties, its low solubility and poor chemical and metabolic stability hinder further drug development [11]. The low chemical stability of curcumin is demonstrated by its rapid decomposition upon exposure to light or high temperatures (>70 °C) [12]. The compound decomposes into several degradation products, including a bicyclopentadione derivative, vanillin, feruloylmethane, and ferulic acid. The degradation of curcumin presumably results from its autoxidation [13], wherein the starting radical species is formed through hydrogen atom removal from the phenolic moiety. The phenolic group with an easily abstractable proton and the keto-enol system display significant hydrogen-donating abilities [14] that could initiate the autoxidative degradation of curcumin. The poor metabolic stability of curcumin can be ascribed to the presence of two phenolic hydroxyl groups that are prone to the formation of glucuronide and sulfate conjugates through the phase II metabolic process [15]. Moreover, the enone moiety of curcumin is accessible to nicotinamide adenine dinucleotide phosphate (NADHP) cytochrome P450 reductase or alcohol dehydrogenases, allowing curcumin to be readily reduced to the hydrogenated metabolites dihydrocurcumin, tetrahydrocurcumin, hexahydrocurcumin, and hexahydrocurcuminol [16].

Medicinal chemists have performed structural optimizations of curcumin to improve its stability and activity. Most of these attempts were focused on modifying the phenolic hydroxyl groups, keto-enol system, and enone moiety of curcumin [17–20]. However, the first two abovementioned structural features are required for radical scavenging activity, which is regarded as vital for curcumin chemopreventive and anti-inflammatory activities [21]. In addition, the enone moiety of curcumin, serving as the electrophilic Michael acceptor, can interact with cysteine or selenocysteine residues in proteins to mediate biological activities [22]. Thus, finding a balance between structural stability and biological potency can become challenging. According to the FDA [23], about 30% of the curcumin clinical trials stalled at stage II, mostly due to discrepancies between in vitro activity and in vivo response. Accordingly, to achieve success in clinical trials, obtaining a more definite picture of signaling pathways and ADME profiling for curcumin is recommended.

We previously developed a series of curcuminoid prodrugs by modifying the phenolic hydroxyl groups of curcumin into a 2,2-bis(hydroxymethyl)propionate group [24]. Among the derivatives, the curcumin 2,2-bis(hydroxymethyl)propionate **1** exhibited stability and solubility superior to those of curcumin. Particularly, **1** showed comparable anticancer activity against MDA-MB-231 cells in vitro and in vivo and had 8 to 10 times more potency than curcumin. Subsequently, the α-position in the central 1,3-diketone moiety (C-4 position) of **1** were alkylated with methyl, ethyl, benzyl, and propargyl groups to give compounds **2**, **3**, **4**, and **5**, respectively. Among them, **2**, **3**, and **5** showed more potent in vitro anti-proliferative activity than **1** against HCT-116 cells; therefore, were recommended them as potential therapeutic agents for colon cancer [24]. In our recent in vitro studies on the enzymatic hydrolysis of **1**, we demonstrated that treating **1** with porcine liver esterase (PLE) cleaves its two ester groups gradually to form curcumin (see supporting information). Compounds **2-5** are ester-based derivatives and structurally correlated to **1**. Therefore, we predicted that a high concentration of esterase, as particularly found in the intestinal tract and mucosa [25], would hydrolyze **2**, **3**, **4**, and **5** into their metabolites **2m**, **3m**, **4m**, and **5m**, respectively (Scheme 2). However, the anticancer activity of **2m–5m** should be evaluated prior to the further in vivo investigations. Compounds **2–5** were substituted with two nonpolar alkyl chains at the C-4 position with the loss of keto-enol tautomerism. The anticancer activity of compound **1** derivatives, which were substituted with two hydrophilic units or two polar side chains at the C-4 position, still remains unclear.

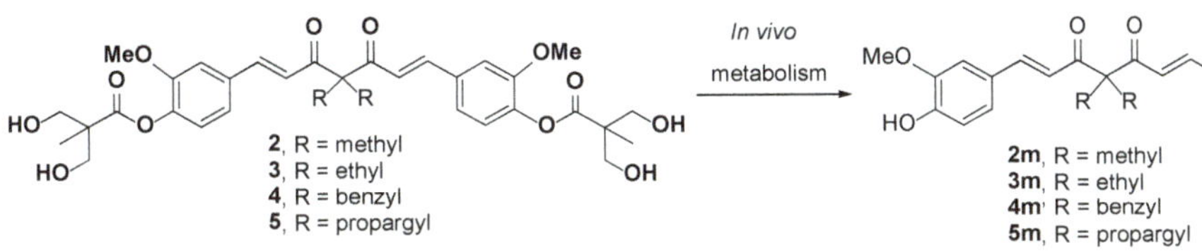

Scheme 2. In vivo enzymatic hydrolysis of **2–5**.

Herein, we report the successful replacement of the acidic α-hydrogens at the C-4 position of compound **1** with polar or nonpolar functional groups to produce compounds **6**, **9**, **10**, **11**, **12**, and **14**. Compounds **2m–6m** were prepared through the chemical hydrolysis of corresponding parent compounds **2–6**. The anticancer activities of **2–6**, **9–12**, **14** and **2m–6m** were evaluated in vitro. A selected compound was submitted for preliminary pharmacokinetic study and antitumor study. The detailed descriptions of the synthetic design, in vitro screening, and in vivo study of the abovementioned compounds are presented as follows.

2. Results and Discussion

2.1. Chemistry

The detailed synthesis of compound **1** derivatives (substituted at the C-4 position with allyl (**6**), hydroxymethyl (**10**), acetoxymethyl (**11**), methoxymethyl (**12**)), novel 2,6-dimethyl curcumin (DMC) derivatives (**13** and **15**), and the ester hydrolysis products of derivatives showing potential anticancer activity (**2m**, **3m**, **4m**, **5m**, **6m**, **16**, and **17**) are depicted in Schemes 3 and 4. As shown in Scheme 3, compound **6** was synthesized according to a procedure established for the synthesis of **2–5**. Compound **7**, prepared by the esterification of curcumin with 2,2,5-trimethyl-1,3-dioxane-5-carboxylic acid, was treated with 1,8-diazabicyclo[5.4.0]undec-7-ene (DBU) and formaldehyde in THF to produce intermediary compound **8**. The latter was used without column chromatographic purification in acid-promoted hydrolysis at room temperature to produce **9** with an overall yield of 6% for two synthetic steps.

Compound **9**, however, proved unstable, slowly decomposing even when kept at 4 °C and presenting an NMR spectrum comprising a messy jungle of peaks after being dissolved in DMSO-d_6 for 1 week at room temperature. The low stability of **9** may have been due to self- or cross-hydrolysis, such that the two hydroxyl groups at the C-4 alkyl chains cleaved the ester linkage of the 2,2-bis(hydroxymethyl)propionate group. Previously, we had found that upon exposure to alcohol for several hours, the ester linkages of **1** were broken under neutral or basic conditions, leading to the conversion of **1** to curcumin and several unknown products. Therefore, we modified the hydroxymethyl group of **9** and were able to produce compounds **10** and **11**. To prepare **10**, compound **9** was acetylated with acetyl chloride in the presence of Et_3N, and the resulting di-acetylation intermediate was hydrolyzed by hydrochloric acid. Compound **11** was obtained with an overall yield of 5% following a similar three-step reaction modified for methylation. The stability of these two products at room temperature may have been due to the conversion of the labile hydroxyl groups of the C-4 alkyl chains of **10** and **11** into non-nucleophilic methyl and acetyl groups, respectively, which prevented degradation.

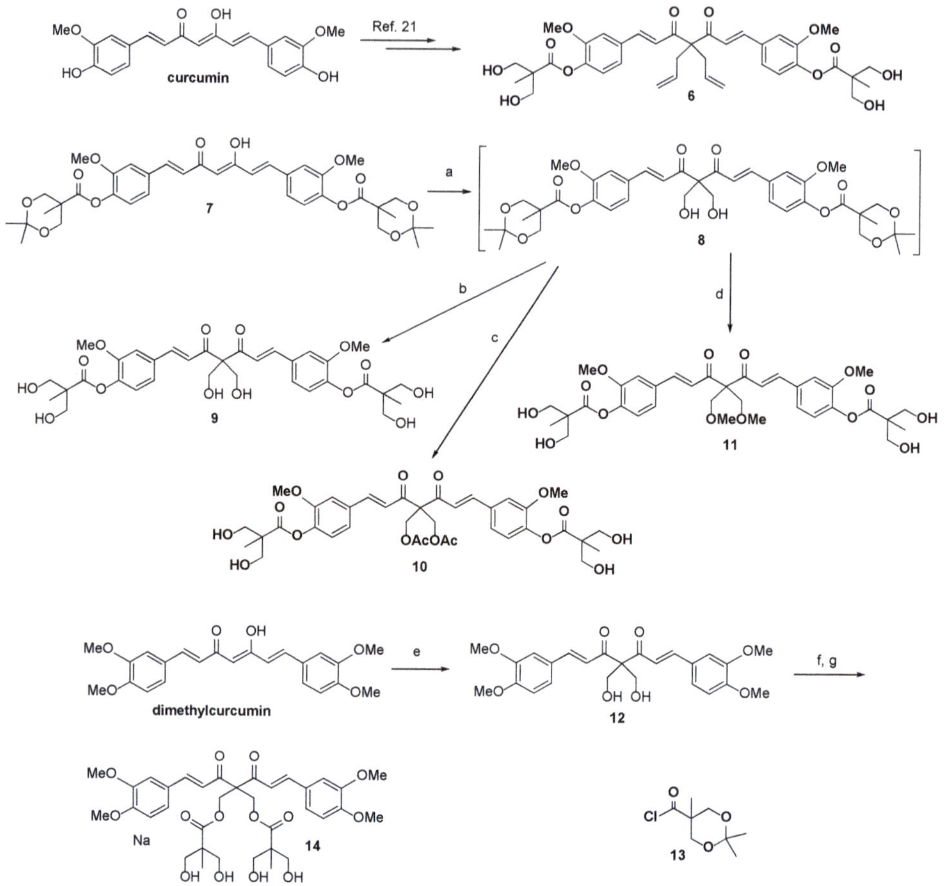

Scheme 3. *Reagents and Conditions:* (**a**) Formaldehyde in THF, DBU, 0 °C to rt, 2 h; (**b**) HCl, MeOH, rt, 1 h, 6% for two steps; (**c**) Et_3N, acetyl chloride, DCM, rt, 12 h; then HCl, MeOH, rt, 1 h, 3% for three steps; (**d**) K_2CO_3, CH_3I, DMF, rt, 6 h; then HCl, MeOH, rt, 1 h, 4% for three steps; (**e**) Formaldehyde in THF, DBU, 0 °C to rt, 2 h, 42%; (**f**) Compound 13, Et_3N, DCM, rt, 12 h; and (**g**) HCl, MeOH, 1 h, 7% for two steps.

Because DMC is a dimethylated derivative of curcumin [26] that exhibits better anticancer activity and stability than curcumin, we synthesized a novel DMC derivative (compound **12**) in which C-4 was substituted with two bulky 2,2-bis(hydroxymethyl)propionate groups. To produce this derivative, commercially available DMC was reacted with DBU and formaldehyde in THF. The reaction of **12** with 2,2,5-trimethyl-1,3-dioxane-5-carbonyl chloride **13** [27] in the presence of Et_3N provided a diol intermediate, which was then hydrolyzed by hydrochloric acid in MeOH to yield compound **14**. In addition, as illustrated in Scheme 4, compounds **2m–6m** were prepared through the NaOMe-mediated hydrolysis of their corresponding parent compounds **2–6**, with high yield.

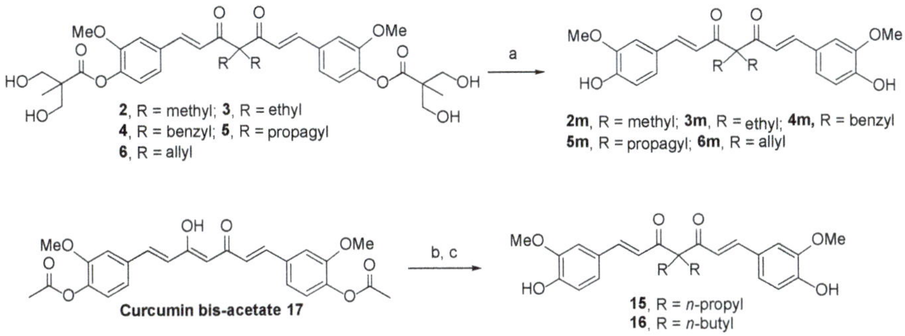

Scheme 4. *Reagents and Conditions*: (**a**) NaOMe, MeOH, rt, 2 h, 82–87%; (**b**) K_2CO_3, alkyl iodide, DMF, rt, 20 h; and (**c**) NaOMe, MeOH, rt, 3 h, 28–33% for two steps.

For SAR analysis, compounds **15** and **16**, which were substituted with two straight alkyl chains with three or four carbons at the C4 position, were prepared from curcumin bisacetate **17** [28] via a two-step procedure comprising K_2CO_3-induced dialkylation and NaOMe-mediated hydrolysis. For the synthesis of four possible phase I metabolites of compound **2**, compound **2m** was subjected to hydrogenation reactions. As shown in Scheme 5, compound **18** is a partially hydrogenated product that can be prepared in 19% yield by treating compound **2m** with $H_{2(g)}$ (1.0 atm, balloon) and a catalytic amount of 10% *w/w* Pd/C in ethyl acetate. When the solvent was changed from ethyl acetate to methanol, the hydrogenation reaction yielded the fully hydrogenated diketone **19** in 52% yield and β-hydroxy ketone **20** in 23% yield. Finally, the reduction of **20** with $NaBH_4$ in MeOH at room temperature (rt) for 2 h provided the desired diol compound **21** in 31% yield.

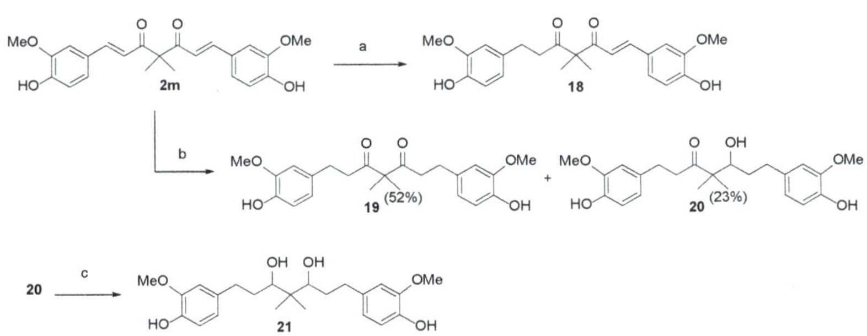

Scheme 5. *Reagents and Conditions*: (**a**) 10% *w/w* Pd/C, ethyl acetate, rt, 1 h, 19%; (**b**) 10% *w/w* Pd/C, MeOH, rt, 2 h; and (**c**) $NaBH_4$, MeOH, 0 °C to rt, 2 h, 31%.

2.2. Structure–Activity Relationship

As mentioned above, we synthesized **2–5** and evaluated their anti-proliferative activity against the MDA-MB-231 and HCT-116 cell lines (72-h treatment). These compounds, along with all newly synthesized compounds, were screened for anti-proliferative activity against the same TNBC and colon cancer cell lines. The results of the in vitro assay are compiled in Table 1.

Table 1. The anti-proliferative effects of curcumin derivatives against MDA-MB-231 and HCT-116 cell lines.

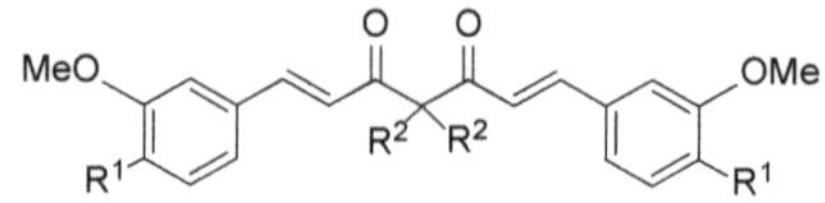

Compound	R^1	R^2	IC_{50} [a] (M)/ 48 h	
			MDA-MB-231	**HCT-116**
Curcumin	**OH**	H	22.05 ± 2.97	26.33 ± 1.38
1	—O–C(O)–C(–OH)(–CH$_3$)(–OH)	H	3.06 ± 0.18	6.25 ± 0.60
2	—O–C(O)–C(–OH)(–CH$_3$)(–OH)	methyl	2.25 ± 0.10	3.10 ± 0.29
3	—O–C(O)–C(–OH)(–CH$_3$)(–OH)	ethyl	3.02 ± 0.15	1.93 ± 0.01
4	—O–C(O)–C(–OH)(–CH$_3$)(–OH)	benzyl	2.45 ± 0.24	7.66 ± 0.61
5	—O–C(O)–C(–OH)(–CH$_3$)(–OH)	propargyl	5.87 ± 0.10	5.81 ± 0.23
6	—O–C(O)–C(–OH)(–CH$_3$)(–OH)	allyl	3.13 ± 0.54	0.92 ± 0.06
9	—O–C(O)–C(–OH)(–CH$_3$)(–OH)	CH_2OH	>100	>100
10	—O–C(O)–C(–OH)(–CH$_3$)(–OH)	CH_2OAc	6.55 ± 0.04	6.53 ± 0.01
11	—O–C(O)–C(–OH)(–CH$_3$)(–OH)	CH_2OMe	84.92 ± 1.67	69.08 ± 3.11
12	OCH_3	CH_2OH	53.50 ± 3.45	58.00 ± 0.60
14	OCH_3	–H_2C–O–C(O)–C(–OH)(–OH)	57.43 ± 2.06	40.10 ± 1.61
2m	OH	methyl	4.17 ± 0.15	2.17 ± 0.16
3m	OH	ethyl	9.12 ± 0.40	8.18 ± 0.57
4m	OH	benzyl	36.52 ± 1.66	9.40 ± 0.02
5m	OH	propargyl	3.10 ± 0.15	1.38 ± 0.06
6m	OH	allyl	6.88 ± 0.36	4.40 ± 0.53
15	OH	propyl	18.17 ± 0.53	10.73 ± 1.63
16	OH	butyl	>100	67.00 ± 3.18

[a] All data presented are means from at least three experiments with standard deviations of the value quoted.

The anticancer effects of **2–5** against both cell lines were still better or at least similar to that of **1**, although the drug treatment duration was reduced to 48 h. The newly synthesized compound **6** had a higher anti-proliferative activity against HCT-116. Compound **9** was less potent against the cell lines than curcumin and **1**, which was used as a reference compound in the current study. We hypothesize that the two polar and hydrophilic side chains at the C-4 position in **9** are the cause for the weak anticancer effects observed for this compound. Another possibility is that **9** decomposed during the screening due to its low stability. Compound **10**, having two acetoxymethyl groups rather than hydroxymethyl groups at the C-4 side chains, showed significant improvement in in vitro anticancer activities compared to **9**; however, it was still less potent than **1**. Compound **11**, substituted with two methoxymethyl groups at the C-4 position, displayed poor anticancer activity with high IC_{50} values for both cell lines used.

After listing the compounds in descending order based on their potency against HCT-116 (**6** > **3** > **2** > **5** > **1** > **10** > **4** >> **11** >> **9**), we concluded that higher polarity and hydrophilicity of the C-4 side chain are detrimental to the anticancer activity of compound **1** derivatives. Alternatively, the DMC derivative 4,4-dihydroxymethyl dimethoxycurcumin (compound **12**) displayed low anti-proliferative activity. When the two hydroxyl moieties of **12** both esterified with 2,2-bis(hydroxymethyl)propionic acid, the resulting compound **14** exhibited activities comparable to those of **12**, which correlates with our previous findings regarding the unfavorable effect of having a hydrophilic moiety at the C-4 position.

Subsequently, the activities of the ester hydrolysis products **2m**–**6m** were explored. Unexpectedly, **5m** displayed substantial improvement in anticancer activity by two- to four-fold relative to that of parent **5**. Compound **2m** exhibited activity similar to that of **2**, whereas **3m**, **4m**, and **6m** showed inferior activity than their corresponding parent compounds **3**, **4**, and **6**. Further screening of **15** and **16** revealed an obvious SAR, such that the anticancer activity of these hydrolysis compounds decreased with the increasing length of the C-4 alkyl side chains. In addition, lower structural volume was seemingly preferred to sterically congested side chains, as demonstrated by the analysis of **2m**, **3m**, **15**, **4m**, and **16**.

Among all examined compounds, **2** and its ester hydrolysis product **2m** possessed significant anticancer activity against MDA-MB-231 and HCT-116; therefore, **2** was chosen for further studies. Notably, the solubility of **2** was more significant than that of curcumin as was evident from the vehicle of **2** for the following in vivo study. The maximum solubility of **2** in the aqueous vehicle comprising 5% ethanol, 10% Tween 80, and 85% saline was approximately 200 mg/mL, while curcumin was almost insoluble in the vehicle. Based on the logic for the metabolism of curcumin [29] and an ester-type prodrug [30], we speculated that the in vivo metabolism of **2** would readily produce **2m**. Subsequently, the latter would be converted into glucuronide **22** and sulfate **23** through the phase II metabolism conjugation of hydrogenated products **18**, **19**, **20**, and **21** via the phase I metabolic pathway (Scheme 6). To verify our assumptions, the preliminary pharmacokinetic evaluation of **2** was executed.

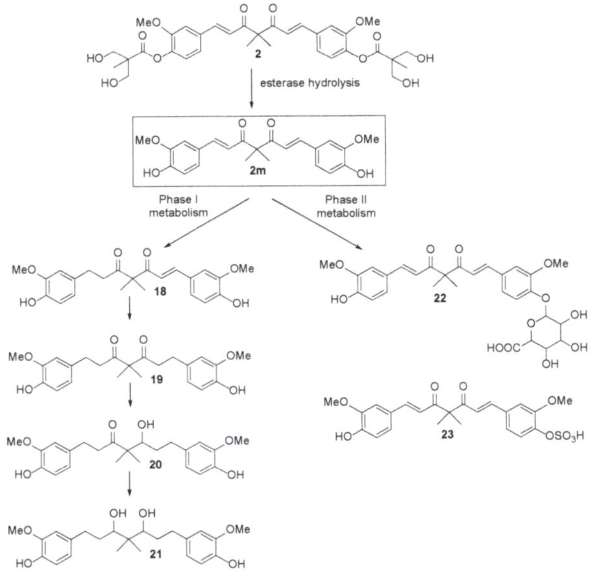

Scheme 6. In vivo metabolism of **2**.

2.3. The Preliminary Pharmacokinetic Evaluation of 2

The preliminary in vivo pharmacokinetic evaluation of **2** was designed to identify the anticipated metabolites and evaluate the plasma stability of **2** and **2m**. The four possible phase I metabolites **18**, **19**, **20**, and **21** were synthesized and subjected to anti-proliferative screening against MDA-MB-231 and HCT-116. Compound **18** exhibited moderate to weak anti-proliferative activity (MDA-MB-231, IC_{50} = 36.48 ± 0.34 μM; HCT-116, IC_{50} = 9.64 ± 0.21 μM) and the others are inactive against both cell lines ($IC_{50} \geq 100$ μM). The assessment of **2** and relevant metabolites was performed in male Sprague-Dawley *rats* after the oral administration of **2** at a dose of 100 mg/kg. The LC signals of **2**, **2m**, and **18–21** was recognized unequivocally by comparison to the signals of standard compounds. The direct analysis of **22** and **23** was not easily attainable due to the low extraction efficiency and high probability of sample loss in LC column. It has been well-documented that curcumin glucuronide and curcumin sulfate can be transformed back into curcumin by the enzyme-mediated hydrolysis reaction [31]. Accordingly, the treatment of **22** and **23** with enzyme that contains sulfatase and glucuronidase is supposed to produce **2m**. In the current study, the amounts of **22** and **23** were not calculated individually but counted as a summary value based on the disparities between the LC signal of **2m** in enzyme-treated and -untreated serum samples.

In practice, LC-MS analysis of one serum sample which was collected at 20 min post-dosing demonstrated the existence of **2**, **2m**, **18**, **19**, **20**, **21**, **22** and **23** (Figure S1 in the Supporting Information). Shown in Figure 1 are the LC signal count-time profiles for **2**, **2m**, **18**, **19**, and the comparison of **2m**, **22**, and **23**, following oral administration. The analysis of **20** and **21** is not illustrated due to their low LC signal response and non-significant anticancer activity. The maximum blood content of **2** was achieved at 20 min, while the second and third highest appeared at 120 min and 360 min, respectively, following administration. The multiple peaks of **2** observed in the profile could be attributed to an enterohepatic circulation cycle.

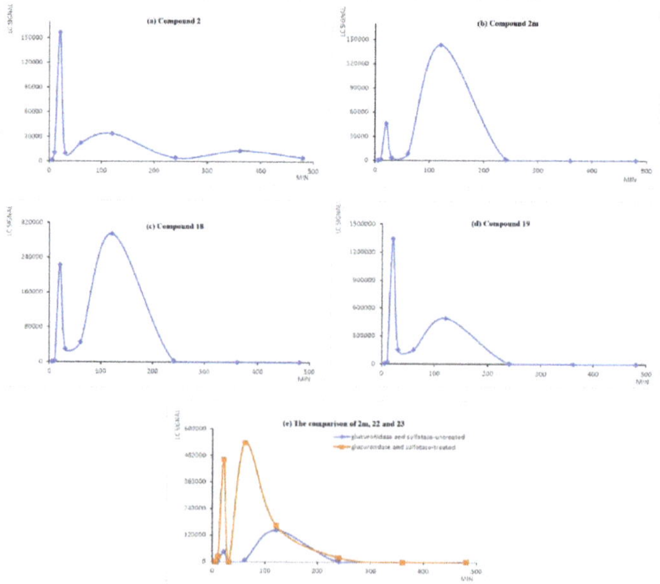

Figure 1. The amount of **2**, **2m**, **18**, **19**, **22**, and **23** in serum at different time points following a single oral administration of **2** at a dose of 100 mg/kg. The blood samples were collected at 5, 10, 20, 30, 60, 120, 240, 360, and 480 min after the administration of **2**. The samples were deproteinated and analyzed by HPLC.

Such behavior, extensively studied in the phenolic drugs, was beneficial in maintaining the circulating concentration of free-form curcumin [32] and may extend the pharmacological effect of drug [33]. The hydrolysis metabolite **2m** emerged from 10 min and reached its maximum blood content at 120 min, indicating that a medium-to-high esterase-mediated biotransformation proceeded in vivo to yield the efficient absorption of **2m**. Subsequently, the phase II metabolic process of **2m** progressed quickly to produce **22** and **23**. The ratio of **2m** to its phase II metabolites (**22** and **23**) is about 1:4 based on the area under the curves in part (e) of Figure 1.

The LC signal count-time profiles of **18** and **19** are analogous to that of **2m**, implying that phase I reduction reaction also occurred rapidly. Compared to non-formulated curcumin with a short elimination period of 28 min in the rat model [34], **2** exhibited superior plasma stability. Otherwise, the active metabolite **2m** is equally potent as **2** with regard to anticancer activity, and the maximum blood content of **2m** appeared after that of **2** could be considered as an extension of drug efficacy of **2**. Thus, the pharmacokinetic outcome was to guide the following in vivo efficacy study of **2**.

2.4. In Vivo Antitumor Efficacy of 2

The therapeutic effect of **2** was evaluated on HCT-116 colon tumor xenograft bearing nude mice. The oral administration of **2** at doses of 50, 100, or 150 mg/kg bw per day were performed after tumors reached approximately 100 mm^3 and continued for 30 consecutive days. 5-Fluorouracil treatment (30 mg/kg bw, i.p., QOD) was used as a positive control. The group treated with compound **2** (100 mg/kg) had a statistically significant reduction in tumor volumes, with a 45% tumor inhibition ratio observed when compared with the cancerous control mice. Positive control group exhibited a capability to cause 40% tumor-growth inhibition. There was a clear dose- and time-dependent inhibitory effect on tumor size in the other two groups, in which at 50 and 150 mg/kg doses the volume of the HCT-116 colon tumor was reduced to 73% and 40% of the control, respectively (Figure 2). The body weight of the mice was recorded prior to dosing every 3 d, and the average body weight of each group is depicted in Figure 3. Otherwise, they were monitored for visible signs of toxicity and behavioral changes 1 h after each administration. No obvious adverse effects were observed between the **2**-administrated and control groups.

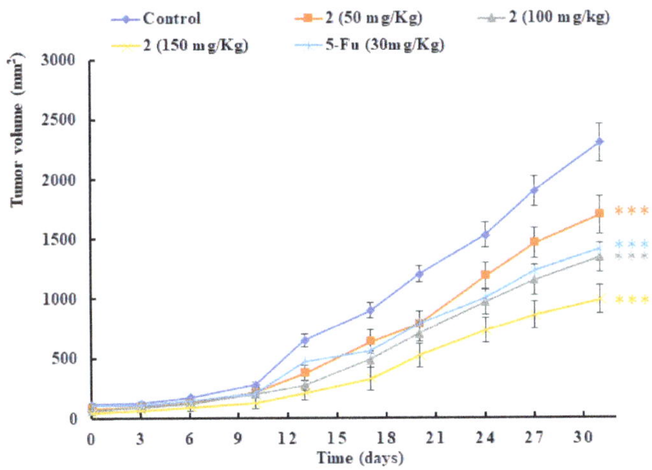

Figure 2. The efficacy study of **2**. In vivo antitumor activity of compound **2** was evaluated in HCT-116 tumor xenograft bearing nude mice. Compound **2** (50, 100, or 150 mg/kg, p.o.) was administered once daily and 5-Fluorouracil was administrated (30 mg/kg, i.p.) every other day. The results are the mean ± SEM of six mice in each group. *** $P < 0.001$ compared to untreated control.

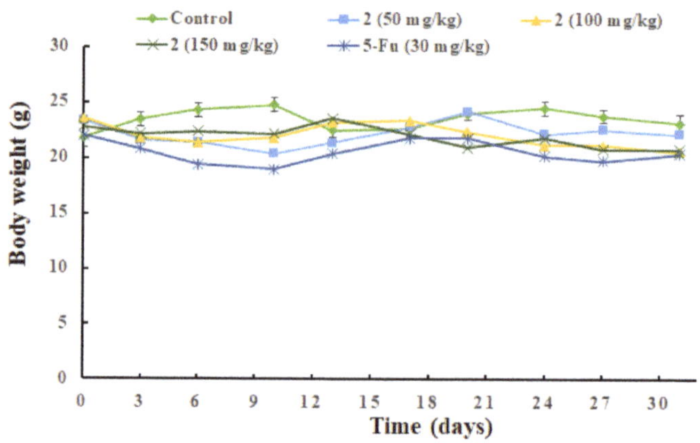

Figure 3. Mean body weight-time profile of **2**- administrated and control groups.

3. Materials and Methods

3.1. General Information

The reactions were performed under an air atmosphere unless otherwise stated. All solvents and reagents were employed as received. Analytical thin layer chromatography (TLC) was performed on SiO_2 60 F_{254} plates and flash column chromatography was carried out using SiO_2 60 (particle size 0.040–0.055 mm, 230–400 mesh), both of which are available from E. Merck (Darmstadt, Germany). Visualization was performed under UV irradiation at 254 nm followed by staining with aqueous potassium permanganate [$KMnO_4$ (3 g) and K_2CO_3 (20 g) in 300 mL of H_2O containing 5 mL of an aqueous solution of NaOH (5%, *w*/*v*)] and charring by heat gun. ^{1}H- and ^{13}C-NMR spectra were recorded on a 500 FT NMR instrument (Bruker, Billerica, MA, USA). Chloroform-*d* and methanol-*d* were used as solvents and TMS (δ = 0.00 ppm) as an internal standard. Chemical shifts are reported as δ values in ppm as referenced to TMS. Multiplicities are recorded as s (singlet), d (doublet), t (triplet), q (quartet), quint (quintet), sext (sextet), sept (septet), dd (doublet of doublets), dt (doublet of triplets), br (broad), m (multiplet). Coupling constants (*J*) are expressed in Hz. LRMS and HRMS were measured by a JMS-HX110 spectrometer (JEOL, Tokyo, Japan) and spectroscopic data were recorded as *m*/*z* values. Melting points were measured using an Electrothermal instrument (Anatec Yanaco Inc., Kyoto, Japan).

3.1.1. *[(1E,6E)-4,4-Diallyl-3,5-dioxohepta-1,6-diene-1,7-diyl]bis(2-methoxy-4,1-phenylene) bis[3-hydroxy-2-(hydroxymethyl)-2-methylpropanoate]* (**6**)

Mp = 195–197 °C; ^{1}H-NMR (DMSO-d$_6$) δ 7.59 (d, *J* = 15.5 Hz, 2H), 7.48 (s, 2H), 7.35 (dd, *J* = 8.2, 1.5 Hz, 2H), 7.12–7.04 (m, 4H), 5.59–5.52 (m, 2H), 5.12–5.04 (m, 4H), 4.82–4.80 (m, 4H), 3.78 (s, 6H), 3.60 (s, 8H), 2.84 (d, *J* = 7.2 Hz, 4H), 1.17 (s, 6H); ^{13}C NMR (DMSO-d$_6$) δ 196.8, 173.3, 151.7, 143.1, 142.1, 133.3, 133.2, 123,9, 122.9, 122.8, 119.5, 113.4, 67.4, 64.0, 56.7, 51.0, 35.0, 17.3; HRMS [ESI]$^+$ calculated for $C_{37}H_{45}O_{12}$: 681.2911 [M + H]$^+$; found: 681.2903.

3.1.2. *[(1E,6E)-4,4-bis(Hydroxymethyl)-3,5-dioxohepta-1,6-diene-1,7-diyl]bis(2-methoxy-4,1-phenylene) bis(2,2,5-trimethyl-1,3-dioxane-5-carboxylate)* (**8**)

To a stirred solution of **7** (1.00 g, 1.47 mmol) in THF (7.3 mL, containing 0.6 M formaldehyde) was added DBU (38.9 mg, 0.04 mmol) at 0 °C. The reaction mixture was stirred at the same temperature for 30 min and then warm to room temperature. After 2 h, the solution was concentrated under vacuum to provide crude compound, which was identified by LRMS [ESI]$^+$ calculated for $C_{39}H_{49}O_{14}$: 741.31 [M + H]$^+$; found: 741.33 and used for next step without purification.

3.1.3. *[(1E,6E)-4,4-bis(Hydroxymethyl)-3,5-dioxohepta-1,6-diene-1,7-diyl]bis(2-methoxy-4,1-phenylene) bis[3-hydroxy-2-(hydroxymethyl)-2-methylpropanoate]* (**9**)

A solution of crude **8** (0.34 g) and 6 N HCl (0.15 mL) in MeOH (1.83 mL) was stirred at room temperature for 1 h. The reaction mixture was concentrated under vacuum to give the crude product, which was purified by flash chromatography on silica gel with MeOH/CH_2Cl_2 (1:20) to afford compound **9** (40 mg) as an orange oil. Yield 6% for two steps. ^{1}H-NMR (DMSO-d_6) δ 7.52 (d, J = 15.6 Hz, 2H), 7.44 (s, 2H), 7.30 (d, J = 8.2 Hz, 2H), 7.08–7.04 (m, 4H), 5.16 (br s, 6H), 4.19 (s, 4H), 3.78 (s, 6H), 3.61 (s, 8H), 1.17 (s, 6H); LRMS [ESI]$^+$ calculated for $C_{33}H_{40}O_{14}$: 661.25 [M + H]$^+$; found: 661.4.

3.1.4. *[(1E,6E)-4,4-bis(Acetoxymethyl)-3,5-dioxohepta-1,6-diene-1,7-diyl]bis(2-methoxy-4,1-phenylene) bis[3-hydroxy-2-(hydroxymethyl)-2-methylpropanoate]* (**10**)

To a stirred solution of crude **8** (0.450 g) in DCM (10 mL) were added NEt_3 (1.0 mL) and acetyl chloride (0.140 g, 1.81 mmol) at 0 °C. The reaction mixture was warmed to room temperature and stirred for 12 h. Then H_2O (10 mL) and diluted $HCl_{(aq)}$ (1 N, 3 mL) were added to quench the reaction. The aqueous layer was separated and extracted with CH_2Cl_2 (3 × 8 mL). The combined organic extracts were washed with brine, dried over $MgSO_4$, filtered and concentrated to give the crude product, which was then treated with $HCl_{(aq)}$ (6 N, 0.2 mL) in MeOH (2 mL) at room temperature. The reaction mixture was stirred at room temperature for 1h and then concentrated under vacuum to give the crude product, which was purified on silica gel with MeOH/CH_2Cl_2 (1:20) to afford **10** (33 mg) as a yellow solid (mp = 194–196 °C). Yield 3% for three steps; ^{1}H-NMR (DMSO-d_6) δ 7.68 (d, J = 15.5 Hz, 2H), 7.51 (s, 2 H), 7.37 (d, J = 8.0 Hz, 2H), 7.20 (d, J = 15.5 Hz, 2H), 7.08 (d, J = 8.5 Hz, 2H), 4.85 (t, J = 5.0 Hz, 8H), 3.80 (s, 6 H), 3.65-3.59 (m, 8H), 1.97 (s, 6H), 1.19 (s, 6H); ^{13}C-NMR (DMSO-d_6) δ 193.1, 173.3, 170.5, 151.7, 144.6, 142.5, 132.9, 123.9, 122,7, 121.7, 113.6, 64.0, 61.9, 56.7, 51.0, 20.9, 17.3; HRMS [ESI]$^+$ calculated for $C_{37}H_{44}NaO_{16}$: 767.2527 [M + Na]$^+$; found: 767.2521.

3.1.5. *[(1E,6E)-4,4-bis(Methoxymethyl)-3,5-dioxohepta-1,6-diene-1,7-diyl]bis(2-methoxy-4,1-phenylene) bis[3-hydroxy-2-(hydroxymethyl)-2-methylpropanoate]* (**11**)

To a stirred solution of crude **8** (0.453 g) in DMF (1.0 mL) were added K_2CO_3 (0.240 g, 1.73 mmol) and MeI (0.247 g, 1.73 mmol) at 0 °C. The reaction mixture was warmed to room temperature and stirred for 6 h. Then H_2O (10 mL) and diluted $HCl_{(aq)}$ (1 N, 3 mL) were added to quench the reaction. The aqueous layer was separated and extracted with EtOAc (3 × 8 mL). The combined organic extracts were washed with brine, dried over $MgSO_4$, filtered and concentrated to give the crude product, which was then treated with $HCl_{(aq)}$ (6 N, 0.25 mL) in MeOH (2 mL) at room temperature. The reaction mixture was stirred at room temperature for 1h and then concentrated under vacuum to give the crude product, which was purified on silica gel with MeOH/CH_2Cl_2 (1:20) to afford **11** (73 mg) as a yellow solid (mp = 187–188 °C). Yield 5% for three steps; ^{1}H-NMR (DMSO-d_6) δ 7.65 (d, J = 16.0 Hz, 2H), 7.49 (s, 2 H), 7.31 (d, J = 10.0 Hz, 2H), 7.08 (d, J = 8.1 Hz, 2H), 6.69 (d, J = 16.0 Hz, 2H), 4.85 (t, J = 5.5 Hz, 4 H), 3.80 (s, 6H), 3.74 (s, 6 H), 3.66-3.60 (m, 8H), 1.20 (s, 6H); ^{13}C-NMR (DMSO-d_6) δ 193.1, 173.4, 167.1, 151.6, 144.4, 141.8, 133.2, 123.8, 121,9, 118.4, 112.6, 64.0, 56.3, 51.9, 51.0, 17.3; HRMS [ESI]$^+$ calculated for $C_{35}H_{44}NaO_{14}$: 711.2629 [M + Na]$^+$; found: 711.2633.

3.1.6. *(1E,6E)-1,7-bis(3,4-Dimethoxyphenyl)-4,4-bis(hydroxymethyl)hepta-1,6-diene-3,5-dione* (**12**)

To a stirred solution of DMC (1.270 g, 3.20 mmol) in THF (20 mL, containing 0.6 M formaldehyde) was added DBU (15 mg, 0.10 mmol). The reaction mixture was stirred at 0 °C for 30 min. The reaction mixture was concentrated under vacuum and then purified by flash chromatography on silica gel with MeOH/CH_2Cl_2 (1:32) to afford **12** (558 mg, 42% yield) as a yellow solid; mp = 174–176 °C; ^{1}H-NMR (DMSO-d_6) δ 7.49 (d, J = 15.5 Hz, 2H), 7.48 (d, J = 2.0 Hz, 2H), 7.24 (d, J = 8.5 Hz, 2H), 6.97–6.91 (m, 4H), 4.78 (t, J = 4.5 Hz, 2H), 4.19 (d, J = 4.5 Hz, 4H), 3.79 (s, 12H); ^{13}C-NMR (DMSO-d_6) δ 196.2, 151.7, 149.1, 142.8, 127.4, 123.8, 120,9, 112.1, 111.3, 71.6, 60.1, 56.1, 56.0; HRMS [ESI]$^+$ calculated for $C_{25}H_{28}NaO_8$: 479.1682 [M + Na]$^+$; found: 479.1677.

3.1.7. *2,2-bis[(E)-3-(3,4-Dimethoxyphenyl)acryloyl]propane-1,3-diylbis[3-hydroxy-2-(hydroxymethyl)-2-methylpropanoate]* (**14**)

To a solution of **12** (0.170 g, 0.37 mmol) in CH_2Cl_2 (4 mL) was added NEt_3 (0.15 mL, 1.10 mmol), DMAP (0.110 g, 0.92 mmol) and 2,2,5-trimethyl-1,3-dioxane-5-carbonyl chloride **13** (0.210 g, 1.10 mmol) sequentially at room temperature. The reaction mixture was stirred at the same temperature for 12h and then concentrated under vacuum to give the crude product, which without purification was treated with $HCl_{(aq)}$ (6 *N*, 0.1 mL) in MeOH (0.5 mL). The resulting mixture was stirred at room temperature for 1h and then concentrated under vacuum to give the crude product, which was purified by flash chromatography on silica gel with MeOH/CH_2Cl_2 (1:19) to afford compound **14** (16 mg, 7% for two steps) as a yellow solid; mp > 300 °C; ^{1}H-NMR (DMSO-d_6) δ 7.64 (d, *J* = 15.5 Hz, 2H), 7.34–7.30 (m, 4H), 7.03 (d, *J* = 15.0 Hz, 2H), 6.99 (d, *J* = 8.0 Hz, 2H), 4.74 (s, 4H), 4.63 (t, *J* = 5.5 Hz, 4H), 3.79 (s, 12H), 3.40-3.37 (m, 8H), 0.95(s, 6H); ^{13}C-NMR (DMSO-d_6) δ 193.3, 174.6, 152.1, 149.4, 145.0, 127.1, 124.7, 119,5, 112.0, 111.4, 66.9, 63.9, 62.3, 56.1, 56.1, 50.7, 17.0; HRMS [ESI]$^+$ calculated for $C_{35}H_{44}NaO_{14}$: 711.2629 [M + Na]$^+$; found: 711.2620.

3.1.8. General Procedure for the Synthesis of Compounds **2m–6m**

The general procedure is illustrated immediately below with compound **2m** as a specific example.

(1E,6E)-1,7-bis(4-Hydroxy-3-methoxyphenyl)-4,4-dimethylhepta-1,6-diene-3,5-dione (**2m**)

To a stirred solution of compound **2** (0.450 g, 0.72 mmol) in CH_3OH (6 mL) were added NaOMe (0.116 g, 2.16 mmol) at room temperature. The reaction mixture was stirred at the same temperature for 2 h and then H_2O (5 mL) and diluted $HCl_{(aq)}$ (1 *N*, 1 mL) were added to quench the reaction. The aqueous layer was separated and extracted with CH_2Cl_2 (3 × 6 mL). The combined organic extracts were washed with brine, dried over $MgSO_4$, filtered and concentrated to give the crude product, which was then purified by flash chromatography on silica gel with EtOAc/*n*-hexane/CH_2Cl_2 (1:2:1) to afford **2m** (0.245 g, 86% yield) as an orange solid; mp = 104–106 °C; ^{1}H-NMR ($CDCl_3$) δ 7.69 (d, *J* = 15.5 Hz, 2H), 7.12 (dd, *J* = 8.0, 1.5 Hz, 2H), 7.01 (s, 2H), 6.91 (d, *J* = 8.5 Hz, 2H), 6.65 (d, *J* = 15.5 Hz, 2H), 5.97 (s, 2H), 3.94 (s, 6H), 1.48 (s, 6H); ^{13}C-NMR ($CDCl_3$) δ 198.2, 148.6, 146.8, 144.5, 126.8, 124.1, 119.1, 114.8, 109.8, 60.8, 56.0, 14.1; HRMS [ESI]$^+$ calculated for $C_{23}H_{24}NaO_6$: 419.1471 [M + Na]$^+$; found: 419.1480

(1E,6E)-4,4-Diethyl-1,7-bis(4-hydroxy-3-methoxyphenyl)hepta-1,6-diene-3,5-dione (**3m**)

Yield: 87%; mp = 114–115 °C; ^{1}H-NMR ($CDCl_3$) δ 7.64 (d, *J* = 15.5 Hz, 2H), 7.05 (dd, *J* = 8.3, 1.5 Hz, 2H), 6.95 (d, *J* = 2.0 Hz, 2H), 6.85 (d, *J* = 8.0 Hz, 2H), 6.59 (d, *J* = 15.5 Hz, 2H), 5.88 (s, 2H), 3.88 (s, 6H), 2.06 (q, *J* = 7.5 Hz, 4H), 0.71 (t, *J* = 7.5 Hz, 6H); ^{13}C-NMR ($CDCl_3$) δ 197.9, 148.5, 146.7, 144.0, 126.8, 124.2, 119.6, 114.7, 109.7, 69.0, 56.1, 21.6, 7.7; HRMS [ESI]$^+$ calculated for $C_{25}H_{28}NaO_6$: 447.1784 [M + Na]$^+$; found: 447.1788.

(1E,6E)-4,4-Dibenzyl-1,7-bis(4-hydroxy-3-methoxyphenyl)hepta-1,6-diene-3,5-dione (**4m**)

Yield: 82%; mp = 134–135 °C; ^{1}H-NMR ($CDCl_3$) δ 7.72 (d, *J* = 15.0 Hz, 2H), 7.26–7.21 (m, 6H), 7.15–7.13 (m, 4H), 7.07–7.05 (m, 2H), 6.90–6.89 (m, 4H), 6.58 (dd, *J* = 15.5, 1.0 Hz, 2H), 5.31 (s, 2H), 3.89 (s, 6H), 3.42 (s, 4H); ^{13}C-NMR ($CDCl_3$) δ 196.9, 148.7, 146.8, 144.0, 136.6, 130.4, 128.1, 126.8, 126.6, 124.3, 120.8, 114.8, 109.8, 70.1, 56.1, 37.7; HRMS [ESI]$^+$ calculated for $C_{25}H_{32}NaO_6$: 571.2097 [M + Na]$^+$; found: 571.2090.

(1E,6E)-1,7-bis(4-Hydroxy-3-methoxyphenyl)-4,4-di(prop-2-yn-1-yl)hepta-1,6-diene-3,5-dione (**5m**)

Yield: 83%; mp = 201–203 °C; ^{1}H-NMR ($CDCl_3$) δ 7.75 (d, *J* = 15.0 Hz, 2H), 7.13 (dd, *J* = 8.0, 1.5 Hz, 2H), 7.03 (d, *J* = 2.0 Hz, 2H), 6.92 (d, *J* = 8.0 Hz, 2H), 6.67 (d, *J* = 15.0 Hz, 2H), 5.97 (s, 2H), 3.95 (s, 6H), 3.19 (s, 4H), 2.04 (s, 4H); ^{13}C-NMR ($CDCl_3$) δ 193.5, 148.9, 145.9, 126.5, 124.7, 117.8, 114.8, 109.8, 79.1, 71.9, 66.9, 56.1, 21.2; HRMS [ESI]$^+$ calculated for $C_{27}H_{28}O_6$: 471.1784 [M + Na]$^+$; found: 471.1783.

(1E,6E)-4,4-Diallyl-1,7-bis(4-hydroxy-3-methoxyphenyl)hepta-1,6-diene-3,5-dione (**6m**)

Yield: 85%; mp = 160–162 °C; ^{1}H-NMR ($CDCl_3$) δ 7.65 (d, *J* = 15.5 Hz, 2H), 7.06 (dd, *J* = 8.3, 1.5 Hz, 2H), 6.95 (d, *J* = 2.0 Hz, 2H), 6.85 (d, *J* = 8.0 Hz, 2H), 6.61 (d, *J* = 15.5 Hz, 2H), 5.87 (s, 2H), 5.54–5.49 (m, 2H), 5.07–5.04 (m 4H), 3.88 (s, 6H), 2.78 (d, *J* = 7.5 Hz, 4H); ^{13}C-NMR ($CDCl_3$) δ 194.6, 148.7, 146.8, 144.6, 132.3, 126.7, 119.3, 119.1, 114.7, 109.7, 67.8, 56.1, 34.4; HRMS [ESI]$^+$ calculated for $C_{27}H_{24}NaO_6$: 467.1471 [M + Na]$^+$; found: 467.1477.

3.1.9. General Procedure for the Synthesis of Compounds **15** and **16**

The general procedure is illustrated below with compound **15** as a specific example. To a solution of curcumin bisacetate **17** (0.76 g, 1.68 mmol) in DMF (8 mL) was added K_2CO_3 (0.58 g, 4.20 mmol) and propyl bromide (0.435 g, 3.53 mmol) sequentially at 0 °C. The reaction mixture was stirred at room temeperatre for 20 h and then H_2O (10 mL) and diluted $HCl_{(aq)}$ (1 *N*, 5 mL) were added to quench the reaction. The aqueous layer was separated and extracted with EtOAc (3 × 10 mL). The combined organic layer was dried over $MgSO_4$, filtered and concentrated to give the crude product, which was subjected without any purification to the base-promoted hydrolysis reaction [NaOMe (0.190 g, 3.53 mmol), MeOH (5 mL)]. The reaction was followed by TLC until no starting material was present. The resulting mixture was then concentrated to give the crude product, which was then purified by flash chromatography on silica gel with EtOAc/*n*-hexane/CH_2Cl_2 (1:2:1) to afford compound **15** (0.213 g, 28% yield for two steps) as a yellow solid.

(1E,6E)-1,7-bis(4-Hydroxy-3-methoxyphenyl)-4,4-dipropylhepta-1,6-diene-3,5-dione (**15**)

mp = 137–139 °C; ^{1}H-NMR ($CDCl_3$) δ 7.69 (d, *J* = 15.0 Hz, 2H), 7.11 (dd, *J* = 8.3, 1.5 Hz, 2H), 7.01 (d, *J* = 1.5 Hz, 2H), 6.91 (d, *J* = 8.0 Hz, 2H), 6.65 (d, *J* = 15.5 Hz, 2H), 5.90 (s, 2H), 3.94 (s, 6H), 2.05–2.01 (m, 4H), 1.15–1.08 (m, 4H), 0.95 (t, *J* = 7.5 Hz, 6H); ^{13}C-NMR ($CDCl_3$) δ 197.9, 148.5, 146.7, 144.0, 126.8, 124.1, 119.5, 114.7, 109.7, 68.4, 56.1, 31.8, 16.8, 14.7; HRMS [ESI]$^+$ calculated for $C_{27}H_{32}NaO_6$: 475.2097 [M + Na]$^+$; found: 475.2100.

(1E,6E)-4,4-Dibutyl-1,7-bis(4-hydroxy-3-methoxyphenyl)hepta-1,6-diene-3,5-dione (**16**)

mp = 144–146 °C; ^{1}H-NMR ($CDCl_3$) δ 7.68 (d, *J* = 15.5 Hz, 2H), 7.11 (dd, *J* = 8.3, 1.5 Hz, 2H), 7.01 (s, 2H), 6.91 (d, *J* = 8.2 Hz, 2H), 6.65 (d, *J* = 15.5 Hz, 2H), 6.01 (s, 2H), 3.93 (s, 6H), 2.07–2.03 (m, 4H), 1.37–1.33 (m, 4H), 1.10–1.03 (m, 4H), 0.90 (t, *J* = 7.3 Hz, 6H); ^{13}C-NMR ($CDCl_3$) δ 198.0, 148.5, 146.8, 144.0, 126.9, 124.1, 119.5, 114.7, 109.7, 68.3, 56.1, 29.1, 25.5, 23.2, 13.9; HRMS [ESI]$^+$ calculated for $C_{29}H_{36}NaO_6$: 503.2410 [M + Na]$^+$; found: 503.2400.

3.1.10. *(E)-1,7-bis(4-Hydroxy-3-methoxyphenyl)-4,4-dimethylhept-1-ene-3,5-dione* (**18**)

To a stirred solution of compound **2m** (0.740 g, 1.87 mmol) in EtOAc (10 mL) was added Pd/C (37 mg, 5% *w*/*w*) in one portion. The resulting mixture was hydrogenated under 1 atmosphere of H_2 at room temperature. After 1 h, the mixture was filtered by celite and concentrated to give the crude product, which was purified by flash chromatography on silica gel with EtOAc/*n*-hexane/CH_2Cl_2 (1:2:1) to afford compound **18** (140 mg, 19% yield) as a light-yellow oil and recover compound **2m** (500 mg, 67% yield). ^{1}H-NMR ($CDCl_3$) δ 7.61 (d, *J* = 15.5 Hz, 1H), 7.07 (dd, *J* = 8.5, 1.5 Hz, 1H), 6.95–6.93 (m, 2H), 6.75 (d, *J* = 8.0 Hz, 1H), 6.65–6.62 (m, 2H), 6.49 (d, *J* = 15.5 Hz, 1H), 6.07 (s, 1H), 5.45 (s, 1H), 3.94 (s, 3H), 3.79 (s, 3H), 2.83 (t, *J* = 7.5 Hz, 2H), 2.74 (t, *J* = 7.0 Hz, 2H), 1.38 (s, 6H); ^{13}C-NMR ($CDCl_3$) δ 209.4, 197.7, 148.7, 146.8, 146.3, 144.7, 143.8, 132.6, 126.9, 123.9, 121.0, 118.2, 114.9, 114.2, 111.0, 110.0, 61.7, 56.0, 55.7, 40.7, 29.6, 21.0; HRMS [ESI]$^+$ calculated for $C_{23}H_{26}NaO_6$: 421.1627 [M + Na]$^+$; found: 421.1233.

3.1.11. *1,7-bis(4-Hydroxy-3-methoxyphenyl)-4,4-dimethylheptane-3,5-dione* (**19**) *and 5-hydroxy-1,7-bis(4-hydroxy-3-methoxyphenyl)-4,4-dimethylheptan-3-one* (**20**)

To a stirred solution of compound **2m** (0.885 g, 2.23 mmol) in MeOH (10 mL) was added Pd/C (45 mg, 10% *w*/*w*) in one portion. The resulting mixture was hydrogenated under 1 atmosphere of H_2 at room temperature. After 2 h, the mixture was filtered by celite and concentrated to give the crude product, which was purified by flash chromatography on silica gel with EtOAc/*n*-hexane/ (1:2) to afford compound **19** (465 mg, 52% yield) as a colorless oil and compound **20** (206 mg, 23% yield) as a colorless oil.

Compound **19**: ^{1}H-NMR ($CDCl_3$) δ 6.83 (d, *J* = 8.0 Hz, 2H), 6.65 (d, *J* = 2.0 Hz, 2H), 6.63–6.61 (m, 2H), 5.51 (s, 2H), 3.87 (s, 6H), 2.76 (t, *J* = 7.0 Hz, 4H), 2.58 (t, *J* = 7.0 Hz, 4H), 1.29 (s, 6H); ^{13}C-NMR ($CDCl_3$) δ 208.9, 146.4, 143.9, 132.7, 120.9, 114.3, 114.9, 111.7, 62.4, 55.9, 40.5, 29.4, 21.0; HRMS [ESI]$^+$ calculated for $C_{23}H_{28}NaO_6$: 423.1784 [M + Na]$^+$; found: 423.1780.

Compound **20**: ^{1}H-NMR ($CDCl_3$) δ 6.86–6.83 (m, 2H), 6.72–6.66 (m, 4H), 5.62–5.61 (br s, 2H), 3.89 (s, 3H), 3.87 (s, 3H), 3.71–3.69 (m, 1H), 2.90–2.71 (m, 5H), 2.61–2.51 (m, 2H), 1.71–1.55 (m, 2H), 1.13 (s, 3H), 1.08 (s, 3H); ^{13}C-NMR ($CDCl_3$) δ 216.5, 146.4, 143.9, 143.7, 133.9, 133.1, 120.9, 120.8, 114.4, 114.3, 111.2, 111.1, 55.9, 51.7, 40.3, 33.7, 32.6, 29.4, 21.7, 19.3; HRMS [ESI]$^+$ calculated for $C_{23}H_{30}NaO_6$: 425.4768 [M + Na]$^+$; found: 425.4760.

3.1.12. *1,7-bis(4-Hydroxy-3-methoxyphenyl)-4,4-dimethylheptane-3,5-diol* (**21**)

To a solution of compound **20** (110 mg, 0.27 mmol) in MeOH (3 mL) was added $NaBH_4$ (30 mg, 0.81 mmol) at 0 °C. The resulting mixture was allowed to stir at room temperature for 2 h, and then H_2O (3 mL) was added. The aqueous layer was separated and extracted with CH_2Cl_2 (3 × 6 mL). The combined organic extracts were washed brine, dried over $MgSO_4$, filtered and concentrated to give the crude residue, which was purified by flash chromatography on silica gel with EtOAc/*n*-hexane/ (1:1) to afford compound **21** (34 mg, 31% yield) as a colorless oil. ^{1}H-NMR ($CDCl_3$) δ 6.86 (d, *J* = 8.0 Hz, 2H), 6.74–6.71 (m, 4H), 5.63 (s, 2H), 3.89 (s, 6H), 3.59–3.56 (m, 2H), 3.11 (s, 2H), 2.85–2.71 (m, 2H), 2.61–2.55 (m, 2H), 1.81–1.72 (m, 4H), 0.90 (s, 6H); ^{13}C-NMR ($CDCl_3$) δ 146.5, 143.7, 134.1, 134.0, 120.9, 114.3, 111.1, 78.7, 55.9, 40.3, 33.9, 32.8, 21.0; HRMS [ESI]$^+$ calculated for $C_{23}H_{32}NaO_6$: 427.4928 [M + Na]$^+$; found: 427.4933.

3.2. Biological Assays

3.2.1. In Vitro MTT [3-(4,5-Dimethylthiazaol-2-yl)-2,4-diphenyltetrazolium bromide] Assay

Cell viability was evaluated by measuring the reduction in MTT to yield blue formazan. Cells were cultured in 96-well plates, allowed to attach overnight, and then treated with compounds. After treatment, MTT solution (1 mg/mL) was added to each well, and plates were incubated for another 2 h. Medium was removed, blue formazan was dissolved in DMSO, and the absorbance was read at 570 nm by Multiskan Go spectrophotometer (Thermo Scientific, Madison, WI, USA).

3.2.2. In Vivo Antitumor Activity Assay

Male nu/nu mice (5 weeks old) from National Laboratory Animal Center (Taipei, Taiwan) were maintained under the procedures and guidelines provided by the Institutional Animal Care and Use Committee of the National Health Research Institutes (Taipei, Taiwan). All experiments were supervised under the Institutional Animal Care and Use Committee, China Medical University, Taichung, Taiwan with a protocol number (CMUIACUC-2018-278). HCT-116 colon cancer cells (5 × 10^6 cells per mouse) were suspended in 0.1 mL of Matrigel solution (50% *v*/*v* Matrigel in PBS) and inoculated into the mammary fat pads of the mice. When the tumor size reached 100 mm^3, the tumor-bearing mice were randomly divided into four groups for treating with vehicle (5% ethanol and 10% tween 80 in 85% saline, 0.1 mL) and **2** (p.o. 50, 100 and 150 mg/kg/day body weight) respectively. Compounds was administered via oral route daily for 30 days. 5-Fluorouracil-treated group (i.p. 30 mg/kg) is the positive control. Tumor size and mouse body weight were measured once every 3 days, and tumor volume (mm^3) was calculated using the equation: length × $(\text{width})^2$ × 0.5. At the end of experiments, mice were sacrificed.

3.2.3. The Preliminary Pharmacokinetic Evaluation of 2 in Rat

Drug administration and blood collection

Male Sparague-Dawley rat (420 g) from National Laboratory Animal Center, Taipei, Taiwan were maintained under the regulations of the Institutional Animal Care and Use Committee of the National Health Research Institutes, Taiwan. The animal studies were supervised under the Institutional Animal Care and Use Committee, China Medical University, Taichung, Taiwan with a protocol number (CMUIACUC-2018-278). Rat was fasted for 16 h before dosing. Compound **2** was dissolved in aqueous solution (5% ethanol and 10% Tween80 in 85% saline) and administrated via oral gavage at a dosage of 100 mg/kg. Blood samples were collected at selected time points (5, 10, 20, 30, 60, 120, 240, 360, 480 min) post dose and centrifuged at 10,000× *g* for 15 min to provide serum samples, which were stored at −80 °C until subsequent sample extraction.

3.2.4. Analysis of **2**, **2m**, **18**, **19**, **20**, **21**, **22** and **23** in Serum

Methods for the preparation of **2**, **2m**, **18**, **19**, **20** and **21** serum sample are delineated as follows. 50 μL of acetate buffer (pH 5.0), 50 μL of ascorbic acid (200 mg/mL), and 50 μL of 0.1 N HCl solution was added into 100 μL of serum which was collected at different time points). The resulting mixture was partitioned with 250 μL of ethyl acetate (containing 5.0 μg/mL of butyl paraben as the internal standard). After centrifuging at 10,000× *g* for 15 min, the upper organic layer was separated and dried under nitrogen gas to provide the crude sample, which was diluted with an appropriate volume of acetonitrile for LC-MS analysis.

The amount of **22** and **23** were analyzed by the serum samples before and after treatment with enzyme solution. The procedure for enzyme treatment is shown as follows. 100 μL of serum was treated with enzyme solution (50 μL; containing 1000 units/mL of sulfatase and 39861 units/mL of β-glucuronidase in pH 5.0 acetate buffer), 50 μL of 0.1 N HCl solution and 50 μL of ascorbic acid (200 mg/mL) in the light protected tubes at 37 °C. For optimal hydrolysis efficiency, the incubation time was decided as 2 h based on a previous study [35]. The mixture was then partitioned with 250 μL of ethyl acetate (containing 5.0 μg/mL of butyl paraben as the internal standard). After centrifuging at 10,000× *g* for 15 min, the upper organic layer was separated and dried under nitrogen gas to provide the crude sample, which was diluted with an appropriate volume of acetonitrile for LC analysis.

3.2.5. LC-MS Analysis Method

All samples obtained from in vivo pharmacokinetic studies were assayed using LC-MS for the analysis of **2** and the corresponding metabolites. The Waters ACQUITY UPLC I-Class system (Waters Corp., Milford, MA, USA) was utilized. System control and all of the mass spectrometry data were acquired and analyzed using UNIFI software (Waters Corp.). For sample separation, the LC was equipped with an Inertsil ph-3 (50 × 2.1 mm, 2 μm particle size) RPLC column (GL Sciences Inc. Tokyo, Japan) in which the column temperature was held at 35 °C, injection volume was 7.5 μL, and the flow rate of the mobile phase was 200 μL/min. The elution started from 50% mobile phase A (ultrapure water + 0.1% formic acid) and 50% mobile phase B (100% methanol + 0.1% formic acid), held at 50% B for 0.5 min, raised to 95% B in 5.5 min, held at 95% B for 1 min, and then lowered to 50% B in 1 min. Before sample injection, the column was equilibrated by pumping 50% B for 4 min. The analysis of target compounds for all samples was performed by using a Waters Vion IMS QTof mass spectrometer. Data were acquired in the electrospray ionization (ESI) positive ion MS^E mode with range of *m*/*z* 100–1000 and 0.5 s scan time. Parameters were set as capillary voltage of 2.5 kV, source temperature of 100 °C, desolvation temperature at 250 °C, cone gas maintained at 10 L/h, desolvation gas maintained at 600 L/h. The acquired m/z and isotope pattern were processed by UNIFI software to illustrate chromatogram, and the amounts of compound was calculated with the integrated peak area of signals.

4. Conclusions

Here, structural modifications at the C-4 position of compound **1**, a 2,2-bis(hydroxymethyl)-propionate curcumin prodrug, with polar and nonpolar functional groups were performed to form a series of 4,4-disubstituted analogs of **1** and **DMC**. The in vitro anti-proliferative assay of these derivatives elucidated an obvious SAR in which the enhancement of polarity on C-4 resulted in the erosion of anticancer activity. The ester prodrug characteristics of **2–6** motivated us to synthesize their hydrolysis compounds **2m–6m**, **15**, and **16**, whose anticancer activities were highly dependent on the volume and length of the C-4 alkyl chains. The preliminary pharmacokinetic evaluations indicated that the esterase-mediated hydrolysis of **2** proceeded smoothly to produce **2m**, which in turn was reduced to **18** and **19** as major phase I metabolites. The phase II metabolic pathway was confirmed by the detection of glucuronide **22** and sulfate **23**. However, it is still possible that reduction precedes ester hydrolysis to provide the hydrogenated counterparts of **2**, which possess novel chemical skeletons with unknown biological activity.

Our preliminary pharmacokinetic results also shed light on future directions for the structural modification of the original compounds. Additional modifications have been scheduled to reduce the high propensity towards ester hydrolysis by converting the methyl of 2,2-bis(hydroxymethyl)-propionate group into more bulky alkyl groups. Pharmacokinetic studies have been designed to identify, biologically evaluate, and quantify the possible metabolites. Apart from this, compound **2** and **2m** may behave like **1** which acts on multiple targets and signaling pathways. The differences between the mechanism of action of **2**, **2m** and **1** are not definitively understood. Currently, we are actively investigating the mechanism of action and pharmacokinetic parameters of **2** and **2m**, including $T_{1/2}$, AUC, and C_{max}, and the results will be reported in due course. The animal study showed that compound **2** inhibits tumor growth by 45% at a dose of 100 mg/kg in HCT-116 xenograft nude mice. Thus, twice-a-day (BID) dosing formulations for achieving higher therapeutic efficacy at lower doses should be considered in further development of title curcuminoid derivatives as anticancer drug candidates.

Supplementary Materials: The following are available online. **Scheme S1**: PLE mediated hydrolysis of curcumin 2,2-bis(hydroxymethyl)propionate **1**. **Figure S1**: The HPLC analysis of compound **18**, **19**, **20**, **21**, **22** and **23**. Copies of ^{1}H and ^{13}C NMR spectra of the new products.

Author Contributions: D.-Y.L. performed the preliminary pharmacokinetic analysis and wrote experimental section and supplementary material file. Y.-C.H. designed the preliminary pharmacokinetic study and integrated the experimental data. J.-S.Y. and H.-Y.L. performed the in vivo assay and in vivo animal study. K.-H.L. and T.-Y.C. reviewed and revised the manuscript. S.-C.K. designed the target compounds and interpreted the experimental data. M.-T.H. designed and synthesized targeted compounds. S.-C.K. and M.-T.H. are the responsible researchers, who wrote and edited the manuscript. All authors have read and agreed to the published version of the manuscript.

Funding: We are grateful to Ministry of Science and Technology, Taiwan (MOST 107-2320-B-039-028) for financial support. Partial support was also provided by the "Chinese Medicine Research Center, China Medical University" from The Featured Areas Research Center Program within the framework of the Higher Education Sprout Project by the Ministry of Education (MOE) in Taiwan (CMRC-CHM-5) awarded to S. C. Kuo.

Conflicts of Interest: The authors declare no conflict of interest.

References

1. Strimpakos, A.S.; Sharma, R.A. Curcumin: Preventive and therapeutic properties in laboratory studies and clinical trials. *Antioxid. Redox Signal.* **2008**, *10*, 511–546. [CrossRef]
2. Nabavi, S.F.; Thiagarajan, R.; Rastrelli, L.; Daglia, M.; Sobarzo-Sánchez, E.; Alinezhad, H.; Nabavi, S.M. Curcumin: A natural product for diabetes and its complications. *Curr. Top Med. Chem.* **2015**, *15*, 2445–2455. [CrossRef]
3. Mathew, D.; Hsu, W.L. Antiviral potential of curcumin. *J. Funct. Foods* **2018**, *40*, 692–699. [CrossRef]

4. Sahebkar, A.; Henrotin, Y. Analgesic efficacy and safety of curcuminoids in clinical practice: A systematic review and meta-analysis of randomized controlled trials. *Pain Med.* **2016**, *17*, 1192–1202. [CrossRef] [PubMed]
5. Osawa, T. Nephroprotective and hepatoprotective effects of curcuminoids. *Adv. Exp. Med. Biol.* **2007**, *595*, 407–423. [PubMed]
6. Miriyala, S.; Panchatcharam, M.; Rengarajulu, P. Cardioprotective effects of curcumin. *Adv. Exp. Med. Biol.* **2007**, *595*, 359–377. [PubMed]
7. Shishodia, S. Molecular mechanisms of curcumin action: Gene expression. *Biofactors* **2013**, *39*, 37–55. [CrossRef] [PubMed]
8. Gupta, S.C.; Prasad, S.; Kim, J.H.; Patchva, S.; Webb, L.J.; Priyadarsini, I.K.; Aggarwal, B.B. Multitargeting by curcumin as revealed by molecular interaction studies. *Nat. Prod. Rep.* **2011**, *28*, 1937–1955. [CrossRef] [PubMed]
9. Bordoloi, D.; Roy, N.K.; Monisha, J.; Padmavathi, G.; Kunnumakkara, A.B. Multi-targeted agents in cancer cell chemosensitization: What we learn from curcumin thus far. *Recent Pat. Anti-Canc.* **2016**, *11*, 67–97. [CrossRef] [PubMed]
10. Kunnumakkara, A.B.; Anand, P.; Aggarwal, B.B. Curcumin inhibits proliferation, invasion, angiogenesis and metastasis of different cancers through interaction with multiple cell signaling proteins. *Cancer Lett.* **2008**, *269*, 199–225. [CrossRef]
11. Nelson, K.M.; Dahlin, J.L.; Bisson, J.; Graham, J.; Pauli, G.F.; Walters, M.A. The essential medicinal chemistry of curcumin. *J. Med. Chem.* **2017**, *60*, 1620–1637. [CrossRef] [PubMed]
12. Schieffer, G.W. Pressurized liquid extraction of curcuminoids and curcuminoid degradation products from turmeric (curcuma longa) with subsequent HPLC assays. *J. Liq. Chromatogr. Relat. Technol.* **2002**, *25*, 3033–3044. [CrossRef]
13. Schneider, C.; Gordon, O.N.; Edwards, R.L.; Luis, P.B. Degradation of curcumin: From mechanism to biological implications. *J. Agric. Food Chem.* **2015**, *63*, 7606–7614. [CrossRef] [PubMed]
14. Priyadarsini, K.I.; Maity, D.K.; Naik, G.H.; Kumar, M.S.; Unnikrishnan, M.K.; Satav, J.G.; Mohan, H. Role of phenolic O-H and methylene hydrogen on the free radical reactions and antioxidant activity of curcumin. *Free Radic. Biol. Med.* **2003**, *35*, 475–484. [CrossRef]
15. Sharma, R.A.; Steward, W.P.; Gescher, A.J. Pharmacokinetics and pharmacodynamics of curcumin. *Adv. Exp. Med. Biol.* **2007**, *595*, 453–470. [PubMed]
16. Ireson, C.; Orr, S.; Jones, D.J.; Verschoyle, R.; Lim, C.K.; Luo, J.L.; Howells, L.; Plummer, S.; Jukes, R.; Williams, M.; et al. Characterization of metabolites of the chemopreventive agent curcumin in human and rat hepatocytes and in the rat in vivo, and evaluation of their ability to inhibit phorbol ester-induced prostaglandin E2 production. *Cancer Res.* **2001**, *61*, 1058–1064. [PubMed]
17. Rodrigues, F.C.; Anil Kumar, N.A.; Thakur, G. Developments in the anticancer activity of structurally modified curcumin: An up-to-date review. *Eur. J Med. Chem.* **2019**, *177*, 76–104. [CrossRef]
18. Tomeh, M.A.; Hadianamrei, R.; Zhao, X. A Review of Curcumin and Its Derivatives as Anticancer Agents. *Int. J. Mol. Sci.* **2019**, *20*, 1033. [CrossRef]
19. Di Martino, R.M.C.; De Simone, A.; Andrisano, V.; Bisignano, P.; Bisi, A.; Gobbi, S.; Rampa, A.; Fato, R.; Bergamini, C.; Perez, D.I.; et al. Versatility of the curcumin scaffold: Discovery of potent and balanced dual BACE-1 and GSK-3b inhibitors. *J. Med. Chem.* **2016**, *59*, 531–544. [CrossRef]
20. Noureddin, S.A.; El-Shishtawy, R.M.; Al-Footy, K.O. Curcumin analogues and their hybrid molecules as multifunctional drugs. *Eur. J. Med. Chem.* **2019**, *82*, 11631–11671.
21. Barzegar, A.; Moosavi-Movahedi, A.A. Intracellular ROS protection efficiency and free radical-scavenging activity of curcumin. *PLoS ONE* **2011**, *6*, e26012. [CrossRef] [PubMed]
22. Epstein, J.; Sanderson, I.R.; Macdonald, T.T. Curcumin as a therapeutic agent: The evidence of in vitro, animal and human studies. *Br. J. Nutr.* **2010**, *103*, 1545–1557. [CrossRef] [PubMed]
23. For the clinical progress of curcumin, refer to https://clinicaltrials.gov/.
24. Hsieh, M.T.; Chang, L.C.; Hung, H.Y.; Lin, H.Y.; Shih, M.H.; Tsai, C.H.; Kuo, S.C.; Lee, K.H. New bis(hydroxymethyl) alkanoate curcuminoid derivatives exhibit activity against triple-negative breast cancer in vitro and in vivo. *Eur. J Med. Chem.* **2017**, *131*, 141–151. [CrossRef] [PubMed]
25. Berry, L.M.; Wollenberg, L.; Zhao, Z. Esterase activities in the blood, liver and intestine of several preclinical species and humans. *Drug Metab. Lett.* **2009**, *3*, 70–77. [CrossRef]

26. Shi, Q.; Shih, C.C.; Lee, K.H. Novel anti-prostate cancer curcumin analogues that enhance androgen receptor degradation activity. *Anticancer Agents Med. Chem.* **2009**, *9*, 904–912. [CrossRef]
27. 2,2,5-trimethyl-1,3-dioxane-5-carbonyl chloride 13 was formed by treating the acid counterpart with thionyl chloride. For detailed procedure, see: R. Giri, X. Chen, J.Q. Yu, Palladium-catalyzed asymmetric iodination of unactivated C-H bonds under mild conditions. *Angew Chem. Int. Ed. Engl.* **2005**, *29*, 2112–2115.
28. Basile, V.; Ferrari, E.; Lazzari, S.; Belluti, S.; Pignedoli, F.; Imbriano, C. Curcumin derivatives: Molecular basis of their anti-cancer activity. *Biochem. Pharmacol.* **2009**, *78*, 1305–1315. [CrossRef]
29. Prasad, S.; Tyagi, A.K.; Aggarwal, B.B. Recent developments in delivery, bioavailability, absorption and metabolism of curcumin: The golden pigment from golden spice. *Cancer Res. Treat.* **2014**, *46*, 2–18. [CrossRef]
30. Liederer, B.M.; Borchardt, R. Enzymes involved in the bioconversion of ester-based prodrugs. *J. Pharm. Sci.* **2006**, *95*, 1177–1195. [CrossRef]
31. Kunihiro, A.G.; Luis, P.B.; Brickey, J.A.; Julia, A.; Jen, B.F.; Chow, H.H.S.; Schmeider, C.; Funk, J.L. Beta-glucuronidase catalyzes deconjugation and activation of curcumin-glucuronide in bone. *J. Nat. Prod.* **2019**, *82*, 500–509. [CrossRef]
32. Ozawa, H.; Imaizumi, A.; Sumi, Y.; Hashimoto, T.; Kanai, M.; Makino, Y.; Tsuda, T.; Takahashi, N.; Kakeya, H. Curcumin β-D-glucuronide plays an important role to keep high levels of free-form curcumin in the blood. *Biol. Pharm. Bull.* **2017**, *40*, 1515–1524. [CrossRef] [PubMed]
33. Ge, S.; Tu, Y.; Hu, M. Challenges and opportunities with predicting in vivo phase II metabolism via glucuronidation from in vitro data. *Curr. Pharmacol. Rep.* **2016**, *2*, 326–338. [CrossRef] [PubMed]
34. Yang, K.Y.; Lin, L.C.; Tseng, T.Y.; Wang, S.C.; Tsai, T.H. Oral bioavailability of curcumin in rat and the herbal analysis from Curcuma longa by LC−MS/MS. *J. Chromatogr. B Anal. Technol.Biomed. Life Sci.* **2007**, *853*, 183–189. [CrossRef] [PubMed]
35. Hsieh, Y.W.; Huang, C.Y.; Yang, S.Y.; Peng, Y.H.; Yu, C.P.; Lee Chao, P.D.; Chi, H.Y. Oral intake of curcumin markedly activated CYP 3A4: In vivo and ex-vivo studies. *Sci. Rep.* **2014**, *4*, 6587–6594. [CrossRef] [PubMed]

Sample Availability: Samples of the compounds are not available from the authors.

molecules

MDPI

Article

Comparative Chemical Profiling and Monacolins Quantification in Red Yeast Rice Dietary Supplements by ^{1}H-NMR and UHPLC-DAD-MS

Rabab Hachem [1], Gaëtan Assemat [1], Stéphane Balayssac [1], Nathalie Martins-Froment [2], Véronique Gilard [1], Robert Martino [1] and Myriam Malet-Martino [1,*]

1 Biomedical NMR Group, Laboratoire SPCMIB, UMR CNRS 5068, Université Paul Sabatier, 118 route de Narbonne, 31062 Toulouse CEDEX 9, France; Rabab.Hachem@evotec.com (R.H.); gaetan-ass@hotmail.fr (G.A.); balayssac@chimie.ups-tlse.fr (S.B.); gilard@chimie.ups-tlse.fr (V.G.); rmartino@chimie.ups-tlse.fr (R.M.)

2 Service commun de spectrométrie de masse, Institut de Chimie de Toulouse, Université Paul Sabatier, 118 route de Narbonne, 31062 Toulouse CEDEX 9, France; martins@chimie.ups-tlse.fr

* Correspondence: martino@chimie.ups-tlse.fr; Tel.: +33-6-81-36-32-19

Academic Editor: Ping-Chung Kuo

check for updates

Received: 9 November 2019; Accepted: 6 January 2020; Published: 13 January 2020

Abstract: Red yeast rice dietary supplements (RYR DS) are largely sold in Western countries for their cholesterol-lowering/regulating effect due to monacolins, mainly monacolin K (MK), which is, in fact, lovastatin, the first statin drug on the market. ^{1}H-NMR was used as an easy, rapid and accurate method to establish the chemical profiles of 31 RYR DS and to quantify their monacolin contents. Among all the ^{1}H resonances of the monacolins found in RYR, only those of the ethylenic protons of the hexahydronaphthalenic ring at 5.84 and 5.56 ppm are suitable for quantification because they show no overlap with the matrix signals. The total content in monacolins per capsule or tablet determined in 28 DS (the content in 3 DS being below the limit of quantification of the method, $\approx$ 0.25 mg per unit dose) was close to that measured by UHPLC, as shown by the good linear correlation between the two sets of values (slope 1.00, y-intercept 0.113, r^2 0.986). Thirteen of the 31 RYR DS analyzed (i.e., 42%) did not provide label information on the concentration of monacolins and only nine of the 18 formulations with an indication (i.e., 50%) actually contained the declared amount of monacolins.

Keywords: red yeast rice; ^{1}H-NMR; UHPLC-DAD-MS; mass spectrometry; monacolins

1. Introduction

The *Monascus*-fermented rice is the fermentation product of non-glutinous white rice with the fungus *Monascus*. It has been used for more than a thousand years in East Asian countries for producing rice wine, for flavoring, coloring and preservation of foods as well as a folk medicine for improving food digestion and blood circulation. Several species other than that isolated in 1895 and named *Monascus purpureus* in recognition of its purple colour, have also been widely used in making red wine and red-coloured foods [1]. These *Monascus* fermentation products are called Red Yeast Rice (RYR) in Western countries although the designation "yeast" is incorrect as *Monascus* is a filamentous fungus and not a yeast [2]. Other more accurate denominations like red fermented rice or red mold rice are also used but much more rarely. Today, the usage of *Monascus* rice products as colorant or flavour in foods, and for brewing red rice wine is permitted in many Asian countries but not in Europe. However, RYR extracts are largely sold in Western countries for their cholesterol-lowering effects.

A multitude of fungal secondary metabolites, phytosterols, isoflavonoids, fatty acids, pigments, monacolins and others, are produced during the fermentation process. Monacolins, in particular monacolin K (MK), which is, in fact, lovastatin, the first marketed statin drug, inhibit the activity of 3-hydroxy-3-methylglutaryl coenzyme A (HMG-CoA) reductase. As a result, the endogenous synthesis of cholesterol is reduced and hence the elevated cholesterol level decreases. Many clinical studies demonstrated the efficacy of RYR in the treatment of hypercholesterolemia and its relative safety [3,4]. For similar cholesterol level reduction, the MK amount in clinically tested RYR is markedly lower than that used with prescription statin drugs [4]. This RYR potency is likely explained by the presence of monacolins other than MK and also by the improved dissolution rate and bioavailability of lovastatin when given as RYR [4,5]. However, adverse effects following RYR consumption have been reported; the nature of the symptoms and the targeted organs and systems are similar to those reported for statin drugs [6–9]. Considering the case reports gathered in the WHO Vigibase (82) [8] and those collected by four national health institutions (FDA (164) [8], French Nutrivigilance system (30) [6], Italian Surveillance system (52) [7] and Netherlands Pharmacovigilance Centre Lareb (74) [9]), the more frequent adverse effects are: (i) myalgias and related neuroskeletal complaints from 30% of FDA cases to 43% of Lareb cases, including cases of rhabdomyolysis in the five registries, (ii) gastrointestinal disorders which concern up to 23% of Italian cases, (iii) hepatobiliary disorders that affect from 9% of the WHO cases to 32% of the French cases, including severe adverse reactions as pancreatitis and acute hepatic failure, and (iv) skin and subcutaneous disorders which concern from 8% of French cases to 17% of Italian cases.

Nowadays, RYR is widely used as a cholesterol-lowering agent by patients with a proven or perceived intolerance to statins or by consumers, even without dyslipidemia or increased cardiovascular risk, interested in complementary and alternative medications to influence their lipid levels, as it is a common belief that "natural" products do not have side effects [4]. The RYR products are registered as dietary supplements (DS) and, despite their ever growing popularity, there is no uniform regulation regarding their content in monacolins, especially in MK, nor strict quality control. So, their efficacy and safety are unpredictable. Therefore, the development of analytical methods for the simultaneous determination of MK and other monacolins in RYR products is of great importance.

High performance or Ultra-high performance liquid chromatography (HPLC or UHPLC) with diode array detection (DAD) and/or mass spectrometry (MS) detection are regarded as the gold standard methods for the accurate identification and quantification of a wide range of components in RYR products, including monacolins and pigments [10–14]. Although sensitive and selective, they are usually time-consuming, require standard reference materials for quantitative analysis and may suffer from the occurrence of co-eluting interferences (matrix effects) which is a major drawback for MS quantification [15]. Proton Nuclear Magnetic Resonance (^{1}H-NMR) is recognized as a method of choice for the analysis of complex mixtures (pharmaceuticals, biological media for instance) [16,17]. Indeed, it is highly reproducible, robust, and nonselective, thus allowing an unbiased overview of the sample composition as all the low molecular weight compounds in the solution (provided they bear ^{1}H nuclei and are present at sufficient concentration) are detected simultaneously in a single run. It is also inherently quantitative because the area of each NMR resonance is directly proportional to the number of corresponding nuclei if spectra are recorded in fully relaxed conditions. Thus, at variance with other techniques, the response factor is not dependent on the molecular structure and there is no need for identical reference materials. Moreover, the sample preparation for NMR analysis is very simple as it requires dissolution (for solid products) or dilution (for liquid products) in an adequate deuterated solvent [16]. Lachenmeyer et al. [18] have already used ^{1}H-NMR for the assay of the total quantity of monacolins in five RYR commercial products. They showed that the inhibitory effect of RYR on HMG-CoA reductase was all the more important as their monacolin content was high, but the monacolin contents they determined were not compared to those measured by an orthogonal analytical method.

The purpose of this study was to validate (or not) the ^{1}H-NMR method for an accurate determination of the monacolin content in RYR DS by comparison with the well-established HPLC method and thus to control their quality. Therefore, 31 RYR DS were analyzed using ^{1}H-NMR to establish their spectral signatures and to determine their monacolin contents based on the quantification of selected protons characteristic of the different monacolin chemical structures usually present. An UHPLC analysis with UV-Visible (UV-Vis) and MS detection was performed in parallel on the same 31 RYR DS in order to determine their chemical profiles and to quantify all the monacolins identified.

2. Results and Discussion

2.1. ^{1}H-NMR Analysis

Thirty-one RYR DS were analyzed by ^{1}H-NMR. All samples with their name, origin, form, batch number, expiration date and RYR extract content are listed in Table 1.

2.1.1. Qualitative ^{1}H-NMR Analysis

Four characteristic spectra are illustrated in Figure 1 (the ^{1}H-NMR spectra of all RYR DS analyzed are shown in Figure S1). The monacolin resonances were identified by comparing the ^{1}H-NMR spectra of RYR DS with those of standard monacolins whose chemical structures are respectively characteristic of monacolins in lactone form (MK and compactin (CP)), monacolins in hydroxyl acid form (MKA), dehydromonacolins (DeMK) and dihydromonacolins (DiMK), the main monacolin derivatives found in RYR (see Figure 2 for chemical structures) [10,12,13]. Their ^{1}H-NMR assignments are given in Table 2. A complete one-dimensional (1D) and two-dimensional (2D) description of the ^{1}H and ^{13}C-NMR signals of standard MK, MKA, CP and DiMK is presented in Table S1.

The resonances of the hexahydronaphthalene ring ethylenic protons (H5, H6, H4 at ≈ 6.01, 5.84 and 5.56 ppm, respectively) are representative of all the monacolins usually found in RYR, except dihydromonacolins as the δ of the ethylenic protons H5 and H6 of DiMK resonated at 5.42 and 5.69 ppm, respectively (Figures 1 and 3, Table 2). The H1 signal (q) at ≈ 5.33 ppm is characteristic of monacolins with a hexahydronaphthalene ring and an ester group (O-CO-R) in position 1. Indeed, for the monacolins with an OH (monacolin J (MJ) and derivatives) or H substituent (monacolin L (ML) and derivatives) instead of ester (Figure 2), the H1 proton(s) resonated respectively at ≈ 4.24 ppm or 1.17 and 1.77 ppm [1,19]. The H20 multiplet at ≈ 4.60 ppm is characteristic of all the monacolins in lactone form including dihydromonacolins but not of monacolins in hydroxyl acid form and dehydromonacolins (Table 2, Figure 3). The H22 signal at ≈ 4.25 ppm is also characteristic of all the monacolins in lactone form (Table 2, Figure 3).

All these monacolin resonances mainly arise from MK, the main monacolin present in RYR DS [1] but other characteristic signals of monacolins were also identified: dehydromonacolins (H22 at 7.03 ppm), dihydromonacolins (H6 at 5.69 ppm) and monacolins in hydroxyl acid form (H22 at 4.05 ppm and H20 at 3.63 ppm) (Figure 1B for some signals).

The singlet of the H1 of citrinin at 5.94 ppm and 8.45 ppm (see footnote 4 of Table 2 for explanation) was never observed, which is not surprising due to the expected very low amount of this compound even if it could have been present [12,13].

Beside the assignments of the different monacolin families, ^{1}H characteristic signals of many other compounds mentioned or not on the label of the RYR DS were detected (Table 2). Fatty acids, both saturated (SFA) and non-conjugated unsaturated (UFA), were found in all the formulations (Figure 1 and Table 3) as it has been reported that they represent ≈ 3% of RYR extracts, each group in approximately identical proportion (≈1.4%) [1]. The presence of glycerol and glucose was observed in 25 and 24 samples, respectively. The singlet at ≈5.50 ppm characteristic of the H4 of monascin and/or other pigments with the same skeleton (ankaflavin, monascuspiloin, monaphilones A or B) (Figure 2) was observed in 20 formulations. Some other compounds, generally mentioned on the label of the RYR DS, were detected in few samples: sorbitol and piperine in two samples as well as carnitine, vitamin B3, vitamin C, chlorogenic acid and isopropyl alcohol in one sample (Tables 2 and 3).

Table 1. Red Yeast Rice (RYR) dietary supplements investigated in this study.

Number	Formulation Name (Origin [1])	Batch Number	Expiration Date	Form	RYR Extract Content Per Capsule or Tablet on the Label (mg)
1	Arkopharma (HFS)	C02194A	04/2016	Capsule	175
2	Arterin (I)	1983-1	07/2016	Tablet	670
3	B.concept (I)	13DN8	09/2016	Capsule	186
4	Belle & Bio (I)	B524A	06/2015	Capsule	250
5	Blue bonnet (I)	31201702	01/2017	Capsule	600
6	Boutique nature (I)	32443A	03/2016	Capsule	222.3
7	Doctor's best (I)	ML847	01/2017	Tablet	1200
8	Ephyto (I)	B1204059CZ	01/2015	Tablet	600
9	Fushi (I)	EPAN110412	08/2015	Capsule	600
10	Hanoju (I)	20130612	06/2016	Capsule	450
11	Health Ace (I)	106012W9	03/2015	Capsule	650
12	Health Spark (I)	1060012/A1	10/2014	Capsule	650
13	Liposterol (HFS)	03023	05/2016	Tablet	600
14	MRM (I)	130736	08/2016	Capsule	600
15	Nat et form (I)	14387J	10/2016	Capsule	600
16	Natrol (I)	2055200	10/2015	Tablet	400
17	Nature algues (I)	G0112112602 PF01G	11/2016	Capsule	250
18	Nature's plus (I)	1249540	09/2016	Tablet	600
19	Nature's way (I)	20012472	05/2016	Capsule	600
20	Naturland (HFS)	C02196A	04/2016	Capsule	175
21	Now (I)	16117750521	06/2015	Capsule	600
22	Nutrisanté (I)	961PAA	05/2016	Capsule	222.2
23	Pharma nature (HFS)	1009.211	12/2014	Capsule	300
24	Phytalessence (HFS)	10134	10/2016	Tablet	600
25	Phytoreponse (I)	B1210229CZ	10/2015	Tablet	600
26	Rizocol (I)	9545	06/2016	Tablet	335
27	Santé verte (HFS)	L066X2	03/2016	Tablet	600
28	Solaray (I)	171306	12/2016	Capsule	600
29	Solgar (I)	747369-02	07/2016	Capsule	600
30	Tradition soleil levant (I)	89853	07/2016	Capsule	208
31	Vit'All + (HFS)	A7248-11	11/2013	Tablet	600

[1] Dietary supplements bought on internet web sites (I) or in health food stores (HFS).

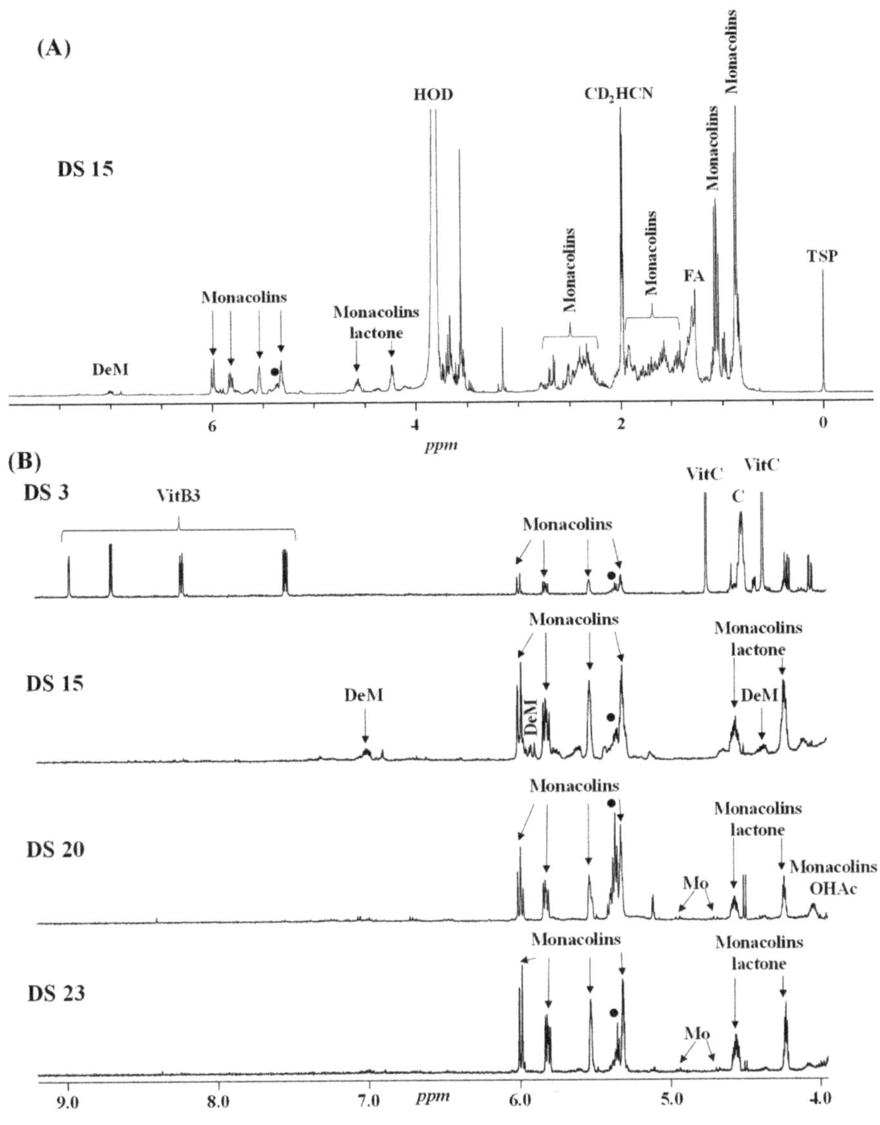

Figure 1. ^{1}H-NMR spectra of selected RYR dietary supplements (DS) recorded in $CD_3CN{:}D_2O$ (80:20). Entire spectrum of DS **15** (**A**) and enlarged downfield region (4–9 ppm) (**B**) of the DS **3**, **15**, **20** and **23**. DeM: dehydromonacolins, Monacolins lactone: monacolins in lactone form, Monacolins OHAc: monacolins in hydroxyl acid form, Mo: monascin and other pigments with the same skeleton, FA: fatty acids (saturated and unsaturated), ●: non-conjugated unsaturated fatty acids, C: carnitine, Vit: vitamin, TSP: sodium 2,2,3,3-tetradeutero-3-(trimethylsilyl) propanoate, set at 0 ppm.

Monacolins		R
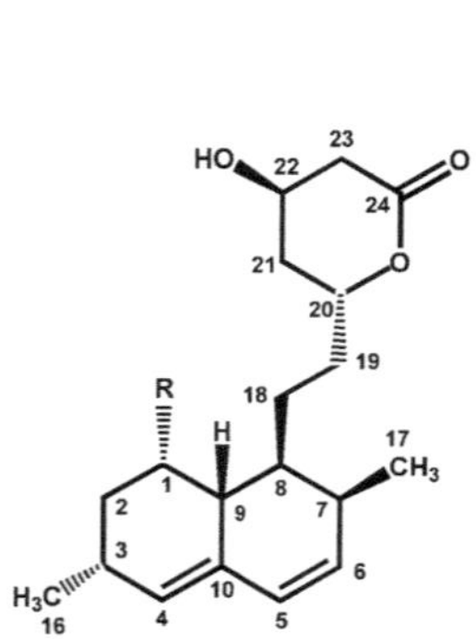	Monacolin K (MK) (lovastatin)	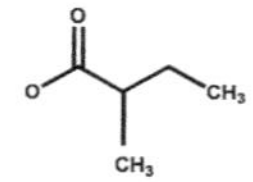
	Monacolin L (ML)	H
	Monacolin J (MJ)	OH
	Monacolin X (MX)	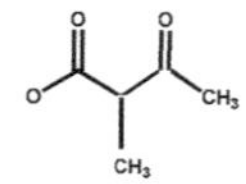
	Monacolin M (MM)	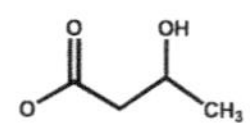
	Monacolin N (MN)	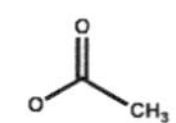
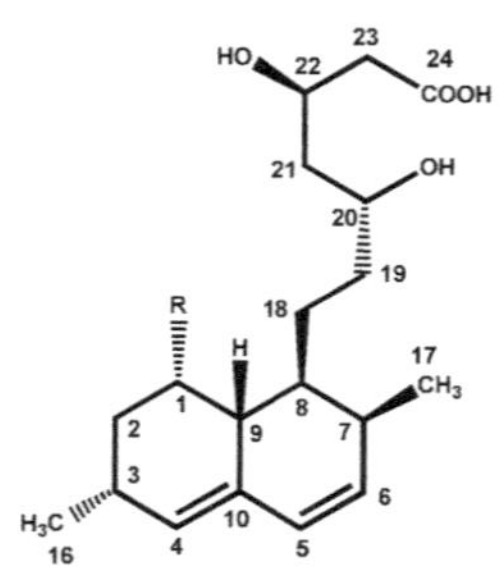	MK hydroxyl acid form (MKA)	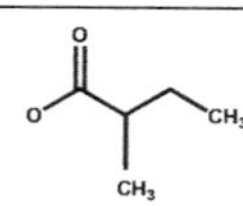
	ML hydroxyl acid form (MLA)	H
	MJ hydroxyl acid form (MJA)	OH

Azaphilone Pigments		R
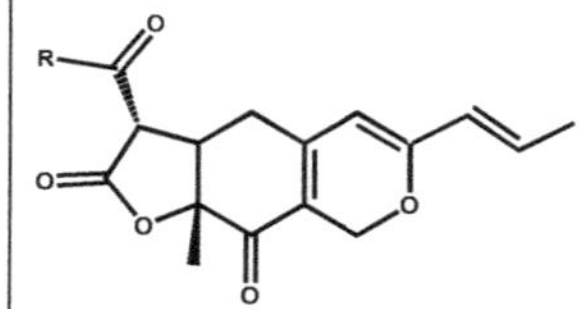	Monascin	n-C_5H_{11}
	Ankaflavin	n-C_7H_{15}
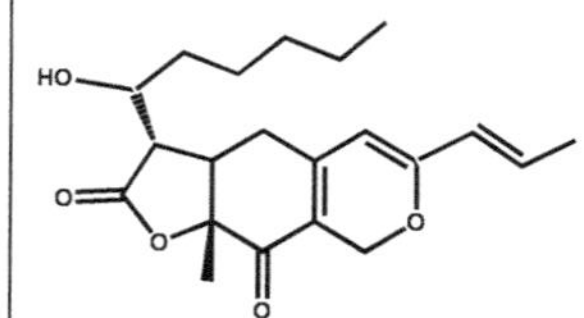	Monascuspiloin (dihydromonascin)	-
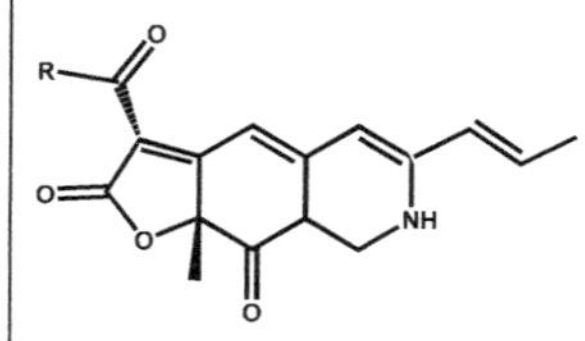	Rubropunctamine	n-C_5H_{11}
	Monascorubramine	n-C_7H_{15}
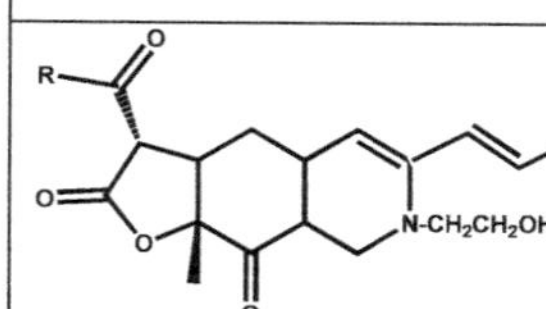	7-(2-hydroxyethyl)-monascorubramine (PP-R)	n-C_7H_{15}

Figure 2. *Cont.*

Structure	Compound	R	Structure	Compound	R
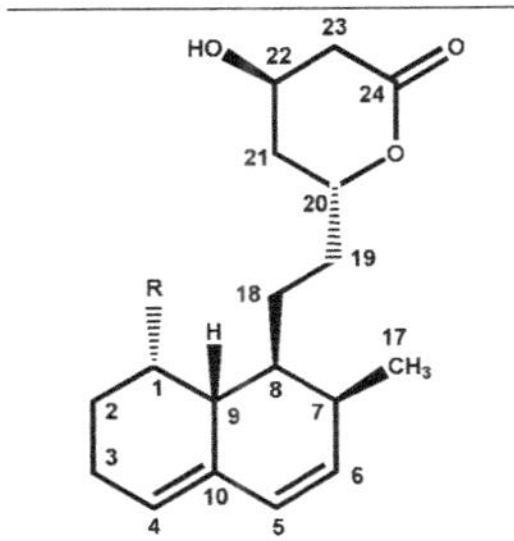	Compactin (CP) (mevastatin)	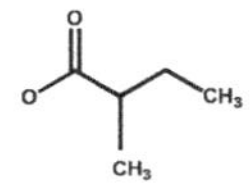	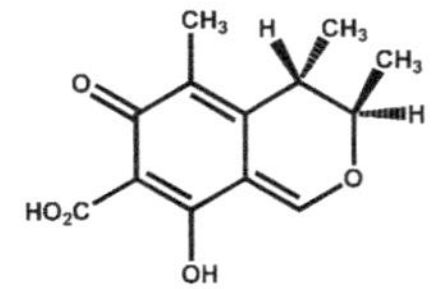	Citrinin	-
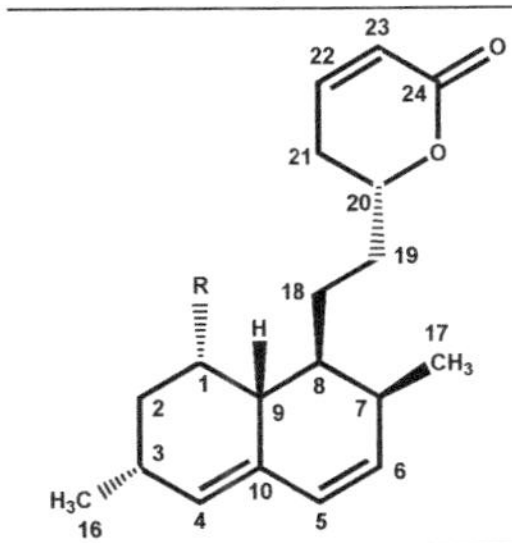	Dehydromonacolin K (DeMK)	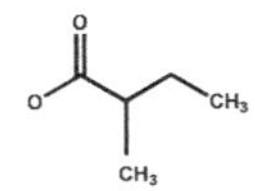	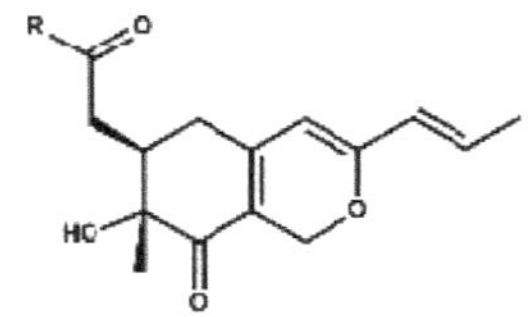	Monaphilone A	n-C_7H_{15}
	Dehydromonacolin L (DeML)	H		Monaphilone B	n-C_5H_{11}
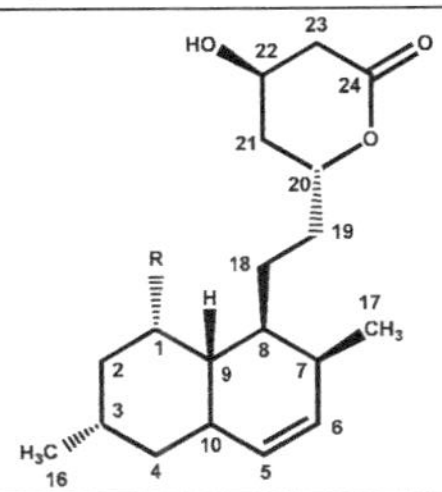	Dihydromonacolin K (DiMK)	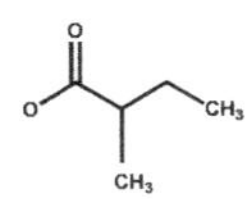	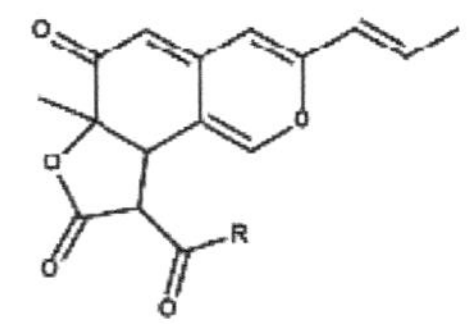	Monasfluore A	n-C_5H_{11}
				Monasfluore B	n-C_7H_{15}

Figure 2. Chemical structures of monacolins and pigments discussed in this study.

Table 2. Structures and ^{1}H-NMR characteristics (solvent: CD_3CN:D_2O 80:20) of standard monacolins and other compounds identified in this study.

Compound	Structure	^{1}H-NMR [1] δ (ppm) (Multiplicity [2], *J* (Hz), Number of Protons, Attribution)
Monacolin K lactone form (MK)	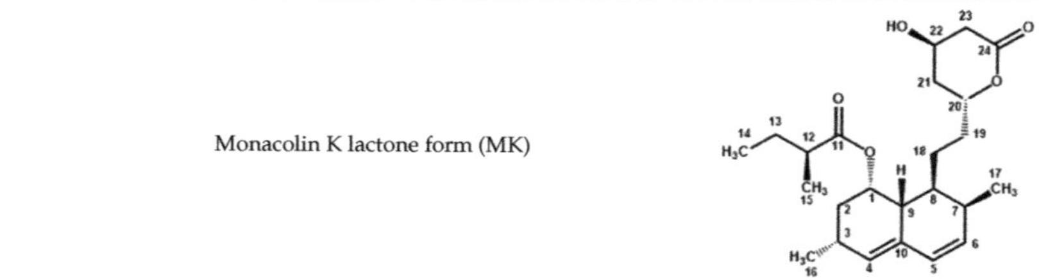	6.01 (d, *J* = 9.6, 1H, H-5), 5.84 (dd, *J* = 6.1, 9.6, 1H, H-6), 5.56 (app t, *J* = 2.8, 1H, H-4), 5.35 (q, *J* = 3.2, 1H, H-1), 4.59 (m, 1H, H-20), 4.25 (app quint, *J* = 3.9, 1H, H-22), 2.69 (Ad, *J* = 4.9, 17.6, 1H, H-23), 2.51 (Bdd, *J* = 1.7, 3.8, 17.6, 1H, H-23), 2.45 (m, 1H, H-3), 2.42 (m, 1H, H-7), 2.37 (m, 1H, H-9), 2.35 (m, 1H, H-12), 1.96 (m, 2H, H-2), 1.90 and 1.71 (two m, 2H, H-21), 1.81 and 1.37 (two m, 2H, H-19), 1.69 (m, 1H, H-8), 1.62 (app qd, *J* = 7.4, 13.6, 1H, H-13), 1.46 (m, 2H, H-13 and H-18), 1.36 (m, 1H, H-18), 1.08 (d, *J* = 6.9, 3H, H-15), 1.06 (d, *J* = 7.4, 3H, H-16), 0.89 (d, *J* = 6.9, 3H, H-17), 0.88 (t, *J* = 7.5, 3H, H-14)
Monacolin K hydroxyl acid form (MKA)	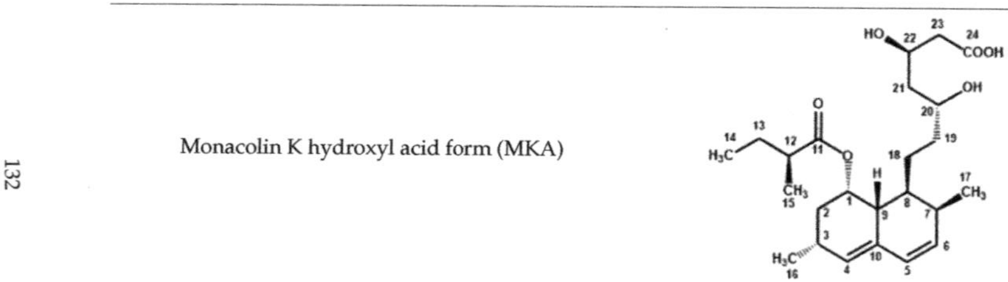	5.99 (d, *J* = 9.6, 1H, H-5), 5.83 (dd, *J* = 6.1, 9.6, 1H, H-6), 5.53 (app t, *J* = 2.8, 1H, H-4), 5.33 (q, *J* = 3.2, 1H, H-1), 4.05 (m, 1H, H-22), 3.63 (app hept, *J* = 4.1, 1H, H-20), 2.42 (m, 1H, H-3), 2.39 (m, 1H, H-7), 2.35 (m, 1H, H-9), 2.33 (m, 2H, H-12 and H-23), 2.16 (dd, *J* = 8.7, 15.2, 1H, H-23), 1.93 (m, 2H, H-2), 1.63 (m, 1H, H-8), 1.59 and 1.45 (two m, 2H, H-13), 1.57 and 1.51 (two m, 2H, H-21), 1.53 and 1.15 (two m, 2H, H-19), 1.32 (m, 2H, H-18), 1.08 (d, *J* = 6.9, 3H, H-15), 1.05 (d, *J* = 7.4, 3H, H-16), 0.87 (d, *J* = 6.9, 3H, H-17), 0.86 (t, *J* = 7.4, 3H, H-14)
Compactin (CP) = Mevastatin	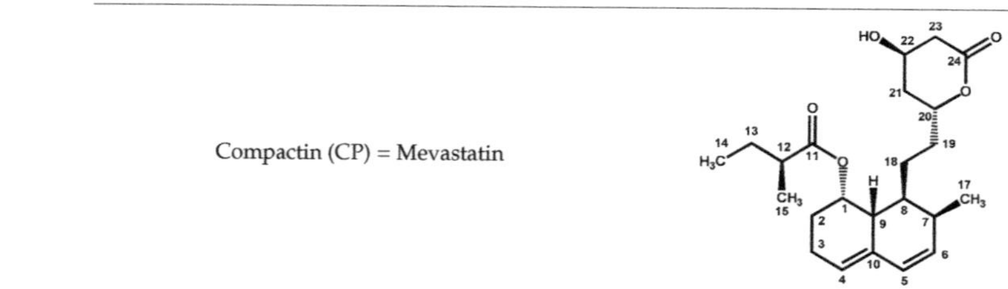	6.00 (d, *J* = 9.7, 1H, H-5), 5.79 (dd, *J* = 6.0, 9.7, 1H, H-6), 5.57 (m, 1H, H-4), 5.30 (m, 1H, H-1), 4.60 (m, 1H, H-20), 4.25 (app quint, *J* = 3.9, 1H, H-22), 2.69 (Ad, *J* = 4.8, 17.6, 1H, H-23), 2.51 (Bdd, *J* = 1.7, 3.6, 17.6, 1H, H-23), 2.42 (m, 2H, H-7 and H-9), 2.38 (m, 1H, H-12), 2.15 (m, 2H, H-3), 2.08 (m, 1H, H-2), 1.90 (m, 1H, H-21), 1.81 and 1.37 (two m, 2H, H-19), 1.72 (m, 2H, H-2 and H-21), 1.68 (m, 1H, H-8), 1.62 (app qd, *J* = 7.6, 13.6, 1H, H-13), 1.48 and 1.39 (two m, 2H, H-18), 1.46 (m, 1H, H-13), 1.11 (d, *J* = 7.0, 3H, H-15), 0.90 (d, *J* = 7.0, 3H, H-17), 0.89 (t, *J* = 7.5, 3H, H-14)

Table 2. *Cont.*

Compound	Structure	^{1}H-NMR [1] δ (ppm) (Multiplicity [2], *J* (Hz), Number of Protons, Attribution)
Dehydromonacolin K (DeMK)	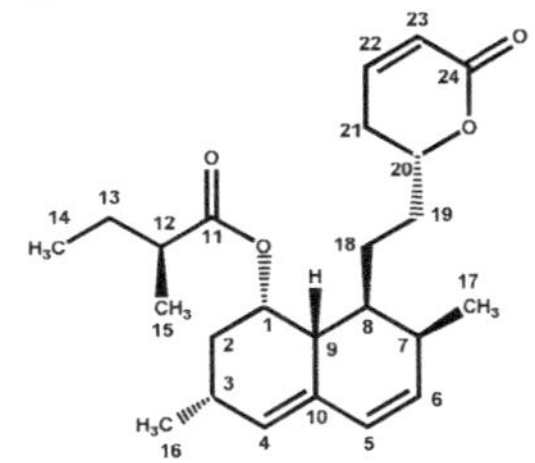	7.03 (ddd, *J* = 2.5, 6.0, 9.8, 1H, H-22), 6.02 (d, *J* = 9.6, 1H, H-5), 5.97 (ddd, *J* = 1.0, 2.7, 9.8, 1H, H-23), 5.84 (dd, *J* = 6.1, 9.6, 1H, H-6), 5.56 (app t, *J* = 2.8, 1H, H-4), 5.33 (q, *J* = 3.2, 1H, H-1), 4.42 (m, 1H, H-20), 2.44 (m, 1H, H-3), 2.43 (m, 1H, H-7), 2.42 and 2.30 (two m, 2H, H-21), 2.36 (m, 1H, H-9), 2.35 (m, 1H, H-12), 1.96 (m, 2H, H-2), 1.90 and 1.41 (two m, 2H, H-19), 1.70 (m, 1H, H-8), 1.62 (app qd, *J* = 7.5, 13.8, 1H, H-13), 1.46 (m, 1H, H-13), 1.45 and 1.39 (two m, 2H, H-18), 1.09 (d, *J* = 6.9, 3H, H-15), 1.07 (d, *J* = 7.4, 3H, H-16), 0.89 (d, *J* = 6.9, 3H, H-17), 0.88 (t, *J* = 7.5, 3H, H-14)
Dihydromonacolin K (DiMK)	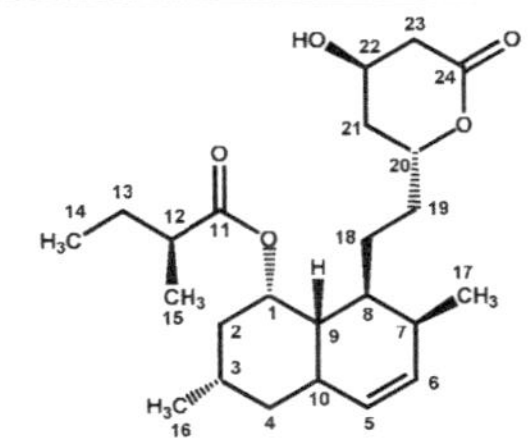	5.69 (ddd, *J* = 2.7, 5.0, 9.8, 1H, H-6), 5.42 (d, *J* = 9.8, 1H, H-5), 5.15 (q, *J* = 2.7, 1H, H-1), 4.58 (m, 1H, H-20), 4.25 (app quint, *J* = 3.8, 1H, H-22), 2.68 (Ad, *J* = 4.9, 17.7, 1H, H-23), 2.50 (Bdd, *J* = 1.7, 3.7, 17.7, 1H, H-23), 2.46 (m, 1H, H-10), 2.39 (m, 1H, H-12), 2.34 (m, 1H, H-7), 2.05 (m, 1H, H-3), 1.88 and 1.72 (two m, 2H, H-21), 1.81 and 1.37 (two m, 2H, H-19), 1.80 (m, 2H, H-2), 1.66 (m, 1H, H-8), 1.64 (m, 1H, H-13), 1.61 and 1.35 (two m, 2H, H-4), 1.49 (app qd, *J* = 7.4, 13.6, 1H, H-13), 1.35 and 1.30 (two m, 2H, H-18), 1.12 (d, *J* = 7.0, 3H, H-15), 1.11 (d, *J* = 7.6, 3H, H-16), 0.90 (t, *J* = 7.5, 3H, H-14), 0.86 (d, *J* = 7.0, 3H, H-17)
Monascin [3]	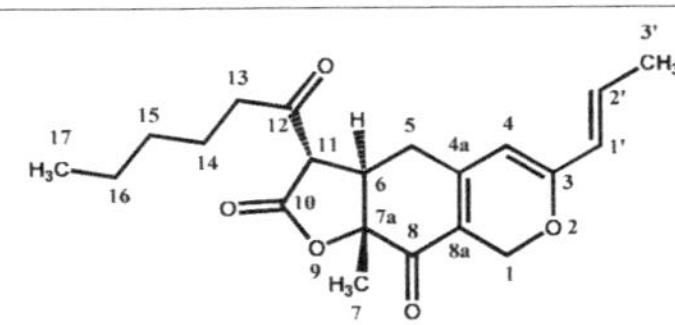	6.50 (qd, *J* = 7.0, 15.5, 1H, H-2′), 6.05 (qd, *J* = 1.7, 15.5, 1H, H-1′), 5.51 (s, 1H, H-4), 4.97 (At, *J* = 0.9, 12.7, 1H, H-1), 4.72 (Bt, *J* = 1.4, 12.7, 1H, H-1), 4.14 (d, *J* = 13.3, 1H, H-11), 3.15 (m, 1H, H-6), 2.88 (At, *J* = 7.3, 18.2, 1H, H-13), 2.67 (Bt, *J* = 7.3, 18.2, 1H, H-13), 2.63 (app br d, *J* = 7.7, 2H, H-5), 1.87 (dd, *J* = 1.7, 7.0, 3H, H-3′), 1.59 (quint, *J* = 7.3, 2H, H-14), 1.43 (s, 3H, H-7), 1.32 (m, 4H, H-15, H-16), 0.91 (t, *J* = 7.2, 3H, H-17)
Citrinin [4]	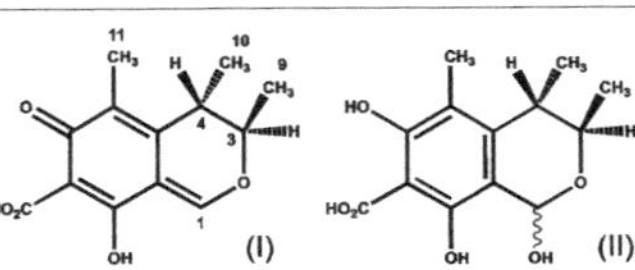	8.45 (br s, H-1 (I)), 5.94 (s, H-1 (II major), 5.90 (br s, H-1 (II minor)), 4.92 (br signal, H-3 (I)), 4.09 (quint, *J* = 6.6, H-3 (II major and minor)), 3.15 (br signal, H-4 (I)), 2.76 (br signal, H-4 (II minor)), 2.67 (quint, *J* = 6.8, H-4 (II major)), 2.03 (s, H-11 *), 2.01 (s, H-11 *), 2.00 (s, H-11 *), 1.30 (d, *J* = 6.7, H-9 *), 1.27 (d, *J* = 6.4, H-9 *), 1.20 (d, *J* = 6.9, H-10 *), 1.17 (very br signal, H-10 *)

Table 2. *Cont.*

Compound	Structure	^{1}H-NMR [1] δ (ppm) (Multiplicity [2], J (Hz), Number of Protons, Attribution)
Piperine	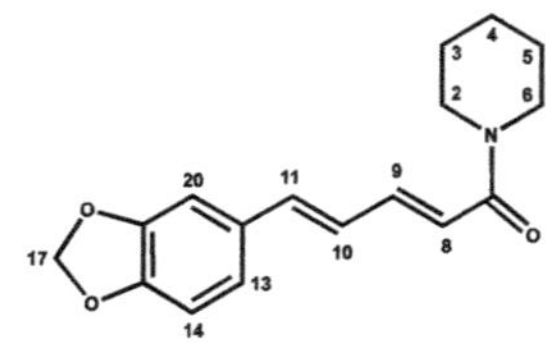	7.28 (ddd, J = 2.7, 7.3, 14.7, 1H, H-9), 7.11 (d, J = 1.6, 1H, H-20), 6.99 (dd, J = 1.7, 8.1, 1H, H-13), 6.87 (d, J = 8.0, 1H, H-14), 6.83 (m, 2H, H-10, H-11), 6.60 (d, J = 14.7, 1H, H-8), 6.00 (s, 2H, H-17), 3.55 (m, 4H, H-2, H-6), 1.65 (m, 2H, H-4), 1.56 (m, 4H, H-3, H-5)
Carnitine	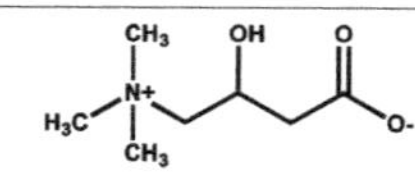	4.56 (br q, J = 7.1, 1H, CH-OH), 3.39 (m, 2H, CH_2-N^+), 3.17 (s, 9H, $(CH_3)_3$-N^+), 2.58 (m, 2H, CH_2-COO^-)
Chlorogenic acid	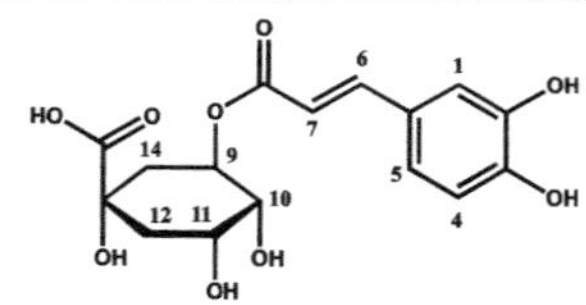	7.61 (d, J = 15.9, 1H, H-6), 7.16 (d, J = 2.0, 1H, H-1), 7.06 (dd, J = 2.0, 8.2, 1H, H-5), 6.88 (d, J = 8.2, 1H, H-4), 6.34 (d, J = 15.9, 1H, H-7), 5.28 (ddd, J = 4.6, 9.4, 10.5, 1H, H-9), 4.18 (app q, J = 3.5, 1H, H-10), 3.75 (dd, J = 3.2, 9.4, 1H, H-11), 2.26–1.97 (m, 4H, H-12, H-14)
Saturated fatty acids	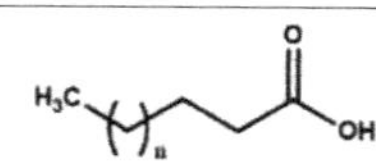	2.20 (t, J = 7.5, 2H, CH_2-COOH), 1.59 (m, 2H, CH_2-CH_2-COOH), 1.30 (m, 2nH, $(CH_2)n$), 0.90 (t, J = 6.3, 3H, CH_3)
α/β Glucose	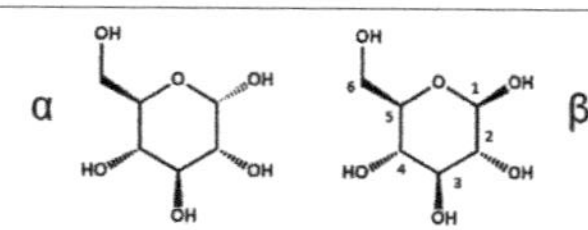	5.14 (d, J = 3.7, 1H, H-1α), 4.52 (d, J = 7.9, 1H, H-1β), 3.85-3.60 (m, 7H, H-3α, H-4α/β, H-6α/β), 3.45–3.28 (m, 4H, H-2α, H-3β, H-5α/β), 3.15 (t, J = 8.5, 1H, H-2β)

Table 2. *Cont.*

Compound	Structure	^{1}H-NMR [1] δ (ppm) (Multiplicity [2], J (Hz), Number of Protons, Attribution)
Glycerol	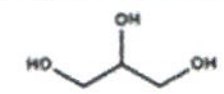	3.66 (tt, J = 4.6, 6.3, 1H, CH), 3.55 (Ad, J = 4.6, 11.5, 2H, CH_2), 3.48 (Bd, J = 6.2, 11.5, CH_2)
Isopropyl alcohol	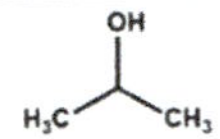	3.93 (hept, J = 6.1, 1H, CH), 1.14 (d, J = 6.1, 6H, CH_3)
Linoleic acid [5]	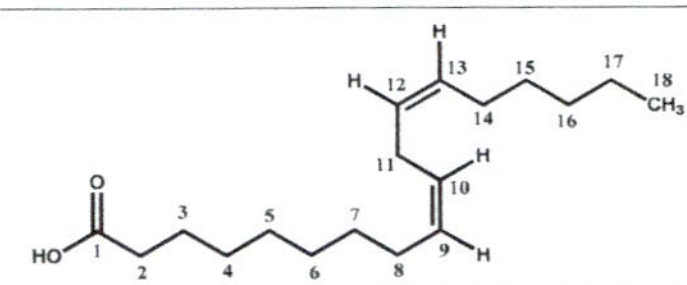	5.36 (m, 4H, H-9, H-10, H-12, H-13), 2.78 (t, J = 6.7, 2H, H-11), 2.27 (t, J = 7.5, 2H, H-2), 2.06 (m, 4H, H-8, H-14), 1.56 (m, 2H, H-3), 1.40-1.27 (m, 14H, H-4, H-5, H-6, H-7, H-15, H-16, H-17), 0.89 (t, J = 6.8, 3H, H-18)
Sorbitol	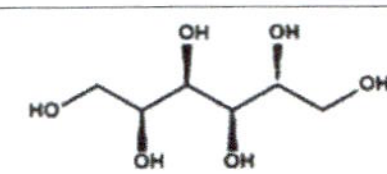	3.82–3.54 (m, 8H)
Vitamin B3 (niacinamide form)	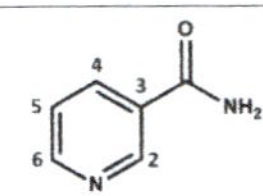	8.99 (d, J = 2.3, 1H, H-2), 8.72 (dd, J = 4.9, 1.6, 1H, H-6), 8.25 (td, J = 8.0, 1.9, 1H, H-4), 7.56 (dd, J = 8.0, 4.9, 1H, H-5)
Vitamin C	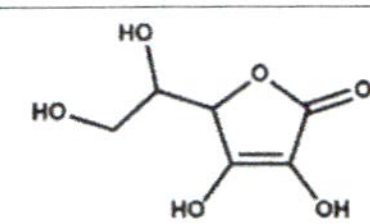	4.76 (d, X part of an ABMX system, J_{XM} = 2.0, 1H, CH-O-), 3.94 (td, M part of an ABMX system, J_{MX} = 1.9, J_{MA} = J_{MB} = 6.6, 1H, CH-OH), 3.66 (AB part of an ABMX system, J_{AB} = 11.3, J_{AM} = J_{BM} = 6.6, 2H, CH_2-OH)

[1] The assignments of ^{1}H-NMR signals of monacolins, citrinin, and monascin were in agreement with the literature data even if the solvents were different: $CDCl_3$ for MK, MKA, CP, DeMK, DiMK and citrinin [1,20–24], DMSO-d6 for monascin [25] and D_2O for citrinin [22]. [2] d: doublet, dd: doublet of doublet, ddd: doublet of doublet of doublet, t: triplet, td: doublet of triplet, tt: triplet of triplet, q: quadruplet, qd: doublet of quadruplet, quint: quintuplet, hept: heptuplet, m: multiplet, A: part A of an AB system, B: part B of an AB system, app: apparent, br: broad. [3] H11 is exchanged with D and its resonance disappears rapidly with time as well as its coupling with H6 whose multiplet becomes a triplet (J = 7.9 Hz). [4] The attributions were done according to the literature [22]. Citrinin exists as the quinone methide (I) in organic NMR solvents but as a diastereoisomeric mixture of hydrates (II) in aqueous solution at physiological pH, one diastereoisomer being major and the other minor. As the solvent used in the present study is a CD_3CN:D_2O (80:20) mixture, the two forms (I) and (II) are observed. * From the literature data, it was not possible to assign the resonances of H9, H10 and H11 to a specific form. [5] As a model of non-conjugated unsaturated fatty acids.

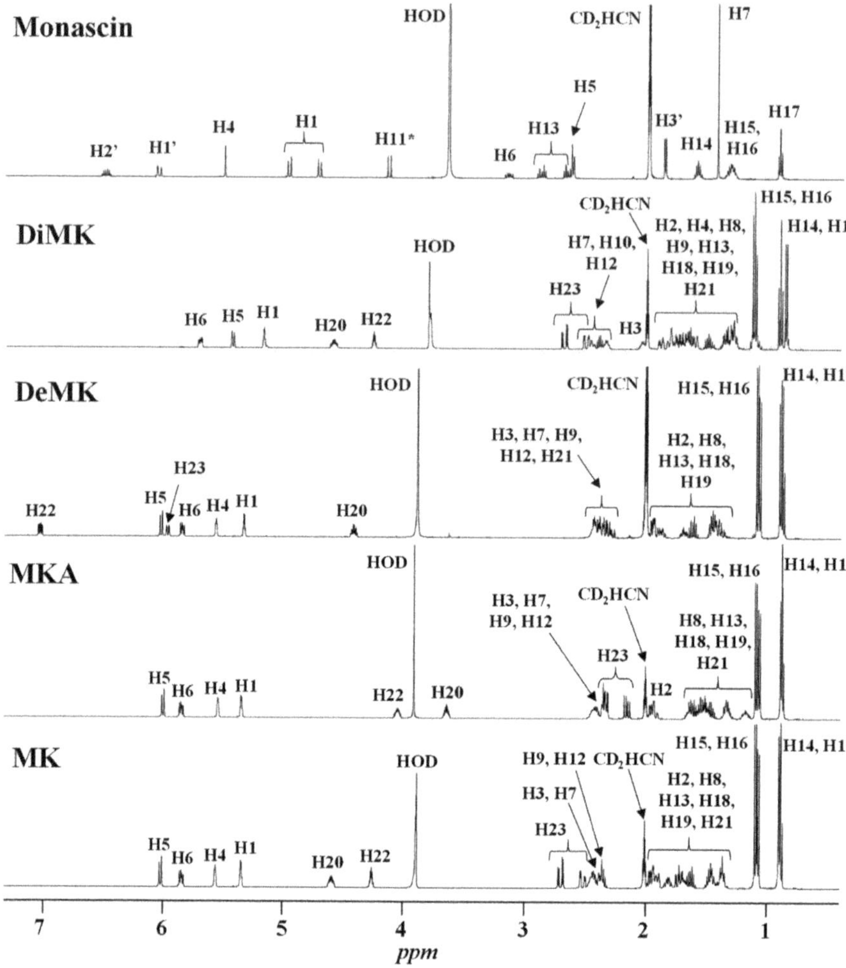

Figure 3. ^{1}H-NMR spectra of standards of monacolin K in lactone form (MK) and in hydroxyl acid form (MKA), dehydromonacolin K (DeMK), dihydromonacolin K (DiMK) and monascin recorded in CD_3CN:D_2O (80:20). The chemical structures of all the compounds and their protons numbering are given in Table 2. (*) The signal of H11 of monascin disappears with time due to exchange with D_2O.

2.1.2. Quantitative ^{1}H-NMR Analysis

With our experimental conditions, the limit of detection (LOD; for a signal-to-noise ratio (SNR) = 3) of monacolin derivatives in real samples was estimated between 4×10^{-5} and 10^{-4} M depending on the multiplicity of the targeted proton signal, corresponding respectively to 0.016 and 0.04 mg of monacolins (considered with the molecular weight of MK (404 g mol^{-1})) in the volume of solvent used for the extraction of 20 to 100 mg of RYR formulation. The limit of quantification (LOQ; SNR = 10) for the same signals was evaluated between 10^{-4} and 3×10^{-4} M, corresponding respectively to 0.04 and 0.12 mg of monacolin derivatives in the amount of powdered RYR extracted. Considering the ratio between the mass of sample extracted and that of the formulation (tablet or powder in the capsule), the LOQ was estimated to be ≈0.25 mg of monacolin per capsule or tablet.

Table 3. ^{1}H-NMR analysis of RYR dietary supplements: quantitative determination of monacolins (mg per capsule or tablet) [1] and list of other compounds identified.

DS Number	Signal at 5.84 ppm [2] H6	Signal at 5.56 ppm [2] H4	Mean Intensity of Signals H6 and H4	Signal at 5.33 ppm [2] H1	Signal at 4.60 ppm [2] H20	Signal at 4.25 ppm [2] H22	Compounds Identified Other than Monacolins [3,4]
1	1.67	1.70	1.685	3.31	1.40	1.34	SFA *, UFA, glycerol, glucose, monascin
2	9.14	9.18	9.16	9.26		7.10	SFA *, UFA, glycerol, glucose
3	3.07	3.07	3.07	2.39			SFA *, UFA, carnitine *, vitamin B3 *, vitamin C *
4	1.02	0.95	0.985	1.64		0.71	SFA, UFA, glucose
5	2.47	2.53	2.50	11.3	2.15		SFA *, UFA, glycerol, glucose, monascin
6	2.92	2.63	2.775	3.27		2.19	SFA *, UFA, glucose, monascin
7	3.76	3.73	3.745				SFA *, UFA, glucose, piperine *
8	8.68	8.79	8.735	8.65	6.89	7.73	SFA, UFA, glycerol, monascin
9	5.14	5.21	5.125	8.03	4.55	4.55	SFA *, UFA, glycerol, glucose, monascin
10	2.10	2.04	2.07	8.67		1.22	SFA, UFA, glycerol, glucose, monascin
11							SFA, UFA, glycerol, glucose, monascin
12							SFA, UFA, glycerol, glucose, monascin
13	23.9	23.9	23.9	21.9	23.9	23.2	SFA *, UFA, glycerol, glucose
14	3.45	3.28	3.365	10.8		2.26	SFA, UFA, glycerol, glucose, monascin
15	11.9	12.3	12.1		9.85		SFA *, UFA, glycerol, monascin
16		0.31					SFA *, UFA, glycerol, glucose, monascin, piperine *
17	1.47	1.51	1.49	4.06	1.28		SFA, UFA, glycerol, monascin
18	8.84	8.19	8.515		6.68	7.04	SFA *, UFA, glycerol, glucose, monascin
19	0.36	0.44	0.40				SFA *, UFA, glycerol, glucose, monascin
20	1.75	1.77	1.76	3.51	1.27	1.49	SFA *, UFA, glycerol, glucose, monascin
21	0.48	0.47	0.475				SFA *, UFA, glycerol

Table 3. *Cont.*

DS Number	Signal at 5.84 ppm [2] H6	Signal at 5.56 ppm [2] H4	Mean Intensity of Signals H6 and H4	Signal at 5.33 ppm [2] H1	Signal at 4.60 ppm [2] H20	Signal at 4.25 ppm [2] H22	Compounds Identified Other than Monacolins [3,4]
22	3.09	3.02	3.055			2.26	SFA *, UFA, glycerol, glucose, monascin
23	10.6	10.7	10.65	10.0	10.0	9.77	SFA *, UFA, glycerol, glucose, monascin
24	11.5	11.8	11.65	9.38		5.97	SFA *, UFA, glucose, sorbitol *
25	10.0	9.72	9.87		5.97	7.26	SFA, UFA, glycerol, monascin
26	4.51	4.54	4.525			3.88	SFA *, UFA, glycerol, glucose, sorbitol
27	4.13	3.84	3.985		2.97	2.90	SFA, UFA, glycerol, glucose, monascin, chlorogenic acid *
28	1.64	1.75	1.695		1.05	1.31	SFA *, UFA, glycerol, glucose
29							SFA *, UFA, glycerol, glucose
30	2.99	3.04	3.015	2.61		2.43	SFA, UFA
31	1.21	1.28	1.245	2.16			SFA *, UFA, glycerol, glucose, monascin, isopropanol

[1] The amounts of monacolins (mg per dosage unit) were calculated from the measured areas using the equation presented in the paragraph 3.4.2. [2] The resonances at 5.84, 5.56 ppm and 5.33 ppm are characteristic of all the monacolins bearing an hexahydronaphthalene ring but the 5.33 ppm signal does not characterize MJ and ML in lactone or in hydroxyl acid form; the resonances at 4.60 and 4.25 ppm are specific of all the monacolins in lactone form including dihydromonacolins. [3] SFA: saturated fatty acids; UFA: non-conjugated unsaturated fatty acids; monascin corresponds to monascin and other pigments with the same skeleton (see Table 2 and Figure 2 for the chemical structures of all these compounds). [4] The asterisk * means that the compound was mentioned on the label of the formulation.

The H5 resonance (6.01 ppm) was not used for the quantification of monacolins as it may also contain part of the signal (ddd) of the H23 of DeMK at 5.97 ppm and can be slightly overlapped with the H1′ signal (qd) of monascin at 6.05 ppm. The H6 (5.84 ppm) and H4 (5.56 ppm) resonances are not hindered by signals of other molecules and they have been quantified respectively in 27 and 28 RYR DS as no monacolin signal was detected in formulations **11**, **12** and **29** whereas both resonances were observed in DS **16** where only the H4 signal could be quantified with an intensity at the LOQ level. The amount of monacolins in lactone form was determined from either H20 or H22 or both signals in only 22 RYR DS because of significant overlap with matrix signals in some formulations. It was possible to quantify the H1 signal (5.33 ppm) in only 17 formulations due to its overlap with the large resonance of the ethylenic protons of non-conjugated UFA at ≈5.36 ppm and maybe also of other compounds. The 2D ^{1}H-^{13}C HSQC-NMR experiment recorded for the DS **15** and illustrated in Figure 4 clearly shows these findings.

The intensity of the DeMK ddd resonance (H22 at 7.03 ppm) was low or very low and overlapped with matrix signals in most sample spectra and could not thus be quantified precisely. The dd resonance of DiMK at 5.69 ppm was not clearly detected in any formulation and could not thus be quantified. The signal intensities of monacolins in hydroxyl acid form (H22 at 4.05 ppm and H20 at 3.63 ppm) would have allowed their quantification in most of the formulations but their substantial overlap with the resonances of matrix compounds did not permit to obtain accurate values without curve deconvolution treatment.

The monacolin contents determined by ^{1}H-NMR are reported in Table 3. For more clarity, the coefficient of variation (CV) and the relative standard deviations (RSD) are given in Table S2. The RSD of the ^{1}H-NMR assays ranged between 1.5% and 13%. The relevance of the ^{1}H-NMR data will be discussed later when compared to those obtained by UHPLC.

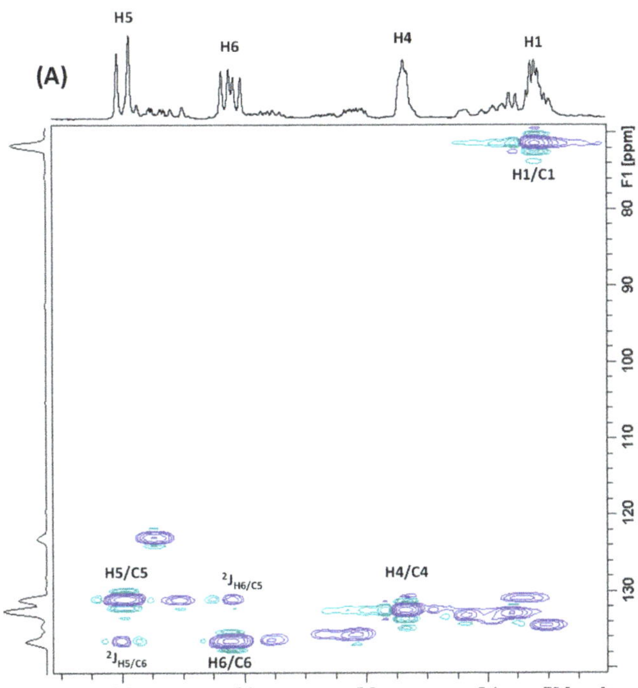

Figure 4. *Cont.*

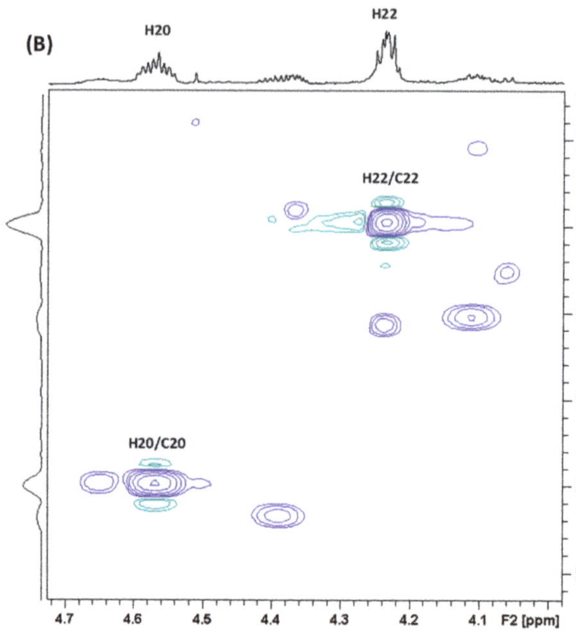

Figure 4. 2D ^{1}H-^{13}C HSQC-NMR spectrum of the dietary supplement **15** recorded in CD_3CN:D_2O (80:20). (**A**) ^{1}H enlarged region 5.2–6.1 ppm. (**B**) ^{1}H enlarged region 3.98–4.72 ppm. The proton numbering of monacolins is given in Table 2.

2.2. Chromatographic Analysis

2.2.1. Identification of Monacolins and Pigments

RYR formulations were traced by UHPLC-DAD-MS. Among the great number of compounds observed, 23 (12 monacolins and 11 azaphilones; see Figure 2 for their chemical structures) were identified by comparing with the literature their elution sequence, their UV-Vis and MS profiles as well as the accurate mass measurements of their parent and fragment ions in High Resolution Electrospray Ionization (HR-ESI) MS and MS/MS [13,26]. Moreover, the identification of MK, CP, MKA, DeMK, DiMK, citrinin and monascin was confirmed by comparing their UV-Vis and MS and MS/MS chromatographic profiles with those of the corresponding standards. The retention time (tR), UV-Vis λ_{max}, accurate masses of parent ion and of its MS/MS major fragment ions for each compound identified are gathered in Table 4. We will not describe the process that allowed the identification of the 23 compounds but we will show the respective contributions of UV-Vis and HR-MS and MS/MS to determine their chemical structure through some selected examples.

The UV-Vis absorption spectra gave some indication on the chemical structure of the compounds detected. For example, all the monacolins with a conjugated hexahydronaphthalene ring displayed the same characteristic mountain-like UV spectrum with three maximum absorptions at $\approx$ 230, 238 and 246 nm (range 229–230, 238–239 and 244–247 nm) (Table 4) in accordance with literature data [1,10,27,28]). Due to the absence of conjugated double bonds, the dihydromonacolins with an octahydronaphthalene ring and so an UV maximum absorption band at $\approx$210 nm [1,20] were not detected at the wavelength used in this study (238 nm). In the same way, the yellow azaphilone pigments with the classical monascin-type chromophore showed specific profiles with high intensity peaks at 227–233 and 386–392 nm and a very low intensity band at 285–291 nm, while the red ones with the rubropunctamine-type chromophore presented specific absorbances at 251–252, 302–307, 412–423 and 525–530 nm (Table 4), all these values being in agreement with literature reports [29–34].

Table 4. Retention times, UV-Vis characteristics and accurate mass measurements of the compounds observed and identified from the RYR dietary supplements analyzed with UHPLC.

Compound Name	Retention Time (min)	UV-Vis λ_{max} (nm)	Formula (Monoisotopic Mass)	Calculated Mass (m/z)		Measured Mass (m/z)	Mass Error (mDa)	Relative Mass Error (ppm)	Major ions of $[M + H]^+$ MS/MS Fragmentation Unless Otherwise Indicated (m/z)
Monacolin J hydroxyl acid (MJA)	0.90	230, 239, 244	$C_{19}H_{30}O_5$ (338.2172)	$[M - H]^-$	337.2015	337.2010	−0.5	−1.5	Not recorded
Citrinin	1.45	243, 326 [1]	$C_{13}H_{14}O_5$ (250.0841)	$[M + H]^+$	251.0919	251.0922	+0.3	+1.2	Not recorded
Monacolin J (MJ)	1.98	230, 239, 246	$C_{19}H_{28}O_4$ (320.1988)	$[M + H]^+$ $[M + Na]^+$ $[M - H]^-$	321.2066 343.1885 319.1909	321.2067 343.1881 319.1910	+0.1 −0.4 +0.1	+0.3 −1.2 +0.3	303.1945, 285.1847, 267.1711, 205.1604, 199.1451, 159.1162
Rubropunctamine	2.70	252, 303, 415, 525	$C_{21}H_{23}NO_4$ (353.1627)	$[M + H]^+$ $[M + Na]^+$ $[M - H]^-$	354.1705 376.1525 352.1549	354.1710 376.1521 352.1551	+0.5 −0.4 +0.2	+1.4 −1.1 +0.6	337.1308, 294.0762, 267.0523, 256.0866, 239.0580
Monacolin N (MN)	3.89	230, 239, 247	$C_{21}H_{30}O_5$ (362.2093)	$[M + H]^+$ $[M + Na]^+$	363.2171 385.1991	363.2159 385.1988	−1.2 −0.3	−3.3 −0.8	345.2052, 285.1865, 267.1750, 243.1711, 199.1490, 173.1335, 159.1173, 143.0856
Monacolin L hydroxyl acid (MLA)	4.05	230, 239, 247	$C_{19}H_{30}O_4$ (322.2144)	$[M + H]^+$ $[M + Na]^+$ $[M - H]^-$	323.2222 345.2042 321.2066	323.2220 345.2037 321.2072	−0.2 −0.5 +0.6	−0.6 −1.5 +1.9	305.2126, 287.1986, 269.1901, 225.1640, 203.1800, 159.1172
Monacolin X (MX)	4.38	230, 239, 246	$C_{24}H_{34}O_6$ (418.2355)	$[M + H]^+$ $[M + Na]^+$	419.2434 441.2253	419.2434 441.2250	0 −0.3	0 −0.7	303.1961, 285.1855, 267.1749, 243.1751, 225.1643, 199.1489, 173.1331, 159.1173, 143.0712
7-(2-hydroxyethyl)-monascorubramine (PP-R)	4.66	251, 302, 423, 530	$C_{25}H_{31}NO_5$ (425.2202)	$[M + H]^+$ $[M + Na]^+$	426.2280 448.2100	426.2265 448.2092	−1.5 −0.8	−3.5 −1.8	Not recorded
Monascorubramine	5.55	251, 307, 412, 530	$C_{23}H_{27}NO_4$ (381.1940)	$[M + H]^+$ $[M + Na]^+$ $[M - H]^-$	382.2018 404.1838 380.1862	382.2018 404.1842 380.1867	0 +0.4 +0.5	0 +1.0 +1.3	365.1624, 294.0765, 267.0528, 250.0870, 239.0582
Monacolin K hydroxyl acid (MKA)	5.64	229, 238, 246	$C_{24}H_{38}O_6$ (422.2669)	$[M + H]^+$ $[M + Na]^+$ $[M - H]^-$ $[M - H + CO]^-$	423.2747 445.2566 421.2590 449.2539	ND [2] 445.2561 421.2589 449.2540	−0.5 −0.1 +0.1	−1.1 −0.2 +0.2	MS/MS $[M - H]^-$ 319.1902, 101.0602, 85.0286
Monascuspiloin (dihydromonascin)	5.92	233, 291, 390	$C_{21}H_{28}O_5$ (360.1937)	$[M + H]^+$ $[M + Na]^+$	361.2015 383.1834	361.2009 383.1836	−0.6 +0.2	−1.7 +0.5	345.2045, 300.2883, 261.1133, 215.1081, 187.1134

Table 4. *Cont.*

Compound Name	Retention Time (min)	UV-Vis λ_{max} (nm)	Formula (Monoisotopic Mass)	Calculated Mass (*m*/*z*)		Measured Mass (*m*/*z*)	Mass Error (mDa)	Relative Mass Error (ppm)	Major ions of $[M + H]^+$ MS/MS Fragmentation Unless Otherwise Indicated (*m*/*z*)
Monacolin L (ML)	6.09	230, 238, 246	$C_{19}H_{28}O_3$ (304.2039)	$[M + H]^+$ $[M + Na]^+$	305.2117 327.1936	305.2116 327.1935	−0.1 −0.1	−0.3 −0.3	287.2004, 269.1902, 251.1179, 225.1649, 203.1797, 201.1643, 199.1488, 173.1333, 159.1174, 145.1015
Compactin (CP) (mevastatin)	6.18	230, 238, 246	$C_{23}H_{34}O_5$ (390.2406)	$[M + H]^+$ $[M + Na]^+$	391.2484 413.2304	391.2488 413.2305	+0.4 +0.1	+1.0 +0.2	289.1793, 271.1693, 253.1588, 229.1587, 211.1481, 185.1324, 159.1168
Monasfluore A	6.28	379 [3]	$C_{21}H_{24}O_5$ (356.1624)	$[M + H]^+$ $[M + Na]^+$	357.1702 379.1521	357.1701 379.1522	−0.1 +0.1	−0.3 +0.3	Not recorded
Monaphilone B	7.23	227, 285, 386	$C_{20}H_{28}O_4$ (332.1988)	$[M + H]^+$ $[M + Na]^+$	333.2066 355.1885	333.2062 355.1880	−0.4 −0.5	−1.2 −1.4	287.2012, 217.1234, 201.0911, 189.1385, 173.0970
Monacolin K (MK) (lovastatin)	7.79	229, 238, 246	$C_{24}H_{36}O_5$ (404.2563)	$[M + H]^+$ $[M + Na]^+$ $[2M + Na]^+$	405.2641 427.2460 831.5023	405.2642 427.2461 831.5026	+0.1 +0.1 +0.3	+0.3 +0.3 +0.4	303.1995, 285.1855, 267.1753, 243.1751, 225.1652, 199.1490, 173.1328, 159.1176, 143.0854
Monascin	8.19	232, 291, 392	$C_{21}H_{26}O_5$ (358.1781)	$[M + H]^+$ $[M + Na]^+$	359.1859 381.1678	359.1854 381.1684	−0.5 +0.6	−1.4 +1.6	343.2274, 315.2317, 261.1112, 215.1073, 187.1121
Dihydromonacolin K (DiMK)	9.78	ND [4]	$C_{24}H_{38}O_5$ (406.2719)	$[M + H]^+$ $[M + Na]^+$	407.2798 429.2617	407.2796 429.2618	−0.2 +0.1	−0.5 +0.2	305.2115, 287.2015, 269.1899, 227.1797, 203.1802
Monasfluore B	9.97	378 [3]	$C_{23}H_{28}O_5$ (384.1937)	$[M + H]^+$ $[M + Na]^+$	385.2015 407.1835	ND [2] 407.1839	+0.4	+1.0	Not recorded
Dehydromonacolin L (DeML)	10.15	230, 238, 246	$C_{19}H_{26}O_2$ (286.1933)	$[M + H]^+$ $[M + Na]^+$	287.2011 309.1830	287.2007 ND [2]	−0.4	−1.4	269.1896, 225.1636, 201.1645, 199.1483, 173.1329, 159.1171, 145.1013
Monaphilone A	11.15	230, 289, 389	$C_{22}H_{32}O_4$ (360.2301)	$[M + H]^+$ $[M + Na]^+$	361.2379 383.2198	361.2380 ND [2]	+0.1	+0.3	287.2020, 217.1322, 189.1390, 173.0952
Dehydromonacolin K (DeMK)	11.42	230, 238, 246	$C_{24}H_{34}O_4$ (386.2457)	$[M + H]^+$ $[M + Na]^+$ $[2M + Na]^+$	387.2535 409.2355 795.4812	387.2532 409.2356 795.4821	−0.3 +0.1 +0.9	−0.8 +0.3 +1.1	345.2037, 285.1851, 267.1747, 249.1625, 199.1484, 173.1325, 159.1167, 143.0858
Ankaflavin	11.75	232, 290, 391	$C_{23}H_{30}O_5$ (386.2093)	$[M + H]^+$ $[M + Na]^+$	387.2172 409.1991	387.2173 409.1994	+0.1 +0.3	+0.3 +0.7	359.2238, 315.2325, 261.1117, 215.1076, 187.1125

[1] The UV absorption bands at 243 and 326 nm (low broad) match well with the UV profile reported in the literature for UHPLC-DAD analytical conditions similar to those used in our study (λ_{max} at 240, 282 (shoulder) and 333 (low broad) nm) [35]. [2] ND: not detected. [3] This low and broad UV absorption band is in agreement with the mean of the λ_{max} of the two enlarged and poorly separated peaks at 371 and 386 nm reported in the literature [36]. [4] ND: not detected at the wavelength used in this study (238 nm).

Nevertheless, the unambiguous determination of compound identity required the additional use of HR-MS and HR-MS/MS. For instance, the UV absorption spectrum of the compound eluting at tR 3.89 min with λ_{max} at 230, 239 and 247 nm was characteristic of monacolins. Its protonated molecular ion $[M + H]^+$ at *m/z* 363.2159 and sodium adduct ion $[M + Na]^+$ at *m/z* 385.1988 were in agreement with respectively the calculated masses 363.2171 of $C_{21}H_{31}O_5$ (relative mass error (RME) −3.3 ppm) and 385.1991 of $C_{21}H_{30}O_5Na$ (RME −0.8 ppm) suggesting a $C_{21}H_{30}O_5$ molecular formula (Table 4). Moreover, the MS/MS product ions obtained were characteristic of the fragmentation pathway of monacolins (Table 4) [13,14,26]. So, the compound detected was a monacolin derivative that might be monacolin N (MN) or dehydromonacolin N hydroxyl acid (DeMNA) recently described by Li et al. [26]. We propose that it was MN based on the fact that its tR at −0.16 min with respect to monacolin L in hydroxyl acid form (MLA) was in agreement with the value given by Li et al. for MN (−0.22 min) and not for DeMNA (+1.72 min). Moreover, DeMNA was detected only at trace level using nanoflow HPLC-Chip-MS whereas MN was also detected by conventional HPLC-MS which is much less sensitive [26].

The identification of the compound eluting at 9.78 min is another example of the interest of coupling tR, UV-Vis profile and HR-MS spectra. This compound was not detected at 238 nm but only by its UHPLC-MS profile exhibiting a molecular ion $[M + H]^+$ at *m/z* 407. Among the known monacolins, two have a nominal mass of 406: DiMK ($C_{24}H_{38}O_5$) with an exact mass for $[M + H]^+$ at *m/z* 407.2798 and monacolin M (MM; $C_{23}H_{34}O_6$) with an exact mass for $[M + H]^+$ at *m/z* 407.2434. The measured mass of the $[M + H]^+$ ion at *m/z* 407.2796 indicated that the compound was DiMK. Furthermore, its MS/MS fragmentation pattern gave the same ions than MK but with 2 more *m/z* units (305, 287, 269 ... compared to 303, 285, 267 ...) confirming that a double bond of the fused ring was hydrogenated (Table 4), whereas MS/MS fragmentation behavior of MM is identical to that of MK [26]. Also, MM eluted between MJ and MLA with a relative tR versus MJ of 1.2–1.3 [13,26] and hence, in our UHPLC conditions, MM would be expected at 2.4–2.6 min and not at 9.78 min. The incorrect assignment of this compound to MM by Li et al. in 2004 [10] was based on the mountain-like UV spectrum centred at 238 nm characteristic of classical monacolins and on the MS $[M + H]^+$ ion at *m/z* 407, but the UV absorption and the $[M + H]^+$ peak did not correspond to the same compound. Indeed, the compound with an UV absorbance at 238 nm was very probably dehydromonacolin L (DeML), a monacolin analogue described for the first time in 2011 [37] whose tR is close to that of DiMK (Table 4) [26].

Monacolins and pigments identified in each RYR formulation are listed in Table 5. UHPLC UV-Vis and MS profiles of five characteristic RYR DS, one with only monacolins (formulation **13**), two with mainly monacolins (formulations **10** and **30**) and two with more pigments than monacolins (formulations **12** and **16**) are illustrated in Figure 5 (the UHPLC UV-Vis and MS profiles of nine other RYR DS analyzed are shown in Figure S2). MK was detected in all the samples analyzed and MKA and DeMK in respectively 28 and 29 DS. MLA, ML, CP, DiMK and DeML were found in 18–22 DS, MJ in 11 and MN in 8. The other monacolins were observed only in few samples: one for monacolin J in hydroxyl acid form (MJA) and two for monacolin X (MX). MM cannot be found in any RYR DS. The mycotoxin citrinin was only detected at trace level in two samples. Among the yellow azaphilone pigments, monascin, monascuspiloin and monaphilone B were identified respectively in 18, 15 and 21 of the RYR DS analyzed, whereas ankaflavin, monaphilone A, monasfluore A and B were found in few samples (6, 4, 5 and 3 respectively). The red azaphilone pigments, rubropunctamine, monascorubramine and PP-R, were identified in 11, 4 and 2 samples, respectively.

2.2.2. Quantitative Analysis of Monacolins

The LOD and LOQ of standard lovastatin, established for SNR 3 and 10, were respectively 0.4×10^{-3} mg mL^{-1} and 1.5×10^{-3} mg mL^{-1} corresponding to 0.4 µg and 1.5 µg in the 10 to 100 mg of powdered.

Table 5. Chemical fingerprints of the RYR dietary supplements analyzed by UHPLC and quantitative determination of the monacolins identified (mg per capsule or tablet).

DS	Monacolins [1]												Total M [2]	MK+MKA (%)	MK/MKA	%DiMK/TotalM	Azaphilones [3]
	MJA	MJ	MN	MLA	MX	MKA	ML	CP	MK	DiMK	DeML	DeMK					
1	-	-	-	-	-	0.49	-	-	1.33	-	-	0.02	1.84	1.82 (98.9)	2.7	-	Mocusp, Mop B, Mo
2	-	-	-	0.04	-	0.82	0.20	0.03	6.77	0.21	0.05	0.68	8.80	7.59 (86.3)	8.3	2.4	
3	-	-	-	-	-	0.23	0.04	0.01	2.16	0.04	0.01	0.13	2.62	2.39 (91.2)	9.4	1.5	
4	-	0.01	0.02	0.01	-	0.10	0.04	0.03	0.48	0.11	0.02	0.29	1.11	0.58 (52.3)	4.8	9.9	Mocusp
5	-	0.01	0.02	0.02	-	0.50	0.05	0.04	1.80	0.20	0.02	0.38	3.04	2.30 (75.7)	3.6	6.6	Rubro, Moco, Mocusp, Mop B, Mo, Mof B, Mop A, Anka
6	-	-	-	0.03	-	0.31	0.08	0.03	2.05	0.11	0.03	0.27	2.91	2.36 (81.1)	6.6	3.8	Mop B
7	-	0.06	-	0.06	-	0.37	0.19	0.08	2.22	0.20	0.06	0.85	4.09	2.59 (63.3)	6.0	4.9	
8	-	0.02	0.09	0.12	-	2.13	0.18	0.28	6.26	0.59	-	0.16	9.83	8.39 (85.4)	2.9	6.0	Cit, Rubro, Mocusp, Mof A, Mop B, Mo
9	-	-	-	-	-	0.13	-	-	4.17	-	-	0.28	4.58	4.30 (93.9)	32.1	-	Mo, Anka
10	-	0.02	-	0.01	-	0.96	0.02	0.04	0.85	0.10	0.01	0.31	2.32	1.81 (78.0)	0.9	4.3	Cit, Rubro, Mocusp, Mop B
11	-	-	-	-	-	-	-	-	0.07	-	-	-	0.07	0.07 (100)	-	-	Rubro, Moco, Mop B, Mo, Mop A, Anka
12	-	-	-	-	-	-	-	-	0.08	-	-	-	0.08	0.08 (100)	-	-	Rubro, PP-R, Moco, Mop B, Mo, Mof B, Mop A, Anka
13	-	-	-	-	-	0.62	-	-	23.18	-	-	0.04	23.84	23.80 (99.8)	37.4	-	
14	0.03	0.05	0.07	0.04	-	0.27	0.18	0.08	2.19	0.23	0.06	0.56	3.76	2.46 (65.4)	8.1	6.1	Mocusp, Mop B, Mo
15	-	-	-	0.19	-	1.33	0.86	0.58	8.27	1.16	0.16	1.44	13.99	9.60 (68.6)	6.2	8.3	Rubro, Mocusp, Mof A, Mop B, Mo
16	-	-	-	-	-	-	0.02	0.02	0.33	0.01	-	0.03	0.41	0.33 (80.5)	-	2.4	Rubro, PP-R, Moco, Mop B, Mo, Mof B, Mop A, Anka
17	-	-	-	0.03	-	0.16	0.12	0.16	1.09	0.13	0.02	0.15	1.86	1.25 (67.2)	6.8	7.0	Mocusp, Mof A, Mop B, Mo
18	-	-	-	-	-	0.98	-	-	6.44	0.13	-	0.38	7.93	7.42 (93.6)	6.6	1.6	Mocusp, Mop B, Mo
19	-	0.01	-	0.01	-	0.05	0.03	0.01	0.25	0.02	0.01	0.11	0.50	0.30 (60.0)	5.0	4.0	
20	-	-	-	-	-	0.48	-	-	1.29	-	-	0.02	1.79	1.77 (98.9)	2.7	-	Mocusp, Mop B, Mo
21	-	0.01	0.01	0.01	-	0.07	0.02	-	0.24	0.02	0.01	0.12	0.51	0.31 (60.8)	3.4	3.9	
22	-	-	-	0.02	-	0.68	0.03	0.08	1.77	0.17	0.01	0.37	3.13	2.45 (78.3)	2.6	5.4	Rubro, Mocusp, Mop B, Mo
23	-	-	-	0.04	0.01	0.68	0.05	0.10	8.47	0.17	0.03	0.23	9.78	9.15 (93.6)	12.5	1.7	Rubro, Mocusp, Mop B, Mo
24	-	-	-	-	-	5.50	-	-	5.21	-	-	0.07	10.78	10.71 (99.4)	0.95	-	
25	-	0.09	0.10	0.11	0.04	3.56	0.14	0.29	5.69	0.67	0.03	0.56	11.28	9.25 (82.0)	1.6	5.9	Rubro, Mocusp, Mof A, Mop B, Mo
26	-	-	-	0.03_5	-	0.39	0.10	0.04	3.52	0.15	0.03_5	0.48	4.75	3.91 (82.3)	9.0	3.2	Mocusp
27	-	-	-	-	-	0.61	-	-	2.68	-	-	0.09	3.38	3.29 (97.3)	4.4	-	Mop B, Mo
28	-	0.03	0.04	0.04	-	0.41	0.07	0.05	0.98	0.15	0.02	0.32	2.11	1.39 (65.9)	2.4	7.1	Mop B
29	-	-	-	-	-	0.01	-	-	0.08	-	-	0.04	0.13	0.09 (69.2)	8.0	-	
30	-	0.02	0.03	0.02	-	0.37	0.05	0.02	2.19	0.07	0.02	0.30	3.09	2.56 (82.8)	5.9	2.3	Mop B
31	-	-	-	-	-	0.31	0.04	0.03	0.85	0.07	0.01	0.12	1.43	1.16 (81.1)	2.7	4.9	Rubro, Mop B, Mo, Anka

[1] Monacolins are listed in increasing order of elution times. MJ, MN, MX, ML, MK: monacolins J, N, X, L, K; MJA, MLA, MKA: monacolins J, L, K in hydroxyl acid forms; CP: compactin; DiMK: dihydromonacolin K; DeML, DeMK: dehydromonacolins L,K. [2] TotalM: total monacolins. [3] Azaphilones are listed in increasing order of elution times. Anka: ankaflavin; Cit: citrinin; Mo: monascin; Moco: monascorubramine; Mocusp: monascuspiloin; Mof A, Mof B: monasfluore A, monasfluore B; Mop A, Mop B: monaphilone A, monaphilone B; PP-R: 7-(2-hydroxyethyl)-monascorubramine; Rubro: rubropunctamine.

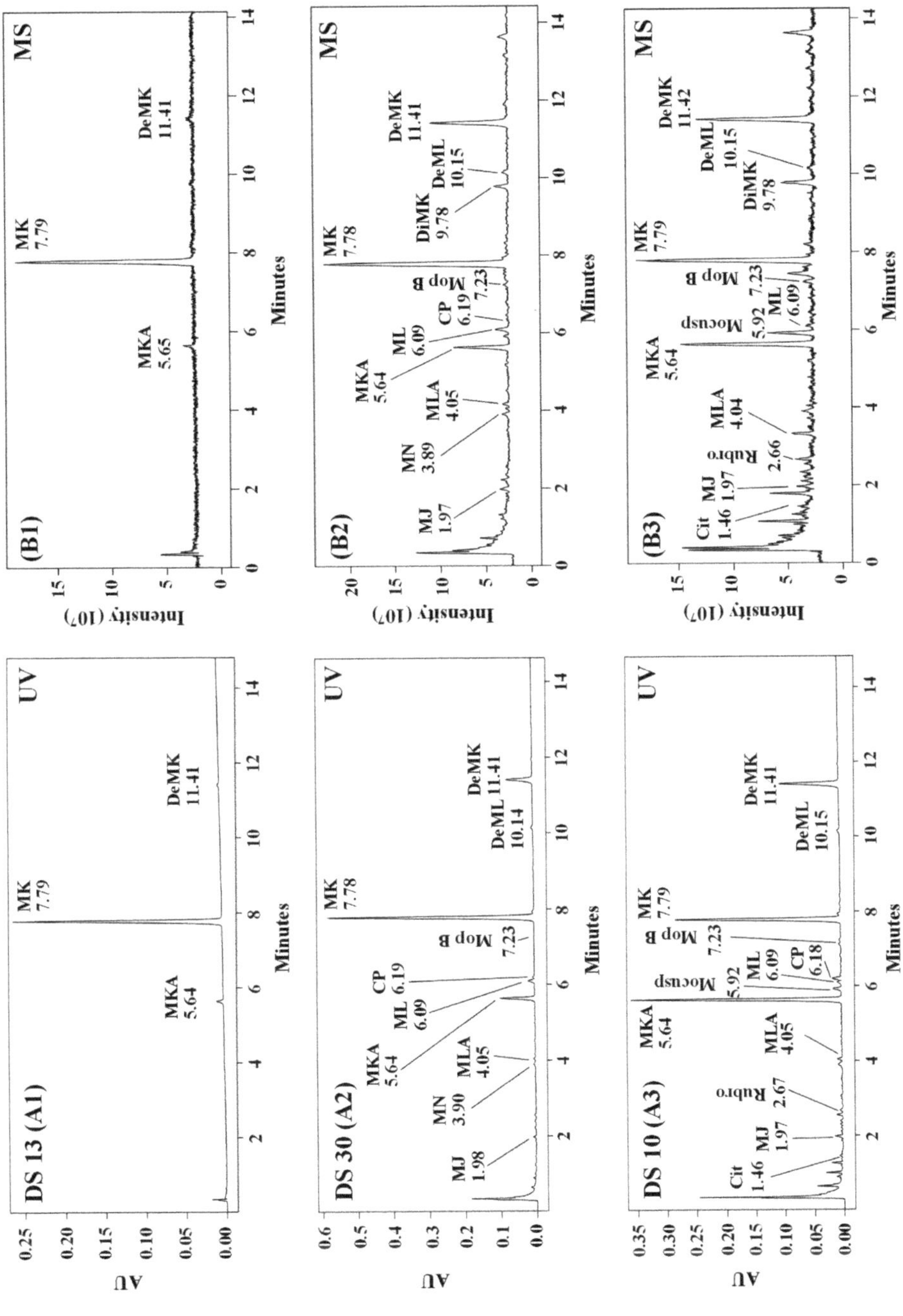

Figure 5. Cont.

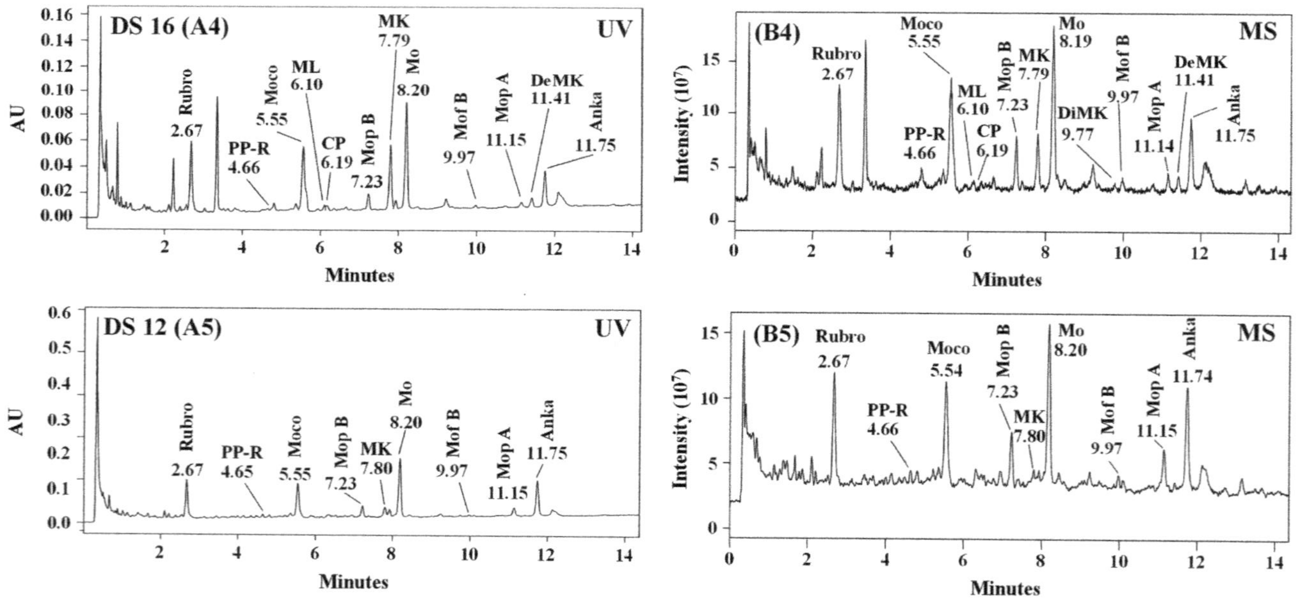

Figure 5. UHPLC chromatograms with UV detection at 238 nm (**A1–A5**) and full scan MS profile in positive ESI mode (**B1–B5**) of five characteristic RYR DS. MJ, MN, ML, MK: monacolins J, N, L, K; MLA, MKA: monacolins L, K in hydroxyl acid forms; CP: compactin; DiMK: dihydromonacolin K; DeML, DeMK: dehydromonacolins L,K; Anka: ankaflavin; Cit: citrinin; Mo: monascin; Moco: monascorubramine; Mocusp: monascuspiloin; Mof B: monasfluore B; Mop A and Mop B: monaphilone A and monaphilone B; PP-R: 7-(2-hydroxyethyl)-monascorubramine; Rubro: rubropunctamine.

RYR formulation extracted. Considering the ratio between the mass of sample extracted and that of the formulation (tablet or powder in the capsule), the LOQ was estimated to be ≈0.9 μg of lovastatin or other monacolins per capsule or tablet. The precision of the method was acceptable with a RSD for the replicates less than 2% (range 0.1%–1.9%). The overall average value of RSD was 5% ranging between 2% and 11%.

The amounts of the 12 monacolins identified in the 31 RYR DS are summarized in Table 5. There is a marked variability in total monacolins (TotalM), MK and MK + MKA contents per capsule or tablet with values ranging respectively between 0.07 and 23.84 mg, 0.07 and 23.18 mg and 0.07 and 23.80 mg. MK is by far the main monacolin in the 31 RYR analyzed representing 68% of TotalM (range 37–100%) and the sum MK + MKA representing 82% (range 52%–100%). As MKA is the active form of MK, the European Food Safety Authority (EFSA) considers that the effective MK content in RYR DS corresponds to the sum of both lactone and hydroxyl acid forms [38], which allows to get rid of the great variability of the ratio MK/MKA. Indeed, this ratio was comprised between 0.9 and 37 in 28 of the RYR DS analyzed in the present study and could not be determined for the three others due to the absence of MKA. These values are in agreement with literature data which report ranges from 0.4 to 85 [10,12,13,39] while the ratio varies between 1.5 and 2 in properly prepared RYR [1]. Minor monacolins represent on average 18% of TotalM, ranging from 0 to 47%. All these results fully confirm previous studies that showed that monacolin contents vary considerably in RYR end-use products, as they depend on the yeast strain and the fermentation process. For example, the HPLC analysis of 15 commercial tablets or capsules showed that TotalM ranges between 0.31 and 11.15 mg per 600 mg of RYR [39–42]. The literature review of the monacolins quantification established that MK represents 57% of TotalM (n = 26; range 0%–99%), MK + MKA 83% (n = 41; range 32%–100%), and the minor monacolins 17% (n = 41; range 0%–68%) [1,10,39–43].

Our own results and all the literature data show that to improve the effectiveness and safety of RYR products in lowering/regulating cholesterol levels, a precise quantification of their monacolin content should be mandatory. Indeed, the MK content indicated for some formulations (18 over the 31 tested) is not sufficient to know the effectiveness of RYR DS as the amount of MKA, which is largely variable, is not specified. However, it is unclear if the term MK used by the manufacturers refers to the sole MK or to the sum MK + MKA. The chemical structures of MK and MKA being different, this should be mentioned. Moreover, the content of minor monacolins should be taken into account as they represent at least ≈ 14% of TotalM in 65% of the RYR tested in our study (Table 5) and in 71% of the 41 RYR products reported in the literature [1,10,39–43]. Indeed, if MK after its transformation by liver metabolism into MKA has the highest HMG-CoA reductase inhibitory activity, the other secondary monacolins such as CP, MJ, ML or DiMK are also active although at a lower or much lower extent (except for DiMK). Relative to MK or DiMK activity stated at 1, those of the other monacolins are 0.5 for CP and dihydroCP, 0.15 for ML, 0.2 for dihydroML, MX and dihydroMX, and 0.04 for MJ [44]. The dehydromonacolins, often considered as inactive, also present a low activity (0.2 for DeMK [45]).

2.3. Comparison of ^{1}H-NMR and UHPLC Results

^{1}H-NMR makes it possible to highlight the overall profile of the compounds present in the RYR DS in a single analysis of the crude extract. Indeed, not only all the monacolins, but also pigments (monascin and other pigments having the same skeleton), fatty acids (SFA and UFA), polyols (glycerol, sorbitol), glucose as well as various other products most often added to the formulation (piperine, carnitine, vitamins) are detected. The main drawback of the method is that the resonances identified are in most cases characteristic of a family of compounds and not of a specific compound as for example, monacolins with a hexahydronaphthalene ring and not only MK, or monascin and other pigments with the same skeleton and not the sole monascin. In contrast, the UHPLC-DAD-MS analysis allows a precise identification of the compounds, but like any separation technique, it requires much more time for implementation and especially the operating conditions used target certain structures of compounds, for example, here, monacolins and pigments.

As the ^{1}H-NMR resonances quantified are specific of monacolin families and not of each monacolin, the monacolin amounts determined by ^{1}H-NMR must be compared to those obtained by UHPLC-DAD-MS for the same group of monacolins. The relationships between the amounts measured from the quantification of the resonances at 5.84 ppm and 5.56 ppm in 27 and 28 formulations respectively, both characteristic of all the monacolin structures except dihydromonacolins (TotalM-DiMK (because only this dihydromonacolin was detected by UHPLC)) (Table 3), and those obtained by UHPLC-DAD-MS for the same set of monacolins (Table 5) were very good, as demonstrated by slopes near 1, y-intercepts close to zero and correlation coefficients (r^2) greater than 0.99 for the two linear regression equations as well as p-values of 0.94 and 0.66 (Table S3). A similar very good correlation between NMR and UHPLC values was obtained when considering the mean concentrations measured from the two NMR resonances (Table S3).

The comparison of the contents of monacolins in lactone form determined from the quantification of their characteristic resonances at 4.60 ppm and/or 4.25 ppm in 22 RYR DS to those obtained by UHPLC-DAD-MS for the same panel of monacolins also showed results in good agreement. Indeed, the linear regression equation of the ^{1}H-NMR and UHPLC values displayed a slope of 0.983, a y-intercept of 0.013 and a correlation coefficient of 0.996, and the *p*-value was 0.50 (Table S3). The resonance at 5.33 ppm is characteristic of all the monacolins except MJ, MJA, ML, MLA and DiMK. Its quantification, which could only be performed on 17 RYR DS, led to values in agreement within ±11% with those obtained by UHPLC for the same set of monacolins for only six (35%) of them (Tables 3 and 5). The very weak relationship between the concentrations measured by the two methods was demonstrated by the low correlation coefficient ($r^2 = 0.723$) and the y-intercept far from zero of the linear regression equation as well as a *p*-value of 0.02 representative of a significant difference between ^{1}H-NMR and UHPLC data (Table S3). Therefore, for the RYR DS analyzed in this study, this resonance is unsuitable for quantification, in contrast to what was observed in a previous ^{1}H-NMR assay of five formulations marketed for German consumers [18].

In the same way, the resonances at 4.05 and 3.63 ppm, characteristic of the monacolins in hydroxyl acid form, cannot be accurately quantified due to their strong overlap with matrix signals. Indeed, all the assays tentatively performed led to values considerably higher than those obtained by UHPLC for the same type of monacolins (data not shown).

From all the ^{1}H resonances characteristic of the various monacolin chemical structures, only those involving the hexahydronaphthalene ring (at 5.84 and 5.56 ppm) and the lactone ring (at 4.60 and 4.25 ppm) are appropriate for quantification without requiring curve deconvolution algorithms. It should be noted that the quantification of the resonances at 5.84 and 5.56 ppm was hindered by the LOQ of the technique for three formulations (28/31 formulations were quantified), and that of resonances at 4.60 or/and 4.25 ppm was additionally hampered by their overlap with matrix signals in some cases (only 22/31 formulations were quantified). To confirm that the ^{1}H-NMR signals at 5.84 and 5.56 ppm can be used to quantify the monacolin content in the RYR formulations, we compared the data obtained to TotalM determined by UHPLC (Table S3). The good relationship between the two sets of values (linear regression equation with a slope of 1.00, a y-intercept at 0.113 and a correlation coefficient of 0.986 as well as a t-test *p*-value of 0.25) was not unexpected as DiMK (the sole dihydromonacolin detected in this study by UHPLC) represented ≈2.8% of the TotalM (range 0–9.9%) (Table 5). The presence of DiMK was not surprising as it is produced in small quantities during the fermentation process of rice with *Monascus* [1,23]. In conclusion, ^{1}H-NMR can be considered as a convenient method to determine the TotalM content in RYR DS.

2.4. Quality Control Issues

2.4.1. About Monacolin Labelling of RYR DS

Only 18 of the 31 RYR DS (58%) tested specified a monacolin(s) content on their label: 14 indicated a level of MK, 1 the sum MK + MKA, and the label was imprecise for 3 DS, 1 indicating "*Monascus purpureus*" and 2 "monacolin" (Table 6). The amount of MK measured was in the range 90%–110% of declared amount for only 3 of the 14 DS mentioning a quantity of MK (DS **13**, **23** and **27**). If we hypothesize that the label "MK" includes MK + MKA, three additional DS (**15**, **22** and **24**) were in the range 90%–110%. If we consider that the claimed dose corresponds to TotalM for the three DS indicating "*Monascus purpureus*" (DS **4**) or "monacolin" (DS **25** and **30**), the measured amount was between 90% and 110% (^{1}H-NMR and UHPLC mean amount for DS **25**) (Table 6). So, 50% of the formulations analyzed contained less than the declared amount of monacolin(s). It is also interesting to note that three DS (**10**, **11** and **12**) contain a very small amount of MK (or MK + MKA) compared to the advertised dose (Table 6). It can thus be concluded that the monacolin label information is not reliable for many RYR products. This deviation between labelled and measured contents has already been reported by Mornar et al. [12] in five DS.

2.4.2. About the Variability of Monacolins Consumed Per Day in RYR DS

There is great variability in the daily consumption of monacolin amounts in terms of MK, MK + MKA, and TotalM (or TotalM-DiMK) calculated from the UHPLC and ^{1}H-NMR data taking into account the recommended serving(s) per day indicated on the labels of the RYR products (Table 6). The amounts of MK, MK + MKA and TotalM consumed daily ranged respectively between 0.08 and 46.2 mg, 0.08 and 47.6 mg, 0.08 and 47.7 mg (UHPLC values) and 0.6 (the very low monacolin content in 3 samples (DS **11**, **12** and **29**) being undetected) and 47.8 mg (NMR values). Although the range of monacolins consumed is the same considering MK, MK + MKA or TotalM, this is not true for all formulations. Indeed, MK represented less than 60% of TotalM in 45% (14/31) of the formulations analyzed and MK + MKA accounted for less than 80% of TotalM in ≈40%, emphasizing the helpfulness of the NMR method which allows the content in all monacolins to be easily determined.

About half of the RYR DS analyzed (48%, 15/31) contained more than 7 mg of MK + MKA or 8 mg of TotalM per recommended daily serving, whereas for ≈45% (14/31), the daily intake of MK + MKA or of TotalM was less than ≈ 3 or 4 mg respectively. Although only single batch formulations have been analyzed and the monacolin content may vary from batch to batch, the question of how effective these levels of monacolin can be for lowering/regulating cholesterol levels can be raised. In 2011, the EFSA concluded that a cause and effect relationship has been established between the consumption of 10 mg per day of MK (sum of lactone and hydroxyl acid forms) from RYR DS and the maintenance of normal low-density lipoproteins-cholesterol (LDL-C) levels [38]. However, several clinical trials using RYR products at daily MK doses well below 10 mg (around 5 mg and even 3 mg) without any other cholesterol-lowering agents such as berberine, policosanol or garlic, showed a statistically significant reduction in LDL-C compared to placebo but these low values either corresponded to MK in the sole lactone form [40,42] or it was not indicated whether they included the hydroxyl acid form of MK [3,4,46]. Nevertheless, when the RYR composition in monacolins was available, the amounts of MK + MKA and TotalM were always at least around 7 mg and 10 mg respectively [40,42]. From these data, it can be concluded that a daily dietary intake of ≈7 mg of MK + MKA or ≈8 mg of TotalM is sufficient to induce a cholesterol-lowering/regulating effect. On the other hand, the very low daily intake of monacolins found in ≈ 45% of the RYR DS analyzed might not lead per se to a significant reduction/regulation of cholesterol levels.

Table 6. Comparison of the percentages of MK, MK + MKA and total monacolins per pill to monacolin labelling; amounts of MK, MK + MKA and total monacolins per recommended daily serving.

N°	Amount of Monacolin(s) Claimed Per Pill (mg) [1]	% Measured/Claimed Per Pill				Serving(s) Per Day [4]	MK Amount (mg/day) [5]	MK + MKA Amount (mg/day) [5]	Total Monacolins (mg/day) [5]	
		MK [2]	MK + MKA [2]	Total Monacolins [3]					UHPLC [3]	NMR [3]
				UHPLC	NMR					
1	MK 2.6	51	70	71	65	1	1.3	1.8	1.8	1.7
2	MK + MKA 10.05	68	76	88	91	1	6.8	7.6	8.8	9.2
3	MK 2.8	77	85	94	110	3	6.5	7.2	7.9	9.2
4	*M. purpureus* 1.0 [6]	48	58	109	99	3	1.4	1.7	3.3	3.0
5						1	1.8	2.3	3.0	2.5
6	MK 3.33	62	71	87	83	3	6.2	7.1	8.7	8.3
7						1	2.2	2.6	4.1	3.8
8						2	12.5	16.8	19.7	17.5
9						2	8.3	8.6	9.2	10.3
10	MK 9	9	20	26	23	4	3.4	7.2	9.3	8.3
11	MK 2.6	3	3	3		4	0.3	0.3	0.3	
12	MK 2.6	3	3	3		1	0.08	0.08	0.08	
13	MK 24.0	97	99	99	100	2	46.2	47.6	47.7	47.8
14						2	4.4	4.9	7.5	6.7
15	MK 9.6	86	100	146	126	1	8.3	9.6	14.0	12.1
16						2	0.7	0.7	0.8	0.6
17						2–3	2.2–3.3	2.5–3.8	3.7–5.6	3.0–4.5
18						1	6.4	7.4	7.9	8.5
19						2–4	0.5–1.0	0.6–1.2	1.0–2.0	0.8–1.6
20	MK 2.6	50	68	69	68	1	1.8	1.8	1.8	1.7
21						2–4	0.5–1.0	0.6–1.2	1.0–2.0	0.95–1.9
22	MK 2.33	76	105	134	131	3	5.3	7.4	9.4	9.2
23	MK 9	94	102	109	118	1	8.5	9.2	9.8	10.6
24	MK 10.0	52	107	108	117	1	5.2	10.7	10.8	11.7
25	Monacolin 9.6	59	96	118	103	1	5.7	9.3	11.3	9.9
26	MK 5.0	70	78	95	91	1–2	3.5–7.0	3.9–7.8	4.8–9.5	4.5–9.1
27	MK 2.6	103	126	130	153	1	2.7	3.3	3.4	4.0
28						1	1.0	1.4	2.1	1.7
29						2–4	0.2–0.3	0.2–0.4	0.25–0.5	
30	Monacolin 3.3	66	77	93	91	3	6.6	7.7	9.3	9.0
31						1	0.9	1.2	1.4	1.2

[1] The exact monacolin labelling is indicated for each formulation. [2] Percentages of MK and MK + MKA were calculated from amounts determined by UHPLC-DAD-MS (see Table 5). [3] Percentages of total monacolins determined by UHPLC-DAD-MS and ^{1}H-NMR correspond, respectively, to TotalM and TotalM-DiMK (see Tables 3 and 5). [4] Serving(s) recommended per day on the label. [5] Amounts of MK, MK + MKA and total monacolins per recommended daily serving. [6] The exact label is: 750 mg of organic RYR titrated at 0.4% of *Monascus purpureus* for three pills.

3. Materials and Methods

3.1. Red Yeast Rice Dietary Supplements

Thirty-one RYR DS bought mainly on internet web sites (24) but also from local health food stores (7) between September 2013 and June 2014 were analyzed before their expiration date with UHPLC and ^{1}H-NMR techniques. RYR products were formulated as capsules (20) or tablets (11). All samples with their name, origin, form, batch number, expiration date and RYR extract content are listed in Table 1.

3.2. Chemicals and Reagents

Authentic standards of lovastatin (MK), citrinin and monascin were purchased from Sigma Aldrich (St. Louis, MO, USA) and those of DeMK, CP and DiMK from TRC (North York, ON, Canada). All other chemicals and reagents used as well the NMR reference for internal chemical shift and quantification (sodium 2,2,3,3-tetradeutero-3-(trimethylsilyl) propanoate (TSP)) were supplied from Sigma Aldrich (St. Louis, MO, USA). Deuterated solvents were obtained from Euriso-Top (91194 Saint Aubin, France). MKA was prepared by hydrolyzing a solution of standard lovastatin in acetonitrile:water (CH_3CN:H_2O) or deuterated acetonitrile:deuterated water (CD_3CN:D_2O) 80:20 *v*/*v* (3.7 mg mL^{-1}) with a 1M NaOH or NaOD solution under the optimized conditions described in literature [12]. The complete conversion of the lactone form (MK) to its acidic form (MKA) was confirmed by HPLC-MS as the $[M + H]^+$ peak of MK at *m*/*z* 405 disappeared and the peak of MKA at *m*/*z* 423 was sole detected. Demineralized water was obtained with a Milli-Q system Purelab flex Veolia Waters.

3.3. Choice of Extraction Solvent and Preparation of Samples for Analyses

The extraction of monacolins from RYR bulk powders with various solvents such as CH_3OH, ethanol: water mixtures, CH_3CN or ethyl acetate has been extensively described, the best results being obtained with CH_3OH or ethanol:water 75:25 [11–13,47,48]. Two extraction solvents were tested in this study, CH_3OH and CH_3CN:H_2O 80:20 (because the MK solubility in CH_3CN is higher than in ethanol, 28 and 16 mg mL^{-1} respectively) and led to the almost same extraction recovery for all compounds, as measured by UHPLC-DAD. In this study, all sample extractions were performed with the mixture CH_3CN:H_2O (or CD_3CN:D_2O).

For the qualitative ^{1}H-NMR analysis, around 100 mg of the powdered RYR samples were mixed with 1 mL of CD_3CN:D_2O (80:20 *v*/*v*) under vortex agitation for 1 min and then sonicated for 10 min. The suspension was then centrifuged (5 min, 3000 rpm) and 700 μL of the supernatant analyzed. TSP as internal chemical shift (δ) reference was added before NMR recording.

For the quantitative ^{1}H-NMR analysis, between 20 and 100 mg of powdered sample was exactly weighed and mixed with 1 mL of CD_3CN:D_2O (80:20 *v*/*v*) under magnetic stirring during 20 min, then sonicated for 10 min. After centrifugation (5 min, 3000 rpm), 30 μL of a 10.0 mM solution of TSP were added to 800 μL of supernatant and the resulting solution was analyzed. The final concentration of TSP was 0.36 mM.

The efficacy of this single-step extraction procedure was demonstrated in samples with low and high contents of monacolins. Around 30, 85 and 90 mg accurately weighed of three different powdered samples (respectively DS **13**, **22** and **5**) were extracted as described above, except that after centrifugation, the whole supernatant was carefully collected and analyzed by ^{1}H-NMR. The exactly weighed residual wet pellet was re-extracted with the same protocol than above and the supernatant analyzed using ^{1}H-NMR. In the second extraction, the amounts of monacolins found were respectively ≈ 4.7%, 8.0% and 14.0% of that measured in the first extract. However, the solvent present in the wet pellet represented respectively at least (as the total weight of the powdered sample used was subtracted from the pellet weight) ≈ 5.4%, 8.3% and 14.0% of the initial solvent weight (1 mL = 905.48 mg). The amount of monacolins extracted was thus directly proportional to the amount of solvent remaining in the pellet. So, all the monacolins were dissolved in the solvent (1 mL) used in the single-step extraction procedure and were thus quantified.

For the quantitative analysis by UHPLC, between 10 and 100 mg of each powdered sample was exactly weighed and mixed with 1 mL of CH_3CN:H_2O (80:20 *v*/*v*) under magnetic stirring during 20 min, then sonicated for 10 min. The suspension was then centrifuged (5 min, 3000 rpm). The supernatant was filtered through a 0.45 µm pore size filter before the injection.

3.4. ^{1}H-NMR Analysis

3.4.1. Recording Conditions

For the qualitative analysis, the 1D ^{1}H-NMR experiments were performed on a Bruker Avance 500 spectrometer (Bruker Biospin AG, Fallanden, Switzerland) equipped with a 5 mm ^{1}H-optimized triple resonance NMR inverse helium-cooled probe (TCI CryoProbe) at 298K (standard ^{1}H sensitivity for 0.1% ethylbenzene in $CDCl_3$: SNR 4200). Typical acquisition parameters were as follows: spectral width 8000 Hz, 32K data points, pulse width 10.0 µs (flip angle 90°), acquisition time 2.04 s, relaxation delay 1 s, number of scans 16, corresponding to a recording time of 0.8 min.

For the quantitative 1D ^{1}H-NMR experiments, analyses were acquired on a Bruker Avance 400 spectrometer equipped with a TBO (Triple resonance Broadband Observe) probe at 298K (standard ^{1}H sensitivity for 0.1% ethylbenzene in $CDCl_3$: SNR 266). The spectra were recorded with the following parameters: spectral width 8000 Hz, 64K data points, pulse width 3.7 µs (flip angle ≈ 30°), acquisition time 5.12 s, and relaxation delay 4.88 s. The number of scans was 64 or 512 for recording times of ≈11 or 85 min respectively. Three different samples of the same DS were analyzed.

T_1 relaxation times of the methyl protons of TSP and those of the characteristic monacolin protons used for quantification (ethylenic protons of the hexahydronaphthalene ring (H6 and H4) and of the unsaturated lactone of dehydromonacolin derivatives (H22), and CH-OH proton of the lactone ring (H22)) (see Table 2 for proton numbering) were measured in RYR DS by the inversion-recovery pulse sequence method. Twenty delays from 0.001 to 30 s were used with an acquisition time of 2.56 sec and a relaxation delay of 30 s. The T_1s found were between 1.6 and 2.2 s for monacolin derivatives and 4.5 s for TSP. All the ^{1}H resonances were thus considered as fully relaxed since more than 98% of the signal intensity of the TSP protons was recovered in the recording conditions used.

The concentrations were calculated by comparing the signal areas of convenient protons of targeted compounds with those of TSP, the area of each NMR peak being directly proportional to the number of corresponding nuclei in fully relaxed recording conditions.

Data were processed with the Bruker TopSpin software 2.1 or 3.1 with one level of zero-filling and Fourier transformation after multiplying FIDs by an exponential line-broadening function of 0.3 Hz. Assignment of signals of MK, MKA, CP and DiMK was achieved with 1D ^{13}C-NMR and 2D experiments (gCOSY, gHSQC and gHMBC).

3.4.2. Quantification

The quantification was performed on the characteristic ^{1}H-NMR signals (CH=, CH-OH and CH-OR) of the different monacolin families (monacolins in lactone and hydroxyl acid forms, dehydromonacolins and dihydromonacolins). The amount of compound (mg) per dosage unit is calculated from the following equation:

$$Q = \frac{A_i}{A_{TSP}} \times \frac{Nb_{TSP}}{Nb_i} \times C_{TSP} \times \frac{V_2}{V_1} \times V_t \times M \times \frac{m_t}{m} \quad (1)$$

with A_i and A_{TSP} being the integrated areas of the characteristic NMR signal(s) of the monacolins to be quantified (*i*) and of the TSP signal respectively, Nb_i and Nb_{TSP} the number of protons contributing to the signal of the analyte and of TSP (9 protons), C_{TSP} the concentration of TSP in the solution analyzed, V_1 the volume of supernatant analyzed, V_2 the volume of solution analyzed (V_1 + V_{TSP} (30 µL)), V_t the volume used to dissolve the sample (1 mL), M the molecular weight of the analyte, m_t and m the weights of the dosage unit (tablet or powder from capsule) and of the sample analyzed. As NMR did not allow to characterize individually all the monacolins present in RYR, the molecular weight considered was that of MK, the main one. This led to an approximation less than ±5% for MKA, DeMK, MX, DiMK and CP but higher for other monacolins, +10% for MN, +16% for MJA, ≈ +20% for MJ and MLA, +25% for ML and +29% for DeML (see Figure 2 for chemical structures). Nevertheless, all the monacolins whose molecular weight differs of more than 5% relative to MK, are very minor. From their amounts determined by UHPLC, we found that the approximation on their molecular weights led to an increase of the total content of monacolins determined by NMR of less than 3% (<1% for 22 DS, between 1 and 2.2% for 8 others, and 2.6% for one formulation).

3.5. UHPLC-DAD-MS Analysis

3.5.1. Instrumentation

The liquid chromatographic system was a Waters Acquity UPLC-DAD-SQD (Ultra Performance Liquid Chromatography with Diode Array Detector and Single Quadrupole Detector). It consists of the following modular components: a binary pump, an automatic sample injector with two 48-well trays, a column oven, a diode array detector and a simple quadrupole detector with ESI. All the analyses were performed using ESI ionization with the following settings: positive mode, electrospray source temperature 135 °C, desolvatation temperature 300 °C, capillary voltage 2.8 kV, cone voltage 3 V, extractor voltage 2.0 V and RF lens voltage 0.1 V. The full scan mass spectra were acquired over a range of *m*/*z* 100–1000. The separations were achieved on a Kinetex C18 (100 mm × 2.1 mm, particle size 1.7 µm) column. The mobile phase consisted of water with 0.02% formic acid (solvent A) and acetonitrile with 0.02% formic acid (solvent B) at a flow rate of 0.6 mL min^{-1}. The temperature of the column oven was set at 40 °C. The chromatographic analysis consisted of an isocratic elution with a 65% A/35% B mixture for 0.5 min followed by a linear gradient program: from 65% A/35% B to 35% A/65% B between 0.5 and 15 min and finally to 0% A/100% B over 3 min. After each run, the percentage of solvents ramped to their initial composition in 1 min and then the column was re-equilibrated for 2 min. The UV detection and quantification were performed at 238 nm and UV-Vis spectra were recorded within a range of 200–800 nm. The data acquisition and processing were done with the Empower 2 software.

3.5.2. Validation Procedure

The UHPLC-UV method described was validated in terms of system suitability, linearity, precision, sensitivity (LOD, LOQ) and specificity.

The system suitability tests to ensure the reproducibility of the chromatographic system were performed by injecting six times 1 µL of a solution of standard lovastatin. The RSD found was 0.2% for a 0.42 mg mL^{-1} solution and 1.5% for a 4.2×10^{-3} mg mL^{-1} solution and was acceptable as it was less than 2% [13].

The linearity of the UHPLC-UV assay was tested for eight concentration levels of standard lovastatin in the range 4.18×10^{-3}–0.42 mg mL^{-1} in CH_3CN:H_2O (80:20 *v*/*v*). The correlation coefficient value ($r^2 = 0.9993$) of the calibration curve obtained by plotting the peak areas versus concentrations indicated satisfactory linearity of the method in the range studied. All the monacolins with a characteristic UV maximum absorption peak at ≈238 nm were quantified using the calibration curve established for standard lovastatin and their amounts in mg calculated considering their respective molecular weights. DiMK was not detected at this wavelength and was thus quantified by MS from its $[MH]^+$ peak area by comparison to that of MK whose amount was previously determined by UV, considering that their ionization efficiencies were similar.

Each solution of extracted RYR sample was analyzed three times. Two independently prepared samples were analyzed for 18 RYR DS and three for the other 13.

The specificity of the method, under the conditions described above, was verified using the chromatographic peak purity tool included in the Empower 2 software and showed no co-elution between peaks of monacolins and those of the complex matrix.

3.6. High-Resolution Mass Spectrometry (HR-MS) Experiments

HR-MS and HR-MS/MS were performed on a Waters Xevo G2 QTOF mass spectrometer (Waters, Manchester, UK) with ESI in positive mode (ESI^+) and in few cases in negative mode (ESI^-). For both modes, the instrument parameters were as follows: for MS analysis, cone voltage 30 or 50 V, source temperature 110 °C, desolvatation temperature 300 °C, cone gas flow rate 20 L h^{-1}, scan range *m*/*z* 50–1200; for MS/MS analysis, three collision energies were used 15, 25 and 35 V with the cone voltage maintained at 30 V or 50 V and the spectra were acquired with a mass precursor ion selection of 3 Da. All analyses were performed using the lockspray, which ensured accuracy and reproducibility. Leucine enkephalin (1 ng μL^{-1}) introduced by a lockspray at 3 μL min^{-1} was used as the lockmass generating reference ions at *m*/*z* 556.2771 or 554.2615 in positive or negative mode respectively. The MassLynx software was used to calculate the most probable chemical formula and the theoretical exact mass of the molecular and fragment ions by comparison with their measured accurate ionic masses.

3.7. Statistical Analysis

Comparison of quantitative determination of the different monacolin sets obtained by ^{1}H-NMR and UHPLC was performed (i) by simple linear regression and (ii) with the Wilcoxon signed-rank test, a non-parametric test that can be used as an alternative to the paired Student's t-test for matched pairs when the population cannot be assessed to be normally distributed; the *p*-values > 0.05 (95% confidence level) were considered as proofs of no significant difference between the values measured by the two methods.

4. Conclusions

This work demonstrated that ^{1}H-NMR is a powerful method to establish rapidly (<1 min) the spectral signature of RYR DS and to afford the quantitative determination of their total monacolin content in a reasonable recording time (from ≈10 to 90 min depending on their concentration in the solution analyzed). The method was validated against UHPLC-DAD-MS, the gold standard technique for a detailed determination of all the monacolins existing in RYR products. Indeed, the quantification of the ^{1}H resonances of the hexahydronaphthalene ring at 5.84 and 5.56 ppm, characteristic of all the monacolin families except the dihydomonacolins present in very low amounts, led to values in close agreement with those of all the monacolins, including dihydromonacolins, measured by UHPLC. These two resonances did not overlap with matrix signals in the RYR DS tested and their quantification is only limited by the sensitivity of the method (LOQ ≈ 0.25 mg per capsule or tablet in our recording conditions). Therefore, ^{1}H-NMR can be considered as an accurate method for determining the total monacolin content of RYR DS, not only those of the main ingredients MK and MKA, but also of minor monacolins which are also active HMG-CoA reductase inhibitors although generally less potent, accounting for a non-negligible amount (mean of ≈ 18%) of TotalM. Taking into account the easy preparation of the RYR DS for the analysis (a simple extraction of capsule/tablet powder by an adequate deuterated solvent), ^{1}H-NMR can be proposed as a high-throughput (thanks to commonly used sample changers) screening method for quality control of RYR formulations on the market. Indeed, the product labels are often incomplete and inappropriate: 42% of the RYR DS analyzed in this study did not provide information on the concentration of monacolins; when the label mentioned MK, it was not specified whether the amount of MK included or not MKA; the amount of the other minor monacolins was never indicated. The deviation between labelled and measured contents must also be emphasized: only 50% of the labelled formulations actually contained the declared amount of monacolin(s). In conclusion, health authorities should impose to manufacturers a strict control of the quality of the RYR DS to assess the ability of the marketed RYR products in reducing/regulating cholesterol level.

Supplementary Materials: The following are available online, Table S1: ^{1}H and ^{13}C-NMR data (solvent: CD_3CN:D_2O 80:20) of monacolin K in lactone (MK) and hydroxyl acid (MKA) forms, compactin and dihydromonacolin K. Table S2: ^{1}H-NMR quantitative determination of monacolins in RYR dietary supplements. Table S3: Comparison of monacolin amounts measured in RYR dietary supplements by ^{1}H-NMR and UHPLC. Figure S1: ^{1}H-NMR spectra (CD_3CN:D_2O (80:20)) of all the dietary supplements analyzed in the present study. Upper part: entire spectrum, lower part: enlarged downfield region (4–9 ppm). Figure S2: UHPLC chromatograms with UV detection at 238 nm (A) and full scan MS profile in positive ESI mode (B) of 9 RYR dietary supplements.

Author Contributions: Conceptualization, M.M.-M.; methodology, R.H., G.A., S.B., N.M.-F.; validation, R.H., G.A., S.B.; formal analysis, S.B.; investigation, R.H., G.A., S.B., N.M.-F.; writing—original draft preparation, R.H., V.G.; writing—review and editing, R.M., M.M.-M.; visualization, V.G., S.B., M.M.-M.; supervision, M.M.-M.; funding acquisition, M.M-M., V.G. All authors have read and agreed to the published version of the manuscript.

Funding: This research was funded by the French National Agency for the Safety of Medicines and Health Products (Agence Nationale de Sécurité du Médicament et des produits de santé: ANSM), grant AAP-2012-082, (convention ANSM/UPS n°2012S071).

Acknowledgments: The authors wish to thank Isabelle Fabing from the SPCMIB lab and the Integrated Screening Platform of Toulouse for support with chromatography.

Conflicts of Interest: The authors declare no conflict of interest. The funders had no role in the design of the study; in the collection, analyses, or interpretation of data; in the writing of the manuscript, or in the decision to publish the results.

References

1. Ma, J.; Li, Y.; Ye, Q.; Li, J.; Hua, Y.; Ju, D.; Zhang, D.; Cooper, R.; Chang, M. Constituents of red yeast rice, a traditional Chinese food and medicine. *J. Agric. Food Chem.* **2000**, *48*, 5220–5225. [CrossRef]
2. Wang, T.-H.; Lin, T.-F. *Monascus* rice products. *Adv. Food Nutr. Res.* **2007**, *53*, 123–159.
3. Li, Y.; Jiang, L.; Jia, Z.; Xin, W.; Yang, S.; Yang, Q.; Wang, L. A meta-analysis of red yeast rice: An effective and relatively safe alternative approach for dyslipidemia. *PLoS ONE* **2014**, *9*, e98611. [CrossRef]
4. Gerards, M.C.; Terlou, R.J.; Yu, H.; Koks, C.H.W.; Gerdes, V.E.A. Traditional Chinese lipid-lowering agent red yeast rice results in significant LDL reduction but safety is uncertain—A systematic review and meta-analysis. *Atherosclerosis* **2015**, *240*, 415–423. [CrossRef] [PubMed]
5. Chen, C.-H.; Yang, J.-C.; Uang, Y.-S.; Lin, C.-J. Improved dissolution rate and oral bioavailability of lovastatin in red yeast rice products. *Int. J. Pharm.* **2013**, *444*, 18–24. [CrossRef] [PubMed]
6. ANSES, OPINION of the French Agency for Food, Environmental and Occupational Health & Safety on the Risks Associated with the Presence of "Red Yeast Rice" in Food Supplements. 2014. Available online: https://www.anses.fr/fr/system/files/NUT2012sa0228EN.pdf (accessed on 9 September 2019).
7. Mazzanti, G.; Moro, P.A.; Raschi, E.; Da Cas, R.; Menniti-Ippolito, F. Adverse reactions to dietary supplements containing red yeast rice: Assessment of cases from the Italian surveillance system. *Br. J. Clin. Pharmacol.* **2017**, *83*, 894–908. [CrossRef] [PubMed]
8. EFSA Panel on Food Additives and Nutrient Sources added to Food (ANS). Scientific opinion on the safety of monacolins in red yeast rice. *EFSA J.* **2018**, *16*, 5368. [CrossRef]
9. Red Yeast Rice—An Update Overview of the Reported ADRs. Netherlands Pharmacovigilance Centre Lareb. Available online: https://www.lareb.nl/media/3204/signals_2019_rode-gistrijst_nvwa-update.pdf (accessed on 9 January 2020).
10. Li, Y.-G.; Zhang, F.; Wang, Z.-T.; Hu, Z.-B. Identification and chemical profiling of monacolins in red yeast rice using high-performance liquid chromatography with photodiode array detector and mass spectrometry. *J. Pharm. Biomed. Anal.* **2004**, *35*, 1101–1112. [CrossRef]
11. Wu, C.-L.; Kuo, Y.-H.; Lee, C.-L.; Hsu, Y.-W.; Pan, T.-M. Synchronous high-performance liquid chromatography with a photodiode array detector and mass spectrometry for the determination of citrinin, monascin, ankaflavin, and the lactone and acid forms of monacolin K in red mold rice. *J. AOAC Int.* **2011**, *94*, 179–190.
12. Mornar, A.; Sertic, M.; Nigovic, B. Development of a rapid LC/DAD/FLD/MS^n method for the simultaneous determination of monacolins and citrinin in red fermented rice products. *J. Agric. Food Chem.* **2013**, *61*, 1072–1080. [CrossRef]

13. Avula, B.; Cohen, P.A.; Wang, Y.H.; Sagi, S.; Feng, W.; Wang, M.; Zweigenbaum, J.; Shuangcheng, M.; Khan, I.A. Chemical profiling and quantification of monacolins and citrinin in red yeast rice commercial raw materials and dietary supplements using liquid chromatography-accurate QToF mass spectrometry: Chemometrics application. *J. Pharm. Biomed. Anal.* **2014**, *100*, 243–253. [CrossRef] [PubMed]
14. Song, F.; El-Demerdash, A.; Lee, S.-J.S.H.; Smith, R.E. Fast screening of lovastatin in red yeast rice products by flow injection tandem mass spectrometry. *J. Pharm. Biomed. Anal.* **2012**, *57*, 76–81. [CrossRef] [PubMed]
15. Svoboda, P.; Sander, D.; Plachká, K.; Nováková, L. Development of matrix effect-free MISPE-UHPLC-MS/MS method for determination of lovastatin in Pu-erh tea, oyster mushroom, and red yeast rice. *J. Pharm. Biomed. Anal.* **2017**, *140*, 367–376. [CrossRef] [PubMed]
16. Malet-Martino, M.; Holzgrabe, U. NMR techniques in biomedical and pharmaceutical analysis. *J. Pharm. Biomed. Anal.* **2011**, *55*, 1–15. [CrossRef] [PubMed]
17. Simmler, C.; Napolitano, J.G.; McAlpine, J.B.; Chen, S.-N.; Pauli, G.F. Universal quantitative NMR analysis of complex natural samples. *Curr. Opin. Biotechnol.* **2014**, *25*, 51–59. [CrossRef]
18. Lachenmeier, D.W.; Monakhova, Y.B.; Kuballa, T.; Löbell-Behrends, S.; Maixner, S.; Kohl-Himmelseher, M.; Waldner, A.; Steffen, C. NMR evaluation of total statin content and HMGCoA reductase inhibition in red yeast rice (*Monascus* spp.) food supplements. *Chin. Med.* **2012**, *7*, 8. [CrossRef]
19. Zhu, L.; Yau, L.-F.; Lu, J.-G.; Zhu, G.-Y.; Wang, J.-R.; Han, Q.-B.; Hsiao, W.-L.; Jiang, Z.-H. Cytotoxic dehydromonacolins from red yeast rice. *J. Agric. Food Chem.* **2012**, *60*, 934–939. [CrossRef]
20. Albers-Schönberg, G.; Joshua, H.; Lopez, M.B.; Hensens, O.D.; Springer, J.P.; Chen, J.; Ostrove, S.; Hoffman, C.H.; Alberts, A.W.; Patchett, A.A. Dihydromevinolin, a potent hypocholesterolemic metabolite produced by *Aspergillus terreus*. *J. Antibiot.* **1981**, *34*, 507–512. [CrossRef]
21. Haruyama, H.; Kuwano, H.; Kinoshita, T.; Terahara, A.; Nishigaki, T.; Tamura, C. Structure elucidation of the bioactive metabolites of ML-236B (mevastatin) isolated from dog urine. *Chem. Pharm. Bull. (Tokyo)* **1986**, *34*, 1459–1467. [CrossRef]
22. Barber, J.; Cornford, J.L.; Howard, T.D.; Sharplese, D. The structure of citrinin in vivo. *J. Chem. Soc. Perkin Trans.* **1987**, *1*, 2743–2744. [CrossRef]
23. Lankhorst, P.P.; Poot, M.M.; De Lange, M.P.A. Quantitative determination of lovastatin and dihydrolovastatin by means of ^{1}H-NMR spectroscopy. *Pharmacopeial. Forum* **1996**, *22*, 2414–2422.
24. Simpson, J. Lovastatin in chloroform-D. In *NMR Case Studies. Data Analysis of Complicated Molecules*, 1st ed.; Elsevier: Cambridge, MA, USA, 2017; pp. 94–104.
25. Su, N.-W.; Lin, Y.-L.; Lee, M.-H.; Ho, C.-Y. Ankaflavin from *Monascus*-fermented red rice exhibits selective cytotoxic effect and induces cell death on Hep G2 cells. *J. Agric. Food Chem.* **2005**, *53*, 1949–1954. [CrossRef] [PubMed]
26. Li, M.-N.; Li, C.-R.; Gao, W.; Li, P.; Yang, H. Highly sensitive strategy for identification of trace chemicals in complex matrix: Application to analysis of monacolin analogues in *monascus*-fermented rice product. *Anal. Chim. Acta* **2017**, *982*, 156–167. [CrossRef]
27. Endo, A. Monacolin K, a new hypocholesterolemic agent produced by a *Monascus* species. *J. Antibiot.* **1979**, *32*, 852–854. [CrossRef]
28. Endo, A.; Hasumi, K.; Negishi, S. Monacolins J and L, new inhibitors of cholesterol biosynthesis produced by *Monascus ruber*. *J. Antibiot.* **1985**, *38*, 420–422. [CrossRef]
29. Mapari, S.A.; Hansen, M.E.; Meyer, A.S.; Thrane, U. Computerized screening for novel producers of *Monascus*-like food pigments in *Penicillium* species. *J. Agric. Food Chem.* **2008**, *56*, 9981–9989. [CrossRef] [PubMed]
30. Zheng, Y.; Xin, Y.; Guo, Y. Study on the fingerprint profile of *Monascus* products with HPLC–FD, PAD and MS. *Food Chem.* **2009**, *113*, 705–711. [CrossRef]
31. Hsu, Y.-W.; Hsu, L.-C.; Liang, Y.-H.; Kuo, Y.-H.; Pan, T.-M. Monaphilones A-C, three new antiproliferative azaphilone derivatives from *Monascus purpureus* NTU 568. *J. Agric. Food Chem.* **2010**, *58*, 8211–8216. [CrossRef]
32. Chiu, H.-W.; Chen, M.-H.; Fang, W.-H.; Hung, C.-M.; Chen, Y.-L.; Wu, M.-D.; Yuan, G.-F.; Wu, M.-J.; Wang, Y.-J. Preventive effects of *monascus* on androgen-related diseases: Androgenetic alopecia, benign prostatic hyperplasia, and prostate cancer. *J. Agric. Food Chem.* **2013**, *61*, 4379–4386. [CrossRef]
33. Ogihara, J.; Kato, J.; Oishi, K.; Fujimoto, Y. PP-R, 7-(2-hydroxyethyl)-monascorubramine, a red pigment produced in the mycelia of *Penicillium* sp. AZ. *J. Biosci. Bioeng.* **2001**, *91*, 44–47. [CrossRef]

34. Mapari, S.A.S.; Meyer, A.S.; Thrane, U. Colorimetric characterization for comparative analysis of fungal pigments and natural food colorants. *J. Agric. Food Chem.* **2006**, *54*, 7027–7035. [CrossRef] [PubMed]
35. Woo, P.C.Y.; Lam, C.-W.; Tam, E.W.T.; Lee, K.-C.; Yung, K.K.Y.; Leung, C.K.F.; Sze, K.-H.; Lau, S.K.P.; Yuen, K.-Y. The biosynthetic pathway for a thousand-year-old natural food colorant and citrinin in *Penicillium marneffei*. *Sci. Rep.* **2014**, *4*, 6728. [CrossRef] [PubMed]
36. Bijinu, B.; Suh, J.-W.; Park, S.-H.; Kwon, H.-J. Delineating *Monascus* azaphilone pigment biosynthesis: Oxidoreductive modifications determine the ring cyclization pattern in azaphilone biosynthesis. *RSC Adv.* **2014**, *4*, 59405–59408. [CrossRef]
37. Li, X.-M.; Shen, X.-H.; Duan, Z.-W.; Guo, S.-R. A new monacolin analogue from Xuezhikang capsule. *Acta Pharm. Sin.* **2011**, *46*, 564–567.
38. EFSA Panel on Dietetic Products, Nutrition and Allergies (NDA). Scientific Opinion on the substantiation of health claims related to monacolin K from red yeast rice and maintenance of normal blood LDL-cholesterol concentrations (ID 1648, 1700) pursuant to Article 13(1) of Regulation (EC) No 1924/2006. *EFSA J.* **2011**, *9*, 2304. [CrossRef]
39. Gordon, R.Y.; Cooperman, T.; Obermeyer, W.; Becker, D.J. Marked variability of monacolin levels in commercial red yeast rice products: Buyer beware! *Arch. Intern. Med.* **2010**, *170*, 1722–1727. [CrossRef] [PubMed]
40. Heber, D.; Yip, I.; Ashley, J.M.; Elashoff, D.A.; Elashoff, R.M.; Go, V.L. Cholesterol-lowering effects of a proprietary Chinese red-yeast-rice dietary supplement. *Am. J. Clin. Nutr.* **1999**, *69*, 231–236. [CrossRef]
41. Becker, D.J.; Gordon, R.Y.; Morris, P.B.; Yorko, J.; Gordon, Y.J.; Li, M.; Iqbal, N. Simvastatin vs therapeutic lifestyle changes and supplements: Randomized primary prevention trial. *Mayo Clin. Proc.* **2008**, *86*, 758–764. [CrossRef]
42. Becker, D.J.; Gordon, R.Y.; Halbert, S.C.; French, B.; Morris, P.B.; Rader, D.J. Red yeast rice for dyslipidemia in statin-intolerant patients: A randomized trial. *Ann. Intern. Med.* **2009**, *150*, 830–839. [CrossRef]
43. Maesschalck, J.; Gruyters, A.M.; Corthout, J. Levure de riz rouge. Que choisir et quel conseil donner? *J. Pharm. Belg.* **2018**, *4*, 6–17.
44. Endo, A. Compactin (ML-236B) and related compounds as potential cholesterol-lowering agents that inhibit HMG-CoA reductase. *J. Med. Chem.* **1985**, *28*, 401–405. [CrossRef] [PubMed]
45. Phainuphong, P.; Rukachaisirikul, V.; Saithong, S.; Phongpaichit, S.; Bowornwiriyapan, K.; Muanprasat, C.; Srimaroeng, C.; Duangjai, A.; Sakayaroj, J. Lovastatin analogues from the soil-derived fungus *Aspergillus sclerotiorum* PSU-RSPG178. *J. Nat. Prod.* **2016**, *79*, 1500–1507. [CrossRef] [PubMed]
46. Heinz, T.; Schuchardt, J.P.; Möller, K.; Hadji, P.; Hahn, A. Low daily dose of 3 mg monacolin K from RYR reduces the concentration of LDL-C in a randomized, placebo-controlled intervention. *Nutr. Res.* **2016**, *36*, 1162–1170. [CrossRef] [PubMed]
47. Li, Y.-G.; Liu, H.; Wang, Z.-T. A validated stability-indicating HPLC with photodiode array detector (PDA) method for the stress tests of *Monascus purpureus*-fermented rice, red yeast rice. *J. Pharm. Biomed. Anal.* **2005**, *39*, 82–90. [CrossRef] [PubMed]
48. Theunis, M.; Naessens, T.; Verhoeven, V.; Hermans, N.; Apers, S. Development and validation of a robust high-performance liquid chromatographic method for the analysis of monacolins in red yeast rice. *Food Chem.* **2017**, *234*, 33–37. [CrossRef]

Sample Availability: Samples of the formulations are available from the authors.

molecules

MDPI

Article

Traceability of Geographical Origin in *Gentiana straminea* by UPLC-Q Exactive Mass and Multivariate Analyses

Zheng Pan [1,*,†], Feng Xiong [2,†], Yi-Long Chen [3], Guo-Guo Wan [1], Yi Zhang [4], Zhi-Wei Chen [5], Wen-Fu Cao [1] and Guo-Ying Zhou [2,*]

1. College of Traditional Chinese Medicine, Chongqing Medical University, Chongqing 400016, China; wanguoguo1@126.com (G.-G.W.); caowenfu@hospital.cqmu.com.edu.cn (W.-F.C.)
2. Qinghai Provincial Key Laboratory of Qinghai-Tibet Plateau Biological Resources, Northwest Institute of Plateau Biology, Chinese Academy of Science, Xining 810008, China; xiongfeng1994@foxmail.com
3. Chongqing Academy of Chinese Materia Medica, Chongqing 404000, China; zychenyl@163.com
4. Centre for Academic Inheritance and Innovation of Ethnic Medicine, Chengdu University of Traditional Chinese Medicine, Chengdu 611130, China; letter2013@sina.com
5. Chongqing Key Laboratory of Traditional Chinese Medicine for Prevention and Cure of Metabolic Diseases, Chongqing 400016, China; chenzw@cqmu.com.edu.cn

* Correspondence: 102796@cqmu.com.edu.cn (Z.P.); zhougy@nwipb.cas.cn (G.-Y.Z.)

† Zheng Pan and Feng Xiong are co-first authors.

Academic Editor: Ping-Chung Kuo
Received: 25 October 2019; Accepted: 4 December 2019; Published: 6 December 2019

check for updates

Abstract: The root of *Gentiana straminea* Maxim. (Gentianaceae), is officially listed as "Qin-Jiao" in the Chinese Pharmacopoeia for the treatment of rheumatic arthritis, icteric hepatitis, constipation, pain, and hypertension. To establish the geographical origin traceability in *G. straminea*, its chemical profiles were determined by a UPLC-Q exactive mass spectrometer, from which 43 compounds were identified by comparing retention times and mass spectrometry. Meanwhile, a pair of isomers (loganin and secologanol) was identified by mass spectrometry based on their fragmentation pathway. A total of 42 samples from difference habitats were determined by an UPLC-Q exactive mass spectrometer and the data were assayed with multivariate statistical analysis. Eight characteristic compounds were identified to determine the geographical origin of the herb. To estimate the key characteristic markers associated with pharmacological function, the inhibiting activities of nitric oxide (NO) production in lipopolysaccharide (LPS)-induced macrophages were examined. This finding is crucial in realizing the determination of botanical origin and evaluating the quality of *G. straminea*.

Keywords: *Gentiana straminea*; geographical origin; UPLC-Q exactive mass; metabolomics; multivariate analysis

1. Introduction

The geographical origin and the authenticity of herbal material are often related to the safe application of their preparations, since the authenticity and quality parameters are often associated with a particular geographical origin and/or production area [1]. However, some herbal products available for purchase do not come from a fixed origin, and could fall short of quality requirements. Thus, the successful traceability of geographical origin attributes is necessary for ensuring efficacy and biosafety [2]. Metabolites are the end product of the majority of cellular processes, and, as such, are representative of the phenotype of an organism. The secondary metabolites of herbs with different geographical origins exhibit certain differences, therefore, it has been speculated that different geographical origins of medicinal herbs may be identified according to their chemical characteristics by a data mining method based on their chemical profiles [3].

The root of *Gentiana straminea* (*G. straminea*) Maxim. (Gentianaceae) is officially listed as "Qin-Jiao" in the Chinese Pharmacopoeia [4] for the treatment of rheumatic arthritis, icteric hepatitis, constipation, pain, and hypertension [5]. Phytochemical investigations have characterized the plant by the presence of a wide range of compounds, such as iridoids, secoiridoids, flavonoids, triterpenoids, alkaloids, and other types of secondary metabolites [6–8]. Some published methods have focused on the qualitative and quantitative determination of iridoids and secoiridoids in the plant by liquid chromatography or liquid chromatography-mass spectrometry [9,10]. Nevertheless, few of these methods have been aimed at determining the geographical origin of *G. straminea* by characteristic marker components.

In recent years, the authors of this paper successfully identified the characteristic components in different geographical origins of *Lamiophlomis rotata* by ultra-high performance liquid chromatography coupled with time-of-flight mass spectrometry (UPLC-Q/TOF/MS) [11–13]. Here, a comprehensive chemical composition analysis of 42 samples is proposed to evaluate the variability of *G. straminea* from different geographical origins, with sensitive, selective, and accurate UPLC-Q exactive mass spectrometer analysis. The UPLC-Q exactive mass spectrometer data was assayed to identify the characteristic components in *G. straminea* from different geographical regions. Partial least-squares discriminant analysis (PLS-DA) and orthogonal partial least-squares discriminant analysis (OPLS-DA) were employed with a metabolomic approach. To estimate the key characteristic marker associated with pharmacological function, the inhibiting activities of nitric oxide (NO) production in lipopolysaccharide (LPS)-induced macrophages were examined. In summary, the findings of this study imply that the origin of the material should be considered when it is used in traditional prescriptions and medicinal preparations.

2. Results

2.1. Identification of the Constituents in G. straminea by UPLC-Q Exactive Mass Spectrometer

The no. 3 sample from Sichuan province (SC-M-03-r3b) was selected for identification by UPLC-Q mass assay as the sample showing the most peaks during 0–30 min. Forty-six peaks were detected in *G. straminea* from MS and MS^n in negative and positive ion mode by a UPLC-Q exactive mass spectrometer (Figure 1). The mass accuracy for all assigned components was less than 5 ppm compared with the molecular formulas of the published compounds in *G. straminea*. Gentiopicroside, swertiamarin, 6′-*O*-β-D-glucopyranosyl-gentiopicroside, sweroside, loganic acide and loganin were identified by comparison with the retention time and mass fragmentation of reference standards. The SciFinder Scholar and the PubChem databases were searched for the spectral data of other compounds reported previously in the genus *Gentiana* and *G. straminea* to identify the constituents of the herb [14–20]. Forty-three of these were identified by comparing the retention times and mass spectrometry, which has already been summarized [21], including 20 iridoids, 16 secoiridoids, 8 flavonoids, 2 triterpenoids, 2 lignins, 2 alkaloids, and 2 saccharides (Table 1).

Among these compounds, gentiopicroside (a type of secoiridoid) and loganin (a type of iridoid) were officially listed in the Chinese Pharmacopoeia 2015 edition for quality control of the herb [22]. In positive ion mode, secoiridoids and iridoids all showed highly abundant proton and sodium ion adducts, but a relatively lower proportion of potassium, and they also showed highly formate and chlorine ion adducts in negative ion mode. Moreover, small peaks for $[2M + Na]^+$, $[2M + HCOOH - H]^-$ (Supplementary Figures S1 and S2), could be observed in the spectra for positive and negative experiments. All of these mass signals were helpful in the identification of secoiridoid and iridoid components [23,24].

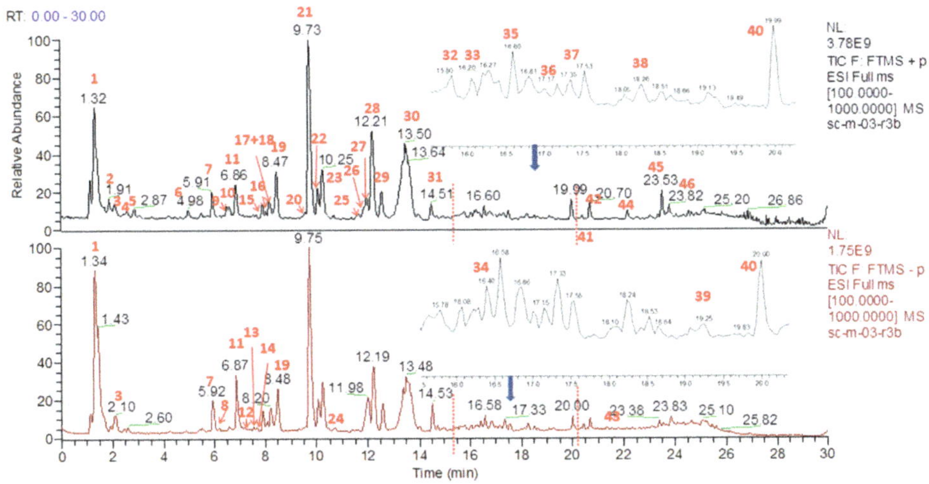

Figure 1. Total ion current chromatograms of substances in the extract of *G. straminea* (the no. 3 sample of Sichuan province) with positive and negative ion modes.

For the first time, a pair of isomers (loganin and secologanol) were identified by mass spectrometry based on their fragmentation pathway. Loganin (a type of iridoid), eluted at 9.95 min, showed fragment ions at *m/z* 413.14157 $[M + Na]^+$ (Figure 2a) and *m/z* 803.29346 $[2M + Na]^+$, with the elemental composition of $C_{17}H_{26}O_{10}Na$ (calculated 413.14240) in positive ion mode. In MS^2, the compound formed product-characterized ions at *m/z* 285.09409 with the neutral loss of $C_6H_8O_3$ (Δm = 128.04831 Da). It also showed ions at *m/z* 185.04211 Da as glucose residue adducts sodium ion, and ions at *m/z* 219.06264 Da indacted the compound with the neutral loss of a glucose and methanol. The proposed fragmentation pathway of the loganin is shown in Figure 3a.

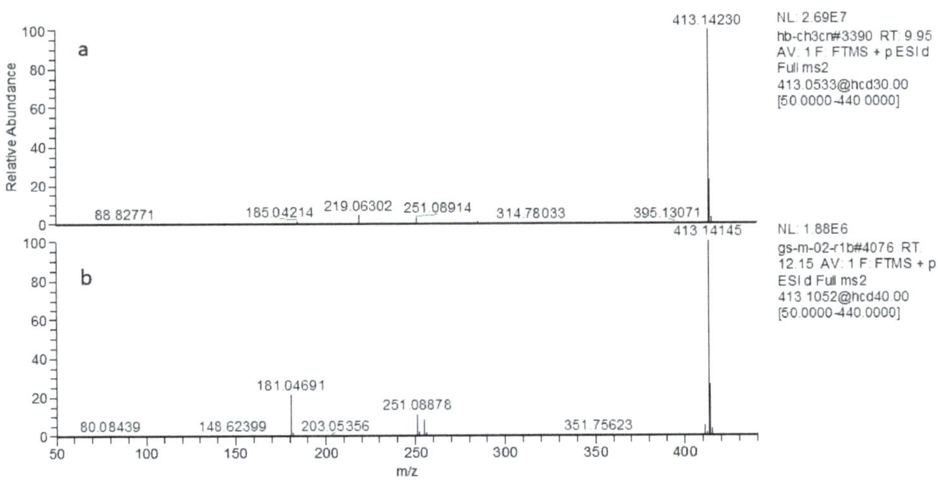

Figure 2. The mass and MS^2 spectra of (**a**) loganin and (**b**) secologanol.

Table 1. The identification of iridoid glycosides of *G. straminea* (the no. 3 sample of Sichuan province) by UPLC-Q exactive mass spectrometer.

Peak No.	RT (min)	Compound	Formula	Calculated (Da)	Selected Ion	Precursor Ion (Da)	Mass Accuracy (ppm)	Class
1	1.33	gentiobiose	$C_{12}H_{22}O_{11}$	342.11622	$[M + Na]^+$	365.10510	−0.25	sugars
2	1.91	gentianose	$C_{18}H_{32}O_{16}$	504.16904	$[M + Na]^+$	527.15784	−0.19	sugars
3	2.12	morroniside	$C_{17}H_{26}O_{11}$	406.14752	$[M + H]^+$	407.13815	−4.22	iridoids
4	2.41	eustomorusside	$C_{17}H_{24}O_{11}$	408.12678	$[M + Na]^+$	431.11603	−0.13	secoiridoids
5	2.66	miserotoxin	$C_9H_{17}O_8N$	267.09542	$[M + Na]^+$	290.08566	0.15	alkaloids
6	4.88	gladiatoside C1	$C_{29}H_{26}O_{12}$	566.14243	$[M + Na]^+$	589.13708	0.82	flavanoids
7	5.85	kingsidic acide	$C_{16}H_{22}O_{11}$	390.11622	$[M + Na]^+$	413.10522	−0.19	iridoids
8	6.20	secologanic acide	$C_{16}H_{24}O_{10}$	374.12130	$[M + Na]^+$	397.11047	−0.16	secoiridoids
9	6.49	6′-*O*-β-ᴅ-glucopyranosyl loganic acide	$C_{22}H_{34}O_{15}$	538.18978	$[M + Na]^+$	561.17889	−0.12	iridoids
10	6.59	6′-*O*-acetylgentiopicroside	$C_{16}H_{24}O_{12}$	398.12130	$[M + Na]^+$	431.15219	−0.76	iridoids
11	6.87	loganic acide	$C_{18}H_{22}O_{10}$	376.13695	$[M + Na]^+$	399.12595	−0.20	iridoids
12	7.10	kushenol I	$C_{16}H_{24}O_{10}$	454.19916	$[M + Na]^+$	477.17078	−3.81	flavanoids
13	7.29	1-*O*-β-ᴅ-gulcopyranosyl-4-epiamplexin	$C_{15}H_{22}O_{10}$	362.12130	$[M + Na]^+$	385.11685	1.49	secoiridoids
14	7.63	macrophylloside A	$C_{27}H_{24}O_9$	492.14204	$[M + H]^+$	493.14243	−1.50	secoiridoids
15	7.91	8-epi-kingsidic acide	$C_{16}H_{22}O_{11}$	390.11622	$[M + Na]^+$	413.10532	−0.17	iridoids
16	8.02	paederotoside	$C_{28}H_{34}O_{15}$	633.18252	$[M + Na]^+$	633.1796	0.46	iridoids
17	8.16	6′-*O*-β-ᴅ-glucopyranosyl gentiopicroside	$C_{22}H_{30}O_{14}$	518.16356	$[M + Na]^+$	541.15302	−0.06	secoiridoids
18	8.29	isovitexin-7-*O*-β-ᴅ-glucopyranoside	$C_{27}H_{30}O_{15}$	594.15848	$[M + Na]^+$	617.14746	−0.13	flavanoids
19	8.51	swertiamarin	$C_{16}H_{20}O_{10}$	374.12130	$[M + Na]^+$	397.11060	−0.13	secoiridoids
20	9.58	2′-acetylswertiamarin	$C_{18}H_{24}O_{11}$	416.13187	$[M + Na]^+$	439.12115	−0.12	secoiridoids
21	9.78	gentiopicroside	$C_{16}H_{20}O_9$	356.11074	$[M + Na]^+$	379.09982	−0.19	secoiridoids
22	9.95	loganin	$C_{17}H_{26}O_{10}$	390.15260	$[M + Na]^+$	413.14166	−0.18	iridoids
23	10.28	sweroside	$C_{16}H_{22}O_9$	358.12639	$[M + Na]^+$	381.11551	−0.18	secoiridoids
24	10.67	coniferin	$C_{16}H_{22}O_8$	342.13147	$[M + Na]^+$	365.11405	−1.98	lignin
25	10.85	Unknown	$C_{15}H_{24}ON_2$	248.18886	$[M + Na]^+$	271.17780	−0.32	alkaloids
26	11.85	Unknown	$C_{18}H_{18}O_{10}$	394.09000	$[M + Na]^+$	417.07639	−0.82	-
27	12.03	Homoorientin	$C_{21}H_{20}O_{11}$	448.10057	$[M + Na]^+$	471.08981	−0.12	flavanoids
28	12.35	Secologanol	$C_{17}H_{26}O_{10}$	390.15260	$[M + Na]^+$	413.14172	−0.16	secoiridoids
29	12.61	qinjiaoside A	$C_{17}H_{24}O_{11}$	404.13187	$[M + Na]^+$	427.12112	−0.13	secoiridoids
30	13.36	8-epikingisde/7-ketologanin	$C_{17}H_{24}O_{10}$	388.13695	$[M + Na]^+$	411.12610	−0.16	iridoids
31	14.54	vitexin/isovitexin	$C_{21}H_{20}O_{10}$	432.10565	$[M + Na]^+$	455.09482	−0.14	flavanoids
32	15.80	olivieroside A	$C_{25}H_{26}O_{11}$	502.14752	$[M + H]^+$	503.13646	−3.75	secoiridoids
33	16.20	pneumonanthoside	$C_{19}H_{30}O_7$	370.19916	$[M + Na]^+$	393.18814	−0.21	lignin
34	16.40	isoorientin-4′-*O*-glucoside	$C_{27}H_{30}O_{16}$	610.15339	$[M + Na]^+$	633.12128	−3.46	flavanoids
35	16.60	saprosmoside H	$C_{34}H_{42}O_{21}S$	818.19394	$[M + Na]^+$	841.21564	3.79	secoiridoids
36	17.17	6-p-coumaroy barlerin	$C_{28}H_{34}O_{14}$	594.19486	$[M + Na]^+$	617.16246	−3.60	iridoids

Table 1. *Cont.*

Peak No.	RT (min)	Compound	Formula	Calculated (Da)	Selected Ion	Precursor Ion (Da)	Mass Accuracy (ppm)	Class
37	17.36	7-*O*-feruloylorientin	$C_{31}H_{28}O_{14}$	624.14791	$[M + Na]^+$	647.13690	−0.13	flavanoids
38	18.26	flavoconomelin	$C_{28}H_{32}O_{15}$	608.17413	$[M + Na]^+$	631.14191	−3.49	flavanoids
39	19.27	alboside III	$C_{22}H_{32}O_{15}$	536.17413	$[M + Na]^+$	559.1639	2.36	secoiridoids
40	19.99	rindoside	$C_{35}H_{42}O_{21}$	798.22187	$[M + Na]^+$	821.21045	−0.15	secoiridoids
41	20.40	unknown	$C_{20}H_{30}O_5$	350.20933	$[M + Na]^+$	373.19589	−0.87	–
42	20.70	triforoside	$C_{35}H_{42}O_{20}$	782.22695	$[M + Na]^+$	805.21564	−0.14	secoiridoids
43	21.29	oliveramine	$C_{20}H_{20}N_2O_4$	352.14231	$[M + Na]^+$	375.12127	−2.89	alkaloids
44	22.15	2-methoxyanofinic acid	$C_{14}H_{16}O_4$	248.10486	$[M + Na]^+$	271.08786	−2.51	phenolic acides
45	23.53	1β,2α,3α,24-tetrahydroxyursa-12,20(30)-dien-28-oic acid	$C_{30}H_{46}O_6$	502.32944	$[M + H]^+$	503.33640	−0.17	triterpenoids
46	23.81	1β,2α,3α,24-tetrahydroxyurs-12-en-28-oic acid	$C_{30}H_{48}O_6$	504.34509	$[M + H]^+$	505.35211	−0.15	triterpenoids

Notably, secoiridoids have always produced fragmentation by Retro–Diels–Alder (RDA) cleavage of the aglycon moiety [24] and these fragmentations are quite different from that of iridoids. Secologanol (a type of secoiridoid), eluted at 12.35 min, showed fragment ions at m/z 413.141172 $[M + Na]^+$ (Figure 2b) and m/z 803.29358 $[2M + Na]^+$ (Supplementary Figure S1b), and after generated $[M + Na - Glu]^+$ at m/z 251.08833 Da, which was identical to the aglycone fragment corresponding to the neutral loss of a glucose unit (Δm = 162 Da). The precursor ion produced characterized ions at m/z 181.04694 with an RDA cleavage reaction of the base skeleton and showed the neutral loss of C_4H_6O (Δm = 70 Da). The proposed fragmentation pathway of the secologanol is shown in Figure 3b.

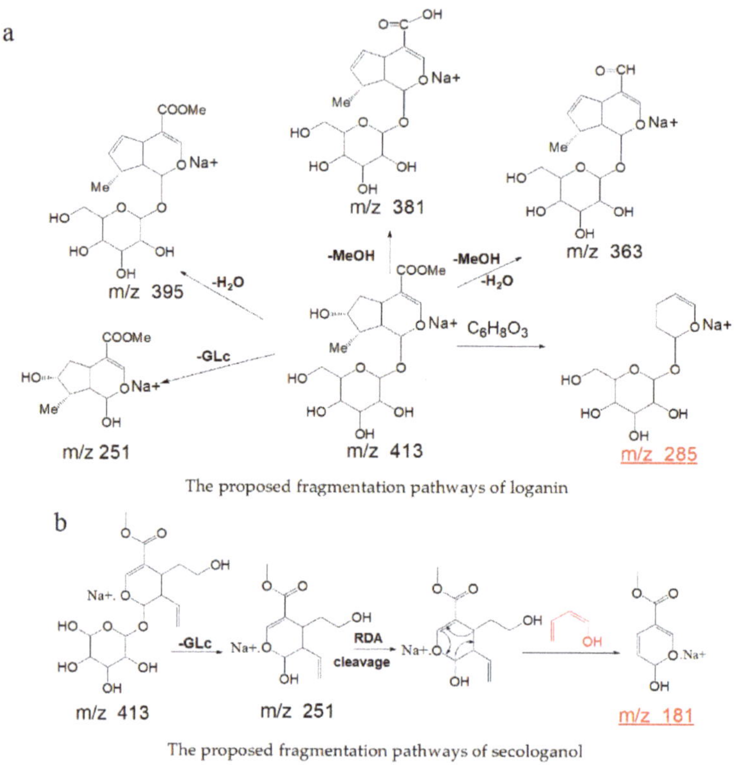

Figure 3. (**a**,**b**) The proposed fragmentation pathways of loganin and secologanol, respectively.

2.2. Multivariate Analysis of the Global Metabolomics Data

To globally evaluate the chemical consistency of *G. straminea* samples of different geographical origins, the UPLC-Q exactive mass datasets were subjected to partial least-squares discriminant analysis (PLS-DA) and orthogonal partial least squared discriminant analysis (OPLS-DA) to highlight differences among the *G. straminea* samples. As shown in Figure 4a, the 42 samples were roughly clustered into three groups by PLS-DA analysis. With OPLS-DA analysis, all of the samples were clearly categorized into three groups in 3D space (Figure 4b), 10 samples (green dots) from Gansu province were assigned to group I, 7 samples (blue dots) from Qinghai were assigned to group II, and 25 samples (red dots) from Sichuan province were assigned to group III.

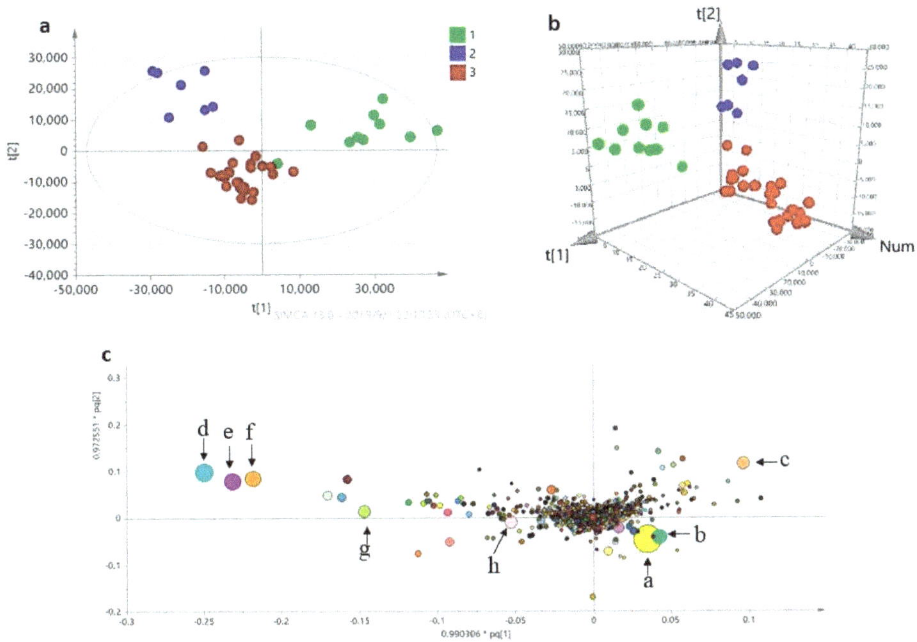

Figure 4. (**a**) Partial least-squares discriminant analysis (PLS-DA) score plot of *G. straminea* from three geographical origins. Green dots: samples from Gansu province, blue dots: samples from Qinghai province, red dots: samples from Sichuan province. (**b**) Orthogonal partial least-squares discriminant analysis (OPLS-DA) in 3D score plot of *G. straminea* from three rgeographical origins. Green dots: samples from Gansu province, blue dots: samples from Qinghai province, red dots: samples from Sichuan province. (**c**) Loading plot of OPLS-DA analysis of *G. straminea*. compound a: gentiopicroside (t_R 9.78 min, *m*/*z* 379.09982), compound b: vitexin (t_R 14.54 min, *m*/*z* 411.12610), compound c: swertiamarin (t_R 8.51 min, *m*/*z* 397.11060), compound d: gentiobiose (t_R 1.33 min, *m*/*z* 365.10510), compound e: sweroside (t_R 10.28 min, *m*/*z* 381.11551), compound f: 2-methoxyanofinic acide (t_R 22.15 min, *m*/*z* 271.08786), compound g: loganic acide (t_R 6.87 min, *m*/*z* 399.12595), and compound h: 1β, α,3α,24-tetrahydroxyursa-12,20 (30)-dien-28-oic acid (t_R 25.53 min, *m*/*z* 503.33640).

In order to identify the most significant discriminatory features between these regions that could act as potential barcodes, an extended statistical analysis was used to provide loading score plots of OPLS-DA (Figure 4c). In this plot, according to the importance of discriminating geographical characteristics, the size and color of these points have been highlighted, as seen in Figure 4c. The eight characteristic compounds were identified as: gentiopicroside (t_R 9.78 min, *m*/*z* 379.09982); vitexin (t_R 14.54 min, *m*/*z* 411.12610); swertiamarin (t_R 8.51 min, *m*/*z* 397.11060); gentiobiose (t_R 1.33 min, *m*/*z* 365.10510); sweroside (t_R 10.28 min, *m*/*z* 381.11551); 2-methoxyanofinic acid (t_R 22.15 min, *m*/*z* 271.08786); 1β,2α,3α,24-tetrahydroxyursa-12,20(30)-dien-28-oic acid (t_R 25.53 min, *m*/*z* 503.33640); and loganic acid (t_R 6.87 min, *m*/*z* 399.12595).

In the loading score plot of OPLS-DA, it is also clearly shown that samples from Gansu province are characterized by a high content of gentiopicroside, vitexin, and loganic acid, while samples from the Sichuan habitat location had a higher relative concentration of swertiamarin, and the populations of the Qinghai province were characterized by high contents of gentiobiose, sweroside, 2-methoxyanofinic acid, and 1β,2α,3α,24-tetrahydroxyursa-12,20(30)-dien-28-oic acid.

2.3. *Anti-Inflammatory Effect of Characterize Components*

With the aid of multivariate statistical analysis, gentiopicroside was confirmed as the most characteristic marker to distinguish the geographical origin of *G. straminea*. Anti-inflammatory activity is directly associated with therapeutic effects on arthritis, thus, to further evaluate the anti-inflammatory pharmacological function of characteristic markers, the inhibiting activities of nitric oxide production were evaluated in the macrophage cell line RAW 264.7 [25].

The cytotoxicity of gentiopicroside and LPS in RAW 264.7 cells were examined using CCK8 assay. As shown in Figure 5a, no significant difference in the viability of RAW 264.7 cells were observed among groups, suggesting that the concentrations of gentiopicroside and LPS used in the present study did not show any significant cytotoxic effects on RAW 264.7 cells. The inhibition effect of NO production induced by LPS in the macrophage-derived RAW 264.7 cells of the compound was assayed. It was found that the level of NO gradually decreased in a concentration-dependent manner in gentiopicroside. At a concentration of 100 μM, the compound significantly inhibited NO generation (83.76 ± 0.57%), with IC_{50} values of 44.2 ± 6.4 μM (Figure 5b).

In addition to gentiopicroside, nitric oxide production was also suppressed by loganic acid, swertiamarin, and vitexin. These compounds possessed the most potent inhibitory activity against NO production with IC_{50} values of 23.13 ± 5.4, 13.65 ± 7.1, and 15.71 ± 6.20 μM, respectively. All the results showed that gentiopicroside, loganic acid, swertiamarin, and vitexin were able to effectively inhibit NO production induced by LPS (1 μg/mL) in a dose-dependent manner in RAW 264.7 cells (Supplementary Figure S3). Specifically, the inhibiting effect against NO production of sweroside was also measured. However, the compound did not show inhibiting activities on LPS-induced NO production in RAW 264.7 macrophages, as has been reported [26], because the cells were incubated with sweroside for only 24 h in the experiment.

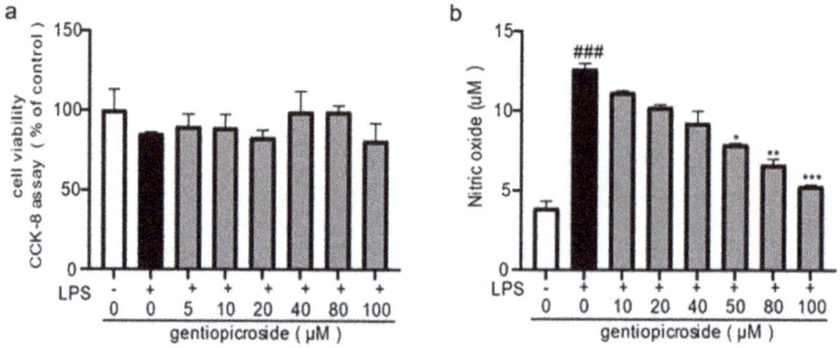

Figure 5. Effects of gentiopicroside on lipopolysaccharide (LPS)-induced NO production in RAW 264.7 cells. (**a**) RAW 264.7 cells were exposed to different concentrations of gentiopicroside (0, 5, 10, 20, 40, 80, and 100 μM) with or without LPS (1 μg/mL) for 24 h. Cell viability was determined by using the CCK-8 method. (**b**) RAW 264.7 cells were incubated with gentiopicroside (0, 5, 10, 20, 40, 80, and 100 μM) with stimulated by LPS (1 μg/mL) for 24 h. Extracellular levels of NO in culture media were measured using commercial Griess reagent. Data were folds of control and expressed as the mean ± SEM of six independent experiments. ### $p < 0.001$ compared with the control group. * $p < 0.05$, ** $p < 0.01$, *** $p < 0.001$, compared with the LPS alone.

3. Conclusions

In the present research, a selective and specificity approach was established to illustrate the chemical composition of 42 samples in *G. straminea* with a UPLC-Q exactive mass spectrometer, and an overall chemical profile of the herb was obtained. The significant differences in metabolite compositions between three geographical origins have been identified with multivariate analyses. The anti-inflammation effects of biomarkers on LPS-induced NO production in RAW264 macrophages were examined. The results suggested that samples from Gansu province have a higher content of gentiopicroside and loganic acid, and showed better anti-inflammatory effects than others. From the legal point of view [27], the result also confirmed that samples of Gansu province have better quality than other samples. Altogether, this finding is crucial in realizing the discrimination of the botanical origin of *G. straminea*, and evaluating the herb quality.

4. Discussion

The objective of the current study is the development of UPLC-Q exactive mass spectrometer methodology to allow qualitative screening of geographical origin traceability in *G. straminea*. Firstly, the chemical profiles of *G. straminea* were determined with a UPLC-Q exactive mass spectrometer, from which 43 compounds were identified by comparing the retention times and mass spectrometry. Meanwhile, a pair of isomers (loganin and secologanol) was identified by mass spectrometry based on their fragmentation pathway. Although Wu, et al. had identified 30 constituents in *G. straminea* with LC-MS [9], the result was also conducive to have a comprehensive understanding of the constituents of *G. straminea*.

Secondly, 42 samples from different habitats were determined by a UPLC-Q exactive mass spectrometer and the data were assayed with multivariate statistical analysis. Gentiopicroside, vitexin, swertiamarin, gentiobiose, sweroside, 2-metho-xyanofinic acide, loganic acide, and 1β,2α,3α,24-tetrahydroxyursa-12,20(30)-dien-28-oic acid were identified as characteristic compounds to identify the geographical origin of the herb. Notably, according to the importance of these characteristic compounds, gentiopicroside was explored as the most characteristic marker to distinguish the geographical origin of *G. straminea*. Additionally, the result also confirmed the rationality of gentiopicroside as the biomarker to determine the quality of *G. straminea*. Moreover, the result indicated that samples from Gansu province would be the most suitable choice for traditional prescriptions and preparations.

It should be emphasized that, according to the Chinese Pharmacopoeia, samples from Gansu province have been shown to have higher gentiopicroside and loganic acide amounts of some compounds than others. However, samples from Sichuan province showed a higher content of swertiamarin, and pharmacological research has revealed that this characteristic compound possesses anti-diabetic and anti-hyperlipidemic effects [28], and inhibits liver fibrosis [29]. Additionally, samples from Qinghai province have shown a higher content of sweroside, which exhibited a hepatoprotective effect [30], protective effects on osteoporosis [31], and aconitine-induced cardiac toxicity effects [32]. In view of the above reasons, it remains a challenge to estimate the herb quality of different populations. Since the herb exhibits various clinical uses in traditional prescriptions, further research should be conducted to better understand its geographical origin and its associated the clinical uses.

5. Materials and Methods

5.1. Plant Materials, Reagents, and Chemicals

Forty-two wild herbs of *G. straminea* were collected around the Qinghai-Tibet plateau in Qinghai, Sichuan, and Gansu provinces during the flowering period (the locations of the samples are provided in Table 2), individuals 10 m apart from each other were sampled randomly throughout the entire range of each location. The herbs were authenticated by Professor Yi Zhang (Chengdu University of Traditional Chinese Medicine, Chengdu, China). The samples were carefully divided into roots, leaves, and inflorescences parts, and dried in the shade. The voucher samples were deposited in the College of Ethnic Medicine (Chengdu University of Traditional Chinese Medicine, Chengdu, China) and the Qinghai Key Laboratory of Qinghai-Tibet Plateau Biological Resources (Northwest Onstitute of Plateau Biology, Chinese Academy of Science, Xining, China).

Table 2. Populations of *G. straminea* from different geographical origin.

Location	Longitude (E)	Latitude (N)	Altitude (m)	No. of Samples
Gansu	101.9308–104.7733	33.36501–34.526	2905–3572	10
Qinghai	96.6487–101.7353	33.7805–33. 9361	3516–3789	7
Sichuan	98.7732–100.5236	31.0219–33.2510	3400–3796	25

Gentiopicrin (CAS:20831-76-9), loganic acid (CAS: 22255-40-9), swertamarin (CAS: 1738839-5); loganin (CAS: 18524-94-2), vitexin (CAS: 3681-93-4), sweroside (CAS: 14215-86-2), and 6′-*O*-β-D-glucopyranosylgentiopicroside (CAS: 115713-06-9) were purchased from Biopurify Phytochemicals Ltd. (Chengdu, China). The purity of all of the standards is higher than 98% (determined by HPLC), and were confirmed by the ^{1}H-NMR spectra to those in the literature [14–20].

HPLC-grade methanol and formic acid were purchased from Merck (Darmstadt, Germany) and Tedia (Fairfield, OH, USA). Deionized water was prepared using a Millipore water treatment system (Bedford, MA, USA). Lipopolysaccharide (LPS, *Escherichia coli* 055: B5) was purchased from Sigma-Aldrich (St. Louis, MO, USA). The Cell Counting Kit-8 was purchased from Dojindo (Kyushu Japan). Griess reagents and dimethyl sulfoxide (DMSO) were purchased from Beyotime (Shanghai, China). All other reagents were of analytical grade.

5.2. Sample Preparation

The dried samples (100 mg of powder each) were made to the same concentration by resuspending in 5 mL of 70% aqueous methanol in an ultrasonic bath for 30 min and cooled at room temperature. The tubes were centrifuged twice at 12,000 rpm at room temperature for 5 min each time. The extraction was repeated three times using fresh aliquots of the solvent. The extracts were transferred to a 5 mL volumetric flask, which was then filled up to the final volume with extraction solvent. The sample solutions were filtered through a 0.22-μm pore size nylon membrane filter before injection into the UPLC. All samples were stored in a refrigerator at a temperature of 4 °C until analysis.

5.3. LC-MS/MS Analysis

The mass spectrometer Thermo Q-Exactive Plus (Thermo Scientific, San Jose, CA, USA) was equipped with heated electrospray ionization (HESI) source. Capillary temperature and vaporizer temperature were set at 330 and 280 °C, respectively, while the electrospray voltage was adjusted at 3.50 kV (operating in both positive and negative mode). Sheath and auxiliary gas were 35 and 15 arbitrary units, with an S lens RF level of 60.

The mass spectrometer was controlled by the Xcalibur 3.0 software (Thermo Fisher Scientific, San Jose, CA, USA). The exact masses of the compounds were calculated using Qualbrowser in Xcalibur 3.0 software. The mass scan range was set in the range of *m*/*z* 100–1000.

The column was a Waters Acquity UPLC BEH C18 column (100 mm × 2.1 mm, 1.7 μm particle size). The mobile phases were (a) water with 0.1% (*v*/*v*) formic acid and (b) methanol with 0.1% (*v*/*v*) formic acid. The optimized elution conditions were as follows: holding at 10% B for 2 min, a linear gradient from 10% to 13% B (all *v*/*v*) (2 to 4 min), 13% to 15% B (4 to 10 min), 15% to 17% B (10 to 15 min), 17% to 21% B (15 to 19 min), 21% to 29% B (19 to 24 min), 29% to 53% B (24 to 29 min), 53% to 75% B (29 to 35 min), 75% to 100% B (35 to 36 min), isocratic 100% B for 1 min, and then back to 7% B in 1 min. The flow rate was 0.3 mL/min. The column temperature was 35 °C. The injection volume was 2 μL.

5.4. Data Processing and Statistical Analysis

The data were processed according to the method described in the references [33]. The MS chromatograms spectra of 42 samples were processed for alignment, data reduction, and normalization by Xcalibur 3.0 software (Thermo Fisher Scientific, San Jose, CA, USA), the data were imported into

Microsoft Excel to carry out peak area normalization after being processed by Compound Discoverer 2.0, and the processed data were exported to SIMCA-P software (ver. 13.0; Umetrics, Umeå, Sweden) for data analysis. A list of the intensities of detected peaks was generated using the retention time (*tR*) and the mass data (*m/z*) pairs to identify each peak. An arbitrary ID was assigned to each *tR–m/z* pair in the order of their UPLC elution to facilitate data alignment. This procedure was repeated for each run. Ions from different samples were considered to be identical when they had the same *tR* (tolerance within 0.01 min) and *m/z* (tolerance within 0.01 Da). If a peak was not detected in a particular sample, that ion intensity was recorded as zero.

5.5. Cell Culture and Cell Viability Measurement

Murine macrophage cell line RAW 264.7 were cultured in Dulbecco's Modified Eagle Medium (DMEM, Hyclone Florida, USA) supplemented with 10% fetal bovine serum (FBS, ExCell Bio Shanghai, China), 1% Penicillin Streptomycin (Gibico California, USA) at 37 °C in a humidified atmosphere of 95% air and 5% CO_2. After spreading at 80–90% confluence, cells were washed with PBS, scraped with fresh culture, and subcultured into 96-well plates at a density of 5.0×10^3 cells/well and incubated with or without LPS (1 μg/mL). The cells were exposed to different concentrations of gentiopicroside, loganic acid, swertiamarin, and vitexin (0, 1, 5, 10, 20, 40, and 50 μM) with or without LPS (1 μg/mL) for 24 h. The optical density was measured at 450 nm using a multi-plate reader (BioTek, Winooski, VT, USA).

5.6. Nitric Oxide (NO) Assay

NO analysis was performed to evaluate inflammatory response and to measure NO release by macrophages. RAW 264.7 cells (1×10^5 cells/well) were seeded in 96-well cell culture plates and allowed to adhere for 12 h. The cells were incubated on swertiamarin and loganic acid (0, 5, 10, 20, 40, 80, and 100 μM), or swertiamarin and vitexin (0, 5, 10, 20, 40, and 50 μM), respectively, with stimulation by LPS (1 μg/mL) for 24 h [34]. NO secretion by LPS-stimulated macrophages was determined by Griess reagents (Beyotime, Shanghai, China) according to the instructions of manufacturer [35]. Absorbance was measured at 540 nm and NO concentration was determined using sodium nitrite as a standard. Three replicates were carried out for each of the different treatments.

Supplementary Materials: The following are available online at http://www.mdpi.com/1420-3049/24/24/4478/s1, Figure S1: the mass spectra of loganin and secologanol in positive ion mode by UPLC-Q Exactive mass spectrometer ($[2M + Na]^+$). Figure S2. the mass spectra of loganin and secologanol in negative ion mode by UPLC-Q Exactive mass spectrometer ($[2M + HCOOH\text{-}H]^-$). Figure S3. Effects of loganic acid, swertiamarin, and vitexin on LPS-induced NO production in RAW 264.7 cells.

Author Contributions: Z.P. and G.-Y.Z. conceived and designed the study, F.X., Y.-L.C., Z.-W.C. and G.-G.W. performed the detailed experiments; Z.P., G.-Y.Z. and W.-F.C.; Funding acquisition, F.X. and Y.Z., samples investigation, Z.P., G.-G.W., and Y.-L.C. wrote the manuscript.

Funding: This research was funded by the National Natural Science Foundation of China (grant number 81973567), and the Chong Qing Science Foundation Project (grant number CSTC2016jcyj A0288 & 2019jcyj-msxmX0180), and the Development Project of Qinghai Provincial Key Laboratory (grant number 2017-ZJ-Y10).

Conflicts of Interest: The authors declare no conflicts of interest associated with this manuscript.

References

1. Corvucci, F.; Nobili, L.; Melucci, D.; Grillenzoni, F.V. The discrimination of honey origin using melissopalynology and Raman spectroscopy techniques coupled with multivariate analysis. *Food Chem.* **2015**, *169*, 297–304. [CrossRef] [PubMed]
2. Risticevic, S.; Carasek, E.; Pawliszyn, J. Headspace solid-phase micro -extractiongas chromatographic–time-of-flight mass spectrometric methodology for geographical origin verification of coffee. *Anal. Chim. Acta* **2008**, *617*, 72–84. [CrossRef] [PubMed]

3. Bentley, J.; Moore, J.P.; Farrant, J.M. Metabolomics as a complement to phylogenetics for assessing intraspecific boundaries in the desiccation-tolerant medicinal shrub *Myrothamnus flabellifolia* (Myrothamnaceae). *Phytochemistry* **2019**, *159*, 127–136. [CrossRef] [PubMed]
4. Zhang, X.X.; Zhan, G.Q.; Jin, M.; Zhang, H.; Dang, J.; Zhang, Y.; Guo, Z.J.; Ito, Y.C. Botany, traditional use, phytochemistry, pharmacology, quality control, and authentication of Radix Gentianae Macrophyllae-A traditional medicine: A review. *Phytomedicine* **2018**, *46*, 142–163. [CrossRef] [PubMed]
5. Wang, Y.M.; Xu, M.; Wang, D.; Yang, C.R.; Zeng, Y.; Zhang, Y.J. Anti-inflammatory compounds of "Qin-Jiao", the roots of *Gentiana dahurica* (Gentianaceae). *J. Ethnopharmacol.* **2013**, *147*, 341–348. [CrossRef]
6. Yang, H.; Liu, J.; Chen, S.; Hu, F.; Zhou, D. Spatial variation profiling of four phytochemical constituents in *Gentiana straminea* (Gentianaceae). *J. Nat. Med.* **2014**, *68*, 38–45. [CrossRef]
7. Kakuda, R.; Iijima, T.; Yaoita, Y.; Machida, K.; Kikuchi, M. Secoiridoid glycosides from *Gentiana scabra*. *J. Nat. Prod.* **2001**, *12*, 1574–1575. [CrossRef]
8. Pan, Y.; Zhao, Y.L.; Zhang, J.; Li, W.Y.; Wang, Y.Z. Phytochemistry and pharmacological activities of the *Genus Gentiana* (Gentianaceae). *Chem. Biodivers.* **2016**, *13*, 107–150. [CrossRef]
9. Wu, J.R.; Zhao, Z.L.; Wu, L.H.; Wang, Z.T. Authentication of *Gentiana straminea* Maxim. and its substitutes based on chemical profiling of iridoids using liquid chromatography with mass spectrometry. *Biomed. Chromatogr.* **2016**, *30*, 2061–2066. [CrossRef]
10. Wei, S.; Zhang, P.; Feng, X.; Kodama, H.; Yu, C.; Chen, G. Qualitative and quantitative determination of ten iridoids and secoiridoids in *Gentiana straminea* Maxim. by LC-UV-ESI-MS. *J. Nat. Med.* **2012**, *66*, 102–108. [CrossRef]
11. Pan, Z.; Fan, G.; Yang, R.P.; Luo, W.Z.; Zhou, X.D.; Zhang, Y. Discriminating *Lamiophlomis rotata* according to geographical origin by ^{1}H-NMR spectroscopy and multivariate analysis. *Phytochem. Anal.* **2015**, *26*, 247–252. [CrossRef] [PubMed]
12. Zhang, D.; Gao, Y.L.; Jiang, S.; Chen, Y.W.; Zhang, Y.; Pan, Z. The similarity and variability of iridoid glycosides profile and antioxidant capacity of aerial and underground parts from *Lamiophlomis rotata* using UPLC-TOF-MS and multivariate analyses. *RSC Adv.* **2018**, *8*, 2459–2468. [CrossRef]
13. Wang, J.; Gao, Y.L.; Chen, Y.L.; Chen, Y.W.; Zhang, Y.; Xiang, L.; Pan, Z. *Lamiophlomis rotata* Identification via ITS2 Barcode and Quality Evaluation by UPLC- QTOF-MS Couple with Multivariate Analyses. *Molecules* **2018**, *23*, 3289.
14. Chen, G.; Wei, S.H.; Yu, C.Y. Secoiridoids from the roots of *Gentiana straminea*. *Biochem. Syst. Ecol.* **2009**, *37*, 766–771. [CrossRef]
15. Xu, M.; Zhang, M.; Zhang, Y.J.; Yang, C.R. New Acylated Secoiridoid Glucosides from *Gentiana straminea* (Gentianaceae). *Helv. Chim. Acta* **2009**, *92*, 321–327. [CrossRef]
16. Jiang, Z.B.; Liu, H.L.; Liu, X.Q.; Shang, J.N.; Zhao, J.; Yuan, C.S. Chemical constituents of *Gentiana macrophylla* Pall. *Nat. Prod. Res.* **2010**, *24*, 1365–1369. [CrossRef]
17. He, Y.M.; Zhu, S.; Ge, Y.W.; Cai, S.Q.; Komatsu, K. Secoiridoid glycosides from the root of *Gentiana crassicaulis* with inhibitory effects against LPS-induced NO and IL-6 production in RAW264 macrophages. *J. Nat. Med.* **2015**, *69*, 366–374. [CrossRef]
18. Tan, R.X.; Kong, L.D.; Wei, H.X. Secoiridoid glycosides and an antifungal anthranilate derivative from *Gentiana tibetica*. *Phytochemistry* **1998**, *47*, 1223–1226. [CrossRef]
19. Zhang, L.; Dang, J.; Mei, L.J.; Shao, Y.; Wang, Q.L.; Liu, X.J. Chemical constituents from roots of *Gentiana straminea* of Tibetan Medicine. *J. Chin. Med. Mater.* **2016**, *39*, 103–106. [CrossRef]
20. Fan, H.; Zang, Y.; Zhang, Y.; Zhang, H.F.; Zhao, Z.; Hu, J.F. Triterpenoids and Iridoid Glycosides from *Gentiana dahurica*. *Helv. Chim. Acta* **2010**, *93*, 2439–2447. [CrossRef]
21. Mandova, T.; Audo, G.; Michel, S.; Grougnet, R. Off-line coupling of new generation centrifugal partition chromatography device with preparative high pressure liquid chromatography-mass spectrometry triggering fraction collection applied to the recovery of secoiridoid glycosides from *Centaurium erythraea* Rafn. (Gentianaceae). *J. Chromatogr. A* **2017**, *1513*, 149–156. [PubMed]
22. Wang, S.F.; Xu, Y.M.; Chen, P.H.; Zhang, Y.F. Structural characterization of secoiridoid glycosides by high performance liquid chromatography/electrospray ionization. *Rapid Commun. Mass Spectrom.* **2014**, *28*, 1569–1579. [CrossRef] [PubMed]

23. Suryawanshi, S.; Mehrotra, N.; Asthana, R.K.; Gupta, R.C. Liquid chromatography/tandem mass spectrometric study and analysis of xanthone and secoiridoid glycoside composition of Swertia chirata, a potent antidiabetic. *Rapid Commun. Mass Spectrom.* **2006**, *20*, 3761–3768. [CrossRef] [PubMed]
24. Han, H.; Zeng, W.L.; He, C.Y.; Annie Bligh, S.W.; Liu, Q.; Yang, L.; Wang, Z.T. Characterization of metabolites of sweroside in rat urine using ultra-high-perfor mance liquid chromatography combined with electrospray ionization quadrupole time-of-flight tandem mass spectrometry and NMR spectroscopy. *J. Mass Spectrom.* **2014**, *49*, 1108–1116. [CrossRef] [PubMed]
25. Deng, Y.T.; Wang, X.S.; Zhao, M.G.; Huang, X.X.; Xu, X.L. Gentiopicroside protects neurons from astrocyte-mediated inflammatory injuries by inhibition of nuclear factor-κB and mitogen-activated protein kinase signaling pathways. *Neuroreport* **2018**, *29*, 1114–1120. [CrossRef] [PubMed]
26. Wang, R.; Dong, Z.; Lan, X.; Liao, Z.; Chen, M. Sweroside Alleviated LPS-Induced Inflammation via SIRT1 Mediating NF-κB and FOXO1 Signaling Pathways in RAW264.7 Cells. *Molecules* **2019**, *24*, 872. [CrossRef]
27. State Pharmacopoeia Commission. *Chinese Pharmacopoeia*; Chemical and Technologic Press: Beijing, China, 2015; pp. 270–271, ISBN 9787506744393.
28. Patel, N.; Tyagi, R.K.; Tandel, N.; Garg, N.; Soni, N. The Molecular Targets of Swertiamarin and its Derivatives Confer Anti-Diabetic and Anti-Hyperlipidemic Effects. *Curr. Drug Targets* **2018**, *19*, 1958–1967. [CrossRef]
29. Li, S.; Wang, Q.; Tao, Y.; Liu, C. Swertiamarin Attenuates Experimental Rat Hepatic Fibrosis by Suppressing Angiotensin II-Angiotensin Type 1 Receptor-ERK Signaling. *J. Pharmacol. Exp. Ther.* **2016**, *395*, 247–255. [CrossRef]
30. Ma, L.Q.; Yu, Y.; Chen, H.; Li, M.; Ihsan, A.; Tong, H.Y.; Huang, X.J.; Gao, Y. Sweroside Alleviated Aconitine-Induced Cardiac Toxicity in H9c2 Cardiomyoblast Cell Line. *Front. Pharmacol.* **2018**. [CrossRef]
31. Ding, Y.; Jiang, H.; Meng, B.; Zhu, B.; Yu, X.; Xiang, G. Sweroside-mediated mTORC1 hyperactivation in bone marrow mesenchymal stem cells promotes osteogenic differentiation. *Cell. Biochem.* **2019**. [CrossRef]
32. Mihailović, V.; Mihailović, M.; Uskoković, A.; Arambašić, J.; Mišić, D.; Stanković, V.; Katanić, J.; Mladenović, M.; Solujić, S.; Matić, S. Hepatoprotective effects of Gentiana asclepiadea L. extracts against carbon tetrachloride induced liver injury in rats. *Food Chem. Toxicol.* **2013**, *52*, 83–90. [CrossRef] [PubMed]
33. Gao, X.X.; Liang, M.L.; Fang, Y.; Zhao, F.; Tian, J.S.; Zhang, X.; Qin, X.M. Deciphering the Differential Effective and Toxic Responses of Bupleuri Radix following the Induction of Chronic Unpredictable Mild Stress and in Healthy Rats Based on Serum Metabolic Profiles. *Front. Pharmacol.* **2018**, *8*, 995. [CrossRef] [PubMed]
34. He, H.Q.; Wu, Y.X.; Nie, Y.J.; Wang, J.; Ge, M.; Qian, F. LYRM03, an ubenimex derivative, attenuates LPS-induced acute lung injury in mice by suppressing the TLR4 signaling pathway. *Acta Pharmacol. Sin.* **2017**, *38*, 342–350. [CrossRef] [PubMed]
35. Somensi, N.; Rabelo, T.K.; Guimarães, A.G.; Quintans-Junior, L.J.; de Souza Araújo, A.A.; Moreira, J.C.F.; Gelain, D.P. Carvacrol suppresses LPS-induced pro-inflammatory activation in RAW 264.7 macrophages through ERK1/2 and NF-kB pathway. *Int. Immunopharmacol.* **2019**, *75*, 105743. [CrossRef] [PubMed]

Sample Availability: Samples of *Gentiana straminea* are available from the authors.

molecules

MDPI

Article

Constituents and Anti-Multidrug Resistance Activity of *Taiwanofungus camphoratus* on Human Cervical Cancer Cells

Hsin-Yi Hung [1], Chin-Chuan Hung [2], Jun-Weil Liang [3], Chin-Fu Chen [4], Hung-Yi Chen [2], Po-Chuen Shieh [5], Ping-Chung Kuo [1,*] and Tian-Shung Wu [1,5,*]

1 School of Pharmacy, College of Medicine, National Cheng Kung University, Tainan 701, Taiwan; z10308005@email.ncku.edu.tw
2 Department of Pharmacy, College of Pharmacy, China Medical University, Taichung 402, Taiwan; cc0206hung@gmail.com (C.-C.H.), hungyi@mail.cmu.edu.tw (H.-Y.C.)
3 Department of Chemistry, National Cheng Kung University, Tainan 701, Taiwan; p77625@hotmail.com
4 Department of Life Sciences, National Cheng Kung University, Tainan 701, Taiwan; chinfu9999@gmail.com
5 Department of Pharmacy, College of Pharmacy and Health Care, Tajen University, Pingtung 907, Taiwan; pochuen@tajen.edu.tw (P.-C.S.)
* Correspondence: z10502016@email.ncku.edu.tw (P.-C.K.); tswu@mail.ncku.edu.tw (T.-S.W.); Tel.: +886-6-2353535-6806 (P.-C.K.); +886-2353535-65333 (T.-S.W.)

Academic Editor: Paula B. Andrade
Received: 16 September 2019; Accepted: 15 October 2019; Published: 16 October 2019

check for updates

Abstract: Resistance to anti-cancer drugs is one of the main factors of treatment failure resulting in high morbidity. Among the reasons of resistance, overexpression of efflux pumps leading to multidrug resistance is an important issue that needs to be solved. *Taiwanofungus camphoratus* has been used as a nutritional supplement to treat various cancers. However, its effects on the resistance to chemotherapeutic agents are still unknown. In this study, we report four new chemical constituents of *T. camphoratus* isolated from an ether extract: camphoratins K (**1**) and N (**2**) and benzocamphorins G (**3**) and I (**4**). Furthermore, we evaluated zhankuic acids A–C for their P-glycoprotein (P-gp) inhibitory effects. The results showed that zhankuic acid A was the most potent P-gp inhibitor compound and (at 20 μM) could reverse drug resistance in human cancer cells, restoring an IC_{50} of 78.5 nM for doxorubicin, of 48.5 nM for paclitaxel, and of 321.5 nM for vincristine, indicating a reversal fold of 48, 38, and 45 times, respectively. This study provides support for the use of *T. camphoratus* in the further development of cancer therapy.

Keywords: *Taiwanofungus camphoratus*; chemoreversing agent; zhankuic acid; P-gp inhibitor

1. Introduction

Treatment failure or metastasis are still the leading causes of death for cancer patients [1]. One of the important factors causing treatment failure is drug resistance, which can be intrinsic or acquired [2,3]. Cancer stem cells [4], tumor microenvironment, and host effects are the main reasons of intrinsic resistance, while efflux pumps, alteration of drug targets, degradation of anticancer drugs, and DNA self-repair cause acquired resistance. In particular, overexpression of efflux pumps leads to multidrug resistance (MDR) [2,3]. In order to increase the efficacy of anticancer drugs, strategies to reverse multidrug resistance are extensively studied, and P-glycoprotein (P-gp) inhibitors have evolved to the fourth generation [5–8]. However, because of their inability to of improve drug efficacy in patients as well as their toxicity, P-gp inhibitors are not yet available in the clinic. Therefore, research is now focusing on natural products, hoping to find safe and effective P-gp inhibitors [5–8].

Taiwanofungus camphoratus, previously named *Ganoderma camphoratum*, *Antrodia cinnamomea*, *or Antrodia camphorate* (Polyporaceae, Aphyllophorales), whose Chinese name is Zhan-Ku or Niu-Chang-Chih, is a kind of fungus parasitic to *Cinnamomum kanehirai* Hay (Lauraceae), found in the inner part of old hollow trunks. Traditionally, it has been used as a kind of medicinal food against intoxication from food, alcohol, and drugs and for its anti-diarrhea, anti-hypertensive, anti-inflammatory, and hepatoprotective effects [9]. Recent studies have revealed that Niu-Chang-Chih exerts immunomodulatory effects [10,11], anti-lung cancer effects [12–14], and tumor-suppressive effects in metastatic patients unresponsive to or unwilling to use chemotherapy [15]. A recent published review summarized the pharmacological effects of this mushroom [16] reporting its anticancer activity against a large variety of cancers, including breast, cervical, ovarian, prostate, bladder, colorectal, pancreatic, liver, and lung cancers, melanoma, leukemia, lymphoma, neuroblastoma, and glioblastoma. Other biological activities include anti-inflammatory, anti-atopic dermatitis, anti-cachexia, immunoregulatory, anti-obesity, anti-diabetic, anti-hyperlipidemic, anti-atherosclerotic, anti-hypertensive, anti-platelet, anti-oxidative, anti-photodamaging, hepatoprotective, renoprotective, neuroprotective, testis protecting, anti-asthmatic, osteogenic, osteoprotective, antiviral, antibacterial, and wound healing properties [16]. The major chemical constituents of this fungus are triterpenoids [17–19] and benzenoids [20,21]. Other components are steroids [22], diterpenoids [23], terpenoids [24], lignans [22], maleic and succinic acid derivatives [25], etc. Although the extract of *T. camphoratus* has been used as a nutritional supplement for treating cancers, the P-gp inhibitory effects of its main constituents are still unknown. Therefore, in addition to reporting new chemical constituents of *T. camphoratus*, this study reveals the P-gp inhibitory effects of zhankuic acids A–C.

2. Results and Discussion

2.1. Purification and Identification of Chemical Constituents

The basswood cultivated fruiting bodies of *T. camphoratus* (3.6 kg) were repeatedly extracted with ether (4 × 10 L) for 3 days. The ether extract was concentrated in vacuo to afford a brown syrup (370 g) and then partitioned between water and ether. The ether layer was chromatographed repeatedly over silica gel, as described in Supplementary Materials. In total, 45 compounds were obtained. Among them, two triterpenoids, camphoratins K (**1**) and N (**2**), and two benzenoids, benzocamphorins G (**3**) and I (**4**), were isolated and characterized from *T. camphoratus* for the first time. Other known isolated compounds were triterpenoids, including methyl antcinate A (**5**), antcins A (**6**), C (**12**), and K (**18**), zhankuic acid A methyl ester (**7**), zhankuic acids A (**8**), B (**11**), C (**9**), and D (**10**), camphoratins E (**13**) and F (**14**), methyl antcinate (**15**), antcamphins A (**19**), B (**16**), and D (**17**); terpenoids, including 1-hydroxy-*p*-menth-3-en-2-one (**20**), nerolidol (**21**), coenzyme Q (**22**), 4-acetylantroquinonol B (**23**); steroids, including ergosterol (**24**), ergosterol peroxide (**25**), camphoratin I (**26**); lignans, including sesamin (**27**) and 4-hydroxysesamin (**28**); and benzenoids, including antrocamphins A (**30**) and B (**29**), benzocamphorins C (**37**), D (**44**), E (**43**), F (**31**), and H (**32**), methyl 3,4,5-trimethoxybenzoate (**33**), methyl 2,3,4,5-tetramethoxy benzoate (**34**), 1-methyl-2,3,4,5-trimethoxy benzene (**35**), 2,3,6-trimethoxy-5-methylphenol (**36**), methyl 2,5-dimethoxy-3,4-methylenedioxybenzoate (**38**), 4,5-dimethoxy-6-methyl-1,3-benzodioxole (**39**), 4,7-dimethoxyl-5-methyl-1,3-benzodioxole (**40**), 2,3-(methylenedioxy)-4-methyl-5-methylphenol (**41**), 2,2,5,5-tetramethoxyl-3,4,3,4-bimethylenedioxyl-6,6-dimethylbiphenyl (**42**), and tetracanyl ferulate (**45**) (see Supplementary Materials for their references).

*2.2. Structural Elucidations of Camphoratins K (**1**) and N (**2**) and Benzocamphorins G (**3**) and I (**4**)*

Camphoratin K (**1**) was isolated as a white powder, and its sodiated molecular formula, $C_{33}H_{54}O_4Na$, was established from a sodium adduct ion peak at *m/z* 537.3917 in high-resolution electrospray ionization mass spectrometric (HR-ESI–MS) analysis. The infrared (IR) absorption bands at 3427, 1714, 1643, 1455, and 891 cm^{-1} were in agreement with the presence of a hydroxyl group, an ester, and a terminal double bond. In its ^{1}H-NMR spectrum, there were proton signals for five methyl singlets at δ 0.80 (6H, s, CH_3-18, 29), 0.97 (3H, s, CH_3-19), 0.99 (3H, s, CH_3-30), 1.01 (3H, s, CH_3-31), 2.04 (3H, s, CH_3-33), two methyl doublets at δ 1.01 (3H, d, J = 6.8 Hz, CH_3-27) and 1.02 (3H, d, J = 6.8 Hz, CH_3-26), and four protons at δ 2.22 (1H, sept, J = 6.8 Hz, H-25), 3.22 (1H, dd, J = 4.4 Hz, 11.6 Hz, H-3), 3.69 (2H, m, H-21), 5.05 (1H, dd, J = 5.8, 9.4 Hz, H-15). The ^{13}C-, DEPT- and HMQC NMR spectra showed 33 carbon signals composed of 8 methyls at δ 15.4, 16.4, 18.3, 19.1, 21.4, 21.8, 21.9, 28.0; 9 methylenes at δ 18.2, 20.8, 26.4, 27.7, 28.2, 30.6, 31.4, 35.5, 36.0; 1 oxygenated methylene at δ 62.0; 2 oxygenated methines at δ 76.0, 78.9, and 1 terminal olefinic carbon at δ 106.4, which indicated a triterpene skeleton. An acetyl group was assigned to link to C-15 from HMBC spectral correlations of H-15 (δ 5.05) to C-16 (δ 36.0) and C-32 (δ 171.1). A terminal olefinic group and an isopropyl group were built up via ^{2}J, ^{3}J-HMBC correlations of CH_3-26 (δ 1.02) to C-24 (δ 156.1), C-25 (δ 33.7), and C-27 (δ 21.9) and H-28 (δ 4.73 and 4.67) to C-23 (δ 31.4), C-24 (δ 156.1), and C-25 (δ 33.7). Other HMBC correlations indicated a sulphurenic acid skeleton with a hydroxyl methylene [H-21 (δ 3.69) to C-17 (δ28.0)/C-20 (δ 43.2)] instead of an acid connected to C-20. This C-21 hydroxyl substitution is novel and rare among all isolates from *T. camphoratus* (Figure 1).

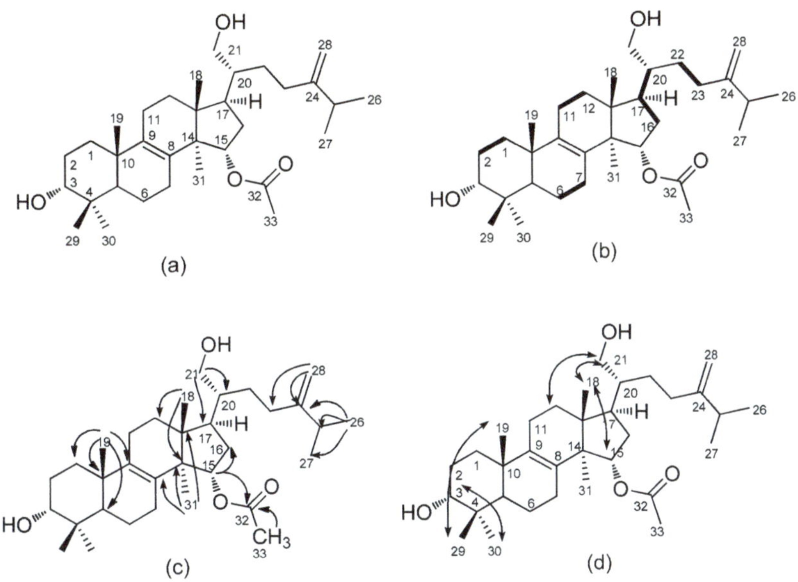

Figure 1. Structure of camphoratin K (**1**) (**a**) and its key COSY (**b**), HMBC (**c**), and NOESY (**d**) correlations.

Camphoratin N (**2**) appeared as a pale yellow solid with sodiated molecular formula $C_{30}H_{42}O_6Na$ (*m/z* 521.2876). The presence of an 8(9)-ene-7,11-dione moiety was proposed along with those of a carboxyl group and a hydroxyl group, according to a UV maximum at 267 nm and IR absorptions at 3495, 1736, 1713, 1674, 1458, and 901 cm^{-1}, respectively. Comparison of its ^{13}C-NMR data with those of antcamphin I [18] indicated the presence of an additional oxygenated methyl group. Assignment of the oxygenated methyl group attached to a terminal carboxylate was based on the HMBC correlation of CH_3-26 (δ 3.67) to C-26 (δ 175.0). Also, 12α-OH and 29α-CH_3 were suggested via the NOE enhancements of H-12 (δ 4.11)/CH_3-18 (δ 0.67) and CH_3-19 (δ 1.54)/H-4 (δ 2.43) (Figure 2). According to a previous study, camphoratin N could include a pair of epimers (25 *S/R*) that have identical NMR data [18].

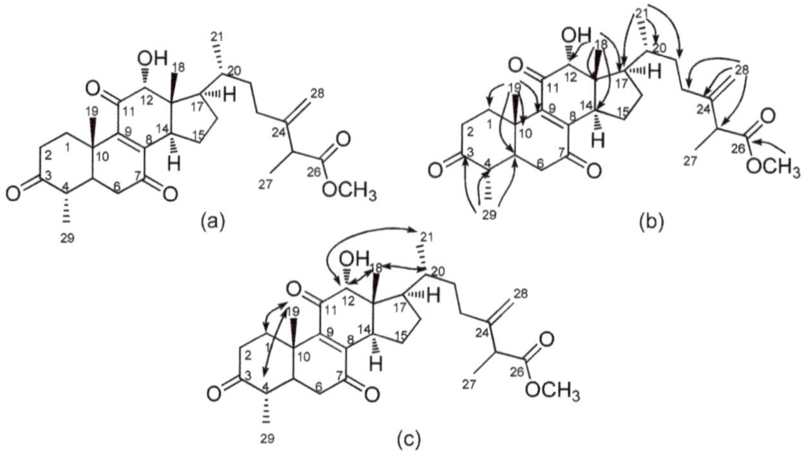

Figure 2. Structure of camphoratin N (**2**) (**a**) and its key HMBC (**b**) and NOESY (**c**) correlations.

Benzocamphorin G (**3**), a colorless syrup, was isolated via thin-layer chromatography and has the pseudomolecular formula of $C_{13}H_{12}O_3Na$, constructed from the sodiated peak at *m/z* 239.0686 in HR-ESI–MS analysis. Characteristic absorption bands in its IR spectrum revealed alkynes (2205 cm^{-1}), conjugated carbonyls (1667 cm^{-1}), and alkenes (1607 cm^{-1}). UV absorption maxima were at 268 and 296 nm. A comparison of its NMR spectra data with those of antrocamphin A [18] indicated similar proton peaks at δ 5.44 (1H, s, terminal alkene), 5.55 (1H, s, terminal alkene), 3.82 (3H, s, OCH_3), 2.22 (3H, s, CH_3-3), and 2.01 (3H, s, CH_3-3′) as well as downfield shift of one aromatic proton (δ 5.99, 1H, s, H-6) and loss of two methoxy signals. ^{13}C- and DEPT-135 NMR spectra revealed a similar pattern to that of antrocamphin A, except for a pair of *ortho*-carbonyls (δ 181.1 and 181.3), a downfield shift of C-6 (δ 107.0), and a loss of two methyl carbons. The positions of the *ortho*-carbonyls were assigned to be at C-1 and C-2 via HMBC correlations of H-6 (δ 5.90)/ with C-1 (δ 183.1), with C-2 (δ 181.3), with C-4 (δ 129.4), with C-5 (δ 158.8); and of CH_3-3 (δ 2.22) with C-2 (δ 181.3), with C-3 (δ 144.6), with C-4 (δ 129.4), with C-1′ (δ 81.7), respectively. Therefore, the structure of benzocamphorin G was established and is shown in Figure 3.

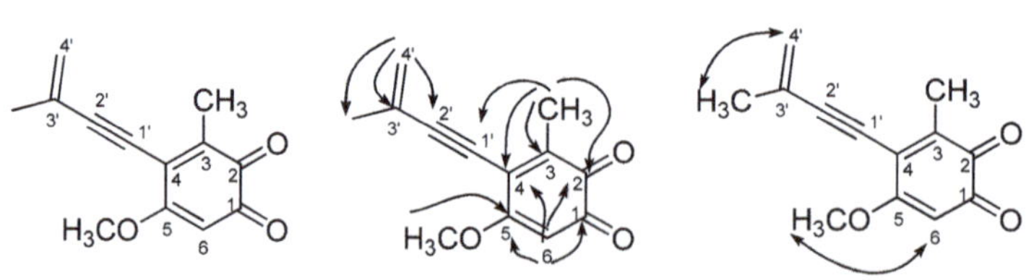

Figure 3. Structure of benzocamphorin G (**3**) (**a**) and its key HMBC (**b**) and NOESY (**c**) correlations.

Benzocamphorin I (**4**) was also isolated via thin-layer chromatography as colorless syrup. HR-ESI–MS analysis indicated its pseudomolecular formula as $C_{18}H_{18}O_8Na$ (*m/z* 385.0898). No alkynes and alkene characteristic peaks were detected in its IR spectrum. Three pairs of proton signals revealed two methyls (δ 1.99, 2.05), two methylenedioxy groups (δ 5.98, 6.00), and two methoxys (δ 3.89, 3.90). Further ^{13}C-, DEPT, and HSQC spectra indicated a benzocamphorin D skeleton [19] with an oxygen linkage between two phenyl groups. However, one methoxy signal was lacking, and an oxygenated aromatic carbon was present compared to benzocamphorin D [19], showing the methoxy group was replaced by a hydroxyl group. Thus, the structure of benzocamphorin I was determined and is shown in Figure 4.

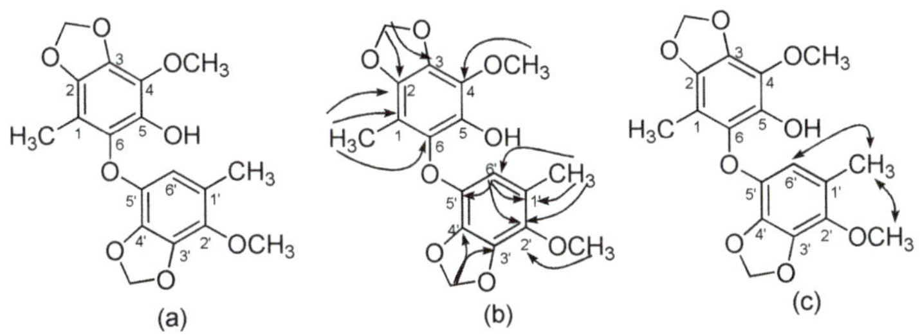

Figure 4. Structure of benzocamphorin I (**4**) (**a**) and its key HMBC (**b**) and NOESY (**c**) correlations.

2.3. *P-gp Inhibitory Effects of the Extract of T. camphoratus*

A pilot study using methanol as a solvent was done to evaluate the P-gp inhibitory effects of the extract of *T. camphoratus* (Figure 5). The methanol extract was further partitioned using water and EtOAc. Therefore, the methanol extract (TAM), the EtOAc layer (TAE), and the water layer (TAW) were tested using human stably P-gp-expressing cells (ABCB1/Flp-InTM-293) in a calcein AM (acetoxymethyl) uptake assay [26]. The increased intracellular calcein fluorescence corresponded to the inhibition level of P-gp efflux function. The methanol extract as well as the EtOAc layer and the water layer exhibited P-gp inhibitory activities at concentrations of 10 and 20 μM. The methanol extract and the EtOAc layer exhibited inhibition in a dose-dependent manner. Moreover, the methanol extract at 20 μM (TAM 20) showed P-gp inhibition comparable to that of the first-generation P-gp inhibitor verapamil at a concentration of 2.5 μM.

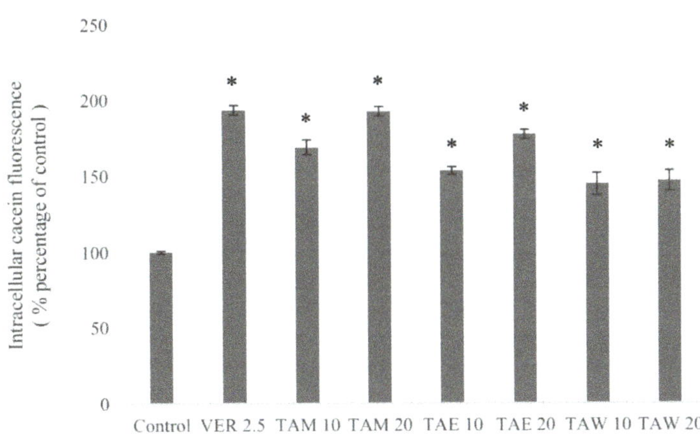

Figure 5. The inhibitory effects of the methanol extract at 10 and 20 μM concentrations (TAM 10, TAM 20), the EtOAc layer (TAE 10, TAE 20), the water layer (TAW 10, TAW 20) and verapamil at 2.5 μM concentration (VER 2.5) on P-glycoprotein (P-gp) in *ABCB1*/Flp-In^{TM}-293 cells. * denotes $p < 0.05$ compared with the intracellular calcein fluorescence in the control group. The numbers, 2.5, 10, 20, indicate the μM concentrations.

2.4. *Zhankuic Acids A–C Inhibited P-gp Efflux Function*

Although four new compounds were isolated, they were in little amount. In order to better understand the main P-gp inhibitory effects of *T. camphoratus*, three of its major components, zhankuic acids (ZAs) A–C, were evaluated for their ability to inhibit P-gp using the calcein AM uptake assay. ZAs A, B, C inhibited P-gp efflux function in a concentration-dependent manner (Figure 6). Among the tested compounds, ZA-A demonstrated the most significant P-gp inhibitory effect.

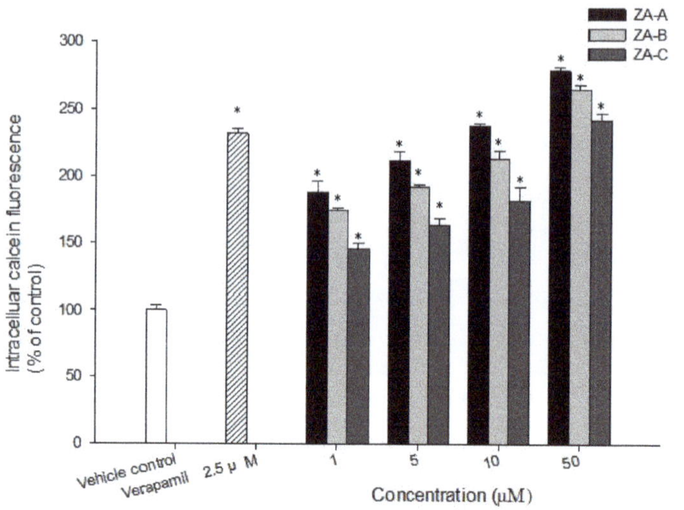

Figure 6. The concentration-dependent inhibitory effects of zhankuic acids (ZAs) A, B, C on P-gp in *ABCB1*/Flp-In^{TM}-293 cells. * denotes $p < 0.05$ compared with the intracellular calcein fluorescence in control group.

2.5. The MDR Reversal Effects of ZAs A, B, C

To examine the MDR reversal effects of ZAs A, B, C, the cytotoxicity of a combination of these triterpenoids and chemotherapeutic drugs was evaluated in HeLaS3 and MDR KBvin cells. The IC_{50} of doxorubicin, paclitaxel, and vincristine in HeLa cells were 104 nM, 4.65 nM, and 41.5 nM, while in KBvin cells they were 3750 nM, 1824 nM, and 14,540 nM, indicating high multidrug resistance of the cells. When ZAs A–C were combined with the chemotherapeutic agents, the IC_{50} of doxorubicin, paclitaxel, and vincristine in MDR KBvin cells were significantly decreased (Table 1). Reversal folds were calculated by dividing the IC_{50} of the individual chemotherapeutic drug by the IC_{50} of the compound–drug combinations. ZA-A possessed the most significant MDR reversal effect among the tested compounds. It (20 μM) reversed drug resistance leading to an IC_{50} of 78.5 nM for doxorubicin, of 48.5 nM for paclitaxel, and of 321.5 nM for vincristine, corresponding to reversal folds of 48, 38, and 45, respectively.

Table 1. The cytotoxic IC_{50} and reversal fold of drug resistance for ZAs A, B, C in combination with chemotherapeutic drugs in HeLaS3 and MDR KBvin cells.

	HeLa		KBvin	
	IC_{50} ± SD (nM)	RF [1]	IC_{50} ± SD (nM)	RF [1]
Doxorubicin	104.5 ± 6.36	1	3750 ± 70.7	1
+Verapamil (2.5 μM)	83.61 ± 3.12 *	1.2	705.21 ± 19.13 *	5.3
+ZA-A (10 μM)	76.000 ± 1.41 *	1.4	420 ± 56.6 *	8.9
+ZA-A (20 μM)	51.500 ± 2.12 *	2	78.5 ± 3.53 *	47.8
+ZA-B (10 μM)	103.000 ± 1.43	1	2050 ± 72.5	1.8
+ZA-B (20 μM)	66.500 ± 4.94 *	1.6	1200 ± 23.5 *	3.1
+ZA-C (10 μM)	101.500 ± 2.12	1	2100 ± 25.3	1.8
+ZA-C (20 μM)	83.000 ± 1.42 *	1.3	1800 ± 45.7 *	2.1
Paclitaxel	4.65 ± 0.21	1	1824 ± 125.87	1
+Verapamil (2.5 μM)	0.95 ± 0.03 *	4.9	75.81 ± 4.95 *	24.1
+ZA-A (10 μM)	1.650 ± 0.07 *	2.8	143.5 ± 4.94 *	12.7
+ZA-A (20 μM)	0.450 ± 0.08 *	10.3	48.5 ± 2.12 *	37.6
+ZA-B (10 μM)	1.900 ± 0.14 *	2.4	228.5 ± 2.23 *	8
+ZA-B (20 μM)	0.750 ± 0.07 *	6.2	141.5 ± 4.78 *	12.9
+ZA-C (10 μM)	4.000 ± 0.28	1.2	253.6 ± 5.16 *	7.2
+ZA-C (20 μM)	3.700 ± 0.56	1.3	221.8 ± 2.54 *	8.2
Vincristine	41.5 ± 0.74	1	14540 ± 719.13	1
+Verapamil (2.5 μM)	37.9 ± 0.64	1.1	370.81 ± 8.34 *	39.2
+ZA-A (10 μM)	6.450 ± 1.06 *	6.4	2187 ± 30.7 *	6.6
+ZA-A (20 μM)	3.450 ± 0.77 *	12	321.5 ± 3.53 *	45.2
+ZA-B (10 μM)	8.350 ± 1.17 *	5	2252 ± 11.31 *	6.5
+ZA-B (20 μM)	5.700 ± 0.98 *	7.3	1355.5 ± 30.41 *	10.7
+ZA-C (10 μM)	31.500 ± 2.47	1.3	2484 ± 55.15 *	5.9
+ZA-C (20 μM)	16.500 ± 0.71 *	2.5	971.5 ± 37.8 *	15

[1] RF: Reversal fold; * $p < 0.05$ compared with substrate drugs transport with the tested compounds.

3. Materials and Methods

3.1. General

The spectroscopic data of the purified compounds including optical rotations ($[\alpha]_D^{25}$), UV, and IR spectra were recorded on a Jasco P-2000 digital polarimeter (Jasco, Tokyo, Japan), a Hitachi U-0080D diode array spectrophotometer (Hitachi, Tokyo, Japan), and a Jasco FT/IR-4100 spectrophotometer (Jasco, Tokyo, Japan), respectively. The mass spectra were collected on a Shimadzu LC-MS 8040 spectrometer (Shimadzu, Kyoto, Japan). The HRMS data were obtained on a JMS-T100LP spectrometer (Jeol, Tokyo, Japan). ^{1}H-, ^{13}C-, and 2D NMR spectra were recorded on the Bruker AV-500 and Avance III-400 NMR spectrometers (Bruker, Billerica, MA, USA). The deuterated solvents were purchased from Sigma-Aldrich (St. Louis, MO, USA). Other chemicals used in this study were provided by Merck KGaA (Darmstadt, Germany). Column chromatography was performed on silica gels in different mesh sizes (70–230 and 230–400 mesh, Kieselgel 60, Merck KGaA, Darmstadt, Germany). Thin-layer chromatography (TLC) was conducted on precoated Kieselgel 60 F 254 plates (Merck KGaA, Darmstadt, Germany). The spots on TLC were detected by UV light or spraying with 10% (*v*/*v*) H_2SO_4 followed by heating at 110 °C for 10 min.

3.2. Plant Materials

The fresh fruiting bodies of *T. camphoratus* were provided by TWHERB Biomedical Co., LTD, Hsinchu, Taiwan (APACC-OG-100-034) in September 2009. The fungus was identified by Dr. Tun-Tschu Chang (Taiwan Forestry Research Institute, Taipei, Taiwan). A voucher specimen (TSWu 2009-001-010) was deposited in the School of Pharmacy, National Cheng Kung University, Tainan, Taiwan.

3.3. Extraction and Isolation

The fruiting bodies of *T. camphoratus* (3.6 kg) were extracted with Et_2O (4 × 10 L) for 3 days. The Et_2O extract was concentrated to afford a brown syrup (370 g) and then partitioned between H_2O and Et_2O. The ether layer was chromatographed on silica gel and eluted with MeOH in chloroform (0–100% of MeOH, gradient) to obtain 10 fractions, (Fractions 1–10) monitored by TLC. Fraction 2 underwent silica gel column chromatography using *n*-hexane–EtOAc (10:1) to obtain compound **3** (7 mg). Fractions 3 and 4 were combined and subjected to silica gel column chromatography, eluted successively with a step gradient of *n*-hexane–EtOAc (3:1 to 1:2) to yield compounds **1** (15 mg) and **4** (7 mg). Fractions 5 and 9 were combined and chromatographed on a column of silica gel, eluted successively with a step gradient of $CHCl_3$–MeOH as eluent to yield compound **2**.

3.3.1. Camphoratin K (**1**)

Colorless powder; $[\alpha]_D^{25}$ +111.2 (c 0.2, MeOH); IR (KBr) ν_{max}: 3427, 2959, 2942, 2885, 1714, 1643, 1455, 1374, 1266, 1249, 1031, 891 cm^{-1}; UV (MeOH) λ_{max}: 243, 253 nm; ESI–MS *m/z* 537 $[M + Na]^+$; HR-ESI-MS *m/z* 537.3917 ($[M + Na]^+$) (Calcd. for $C_{33}H_{44}O_4Na$: 531.3920); ^{1}H-NMR ($CDCl_3$, 400 MHz) δ 5.05 (1H, dd, *J* = 5.8, 9.4 Hz, H-15), 4.73 (1H, s, H-28), 4.67 (1H, s, H-28), 3.67 (3H, s, OCH_3), 3.69 (1H, m, H-21), 3.22 (1H, dd, *J* = 4.4, 11.6 Hz, H-3), 2.22 (1H, sept, *J* = 6.8 Hz, H-25), 2.12 (1H, m, H-16), 2.11 (1H, m, H-23), 2.10 (1H, m, H-7), 2.04 (1H, m, H-11), 2.03 (1H, m, H-7), 2.01 (1H, m, H-20), 1.93 (1H, m, H-12), 1.91 (2H, m, H-11, H-23), 1.73 (1H, m, H-1), 1.63 (2H, m, H-2, H-22), 1.61 (1H, m, H-12), 1.56 (1H, m, H-6), 1.55 (1H, m, H-2), 1.50 (1H, m, H-22), 1.48 (1H, m, H-6), 1.46 (1H, m, H-17), 1.22 (1H, m, H-1), 1.02 (3H, d, *J* = 6.8 Hz, H-26), 1.02 (3H, s, H-31), 1.01 (1H, m, H-5), 1.01 (3H, d, *J* = 6.8 Hz, H-27), 0.99 (3H, s, H-30), 0.97 (3H, s, H-19), 0.80 (3H, s, H-18), 0.80 (3H, s, H-29). ^{13}C-NMR ($CDCl_3$, 100 MHz) δ 171.1 (C-32), 156.1 (C-24), 135.5 (C-9), 132.8 (C-8), 106.4 (C-28), 78.9 (C-3), 76.0 (C-15), 62.0 (C-21), 51.0 (C-13), 50.1 (C-5), 44.4 (C-14), 43.2 (C-20), 38.8 (C-4), 37.1 (C-10), 36.0 (C-16), 35.5 (C-1), 33.7 (C-25), 31.4 (C-23), 30.6 (C-12), 28.2 (C-6), 28.0 (C-17), 28.0 (C-31), 27.7 (C-2), 26.4 (C-11), 21.9 (C-27), 21.8 (C-26), 21.4 (C-33), 20.8 (C-7), 19.1 (C-19), 18.3 (C-31), 18.2 (C-22), 16.4 (C-18), 15.4 (C-29).

3.3.2. Camphoratin N (**2**)

Yellow solid; $[\alpha]_D^{25}$ +194.9 (c 0.1, MeOH); IR (KBr) ν_{max}: 3495, 2944, 2926, 2892, 1736, 1713, 1674, 1458, 1378, 1238, 1200, 1167, 1065, 901 cm^{-1}; UV (MeOH) λ_{max}: 267 nm; ESI–MS *m/z* 521 $[M + Na]^+$; HR-ESI-MS *m/z* 521.2876 ($[M + Na]^+$) (Calcd. for $C_{30}H_{42}O_6Na$: 521.2879); ^{1}H-NMR ($CDCl_3$, 400 MHz) δ 4.91 (1H, s, H-28), 4.87 (1H, s, H-28), 4.11 (1H, s, H-12), 3.67 (3H, s, OCH_3), 3.13 (1H, q, *J* = 7.1 Hz, H-25), 3.03 (1H, dd, *J* = 7.3, 12.4 Hz, H-14), 2.93 (1H, ddd, *J* = 2.7, 7.0, 2.7 Hz, H-1), 2.55 (1H, m, H-6), 2.54 (1H, m, H-2), 2.45 (1H, m, H-6), 2.43 (1H, m, H-4), 2.41 (1H, m, H-2), 2.40 (1H, m, H-15), 2.11 (1H, m, H-23), 1.99 (1H, m, H-16), 1.99 (1H, m, H-5), 1.96 (1H, m, H-23), 1.87 (1H, m, H-17), 1.58 (1H, m, H-22), 1.55 (1H, m, H-16), 1.54 (3H, s, H-19), 1.47 (1H, m, H-1), 1.44 (1H, m, H-15), 1.42 (1H, m, H-20), 1.28 (3H, d, *J* = 7.1 Hz, H-27), 1.20 (1H, m, H-22), 1.04 (3H, d, *J* = 6.6 Hz, H-29), 0.97 (3H, d, *J* = 6.5, H-21), 0.67 (3H, s, H-18). ^{13}C-NMR ($CDCl_3$, 100 MHz) δ 210.7 (C-3), 201.9 (C-11), 200.1 (C-7), 150.3 (C-9), 148.4 (C-24), 145.3 (C-8), 80.4 (C-12), 49.3 (C-13), 48.5 (C-5), 45.6 (C-17), 43.9 (C-4), 41.7 (C-14), 39.0 (C-6), 37.9 (C-10), 37.5 (C-2), 35.3 (C-20), 34.5 (C-1), 33.8 (C-22), 31.2 (C-23), 26.8 (C-16), 23.8 (C-15), 17.9 (C-21), 16.3 (C-19), 11.4 (C-18).

3.3.3. Benzocamphorin G (**3**)

Colorless syrup; IR (KBr) ν_{max}: 2937, 2205, 1667, 1607, 1588, 1450, 1365, 1273, 1230, 1078, 853 cm^{-1}; UV (MeOH) λ_{max}: 268, 296 nm; ESI–MS *m/z* 239 $[M + Na]^+$; HR-ESI–MS *m/z* 239.0686 ($[M + Na]^+$) (Calcd. for $C_{13}H_{12}O_3Na$: 239.0684); ^{1}H-NMR ($CDCl_3$, 400 MHz) δ 5.90 (1H, s, H-2), 5.55 (1H, s, H-4′), 5.44 (1H, s, H-4), 3.82 (3H, s, OCH_3), 2.22 (3H, s, CH_3-5), 2.01 (3H, s, CH_3-3′). ^{13}C-NMR ($CDCl_3$, 100 MHz) δ 183.1 (C-1), 181.3 (C-6), 158.8 (C-3), 144.6 (C-5), 129.4 (C-4), 126.2 (C-3′), 125.1 (C-4′), 108.7 (C-2′), 107.0 (C-2), 81.7 (C-1′), 56.3 (OCH_3), 23.0 (CH_3-3′), 14.5 (CH_3-5).

3.3.4. Benzocamphorin I (**4**)

Colorless syrup; IR (KBr) ν_{max}: 3276, 2936, 2892, 1496, 1457, 1443, 1232, 1115, 1070, 1051, 956 cm^{-1}; UV (MeOH) λ_{max}: 265, 284 nm; ESI-MS *m/z* 385 $[M + Na]^+$; HR-ESI–MS *m/z* 385.0898 ($[M + Na]^+$) (Calcd. for $C_{18}H_{18}O_8Na$: 385.0899); ^{1}H-NMR ($CDCl_3$, 400 MHz) δ 6.00 (2H, OCH_2O-2′,3′), 5.98 (2H, OCH_2O-3′,4′), 5.97 (1H, s, H-6′), 5.17 (1H, s, OH-5), 3.90 (3H, s, OCH_3-4), 3.89 (3H, s, OCH_3-2′), 2.05 (3H, s, CH_3-1′), 1.99 (3H, s, CH_3-1). ^{13}C-NMR ($CDCl_3$, 100 MHz) δ 136.8 (C-3, 137.4 (C-2′), 136.1 (C-6, C-3′), 135.5 (C-5′), 135.0 (C-4), 134.4 (C-2), 133.4 (C-4′), 129.4 (C-5), 116.9 (C-1), 109.5 (C-6′), 101.8 (OCH_2O-3′,4′), 101.8 (OCH_2O-2, 3), 60.1 (OCH_3-4), 60.0 (OCH_3-2), 59.8 (OCH_3-2′), 15.9 (CH_3-1′), 9.4 (CH_3-1).

3.4. *Culture of Cell Lines*

Human stably P-gp-expressing cells (ABCB1/Flp-InTM-293) were established and cultured in DMEM as in a previous study [27]. The human cervical epithelioid carcinoma cell line HeLaS3 was purchased from Bioresource Collection and Research Center (Hsinchu, Taiwan), and the multi-drug resistant human cervical cancer cell line KBvin was kindly provided by Dr. Kuo-Hsiung Lee (University of North Carolina, Chapel Hill, NC, USA). All cancer cell lines were cultured in RPMI-1640 containing 10% FBS, at 37 °C in a humidified atmosphere of 5% CO_2.

3.5. *Calcein AM Uptake Assay*

The calcein AM uptake assay was performed to evaluate the inhibitory effect of the test compounds on human P-gp efflux function. To be brief, 1×10^5 cells/well were seeded in 96-well black plates overnight. Before starting the assay, the cells were washed and pre-incubated with warm Hanks' balanced salt solution (HBSS) for 30 min. Then, the test compounds were added, and incubation was carried out for 30 min. Calcein-AM was added after washing with warm PBS. The BioTek Synergy HT Multi-Mode Microplate Reader was utilized to detect calcein fluorescence (excitation/emission wavelength = 485 nm/528 nm) at 37 °C every 3 min for 30 min. Each experiment was performed at least three times, each in triplicate on different days.

3.6. *SRB Cytotoxicity Assay and Reversal Fold Calculation*

Briefly, after 72 h of treatment with series concentrations of chemotherapeutic drugs with or without the test compounds, 50% trichloroacetic acid (TCA) was added to fix the cells for 30 min. After air-drying, the cells were stained with 0.04% sulforhodamine B (SRB) for 30 min and washed with 1% acetic acid. The bound stain was solubilized in 10 mM Tris base, and the absorbance was measured by a Synergy HT Multi-Mode Microplate Reader (BioTek, Winooski, VT, USA) at 515 nm. Reversal folds were calculated by dividing the IC_{50} of each drug by the IC_{50} of the compound–drug combination treatment.

4. Conclusions

Although numerous anti-cancer drugs are marketed, resistance to cancer treatments is still the top reason for cancer death. Multidrug resistance is largely due to the high expression of efflux pumps. *T. camphoratus*, a medicinal fungus, was reported to exhibit anti-cancer properties, but its effects toward cancer multidrug resistance are unknown. In this study, four new chemical constituents, camphoratins K (**1**) and N (**2**) and benzocamphorins G (**3**) and I (**4**), were reported for the first time, and the main constituents of *T. camphoratus*, zhankuic acids A–C, were found to have P-gp inhibitory effects in a dose dependent manner. In addition, zhankuic acid A (20 μM), the most potent P-gp inhibitor, could effectively reverse MDR in KBvin cells, leading to an IC_{50} of 78.5 nM for doxorubicin, of 48.5 nM for paclitaxel, and of 321.5 nM for vincristine, corresponding to reversal folds of 48, 38, and 45, respectively.

Supplementary Materials: The following are available online. S01: References of the known compounds; Figures S1–S36: NMR spectra of compounds **1–4**.

Author Contributions: Conceptualization, P.-C.K. and T.-S.W.; investigation, H.-Y.H., C.-C.H., J.-W.L.; resources, P.-C.S., P.-C.K., and T.-S.W.; data curation, C.-F.C., H.-Y.C.; writing—original draft preparation, H.-Y.H., C.-C.H.; writing—review and editing, P.-C.K. and T.-S.W. All authors read and approved the final manuscript.

Funding: This research was sponsored by the Ministry of Science and Technology (MOST), Taiwan, granted to T.-S.W.

Acknowledgments: Thanks TWHERB Biomedical Co., LTD for providing plant materials and thanks Taiwan Forestry Research Institute for helping identifying the fungus.

Conflicts of Interest: The authors declare no conflict of interest.

References

1. Institute, N.C. Cancer Statistics. Available online: https://www.cancer.gov/about-cancer/understanding/statistics (accessed on 16 September 2019).
2. Avril, T.; Vauleon, E.; Chevet, E. Endoplasmic reticulum stress signaling and chemotherapy resistance in solid cancers. *Oncogenesis* **2017**, *6*, e373. [CrossRef] [PubMed]
3. Sun, Y. Tumor microenvironment and cancer therapy resistance. *Cancer Lett.* **2016**, *380*, 205–215. [CrossRef] [PubMed]
4. Nimmakayala, R.K.; Batra, S.K.; Ponnusamy, M.P. Unraveling the Journey of Cancer Stem Cells from Origin to Metastasis. *Biochim. Biophys. Acta Rev. Cancer* **2018**. [CrossRef]
5. Sodani, K.; Patel, A.; Kathawala, R.J.; Chen, Z.S. Multidrug resistance associated proteins in multidrug resistance. *Chin. J. Cancer* **2012**, *31*, 58–72. [CrossRef]
6. Joshi, P.; Vishwakarma, R.A.; Bharate, S.B. Natural alkaloids as P-gp inhibitors for multidrug resistance reversal in cancer. *Eur. J. Med. Chem.* **2017**, *138*, 273–292. [CrossRef]
7. Kathawala, R.J.; Gupta, P.; Ashby, C.R., Jr.; Chen, Z.S. The modulation of ABC transporter-mediated multidrug resistance in cancer: A review of the past decade. *Drug Resist. Update* **2015**, *18*, 1–17. [CrossRef]
8. Palmeira, A.; Sousa, E.; Vasconcelos, M.H.; Pinto, M.M. Three decades of P-gp inhibitors: Skimming through several generations and scaffolds. *Curr. Med. Chem.* **2012**, *19*, 1946–2025. [CrossRef]
9. Lu, M.C.; El-Shazly, M.; Wu, T.Y.; Du, Y.C.; Chang, T.T.; Chen, C.F.; Hsu, Y.M.; Lai, K.H.; Chiu, C.P.; Chang, F.R.; et al. Recent research and development of *Antrodia cinnamomea*. *Pharmacol. Ther.* **2013**, *139*, 124–156. [CrossRef]
10. Lin, Y.H.; Kuo, J.T.; Chen, Y.Y.; Kumar, K.J.S.; Lo, C.P.; Lin, C.C.; Wang, S.Y. Immunomodulatory Effects of the Stout Camphor Medicinal Mushroom, *Taiwanofungus camphoratus* (Agaricomycetes)-Based Health Food Product in Mice. *Int. J. Med. Mushrooms* **2018**, *20*, 849–858. [CrossRef]
11. Chen, Y.Y.; Lo, C.P.; Lin, C.C.; Hsieh, Y.H. Effects of *Taiwanofungus camphoratus* on non-specific and specific immune activities in mice. *Mycology* **2018**, *9*, 129–135. [CrossRef] [PubMed]
12. Yang, H.; Zhang, J.; Zhang, H.; Yang, Y.; Liu, Y.; Sun, W.; Wang, W.; Jia, W. Compound of Stout Camphor Medicinal Mushroom, *Taiwanofungus camphoratus* (Agaricomycetes), Induces Protective Autophagy in SPCA-1 Cells through AMPK Inhibition-Independent Blockade of the Akt/mTOR Pathway. *Int. J. Med. Mushrooms* **2018**, *20*, 727–738. [CrossRef] [PubMed]
13. Lai, I.C.; Lai, G.M.; Chow, J.M.; Lee, H.L.; Yeh, C.F.; Li, C.H.; Yan, J.L.; Chuang, S.E.; Whang-Peng, J.; Bai, K.J.; et al. Active fraction (HS7) from *Taiwanofungus camphoratus* inhibits AKT-mTOR, ERK and STAT3 pathways and induces CDK inhibitors in CL1-0 human lung cancer cells. *Chin. Med.* **2017**, *12*, 33. [CrossRef] [PubMed]
14. Wang, W.H.; Wang, L.Y.; Yang, H.R.; Zhang, Y.P.; Zhang, H.N.; Fan, H.; Zhao, X.L.; Zhang, J.S.; Jia, W. Antitumor Effect of By-1 from Spent Broth from Submerged Cultures of Stout Camphor Medicinal Mushroom, *Taiwanofungus camphoratus* (Higher Basidiomycetes), on A549 Adenocarcinoma Cells. *Int. J. Med. Mushrooms* **2017**, *19*, 225–232. [CrossRef] [PubMed]
15. Tai, C.J.; Shi, Y.C.; Tai, C.J.; Kuo, L.J.; Chen, R.J.; Chang, Y.J.; Tzao, C.; Wu, C.H.; Chang, C.C.; Chiou, H.Y.; et al. *Taiwanofungus camphoratus* Combined With Amphotericin B for Metastatic Cancer Patients Unresponsive to or Unwilling to Undergo Chemotherapy: A Pilot Study. *Altern. Ther. Health Med.* **2018**. Available online: https://europepmc.org/abstract/med/29477137 (accessed on 15 October 2019).

16. Wang, C.; Zhang, W.; Wong, J.H.; Ng, T.; Ye, X. Diversity of potentially exploitable pharmacological activities of the highly prized edible medicinal fungus *Antrodia camphorata*. *Appl. Microbiol. Biotechnol.* **2019**, *103*, 7843–7867. [CrossRef]
17. Wu, S.J.; Leu, Y.L.; Chen, C.H.; Chao, C.H.; Shen, D.Y.; Chan, H.H.; Lee, E.J.; Wu, T.S.; Wang, Y.H.; Shen, Y.C.; et al. Camphoratins A–J, potent cytotoxic and anti-inflammatory triterpenoids from the fruiting body of *Taiwanofungus camphoratus*. *J. Nat. Prod.* **2010**, *73*, 1756–1762. [CrossRef]
18. Huang, Y.; Lin, X.; Qiao, X.; Ji, S.; Liu, K.; Yeh, C.T.; Tzeng, Y.M.; Guo, D.; Ye, M. Antcamphins A-L, ergostanoids from *Antrodia camphorata*. *J. Nat. Prod.* **2014**, *77*, 118–124. [CrossRef]
19. Shi, L.S.; Chao, C.H.; Shen, D.Y.; Chan, H.H.; Chen, C.H.; Liao, Y.R.; Wu, S.J.; Leu, Y.L.; Shen, Y.C.; Kuo, Y.H.; et al. Biologically active constituents from the fruiting body of *Taiwanofungus camphoratus*. *Bioorg. Med. Chem.* **2011**, *19*, 677–683. [CrossRef]
20. Wu, M.D.; Cheng, M.J.; Wang, W.Y.; Huang, H.C.; Yuan, G.F.; Chen, J.J.; Chen, I.S.; Wang, B.C. Antioxidant activities of extracts and metabolites isolated from the fungus *Antrodia cinnamomea*. *Nat. Prod. Res.* **2011**, *25*, 1488–1496. [CrossRef]
21. Chen, Y.C.; Chiu, H.L.; Chao, C.Y.; Lin, W.H.; Chao, L.K.; Huang, G.J.; Kuo, Y.H. New Anti-Inflammatory Aromatic Components from *Antrodia camphorata*. *Int. J. Mol. Sci.* **2013**, *14*, 4629–4639. [CrossRef] [PubMed]
22. Chen, J.J.; Lin, W.J.; Liao, C.H.; Shieh, P.C. Anti-inflammatory benzenoids from *Antrodia camphorata*. *J. Nat. Prod.* **2007**, *70*, 989–992. [CrossRef] [PubMed]
23. Chen, C.C.; Shiao, Y.J.; Lin, R.D.; Shao, Y.Y.; Lai, M.N.; Lin, C.C.; Ng, L.T.; Kuo, Y.H. Neuroprotective diterpenes from the fruiting body of *Antrodia camphorata*. *J. Nat. Prod.* **2006**, *69*, 689–691. [CrossRef] [PubMed]
24. Lee, T.H.; Lee, C.K.; Tsou, W.L.; Liu, S.Y.; Kuo, M.T.; Wen, W.C. A new cytotoxic agent from solid-state fermented mycelium of *Antrodia camphorata*. *Planta Med.* **2007**, *73*, 1412–1415. [CrossRef] [PubMed]
25. Wu, M.D.; Cheng, M.J.; Wang, B.C.; Yech, Y.J.; Lai, J.T.; Kuo, Y.H.; Yuan, G.F.; Chen, I.S. Maleimide and maleic anhydride derivatives from the mycelia of *Antrodia cinnamomea* and their nitric oxide inhibitory activities in macrophages. *J. Nat. Prod.* **2008**, *71*, 1258–1261. [CrossRef]
26. Teng, Y.-N.; Chang, C.-S.; Lee, T.-E.; Hung, C.-C. Cordycepin re-sensitizes multidrug resistance cancer cells to chemotherapeutic agents through modulating P-glycoprotein expression and ATPase function. *J. Funct. Foods* **2016**, *26*, 681–690. [CrossRef]
27. Teng, Y.N.; Hsieh, Y.W.; Hung, C.C.; Lin, H.Y. Demethoxycurcumin modulates human P-glycoprotein function via uncompetitive inhibition of ATPase hydrolysis activity. *J. Agric. Food Chem.* **2015**, *63*, 847–855. [CrossRef]

Sample Availability: Samples of all the isolated compounds are available from the authors.

molecules

MDPI

Article

Characterizing Tyrosinase Modulators from the Roots of *Angelica keiskei* Using Tyrosinase Inhibition Assay and UPLC-MS/MS as the Combinatorial Novel Approach

Jia-Hao Lee [1], Hui-Ching Mei [2], I-Chih Kuo [3], Tzong-Huei Lee [4], Yu-Hsin Chen [5] and Ching-Kuo Lee [1,*]

1 School of Pharmacy, Taipei Medical University, Taipei 11031, Taiwan
2 Department of Science Education, National Taipei University of Education, Taipei 10671, Taiwan
3 Department of Microbiology and Immunology, University of British Columbia, Vancouver, BC V6T 1Z3, Canada
4 Institute of Fisheries Science, National Taiwan University, Taipei 10617, Taiwan
5 Taichung District Agricultural Research and Extension Station, Council of Agriculture, Executive Yuan, Taichung 42081, Taiwan
* Correspondence: cklee@tmu.edu.tw; Tel.: +889-2736-1661-6150

Academic Editor: Ping-Chung Kuo
Received: 20 August 2019; Accepted: 8 September 2019; Published: 10 September 2019

check for updates

Abstract: In this study, an in vitro tyrosinase inhibition assay in combination with ultra performance liquid chromatography-orbitrap mass spectrometry (UPLC-orbitrap-MS) was developed for the rapid screening and identification of tyrosinase modulators from roots of *Angelica keiskei*. Of the 15 candidates considered, nine chalcones, xanthoangelols (**1**), B (**2**), D (**3**), E (**4**), G (**5**), H (**6**), 4-hydroxyderricin (**7**), xanthokeismin B (**8**) and (2*E*)-1-[4-hydroxy-2-(2-hydroxy-2-propanyl)-2,3-dihydro-1- benzofuran-7-yl]-3-(4-hydroxyphenyl)-2-propen-1-one (**9**), five coumarins, umbelliferone (**10**), selinidin (**11**), isopimpinellin (**12**), phellopterin (**13**) and xanthyletin (**14**), and one other compound, ashitabaol A (**15**), were distinguished between the test samples and the controls with statistical significance, and the structure of each compound was determined by comparing with in-house standards and the literature. Among these, six compounds, xanthoangelol (**1**), xanthoangelol D (**3**), xanthoangelol H (**6**), 4-hydroxyderricin (**7**), laserpitin (**16**) and isolaserpitin (**17**), were isolated from roots of *A. keiskei*. Of the compounds isolated, compounds **1**, **7** and **16** were subjected to tyrosinase inhibitory assay, and the IC_{50} values were 15.87 ± 1.21, 60.14 ± 2.29 and >100 μM, respectively. The present study indicated that the combination of in vitro tyrosinase inhibition assay coupled with UPLC-MS/MS could be widely applied to the rapid screening of active substances from various natural resources.

Keywords: *Angelica keiskei*; tyrosinase; UPLC-MS/MS; chalcone; coumarins

1. Introduction

Natural products have been used in the field of medicine and cosmetics for centuries. Their potential to treat various skin diseases and to improve skin condition is well-known. As ultraviolet (UV) radiation is a contributing cause for sunburns, wrinkles, premature aging, cancer and reduced immunity against infections, there is an increasing demand for products that provide protection against UV radiation [1]. Tyrosinase is a key enzyme that catalyzes the initial rate-limiting steps of melanin synthesis [2,3]. Abnormal and excessive melanin synthesis is the primary cause for skin disorders including melasma, senile lentigo, freckles and age spots [4]. As a result, finding novel and effective melanogenesis inhibitors has profound importance in controlling melanin production and pigmentation-related disorders [5]. In an attempt to find potent and safe tyrosinase inhibitors, this experiment evaluated the in vitro tyrosinase inhibitory activity of *Angelica keiskei*.

A. keiskei Koidzumi (Japanese name: 'Ashitaba', Umbelliferae) is a hardy perennial herb with exuberant vitality and multi-bioactive components. It originated in Hachijojima, Japan, and now mainly grows along the Pacific coast of Japan. *A. keiskei* is traditionally used as a diuretic, laxative, analeptic and galactagogue [6–8]. It is a dark green leafy vegetable that has been recognized as a medicinally important herb and cultivated in many Asian countries because of its health benefits. The stems and leaves have been consumed commercially as health foods and the roots have also been used as a food additive and medicine to alleviate pain and diabetic symptoms [9]. Various chalcones such as xanthoangelol, 4-hydroxyderricin and coumarins like xanthotoxin and laserpitin have been isolated and characterized from this plant [6,8].

Chalcones and coumarins are the main bioactive compounds in *A. keiskei*. Chalcones have been widely studied and are known to contain antioxidants [10], anticancer agents [11,12] and α-glucosidase inhibitors [13]. Coumarins, on the other hand, were proved to have antioxidants [14], antidepressants [15] and anticancer agents [8,16,17].

An ultra-performance liquid chromatography (UPLC) coupled with the high resolution MS/MS method is the advanced method for the identification and quantification of phenol compounds from plants and food [18]. The full scan analysis reveals the molecular weight (MW) of the unknown compounds, while tandem MS reveals aspects of the chemical structure of the precursor ion via fragmentation. Ion trap mass spectrometers can carry out sequential fragmentations of the precursor ion to form product ions. By using this method, unknown compounds are identified from the exact mass and MS/MS fragmentation. Since compounds are scanned separately and identified, the UPLC-MS/MS approach has increased sensitivity and provides more structural information based on fragmentation pattern of the analyte [19].

The aim of this study was to establish a rapid method by using UPLC-MS/MS to identify the components in *A. keiskei* that regulate tyrosinase activity via statistical analysis.

2. Results

2.1. The Tyrosinase Modulatory Effect of *Angelica keiskei*

According to the current understanding, *A. keiskei* has anti-inflammatory, antimicrobial and antihypertensive characteristics [14], while its effect on tyrosinase activity remains largely unknown [20,21]. With the lack of research in this field, our study focused on the analysis of tyrosinase activity in the presence of different concentrations of *A. keiskei* extracts from leaf and root parts. As shown in Figure 1, the preliminary results from our study showed that at the concentration of 500 μg/mL the tyrosinase activity remained unchanged and similar to blank when treated with *A. keiskei* leaves. On the other hand, a dosage-dependent inhibition of tyrosinase activity was observed when treated with *A. keiskei* roots. This trend of the inhibitory effect started at 200 μg/mL and increased with the higher concentration of *A. keiskei* roots. At the 1000 μg/mL level, conditions even exceeded the inhibitory effect achieved by 10 μM kojic acid (positive control). Since the leaf extract showed no evident change at the 500 μg/mL concentration, the subsequent experiments were performed using the root extract of *A. keiskei* with the purpose of further examining its modulatory effect on tyrosinase activity.

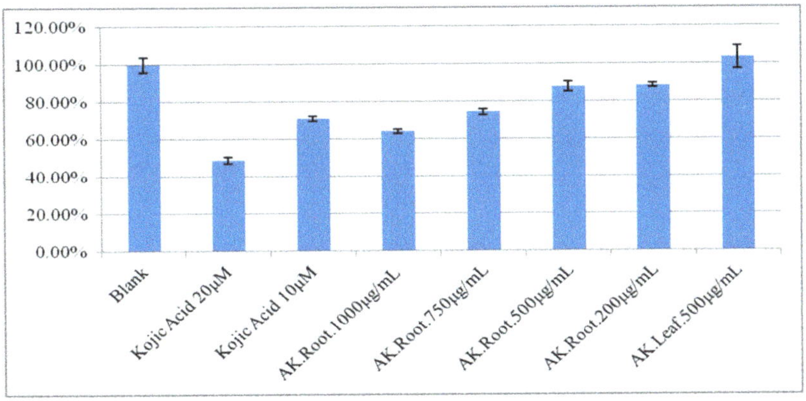

Figure 1. The comparative ratio of tyrosinase activity of the *A. keiskei* Leaf (*AK.* Leaf) at 500 μg/mL and roots (*AK.* Root) at 200, 500, 750 and 1000 μg/mL concentrations using kojic acid (10 and 20 μM) as the positive control.

2.2. Screening Tyrosinase Modulator by UPLC-MS/MS

In order to investigate the compounds of *A. keiskei* roots that modulate the tyrosinase activity in a timely manner, we analyzed the samples using the simple LC-MS protocol optimized by our lab [22]. The base peak chromatograms (BPC) showed that under the ESI (+) MS mode in Figure 2, there were differences in signal intensities for the roots of *A. keiskei* treated with or without tyrosinase at the retention times of 7.93, 10.83, 10.84, 14.86, 14.88, 15.52, 18.56 and 19.04 min (Table 1). The base peak chromatograms (BPC) ESI (−) MS mode are shown in Figure 3; differences in signal intensities were observed at the retention times of 7.65, 10.08, 10.87, 13.20, 13.82, 14.32, 16.35 and 19.95 min (Table 2). Consequently, we suspect that components in the roots of *A. keiskei* might have a modulatory effect on tyrosinase activity.

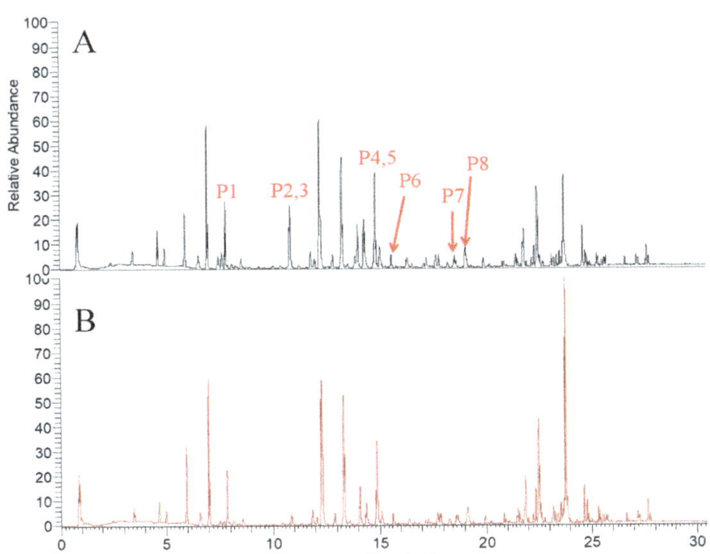

Figure 2. The base peak chromatograms (BPC) of untreated (**A**) and tyrosinase-treated samples (**B**) of *A. keiskei* in positive ion mode.

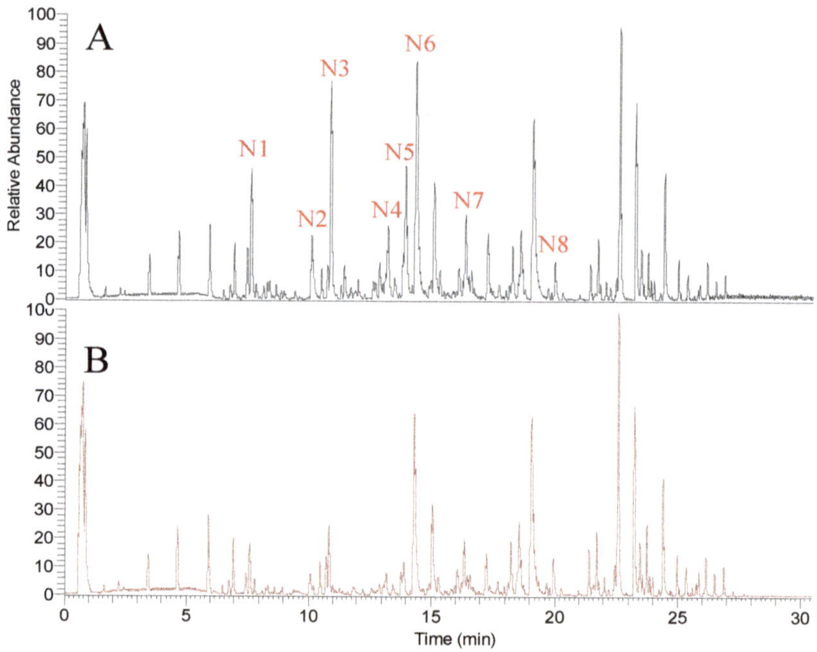

Figure 3. The base peak chromatograms (BPC) of untreated (**A**) and tyrosinase-treated samples (**B**) of *A. keiskei* in negative ion mode.

2.3. Multivariate Analysis to Differentiate Processed A. keiskei

The highly complex UPLC-MS spectra are difficult to visually link to various components; therefore, multivariate data analyses were performed to comprehensively characterize the distinct composition of various metabolites from the untreated samples and the tyrosinase-treated samples. For each condition, there were three biological replicates of *A. keiskei* metabolites ($n = 3$). After performing UPLC-orbitrap-MS-based profiling on individual samples, each dataset was processed using SIEVE software to align and extract meanings. All datasets obtained from the two treatment groups were then analyzed with PCA, OPLS-DA and S-plot functions within the SIMCA-P software to find the candidates of interest.

Due to the similarity in compositions, the differences between untreated (blank) and tyrosinase-treated (test) samples were hard to identify based on the BPC chromatograms (shown in Figures 2 and 3). Henceforth, a two-component PCA score plot of UPLC-MS data was utilized to depict general variations in components among the *A. keiskei* samples (Figure 4). According to the PCA scores plot in Figure 4, the distribution can be readily classified into two clusters, with the untreated (blank) and treated (test) samples clearly separated by principal component 1 (PC1) (Figure 4).

First, OPLS-DA was performed to compare untreated and tyrosinase-treated samples. An S-plot analysis was then used to select the critical variables that allowed differentiation. In the S-plot, each point represents *m*/*z*-RT pairs of molecules. To further reveal the tyrosinase inhibiting effect of *A. keiskei*, an OPLS-DA model was carried out between blank and test groups. The model fit well with R2Y value 0.996 and Q2 value 0.821 in positive mode, and R2Y value 1.0 and Q2 value 0.999 in negative mode. The score plot (Figure 4A,C) showed good separation, confirming the tyrosinase inhibiting effect. The corresponding S-plot was shown in Figure 4B, where coordinates in the lower-left quadrant were metabolites significantly increased in the blank group compared with the test group, while those in the upper-right quadrant represent the decreased ones. In Figure 4D, the components elevated in untreated samples were shown in the upper-right quadrant of the S-plot, while the lower-left quadrant revealed the components elevated in tyrosinase-treated samples.

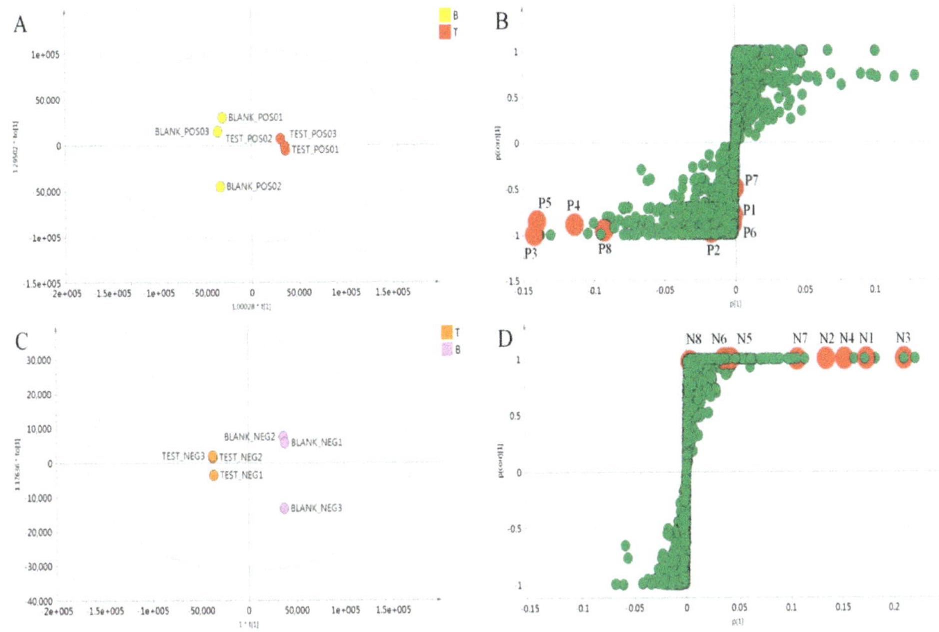

Figure 4. OPLS-DA score plot (**A**) and S-plot (**B**) of positive mode, OPLS-DA score plot (**C**) and S-plot (**D**) of negative mode, mass spectra obtained from untreated (blank) and tyrosinase-treated (test) groups.

2.4. Structural Characterization of Chalcones and Coumarines

To find the regulators for tyrosinase and the indicator component, the acetone extract was dissolved in acetone and coated in silica gel (1.5 g) and was then subjected to medium-pressure liquid chromatography (MPLC) and semi-preparative HPLC purification using the LC/MS-guided isolation approach. The process resulted in the successful isolation of six compounds, **1**, **3**, **6**, **7**, **16** and **17** (Figure 5). By using extensive NMR spectroscopic methods (1D and 2D-NMR), and LC/MS analysis to compare our results with previously reported spectroscopic values, the isolated compounds were determined to be xanthoangelol (**1**) [23], 4-hydroxyderricin (**7**) [24], xanthoangelol D (**3**) [25], xanthoangelol H (**6**) [26], laserpitin (**16**) [23] and isolaserpitin (**17**) [23] (Figure 5). These types of chalcones and coumarins are common amongst the *Angelica* species. In order to further confirm the effectiveness of the screening system, purified compounds, which in the S-plot are different between the blank and test groups, like the compounds **1**, **7** and **16**, were subjected to tyrosinase inhibition assay using kojic acid as the positive control. Results showed that both compounds **1** and **7** inhibited tyrosinase activity with IC_{50} values of 15.87 ± 1.21 μM and 60.14 ± 2.29 μM, respectively, whereas compound 16 had no inhibitory effect (IC_{50} > 100 μM) on tyrosinase. These findings corroborated the results found earlier via the screening system (Tables 1–3).

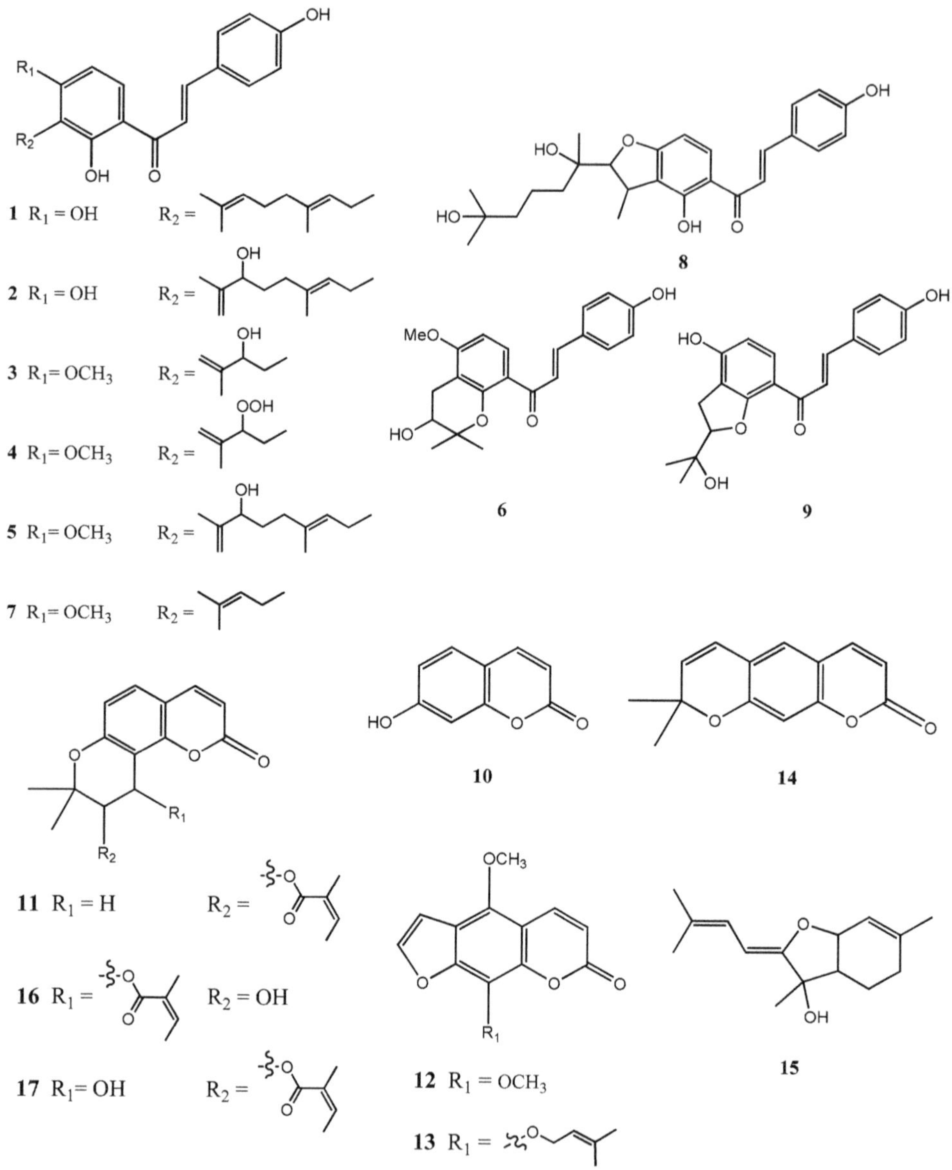

Figure 5. The chemical structure of 17 compounds in *Angelica keiskei* Koidzumi, xanthoangelol (**1**), B (**2**), D (**3**), E (**4**), G (**5**), H (**6**), 4-hydroxyderricin (**7**), xanthokeismin B (**8**), (2*E*)-1-[4-hydroxy-2-(2-hydroxy-2-propanyl)-2,3-dihydro-1-benzofuran-7-yl]-3-(4-hydroxyphenyl)-2-propen-1-one (**9**), umbelliferone (**10**), selinidin (**11**), isopimpinellin (**12**), phellopterin (**13**), xanthyletin (**14**), ashitabaol A (**15**), laserpitin (**16**) and isolaserpitin (**17**).

Table 1. Identification of the different amount of compounds from *A. keiskei* between the blank and test groups by UPLC-MS/MS in positive ion mode.

No.	t_R (min)	Measured *m/z* $[M + H]^+$	Theoretical Formula $[M + H]^+$	Error (ppm)	HCD (eV)	Fragment Ions (*m/z*) Abundance Rate (%)	Identification
P1	7.93	247.0605	$C_{13}H_{11}O_5$	1.8	50	247.0608 (15) 232.0373 (40) 217.0137 (100)	Isopimpinellin
P2	10.8	163.0387	$C_9H_7O_3$	−1.7	50	163.0389 (51) 133.0282 (72) 105.0332 (100)	Umbelliferone
P3	10.84	355.1542	$C_{21}H_{23}O_5$	0.5	50	235.0959 (46) 217.0855 (45) 181.0493 (83) 163.0384 (100)	Xanthoangelol H
P4	14.86	229.0861	$C_{14}H_{13}O_3$	0.9	50	229.0856 (100) 187.0386 (23) 175.0387 (50) 159.0438 (22)	Xanthyletin
P5	14.88	329.1384	$C_{19}H_{21}O_5$	0.2	30	245.0807 (54) 227.0686 (100)	Selinidin
P6	15.52	235.1697	$C_{15}H_{23}O_2$	1.7	50	235.1683 (100) 179.106 (82)	Ashitabaol A
P7	18.56	423.2170	$C_{26}H_{31}O_5$	0.9	30	283.0945 (100) 181.0499 (23) 163.0375 (42) 131.5666 (18)	Xanthoangelol G
P8	19.04	339.1592	$C_{21}H_{23}O_4$	0.2	40	283.0945 (32) 181.0484 (99) 163.0379 (100)	4-Hydroxyderricin

Table 2. Identification of the different amount of compounds from *A. keiskei* between the blank and test groups by UPLC-MS/MS in negative ion mode.

No.	t_R (min)	Measured *m/z* $[M + H]^+$	Theoretical Formula $[M + H]^+$	Error (ppm)	HCD (eV)	Fragment Ions (*m/z*) Abundance Rate (%)	Identification
N1	7.65	369.1320	$C_{21}H_{21}O_6$	−3.4	50	369.1389 (100) 297.0811 (1.3) 119.0517 (0.9)	Xanthoangelol E
N2	10.08	299.0905	$C_{17}H_{15}O_5$	−2.9	40	299.0959 (100) 119.0516 (10)	Phellopterin
N3	10.87	353.1370	$C_{21}H_{21}O_5$	−3.7	50	353.1436 (100) 239.1103 (2.4) 119.0518 (32)	Xanthoangelol H
N4	13.2	339.1216	$C_{20}H_{19}O_5$	−3.1	50	339.1274 (100) 119.0517 (17)	(2*E*)-1-[4-hydroxy-2-(2-hydroxy-2-propanyl)-2,3-dihydro-1-benzofuran-7-yl]-3-(4-hydroxyphenyl)-2-propen-1-one
N5	13.82	439.1737	$C_{25}H_{27}O_7$	−3.2	50	439.1779 (100) 319.1223 (3.4) 119.0525 (2.2)	Xanthokeismin B
N6	14.32	353.1373	$C_{21}H_{21}O_5$	−2.9	50	353.1435 (100) 283.1014 (12) 119.0517 (32)	Xanthoangelol D
N7	16.35	407.1837	$C_{25}H_{27}O_5$	−3.8	50	407.1913 (100) 287.1324 (5.5) 119.0517 (2.2) 243.142 (0.8)	Xanthoangelol B
N8	19.95	391.1892	$C_{25}H_{27}O_4$	−3.0	50	391.1956 (100) 271.137 (11) 119.0518 (4)	Xanthoangelol

Table 3. Effect of kojic acid, 4-hydroderricin, xanthoangelol and laserpitin on mushroom tyrosinase activity.

Compounds	L-tyrosine IC_{50} (μM) (%) [1]
Kojic acid [2]	3.8 ± 0.2
Xathoangelol	15.87 ± 1.21
4-hydroyderricin	60.14 ± 2.29
Lasepitin	>100 μM (No inhibition at 100 μM)

[1] Relative inhibitory activity, [2] Positive control.

3. Materials and Methods

3.1. Reagents and Materials

The fresh *Angelica keiskei* was collected from Ali Mountain, Chiayi County of Taiwan, and authenticated by Dr. Yu-Hsin Chen (Taichung District Agricultural Research and Extension Station, Taichung, Taiwan). The voucher specimens (TMU-LCK-77) were deposited at the School of Pharmacy, Taipei Medical University, Taipei, Taiwan.

All the reagents including phosphate buffer (sodium phosphate monobasic (NaH_2PO_4), sodium phosphate dibasic (Na_2HPO_4)), kojic acid (positive control for the enzyme activity assay), dimethyl sulfoxide (DMSO) and acetone were purchased from Sigma Aldrich (St. Louis, MO, USA). Methanol and acetonitrile (all MS grade), on the other hand, were purchased from Macron Fine Chemicals™ (Radnor, PA, USA). The ultra-pure water was prepared with the Millipore-Q water purification system (Bedford, MA, USA).

3.2. Sample Preparation

The fresh *A. keiskei* was weighed, ground and precisely cut into thin blocks. The root (425 g) and leaf (490 g) specimens of *A. keiskei* were then purified and extracted three times using acetone in a 1:5 ratio to obtain 20.58 g and 18.11 g of crude extract, respectively.

3.2.1. Crude Extract Preparation and Effective Compound Isolation

One gram of *A. keiskei* roots extracts was dissolved in acetone and coated in silica gel (1.5 g). The sample was subjected to medium-pressure liquid chromatography (MPLC, Isolera ONE, Uppsala, Sweden) on a silica SP column (Daiso gel, 50 μm, 10 g, column volume: 25 mL) with a fixed flow rate of 10 mL/min. The step-gradient of purification method was used to obtain the 0%, 20%, 30%, 40%, 60%, 80% and 100% fraction ethyl acetate solvents. A total volume of 100 mL was collected for each of the 6 ethyl acetate (EA) fractions (20 mL/tube). Tube numbers 12 to 15 were loaded on the same silica column for HPLC (250 × 10 mm, Luna 5μ Silica (2) 100 Å). The eluting solvent was ethyl acetate-hexane with a ratio of 20:80. At the flow rate of 4.0 mL/min, we were able to isolate 4-hydroxyderricin (**7**), xanthoangelol (**1**), laserpitin (**16**) and isolaserpitin (**17**). Similarly, HPLC was performed on tube numbers 18 to 23 on the same silica column. EA-hexane with a ratio of 40:60 was used as the eluting solvent. With the flow rate of 4.0 mL/min, we successfully isolated xanthoangelol D (**3**) and xanthoangelol H (**6**). All of the compounds were identified through HR-MS and NMR and their identities were confirmed via comparison with the literature values.

3.2.2. Tyrosinase Assay

The L-tyrosine substrate and the mushroom tyrosinase were purchased from Sigma Aldrich. The test samples were prepared by first dissolving mushroom tyrosinase (250 U/mL) and L-Tyrosine (0.1 mg/mL) in DMSO and diluting the stock to different concentrations using phosphate buffer (66.7 mM at pH 6.8) to obtain a final DMSO concentration of 1%. The 96-well plate was divided into four groups: Blank control (BC), experimental control (EC), sample control (SC) and sample experiment (SE). All wells contained 40 μL of tyrosinase and had a total volume of 200 μL. On top of the enzyme solution, the BC group had phosphate buffer (160 μL) in the wells; EC contained phosphate buffer (80 μL) and L-Tyrosine (80 μL) in each well; SC wells consisted of phosphate buffer (120 μL) and sample solution (40 μL); SE wells, on the other hand, had phosphate buffer (40 μL), L-Tyrosine (80 μL) and sample solution (40 μL) in them. These assay mixtures were incubated at 37 °C for 30 min and measured at 475 nm using the microplate reader (Molecular Devices, CA, US). Percentages of tyrosinase activity were calculated using this formula:

$$\% \text{ Tyrosinase activity} = [(SE - SC) \div (EC - BC)] \times 100\% \quad (1)$$

Note:

SE = absorbance of sample experiment (with tyrosine)
SC = absorbance of sample control (without tyrosine)
EC = absorbance of experimental control (with tyrosine)
BC = absorbance of blank control (without tyrosine)

3.3. UPLC-MS/MS Analysis

UPLC-MS/MS analyses were performed using a linear ion trap-orbitrap mass spectrometer (Orbitrap Elite; Thermo Fisher Scientific, Bremen, Germany) coupled with the online UPLC system (ACQUITY UPLC; Waters, Waters Corp., Manchester UK). The working solution was separated by an ACQUITY UPLC BEH C18 column (100 × 2.1 mm, 1.7 μm; Waters) at 40 °C. The flow rate of mobile phase A (ddH_2O) and B (methanol) was 0.3 mL/min. The sample injected 5 μL. The gradient program was set as 20–30% phase B from 0–1 min and 30–100% phase B from 1–25 min. The column was washed with 100% phase B for 5 min before being re-equilibrated for 5 min. The mass spectrometer was equipped with an electrospray interface controlled by the Xcalibur software (version 2.0, Thermo Fisher Scientific, Bremen, Germany) with two types of operations: The positive ion and negative ion modes. The ESI source was set at these parameters: Spray voltage of 3.5 kV for the positive-ion mode and −3.2 kV for the negative-ion mode; capillary temperature was set at 360 °C and the source heater temperature was maintained at 350 °C. During the analysis, the mass spectrometer performed high-resolution (resolving power, r = 15,000) full scan cycles (*m/z* 100–1000). The analyte profiles were first made by the orbitrap; the MS2 spectra were then recorded in the centroid mode for the five most intense ions. The isolation width was set at the mass-to-charge ratio (*m/z*) of 0.2 and the higher-energy collisional dissociation (HCD) was performed at collision energies of 20, 30, 40 and 50 eV

3.4. In Vitro Tyrosinase Inhibition Assay by UPLC-MS/MS Analysis

The crude extract of *A. keiskei* root (15 mg) was dissolved in 75 μL of DMSO and 525 μL of 50 mM phosphate buffer (pH 6.8) to form the blank solution. Similarly, another 15 mg of the crude extract was dissolved in 75 μL DMSO and 425 μL of phosphate buffer solution (50 mM, pH 6.8) before 100 μL of tyrosinase (716 units/mL) was added as the test solution. Both solutions were stirred and incubated in the 37 °C water bath for 40 minutes in the dark. Subsequently, 900 μL of acetonitrile was added to each solution in order to terminate the enzyme reaction. Both test solutions with the final concentration of 10 mg/mL were filtered through the 0.22 μm filters. The working solutions were then analyzed by UPLC-MS/MS.

3.5. Mass Fragmentation Analyses and Compound Identifications

The MassLynx Mass Spectrometry software version 2.0.w.15 (Waters Corporation, Milford, MA, USA) was used for analyses. We extracted the mass fragment spectrum of each candidate in positive and negate mode, separately, and then transferred the mass fragment spectrum to the mass list. In the last step, via the MassFragment software in the MarkerLynx XS, structures were assigned to our observed fragment ions of small molecule compounds and, further, the reported structures were cross referenced with the literature on *A. keiskei*.

3.6. Statistics

The raw data were aligned and extracted using the SIEVE v2.2 application software from Thermo Fisher Scientific. The experimental target was metabolomics and the minimum intensity for the base peak was 1,000,000. The frames and threshold values were defined as 481,351 for the positive mode and 674,992 for the negative mode. SIMCA-P 13.0.3 software from MKS Umetrics (Umeå, Sweden) was then used to obtain OPLS-DA. The S-plots were also utilized for finding different candidates between blank and test groups.

4. Conclusions

Besides xanthoangelols (**1**), B (**2**), D (**3**), E (**4**), G (**5**), H (**6**), 4-hydroxyderricin (**7**), xanthokeismin B (**8**) and (2*E*)-1-[4-hydroxy-2-(2-hydroxy-2-propanyl)-2,3-dihydro-1-benzofuran-7-yl]-3-(4-hydroxyphenyl)-2-propen-1-one (**9**), umbelliferone (**10**), selinidin (**11**), isopimpinellin (**12**), phellopterin (**13**), xanthyletin (**14**) and ashitabaol A (15) in the positive and negative modes of UPLC-MS/MS, along with the differences between untreated and treated, revealed that there were nearly 15 different tyrosinase modulators found in these two groups. In addition to umbelliferone (**10**) and isopimpinellin (**12**), which are already known to inhibit tyrosinase activity, we also identified compounds **1** and **7** as active components with similar effects. This is the first research to investigate the tyrosinase-modifying effects of *A. keiskei*. We compared the compounds with the published literature and found xanthoangelols (**1**) [20], 4-hydroxyderricin (**7**) [20], umbelliferone (**10**) [27], isopimpinellin (**12**) [28] and phellopterin (**13**) [28] had been reported to have tyrosinase inhibition activity. The compounds from its roots demonstrate promising potential for treating hyperpigmentation and related disorders. Future research can focus on elucidating the in vivo effect and the optimal dosage.

With the aim of evaluating the effectiveness of the novel screening system, we examined the purified compounds to check the candidates from the base peak chromatograms and the tyrosinase inhibitory assay and found out that the results from the two experiments support each other. We therefore conclude this novel screening is highly accurate and effective. Furthermore, in comparison to the traditional protocols, our method reduces the sample and solvent volumes required for UPLC-MS/MS analysis. In addition, it allows researchers to quickly analyze and screen for more candidates with tyrosinase modulatory effects. This strategy can also be used for the rapid development of applications in other screening platforms.

Author Contributions: Investigation, J.-H.L. performed the experiments, data analysis and wrote the manuscript. Resources, Y.-H.C. executed the experiments and identified the plant. Conceptualization, C.-K.L. designed the research experiments and provided the study outline. Methodology, T.-H.L. created models and suggestion of direction. Writing—original draft, I.-C.K. and writing—review and editing, H.-C.M. All authors edited and approved the final manuscript.

Funding: This research was funded by the National Science Council (Taiwan), grant number NSC104-2320-038-020-MY3.

Acknowledgments: We are grateful to S.-H.W. and S.-Y.S. of the Instrumentation Center of Taipei Medical University and the Instrumentation Center of the College of Science, National Taiwan University, respectively, for the NMR and MS data acquisition.

Conflicts of Interest: The authors declare no conflict of interest.

References

1. Korać, R.; Khambholja, K. Potential of herbs in skin protection from ultraviolet radiation. *Pharmacogn. Rev.* **2011**, *5*, 164–173. [CrossRef] [PubMed]
2. D'Mello, S.; Finlay, G.; Baguley, B.; Askarian-Amiri, M. Signaling Pathways in Melanogenesis. *Int. J. Mol. Sci.* **2016**, *17*, 1144. [CrossRef]

3. Pillaiyar, T.; Manickam, M.; Namasivayam, V. Skin whitening agents: Medicinal chemistry perspective of tyrosinase inhibitors. *J. Enzym. Inhib. Med. Chem.* **2017**, *32*, 403–425.
4. Picardo, M.; Maresca, V.; Eibenschutz, L.; De Bernardo, C.; Rinaldi, R.; Grammatico, P. Correlation Between Antioxidants and Phototypes in Melanocytes Cultures. A Possible Link of Physiologic and Pathologic Relevance. *J. Investig. Dermatol.* **1999**, *113*, 424–425. [PubMed]
5. Pillaiyar, T.; Namasivayam, V.; Manickam, M.; Jung, S.-H. Inhibitors of melanogenesis: An updated review. *J. Med. Chem.* **2018**, *61*, 7395–7418. [PubMed]
6. Akihisa, T.; Tokuda, H.; Hasegawa, D.; Ukiya, M.; Kimura, Y.; Enjo, F.; Suzuki, T.; Nishino, H. Chalcones and Other Compounds from the Exudates of Angelica keiskei and Their Cancer Chemopreventive Effects. *J. Nat. Prod.* **2006**, *69*, 38–42. [CrossRef]
7. Kil, Y.-S.; Choi, S.-K.; Lee, Y.-S.; Jafari, M.; Seo, E.-K. Chalcones from Angelica keiskei: Evaluation of Their Heat Shock Protein Inducing Activities. *J. Nat. Prod.* **2015**, *78*, 2481–2487. [CrossRef]
8. Akihisa, T.; Tokuda, H.; Ukiya, M.; Iizuka, M.; Schneider, S.; Ogasawara, K.; Mukainaka, T.; Iwatsuki, K.; Suzuki, T.; Nishino, H. Chalcones, coumarins, and flavanones from the exudate of Angelica keiskei and their chemopreventive effects. *Cancer Lett.* **2003**, *201*, 133–137. [CrossRef]
9. Li, L.; Aldini, G.; Carini, M.; Chen, C.Y.O.; Chun, H.-K.; Cho, S.-M.; Park, K.-M.; Correa, C.R.; Russell, R.M.; Blumberg, J.B.; et al. Characterisation, extraction efficiency, stability and antioxidant activity of phytonutrients in Angelica keiskei. *Food Chem.* **2009**, *115*, 227–232. [CrossRef]
10. Aoki, N.; Muko, M.; Ohta, E.; Ohta, S. C-geranylated chalcones from the stems of Angelica keiskei with superoxide-scavenging activity. *J. Nat. Prod.* **2008**, *71*, 1308–1310. [CrossRef]
11. Kimura, Y.; Baba, K. Antitumor and antimetastatic activities of Angelica keiskei roots, part 1: Isolation of an active substance, xanthoangelol. *Int. J. Cancer* **2003**, *106*, 429–437. [CrossRef] [PubMed]
12. Shin, J.E.; Choi, E.J.; Jin, Q.; Jin, H.G.; Woo, E.R. Chalcones isolated from Angelica keiskei and their inhibition of IL-6 production in TNF-alpha-stimulated MG-63 cell. *Arch. Pharmacal Res.* **2011**, *34*, 437–442. [CrossRef] [PubMed]
13. Luo, L.; Wang, R.; Wang, X.; Ma, Z.; Li, N. Compounds from Angelica keiskei with NQO1 induction, DPPH scavenging and α-glucosidase inhibitory activities. *Food Chem.* **2012**, *131*, 992–998.
14. Caesar, L.; Cech, N. A Review of the Medicinal Uses and Pharmacology of Ashitaba. *Planta Med.* **2016**, *82*, 1236–1245. [PubMed]
15. Chen, Y.; Kong, L.-D.; Xia, X.; Kung, H.-F.; Zhang, L. Behavioral and biochemical studies of total furocoumarins from seeds of Psoralea corylifolia in the forced swimming test in mice. *J. Ethnopharmacol.* **2005**, *96*, 451–459. [CrossRef]
16. Akihisa, T.; Kikuchi, T.; Nagai, H.; Ishii, K.; Tabata, K.; Suzuki, T. 4-Hydroxyderricin from Angelica keiskei roots induces caspase-dependent apoptotic cell death in HL60 human leukemia cells. *J. Oleo Sci.* **2011**, *60*, 71–77.
17. Sumiyoshi, M.; Taniguchi, M.; Baba, K.; Kimura, Y. Antitumor and antimetastatic actions of xanthoangelol and 4-hydroxyderricin isolated from Angelica keiskei roots through the inhibited activation and differentiation of M2 macrophages. *Phytomedicine* **2015**, *22*, 759–767. [CrossRef]
18. Gupta, D.K.; Verma, M.K.; Anand, R.; Khajuria, R.K. Development of a validated UPLC–qTOF-MS/MS method for determination of bioactive constituent from Glycyrrhiza glabra. *J. Pharm. Anal.* **2013**, *3*, 205–210. [CrossRef]
19. Aldini, G.; Regazzoni, L.; Pedretti, A.; Carini, M.; Cho, S.M.; Park, K.M.; Yeum, K.J. An integrated high resolution mass spectrometric and informatics approach for the rapid identification of phenolics in plant extract. *J. Chromatogr. A* **2011**, *1218*, 2856–2864.
20. Arung, E.T.; Furuta, S.; Sugamoto, K.; Shimizu, K.; Ishikawa, H.; Matsushita, Y.-I.; Kondo, R. The inhibitory effects of representative chalcones contained in Angelica keiskei on melanin biosynthesis in B16 melanoma cells. *Nat. Prod. Commun.* **2012**, *7*, 1007–1010. [CrossRef]
21. Asthana, S.; Zucca, P.; Vargiu, A.V.; Sanjust, E.; Ruggerone, P.; Rescigno, A. Structure–Activity Relationship Study of Hydroxycoumarins and Mushroom Tyrosinase. *J. Agric. Food Chem.* **2015**, *63*, 7236–7244. [CrossRef] [PubMed]
22. Chen, C.H.; Chen, P.Y.; Wang, K.C.; Lee, C.K. Rapid identification of the antioxidant constituent of Koelreuteria henryi. *J. Chin. Chem. Soc.* **2010**, *57*, 404–410. [CrossRef]

23. Jayasinghe, L.; Balasooriya, B.A.I.S.; Padmini, W.C.; Hara, N.; Fujimoto, Y. Geranyl chalcone derivatives with antifungal and radical scavenging properties from the leaves of Artocarpus nobilis. *Phytochemistry* **2004**, *65*, 1287–1290. [CrossRef] [PubMed]
24. Kim, D.W.; Curtis-Long, M.J.; Yuk, H.J.; Wang, Y.; Song, Y.H.; Jeong, S.H.; Park, K.H. Quantitative analysis of phenolic metabolites from different parts of Angelica keiskei by HPLC-ESI MS/MS and their xanthine oxidase inhibition. *Food Chem.* **2014**, *153*, 20–27. [CrossRef] [PubMed]
25. Baba, K.; Nakata, K.; Taniguchi, M.; Kido, T.; Kozawa, M. Chalcones from Angelica keiskei. *Phytochemistry* **1990**, *29*, 3907–3910. [CrossRef]
26. Sugamoto, K.; Matsusita, Y.-I.; Matsui, K.; Kurogi, C.; Matsui, T. Synthesis and antibacterial activity of chalcones bearing prenyl or geranyl groups from Angelica keiskei. *Tetrahedron* **2011**, *67*, 5346–5359. [CrossRef]
27. Ashraf, Z.; Rafiq, M.; Seo, S.-Y.; Babar, M.M.; Zaidi, N.-U.-S.S. Design, synthesis and bioevaluation of novel umbelliferone analogues as potential mushroom tyrosinase inhibitors. *J. Enzym. Inhib. Med. Chem.* **2015**, *30*, 874–883. [CrossRef] [PubMed]
28. Orhan, I.; Tosun, F.; Skalicka-Woźniak, K. Cholinesterase, tyrosinase inhibitory and antioxidant potential of randomly selected Umbelliferous plant species and chromatographic profile of Heracleum platytaenium Boiss. and *Angelica sylvestris* L. var. sylvestris. *J. Serb. Chem. Soc.* **2016**, *81*, 357–368. [CrossRef]

Sample Availability: Samples of the compounds are not available from the authors.

molecules

Article

Intake of Molecular Hydrogen in Drinking Water Increases Membrane Transporters, *p*-Glycoprotein, and Multidrug Resistance-Associated Protein 2 without Affecting Xenobiotic-Metabolizing Enzymes in Rat Liver

Hsien-Tsung Yao *, Yu-Hsuan Yang and Mei-Ling Li

Department of Nutrition, China Medical University, 91 Hsueh-shih Road, Taichung 404, Taiwan
* Correspondence: htyao@mail.cmu.edu.tw; Tel.: +886-4-22053366 (ext. 7526); Fax: +886-4-22062891

Academic Editor: Ping-Chung Kuo
Received: 26 June 2019; Accepted: 17 July 2019; Published: 19 July 2019

Abstract: Molecular hydrogen (H_2) has been shown to have antioxidant and anti-inflammatory activities that may reduce the development and progression of many diseases. In this study, hydrogen-rich water (HRW) was obtained by reacting hybrid magnesium–carbon hydrogen storage materials with water. Then, the effects of intake of HRW on the activities of xenobiotic-metabolizing enzymes, membrane transporters, and oxidative stress in rats were investigated. Rats were given HRW ad libitum for four weeks. The results showed that intake of HRW had no significant effect on the activities of various cytochrome P450 (CYP) enzymes (CYP1A1, 1A2, 2B, 2C, 2D, 2E1, 3A, and 4A), glutathione-*S*-transferase, and Uridine 5′-diphospho (UDP)-glucuronosyltransferase. Except for a mild lower plasma glucose concentration, intake of HRW had no effect on other plasma biochemical parameters in rats. *p*-Glycoprotein and multidrug resistance-associated protein (Mrp) 2 protein expressions in liver were elevated after intake of HRW. However, HRW had no significant effects on glutathione, glutathione peroxidase, or lipid peroxidation in liver. The results from this study suggest that consumption of HRW may not affect xenobiotic metabolism or oxidative stress in liver. However, intake of HRW may increase the efflux of xenobiotics or toxic substances from the liver into bile by enhancing *p*-glycoprotein and Mrp2 protein expressions.

Keywords: molecular hydrogen; hydrogen-rich water; xenobiotic-metabolizing enzymes; membrane transporters; liver; oxidative stress

1. Introduction

Xenobiotics, such as drugs or toxic chemicals, can be metabolized and eliminated by xenobiotic-metabolizing enzymes and membrane transporters. The liver is the major tissue responsible for detoxification of xenobiotics. The xenobiotic-metabolizing enzymes include phase I and phase II enzyme systems. The cytochrome P450 (CYP) enzymes are the major phase I enzymes responsible for the metabolism of endogenous molecules (e.g., sterols and fatty acids) and exogenous xenobiotics (e.g., drugs and toxic chemicals), resulting in the formation of more water-soluble and less toxic metabolites [1]. However, some CYP enzymes, such as CYP1A1, 3A, and 2E1, are involved in the bioactivation of chemicals, such as benzo[a]pyrene, aflatoxin B1, and acetaminophen [2–4], and produce electrophile intermediates that may covalently bind to proteins, lipids, and DNA. These enzyme reactions may therefore produce more reactive oxygen species (ROS) and increase oxidative damage to tissues [5]. Uridine 5′-diphospho (UDP)-glucuronosyltransferase (UGT) and glutathione S-transferase (GST) are two important phase II detoxifying enzymes that catalyze the conjugation reactions, resulting in the formation of water-soluble glucuronate and glutathione conjugates to facilitate the excretion of xenobiotics. Induction of phase II detoxifying enzymes and reduction of ROS is most pronounced in the prevention of chemical-induced tissue injuries and carcinogenesis [6]. Phase III membrane transporters include *p*-glycoprotein and multidrug resistance-associated proteins, such as Mrp2/3, which may function to shuttle xenobiotics or their metabolites across cellular membranes and to facilitate the excretion of these compounds from the liver into bile (e.g., *p*-glycoprotein and Mrp2) and blood (e.g., Mrp3) [7–9].

Molecular hydrogen (H_2) acts as an antioxidant by selectively reducing particularly strong oxidants such as the hydroxyl radical (•OH), which can exhibit relatively stronger oxidative activities than other ROS in cells [10]. It is noteworthy that H_2 has no cytotoxicity even at high concentrations. Consumption of hydrogen water (H_2 dissolved in water) is a convenient, easily administered, and safe way to ingest H_2. Hydrogen water can be made by several methods: infusing H_2 gas into water under high pressure, electrolyzing water to producing H_2, and reacting magnesium metal or its hydride with water [11]. Despite the low solubility of H_2 gas in water, which can be only up to 1.6 ppm under 1 atmospheric pressure at room temperature, consumption of hydrogen-rich water (HRW) has been shown to be effective for ameliorating various diseases caused by oxidative stress in animal and clinical studies [11,12]. These findings suggest that H_2 may be a versatile element with therapeutic activity.

The quality of drinking water is an important health and safety information [13]. It is known that dissolved H_2 in drinking water will gradually decrease over time. Until now, in addition to the H_2 concentration, the quality of HRW has not been adequately described. In addition, the effects of intake of HRW on the activity of xenobiotic-metabolizing enzymes and on the membrane transporters involved in the metabolism of various drugs or chemicals are still unknown. In this study, therefore, we first assessed quality parameters of HRW, such as the dissolved H_2 concentration, total dissolved solids (TDS), ions, and water cluster. Then, the effects of intake of HRW on xenobiotic-metabolizing enzymes, membrane transporters, and antioxidant activity in rat liver were investigated.

2. Results

2.1. *Quality of HRW*

The quality of drinking water is an important information for health and safety. The quality parameters of HRW, including dissolved H_2 concentration, oxidation-reduction potential (ORP) value, and physical and chemical properties of the water samples, are shown in Table 1. The fresh HRW contained 1550 ppb H_2 and had a negative ORP value (control water: +293 mv; HRW: −453 mv). HRW also had a higher pH than the control water. Ca^{2+} concentrations were lower and Mg^{2+} concentrations were higher in the HRW. Other parameters of water quality, including TDS, salt, electrolytic conductivity (EC), dissolved oxygen (DO), Na^+, K^+, Cl^-, and $SO4^{2-}$ levels, showed little or no difference between the control water and HRW.

The stability of dissolved H_2 in HRW is shown in Figure 1. A H_2 concentration of around 1500 ppb was maintained in water for 1 h, and the concentration remained around 1300 ppb for 2 h. After that, the H_2 concentration in water gradually decreased over time. The remaining H_2 concentration in HRW was around 200 ppb after the sample was set at 4 °C or 25 °C for 24 h. The ORP value in HRW increased gradually over time and, after 24 h, was around −100 mv. No significant differences in dissolved H_2 or ORP values were observed in the HRW samples either at 4 °C or at 25 °C within 24 h.

It was noteworthy that water cluster size, determined by ^{17}O NMR line-width, was comparable between the control water (55.8 Hz) and the HRW (54.9 Hz) (Figure 2). In a previous study, the water cluster size of tap water was shown to be greater than 110 Hz [14]. These results suggest that, after tap water is filtered through the first two filters, the water cluster size can be lowered by shear force. The final magnetized rod device may have strengthened the hydrogen bond network of the water and, thus, stabilized the water cluster [15].

Table 1. Quality of control water and hydrogen-rich water (HRW) [a].

Parameters	Control	HRW
H_2 (ppb)	0	1550
ORP (mv)	292.6	−452.7
TDS (g/L)	197.3	196.7
Salt (kU.m)	149.3	149.3
EC (ds/m)	197	197
DO (mg/L)	6.1	5.8
pH	7.6	9.3
Ions (ppm)		
Ca^{2+}	28.0	9.1
Mg^{2+}	10.2	22.8
Na^+	4.8	5.4
K^+	1	1
Cl^-	2.9	3.2
SO_4^{2-}	38.5	59.8
Water cluster (Hz)	55.8	54.9

[a] All data were measured at 25 °C. The quality of tap water was as follows: H_2, 0 ppb; ORP: 203.7 mv; TDS: 204.7 g/L; salt: 155 kU.m; EC: 310.7 ds/m; DO: 8.0 mg/L; pH: 8.0; Na^+: 4.8 ppm; K^+: 1 ppm; Ca^{2+}: 51 ppm; Mg^{2+}: 10 ppm; Cl^-: 3.2 ppm; and SO_4^{2-}: 42.0 ppm. Except for water cluster, all of the above parameters were expressed as the mean value of three determinations. In the previous study, the water cluster of tap water determined by the ^{17}O NMR line-width method was 113 Hz [14]. DO, dissolved oxygen; EC, electrolytic conductivity; ORP, oxidation-reduction potential; TDS, total dissolved solids.

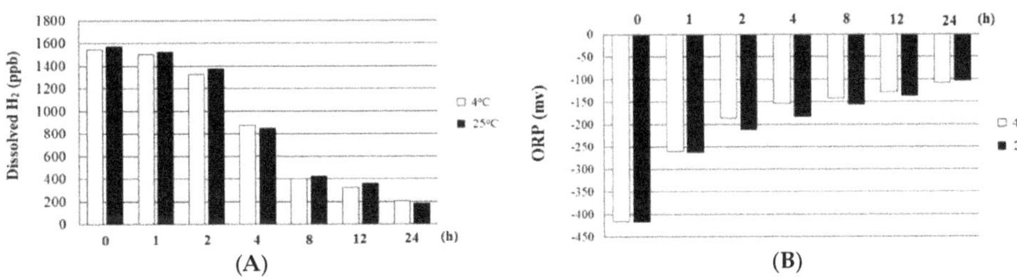

Figure 1. (**A**) Stability of dissolved H_2 and (**B**) the ORP of hydrogen-rich water at 4 °C and 25 °C. ORP, oxidation-reduction potential.

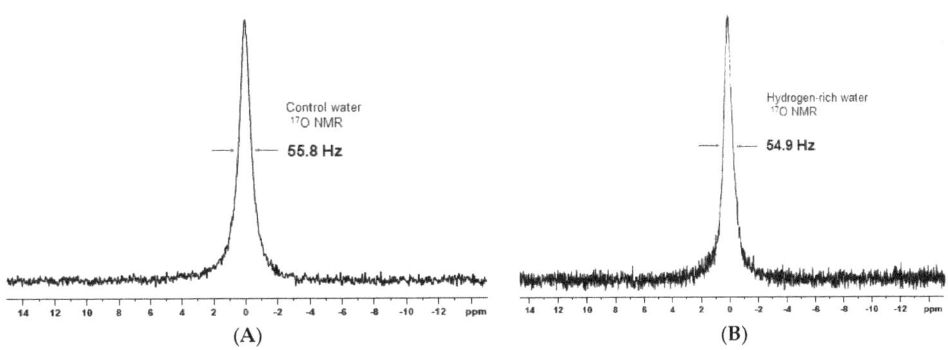

Figure 2. ^{17}O Nuclear magnetic resonance line-width of water samples: (**A**) Control water and (**B**) hydrogen-rich water.

2.2. Body Weight, Tissue Weight, Water Drinking Volume, and Plasma Biochemical Parameters

In this study, no significant effects were found on body weight, liver weight, kidney weight, or food intake in rats that ingested the control water or HRW for four weeks. The volume of drinking water ingested was mildly higher (+10.7%, $p < 0.05$) in the HRW group (81.8 ± 5.1 mL/day) than in the control group (73.9 ± 5.0 mL/day). The plasma biochemical parameters are shown in Table 2. Plasma glucose was mildly lower (−7.7%) in the HRW group than in the control group ($p < 0.05$). There were no significant differences in other plasma parameters including ions between the control and HRW groups.

Table 2. Effects of intake of HRW on plasma biochemical parameters in rats [a].

Parameters	Control	HRW
Total cholesterol (mg/dL)	72.0 ± 8.2	78.9 ± 9.1
Triglyceride (mg/dL)	53.9 ± 20.5	58.2 ± 13.3
Glucose (mg/dL)	155.9 ± 10.3	143.9 ± 8.1 *
BUN (mg/dL)	16.8 ± 2.4	18.8 ± 2.6
Creatinine (mg/dL)	0.31 ± 0.02	0.31 ± 0.03
Uric acid (mg/dL)	2.2 ± 0.6	2.0 ± 0.3
ALT (U/L)	44.6 ± 6.1	44.3 ± 5.1
TBARS (nmol/mL)	0.6 ± 0.1	0.6 ± 0.1
GSH (nmol/mL)	6.4 ± 2.6	6.5 ± 2.4
Na^+ (mg/dL)	140.2 ± 1.6	144.0 ± 5.9
K^+ (mg/dL)	6.1 ± 0.3	6.1 ± 0.3
Cl^- (mg/dL)	1.3 ± 0.1	1.4 ± 0.2
Ca^{2+} (mg/dL)	91.5 ± 1.1	93.3 ± 3.9
Mg^{2+} (mg/dL)	2.9 ± 0.2	3.0 ± 0.2

[a] Results are expressed as the mean ± S.D. of eight rats in each group. ALT, alanine aminotransferase; BUN, blood urea nitrogen; GSH, reduced glutathione; TBARS, thiobarbituric acid reactive substances. * Significantly different from the control group, $p < 0.05$.

2.3. *Activities of Xenobiotic-Metabolizing Enzymes and Membrane Transporters*

Table 3 shows the effects of intake of HRW on the activities of xenobiotic-metabolizing enzymes in the liver. No effects were found in the activity of UGT or GST in rats based on intake of HRW for four weeks. In addition, intake of HRW had no significant effect on CYP-mediated reactions.

Table 3. Changes in hepatic drug-metabolizing enzyme activities in rats [a].

	Control	HRW
Phase I enzymes (pmol/min/mg protein)		
Testosterone 6β-hydroxylase (CYP3A)	682.3 ± 75.6	640.9 ± 168.1
Diclofenac 4-hydroxylase (CYP2C)	155.9 ± 10.3	147.9 ± 8.1
Dextromethorphan *O*-demethylase (CYP2D)	51.9 ± 6.1	45.2 ± 4.6
Nitrophenol 6-hydroxylase (CYP2E1)	121.1 ± 12.3	109.8 ± 15.6
Ethoxyresorufin *O*-deethylase (CYP1A1)	167.8 ± 17.2	171.5 ± 10.1
Methoxyresorufin *O*-demethylase (CYP1A2)	133.2 ± 18.2	140.6 ± 17.4
Pentoxyresorufin *O*-depentylase (CYP2B)	61.9 ± 10.4	62.3 ± 11.3
Lauric acid 12-hydroxylauric acid (CYP4A)	1524 ± 125	1488 ± 132
Phase II enzymes (nmol/min/mg protein)		
Glutathione *S*-transferase (GST)	1072 ± 175	1132 ± 88
UDP-glucurosyltransferase (UGT)	23.4 ± 2.3	25.5 ± 3.0

[a] Results are expressed as the mean ± S.D. of eight rats in each group.

The gene expression of CYP enzymes was also not changed ($p > 0.05$) by intake of HRW (Figure 3).

Immunoblots of liver membrane transporters are shown in Figure 4. Intake of HRW increased ($p < 0.05$) *p*-glycoprotein and Mrp2 protein expressions in liver. However, HRW had no significant effect on Mrp3 protein expression. These results suggest that intake of HRW may increase the efflux of endogenous substances and exogenous xenobiotics from liver into bile by increasing Mrp2 and *p*-glycoprotein protein expression without affecting the hepatic activity of xenobiotic-metabolizing enzymes in rats.

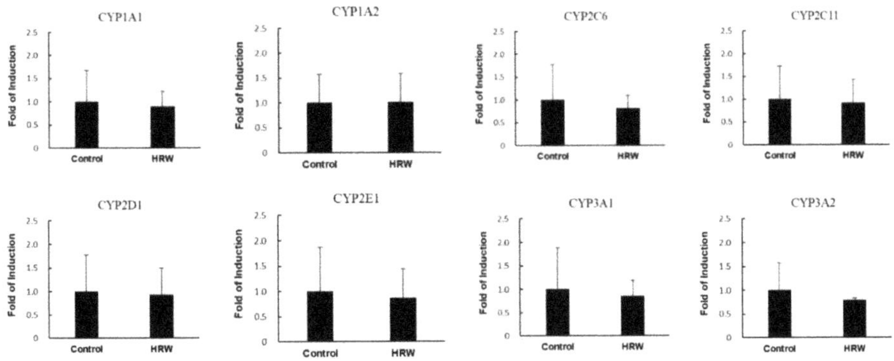

Figure 3. Effects of intake of HRW on the mRNA expression of various CYP enzymes in liver: The results are expressed as the mean ± S.D. of five rats.

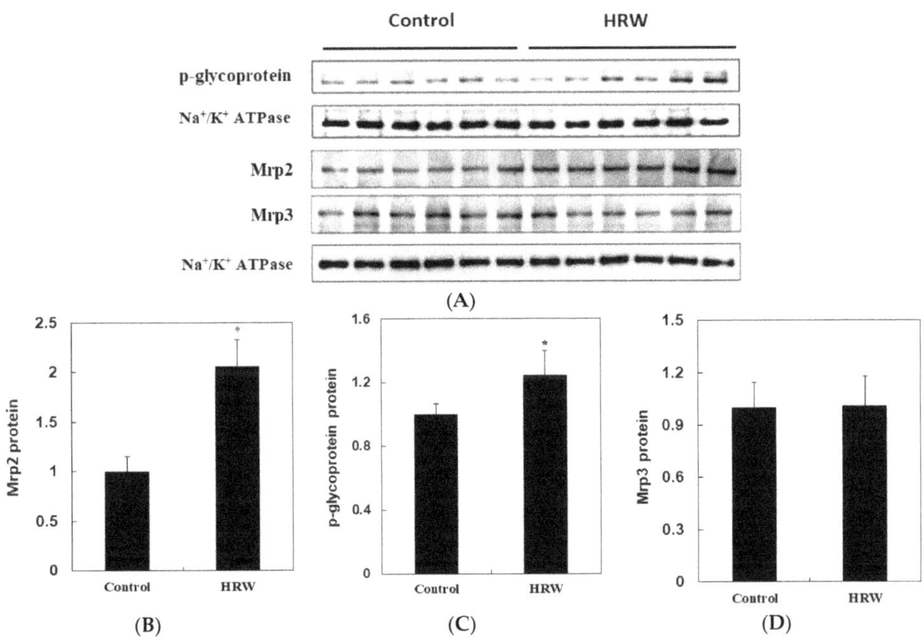

Figure 4. Effects of intake of HRW on *p*-glycoprotein and Mrp2/3 protein expression in liver of rats (**A**). Protein expression was determined by Western blotting. Densitometric analysis for Mrp2 (**B**), *p*-glycoprotein (**C**) and Mrp3 (**D**) protein levels corrected to each internal control is shown. The results are expressed as the mean ± S.D. of six rats. Na^+/K^+ ATPase acts as an internal control. The protein band was quantified by densitometry, and the level of the control was set at 1. * Significantly different from the control group at $p < 0.05$.

2.4. Oxidative Stress

As shown in Table 4, intake of HRW had no significant effects on hepatic GSH or GSSG levels, the GSH/GSSG ratio, or GSH peroxidase activity in rats compared with these animals in the control group. Also, intake of HRW did not change the hepatic TBARS value.

Table 4. Effect of intake of HRW on oxidative stress in liver [a].

	Control	HRW
GSH (nmol/mg protein)	4.1 ± 0.7	5.0 ± 1.4
GSSG (nmol/mg protein)	0.14 ± 0.03	0.16 ± 0.02
GSH/GSSG	29.4 ± 4.2	31.3 ± 10.2
GSH peroxidase (nmol/min/mg protein)	314.3 ± 25.6	302.0 ± 41.3
TBARS (nmol/g protein)	18.8 ± 3.6	18.9 ± 2.1

[a] Results are expressed as the mean ± S.D. of eight rats in each group. GSH, reduced glutathione; GSSG, oxidized glutathione; TBARS, thiobarbituric acid reactive substances.

3. Discussion

Xenobiotic-metabolizing enzymes and membrane transporters are responsible for the detoxification and elimination of xenobiotics from the body. Food components can change xenobiotic metabolism by modifying these enzymes and membrane transporters. This study first showed that intake of HRW had no effect on CYP enzyme activities or antioxidant activity in rat livers. However, intake of HRW increased the efflux pumps of *p*-glycoprotein and Mrp2 in liver. These results indicate that HRW may enhance the excretion of xenobiotics from the liver into bile without altering their metabolism by xenobiotic-metabolizing enzymes. In addition, intake of HRW had no significant effect on oxidative stress in the normal physiologic condition.

CYP-mediated bioactivation and ROS formation may be responsible for chemical or drug toxicity. Increased phase II-conjugated enzymes may facilitate the elimination of xenobiotics [6]. We showed here that intake of HRW did not change the activities of CYP isozymes, UGT, and GST in the liver. Thus, intake of HRW may not change the metabolism of xenobiotics by xenobiotic-metabolizing enzymes in liver.

GSH is the most important biomolecule against ROS-induced tissue injury and can participate in the elimination of xenobiotics through GST [6]. HRW has been shown to be able to scavenge free radicals, especially hydroxyl radical (•OH), and can prevent the progression of various diseases induced by oxidative damage [11,12]. In the present study, the plasma GSH concentration and hepatic GSH, GSH/GSSG, and GSH peroxidase activities were not changed after intake of HRW. In addition, the lipid peroxidation levels in plasma and liver were not affected by HRW. Therefore, in the normal physiologic condition, intake of HRW may not affect antioxidant activity or oxidative stress in liver. These results are consistent with previous results indicating that H_2 does not disturb normal cellular metabolic redox reactions [11].

Membrane transporters are effective pumps for elimination of conjugates of xenobiotics from hepatocytes into bile (e.g., Mrp2 and *p*-glycoprotein) [16] and plasma (e.g., Mrp3) [9]. Therefore, increased protein expressions of Mrp2/3 and *p*-glycoprotein also play roles in detoxification processes. These membrane proteins can be induced and can protect normal tissues from endogenous and exogenous toxic substances [16,17]. In the present study, Mrp2 and *p*-glycoprotein increased in the liver whereas Mrp3 did not change significantly after intake of HRW (Figure 4). Therefore, although the mechanism is still unknown, HRW may enhance the excretion of xenobiotics or endogenous toxic substances from the liver into bile and then increase their fecal excretion by increasing the efflux pumps of Mrp2 and *p*-glycoprotein. On the other hand, in this study, a little change on the concentrations of Ca^{2+}, Mg^{2+}, and $SO4^{2-}$ between the control water and HRW groups did not have any influence on plasma mineral ions. A higher daily water drinking volume was found in the HRW group compared with the control group, suggesting HRW had better palatability (see descriptions in the Results section). Until now, there is a lack of evidence demonstrated that ions in drinking water may change the xenobiotic-metabolizing enzymes and transporters. Therefore, it is suggested that increased Mrp2 and *p*-glycoprotein expression in the liver after intake of HRW may be attributed to the molecular hydrogen.

Calcium and magnesium are important nutrients in the development and maintenance of human health. Supplementation with magnesium ion (Mg^{2+}) from drinking water may provide substantial contributions to total intakes of Mg^{2+} in some populations and may exert beneficial effects on reducing many diseases. A protective effect of Mg^{2+} intake from drinking water has been demonstrated on reducing cerebrovascular disease and cardiovascular mortality in humans [18,19], especially in men with lower dietary magnesium intake [20]. A recent study demonstrated that Mg^{2+} added to drinking water reduces blood glucose levels by inhibiting the gluconeogenesis pathway in rat liver [21]. In this study, HRW contained a higher Mg^{2+} concentration than control water. This may be one of the explanations for the lower plasma glucose level in the HRW group, although H_2 is regarded as an effective element to improve glucose intolerance in diabetic mice and some type 2 patients [22,23]. Therefore, HRW manufactured by reacting water with magnesium–carbon hydrogen storage hybrid materials may result in a higher Mg^{2+} concentration in drinking water and, thus, may complement daily magnesium, especially in populations with magnesium deficiency.

To date, various commercial apparatuses for HRW production (e.g., manufactured by electrolyzing water or water reacted with magnesium-containing materials) are being developed. These HRW products may have a high dissolved H_2 concentration and H_2 stability in drinking water. However, the quality and safety of these HRW products should be a concern (e.g., undesirable flavor and/or unknown reaction products).

In summary, intake of HRW for four weeks may not change xenobiotic-metabolizing enzymes or antioxidant activity in liver. Regular consumption of HRW may enhance detoxification process, at least in part, through an increase in the efflux of toxic substances from the liver into bile. In addition to measuring dissolved H_2 concentration, the present study also evaluated the water quality of HRW on biological function. Because HRW is becoming more popular worldwide, the results of the present study may provide health and safety information on HRW to consumers.

4. Materials and Methods

4.1. Materials

Testosterone, ethoxyresorufin, methoxyresorufin, pentoxyresorufin, resorufin, *p*-nitrophenol, 4-nitrocatechol, NADPH, glutathione, 1-chloro-2,4-dinitrobenzene, lauric acid, 12-hydroxy lauric acid, diclofenac (sodium salt), chlorzoxazone, dextromethophen, 1,1,3,3-tetraethoxypropan, thiobarbituric acid, and heparin were obtained from Sigma (St. Louis, MO, USA). 6-β-Hydroxytestosterone was purchased from Ultrafine Chemicals (Manchester, UK). All other chemicals and reagents were of analytical grade and were obtained commercially.

4.2. HRW Production

As shown in Figure 5, the control water (control group) was filtered from tap water (Taitung, Taiwan) by passage through a calcined ceramic filter, an activated carbon filter, and a magnetized rod (purchased from Japin biotech company, Taitung, Taiwan). The whole water devices are certified by the National Sanitation Foundation (NSF)/American National Standards Institute (ANSI) standards No. 42 (Filters are certified to reduce aesthetic impurities such as chlorine and taste/odor.), No. 53 (Filters are certified to reduce a contaminant with a health effect, which are set in this standard as regulated by the U.S. Environmental Protection Agency (EPA) and Health Canada.), and No. 401 (Treatment systems that have been verified to reduce one or more of 15 emerging contaminants, which can be pharmaceuticals or chemicals not yet regulated by the EPA or Health Canada, from drinking water (http://www.nsf.org/consumer-resources/water-quality/water-filters-testing-treatment/standards-water-treatment-systems)). The HRW was obtained from the same water apparatus except that the tap water was passed through the first two filters and then reacted with the magnesium–carbon hydrogen storage hybrid materials (Kuraray Co., Ltd., Japan) in the third device. The resulting HRW was then passed through an activated carbon filter and magnetized rod at a flow rate of 2 L/min.

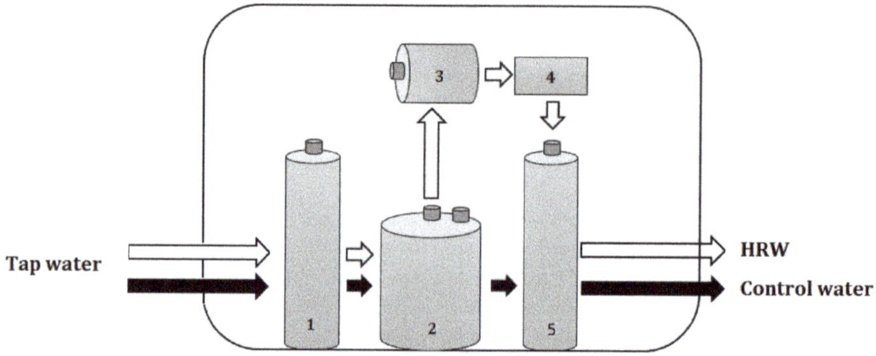

Figure 5. Schemes of the manufacturing process for control water and HRW from tap water: 1. Calcined ceramic filter; 2. activated carbon filter; and 3. magnesium–carbon hydrogen storage hybrid materials. Water reacted with this material and then release stable H_2 gas; 4. activated carbon filter; 5. magnetized rod.

4.3. Determinations of the Quality of HRW

The dissolved H_2 in fresh HRW was measured with an ENH-1000 electrode (TRUSTLEX Inc, Osaka, Japan), and the oxidation-reduction potential (ORP) value was determined by use of an MP-103 electrode (Gondo Electronic Co., Ltd. Taipei, Taiwan). The stability of dissolved H_2 and ORP values in HRW was determined at various time points, including initial (0 h), 1, 2, 4, 8, 12, and 24 h, and at 4 °C and 25 °C, respectively (Figure 1).

The other water quality parameters, including pH, total dissolved solids (TDS), electrolytic conductivity (EC), and dissolved oxygen (DO), in the experimental drinking water were determined by use of electrode equipment with a bench-top water quality meter (Chi Jui Instrument Enterprise Co., Ltd., Taiwan) (Table 1). The concentrations of cations (Na+, K+, Ca^{2+}, and Mg^{2+}) and anions (Cl^- and $SO4^{2-}$) in the water samples were determined by inductively coupled plasma-optical emission spectroscopy (ICP-OES) (HORIBA Jobin Yvon Longjumeau, France). It is known that water quality can be influenced by ions (e.g., Na^+, K^+, Ca^{2+}, Mg^{2+}, Cl^-, NO^{3-}, and $SO4^{2-}$ are commonly found in natural water), pH, and TDS in water and that these variables have an important influence on human health. The above water parameters can affect water clusters by the interaction of water molecules and ions or the formation of clusters of ion–water and water–water, which can change the physical properties of water (e.g., melting point) [14]. To estimate the water clustering of control water and HRW, ^{17}O nuclear magnetic resonance (NMR) line-width was measured to estimate median water cluster size [14]. The wider the ^{17}O NMR line-width, the larger the water cluster size. In this study, water samples were characterized by ^{17}O NMR spectroscopy (Bruker 500 MHz NMR, Varian Inova, Canada). The sample (700 μL) was mixed in a 5-mm NMR spectroscopy tube and analyzed under the following conditions: 67.80 MHz, 0.202-s sampling time, 10162.6-Hz bandwidth, 4096 scans, 90° flip angle, 0.2-s relaxation delay, and room temperature (25 °C).

4.4. Animal Study

To investigate the effect of intake of HRW on the xenobiotic-metabolizing enzymes and membrane transporters in rat livers, sixteen male Wistar rats (aged six weeks) obtained from BioLASCO in Ilan, Taiwan were used. Rats were fed a pelleted laboratory diet with fresh control water or HRW (replaced at 5 p.m. every day) ad libitum for four weeks. The rats were all housed in plastic cages in a room kept at 23 ± 1°C with 60 ± 5% relative humidity and a 12-h light-dark cycle. At the end of the experiment, food was withdrawn for 12 h and the animals were sacrificed by exsanguination via the abdominal aorta while under carbon dioxide (70:30, CO_2/O_2) anesthesia. Heparin was used as the anticoagulant, and the plasma was separated from the blood by centrifugation (1750× *g*) at 4 °C for 20 min. Plasma concentrations of total cholesterol, triglyceride, alanine aminotransferase (ALT), glucose, blood urine nitrogen (BUN), creatinine, uric acid, and ions were measured immediately by use of a serum autoanalyzer (DiaSYS Diagnostic system, Germany). The liver and kidney samples from each animal were weighed and stored at −80 °C.

This study was approved (No: 2017-056) by the Institutional Animal Care and Use Committee (IACUC) of China Medical University, Taiwan. The animals were maintained in accordance with the guidelines for the care and use of laboratory animals as issued by the IACUC ethics committee.

4.5. Preparation of Liver Microsomes

The frozen liver was homogenized (1:4, *w*/*v*) in ice-cold 0.1 M phosphate buffer (pH 7.4) containing 1 mM ethylenediaminetetraacetic acid (EDTA). The homogenates were centrifuged at 10,000× *g* for 15 min at 4 °C. The supernatants were then centrifuged at 105,000× *g* for 60 min. The resulting microsomal pellets were suspended in a 0.25 M sucrose solution containing 1 mM EDTA and were stored at −80 °C until use. The microsomal protein concentration was determined by using a BCA protein assay kit (Pierce, Rockford, IL, USA).

4.6. Xenobiotic-Metabolizing Enzyme Activity Assays

The CYP enzyme activities, including methoxyresorufin *O*-demethylation (CYP1A2), ethoxyresorufin *O*-deethylation (CYP1A1), pentoxyresorufin *O*-depentylation (CYP2B), diclofenac 4-hydroxylation (CYP2C), dextromethorphan *O*-demethylation (CYP2D), *p*-nitrophenol 6-hydroxylation (CYP2E1), testosterone 6β-hydroxylation (CYP3A), and lauric acid 12-hydroxylation (CYP4A), were determined by the high performance liquid chromatography (HPLC)/mass spectrometric (MS) method [24]. Enzyme activities were expressed as pmol of metabolite formation/min/mg protein. Microsomal UGT activity was determined by using *p*-nitrophenol as the substrate, and the rate of formation of *p*-nitrophenol glucuronic acid was measured by HPLC/MS (Agilent, USA) [25]. GST was measured by the spectrophotometric method [26].

4.7. Determinations of GSH, GSH Peroxidase Activity, and Lipid Peroxidation

Plasma and liver homogenate were used to determine the reduced (GSH) or oxidized glutathione (GSSG) content by the HPLC/MS method as reported previously [27]. Glutathione peroxidase activity was determined spectrophotometrically according to the method of Mohandas et al. [28]. Lipid peroxidation, as measured by thiobarbituric acid reactive substances (TBARS), in plasma and tissues were assessed by the modified method of Mihara and Uchiyama [29]. 1,1,3,3-Tetramethoxypropane (Sigma, USA) was used to determine the concentrations of TBARS in the samples. Fluorescence was measured at excitation and emission wavelengths of 515 nm and 553 nm, respectively.

4.8. Determination of p-Glycoprotein and Mrp 2/3

Crude membrane from liver was prepared according to the method of Aleksunes et al. [30]. Each gram of liver was homogenized with 4 mL of sucrose-Tris buffer (0.25 M sucrose, 10 mM Tris-HCl, pH 7.4) containing 50 g/mL of aprotinin. The homogenate was then centrifuged at 100,000× *g* for 60 min at 4 °C. The resulting pellet was resuspended in sucrose-Tris buffer and was used for determinations of *p*-glycoprotein and Mrp2/3 by Western blot.

4.9. Western Blot Analysis

Equal amounts of proteins from membranes of each group were separated by SDS-PAGE and transferred to nitrocellulose membranes. The *p*-glycoprotein was purchased from Calbiochem (Darmstadt, Germany). Mrp2/3 antibodies were purchased from Abcam (Cambridge, UK). The western blot analysis was performed as described previously [24].

4.10. Reverse Transcription Polymerase Chain Reaction (RT-PCR) Analysis

Total RNA was extracted from homogenized liver tissue by using TRIZOL reagent (Invitrogen, Carlsbad, CA, USA) according to the manufacturer's instructions. Total RNA (1 μg) was reverse-transcribed into first-strand cDNA by using 200 units of MMLV-RT (Promega) in a total volume of 20 μL. For real-time PCR, a SYBR system with self-designed primers and 12.5 ng cDNA was used. The self-designed primers were as follows: CYP1A1 forward: GGTTCTGGATACCCAGCTGAC; reverse: TGTGGCCCTTCTCAAATGTCC, CYP1A2 forward: GCTGTGGACTTCTTTCCGGT; reverse: TGTCCTGGATACTGTTCTTGTTGA, CYP2C6 forward: TCCTGCTGAAGTGTCCAGAA; reverse: TGCAAGGGCTGCGATGTTT, CYP2C11 forward: TGAAGGACATCGGCCAATCA; reverse: CCCA TGCAACACCACAAAGG, CYP2D1 forward: ACCCATGGCTTCTTTGCTTTTC; reverse: GTCCTTGC TCCCGTACCAC, CYP3A1 forward: CTCAAGGAGATGTTCCCTGTCA; reverse: CAGGTTTGCCTT TCTCTTGCC, CYP3A2 forward: CCATCCACATCTGGTGGTCT; and reverse: TCAAAGGACGAG GACATGGTT. Amplification using 40 cycles of 2 steps (95 °C for 15 s and 60 °C for 1 min) was performed on an ABI Prism 7900HT sequence detection system (Foster City, CA, USA).

4.11. Statistical Analysis

Statistical differences among groups were calculated by using one-way ANOVA (SAS Institute, Cary, NC, USA). The differences were considered to be significant at $p < 0.05$ as determined by independent-sample t tests.

5. Conclusions

Intake of HRW for four weeks may not change xenobiotic-metabolizing enzymes and antioxidant activity in liver. Regular consumption of HRW may enhance detoxification process possibly through an increase in the efflux of toxic substances from the liver into bile.

Author Contributions: Conceptualization, H.-T.Y.; methodology, H.-T.Y.; software, Y.-H.Y.; validation, H.-T.Y.; formal analysis, Y.-H.Y. and M.-L.L.; investigation, Y.-H.Y. and M.-L.L.; resources, H.-T.Y.; data curation, Y.-H.Y.; writing—original draft preparation, H.-T.Y.; writing—review and editing, H.-T.Y.; visualization, Y.-H.Y.; supervision, H.-T.Y.; project administration, H.-T.Y.; funding acquisition, H.-T.Y. All authors approved the final submitted version.

Funding: This research was financially supported by grant-aid (MOST 106-2320-B-039-041-MY3) of the Ministry of Science and Technology, Taiwan.

Conflicts of Interest: The authors have no conflicts of interest to report.

References

1. Rendić, S. Summary of information on human CYP enzymes: Human P450 metabolism data. *Drug Metab. Rev.* **2002**, *34*, 83–448. [CrossRef] [PubMed]
2. James, L.P. Acetaminophen-induced hepatotoxicity. *Drug Metab. Dispos.* **2003**, *31*, 1499–1506. [CrossRef] [PubMed]
3. Gonzalez, F.J.; Gelboin, H.V. Role of Human Cytochromes P450 in the Metabolic Activation of Chemical Carcinogens and Toxins. *Drug Metab. Rev.* **1994**, *26*, 165–183. [CrossRef] [PubMed]
4. Arlt, V.M.; Krais, A.M.; Godschalk, R.W.; Riffo-Vasquez, Y.; Mrizova, I.; Roufosse, C.A.; Corbin, C.; Shi, Q.; Frei, E.; Stiborova, M.; et al. Pulmonary Inflammation Impacts on CYP1A1-Mediated Respiratory Tract DNA Damage Induced by the Carcinogenic Air Pollutant Benzo[a]pyrene. *Toxicol. Sci.* **2015**, *146*, 213–225. [CrossRef] [PubMed]
5. Kondraganti, S.R.; Jiang, W.; Jaiswal, A.K.; Moorthy, B. Persistent induction of hepatic AND pulmonary phase II enzymes by 3-methylcholanthrene in rats. *Toxicol. Sci.* **2008**, *102*, 337–344. [CrossRef] [PubMed]
6. Krajka-Kuźniak, V. Induction of phase II enzymes as a strategy in the chemoprevention of cancer and other degenerative diseases. *Postępy Hig. Med. Doświadczalnej* **2007**, *61*, 627–638.

7. Kong, L.L.; Shen, G.L.; Wang, Z.Y.; Zhuang, X.M.; Xiao, W.B.; Yuan, M.; Gong, Z.H.; Li, H. Inhibition of *p*-Glycoprotein and Multidrug Resistance-Associated Protein 2 Regulates the Hepatobiliary Excretion and Plasma Exposure of Thienorphine and Its Glucuronide Conjugate. *Front. Pharmacol.* **2016**, *7*, 242. [CrossRef]
8. Liu, Y.T.; Chen, Y.H.; Uramaru, N.; Lin, A.H.; Yang, H.T.; Lii, C.K.; Yao, H.T. Soy isoflavones reduce acetaminophen-induced liver injury by inhibiting cytochrome P-450-mediated bioactivation and glutathione depletion and increasing urinary drug excretion in rats. *J. Funct. Foods* **2016**, *26*, 135–143. [CrossRef]
9. Slitt, A.L.; Cherrington, N.J.; Maher, J.M.; Klaassen, C.D. Induction of multidrug resistance protein 3 in rat liver is associated with altered vectorial excretion of acetaminophen metabolites. *Drug Metab. Dispos.* **2003**, *31*, 1176–1186. [CrossRef]
10. Setsukinai, K.; Urano, Y.; Kakinuma, K.; Majima, H.J.; Nagano, T. Development of novel fluorescence probes that can reliably detect reactive oxygen species and distinguish specific species. *J. Biol. Chem.* **2003**, *278*, 3170–3175. [CrossRef]
11. Ohta, S. Molecular hydrogen as a preventive and therapeutic medical gas: Initiation, development and potential of hydrogen medicine. *Pharmacol. Ther.* **2014**, *144*, 1–11. [CrossRef]
12. Cejka, C.; Kubinova, S.; Cejkova, J. The preventive and therapeutic effects of molecular hydrogen in ocular diseases and injuries where oxidative stress is involved. *Free Radic. Res.* **2019**, *53*, 237–247. [CrossRef]
13. World Health Organization. *Guidelines for Drinking-Water Quality: Fourth Edition Incorporating the 1st Addendum*; World Health Organization: Geneva, Switzerland, 2017.
14. Yan, Y.; Ou, X.X.; Zhang, H.P.; Shao, Y. Effects of nano-materials on 17O NMR line-width of water clusters. *J. Mol. Struct.* **2013**, *1051*, 211–214. [CrossRef]
15. Szcześ, A.; Chibowski, E.; Hołysz, L.; Rafalski, P. Effects of static magnetic field on water at kinetic condition. *Chem. Eng. Process. Process. Intensif.* **2011**, *50*, 124–127. [CrossRef]
16. Ayrton, A.; Morgan, P. Role of transport proteins in drug absorption, distribution and excretion. *Xenobiotica* **2001**, *31*, 469–497. [CrossRef]
17. Ghanem, C.I.; Gómez, P.C.; Arana, M.C.; Perassolo, M.; Ruiz, M.L.; Villanueva, S.S.; Ochoa, E.J.; Catania, V.A.; Bengochea, L.A.; Mottino, A.D. Effect of acetaminophen on expression and activity of rat liver multidrug resistance-associated protein 2 and *p*-glycoprotein. *Biochem. Pharmacol.* **2004**, *68*, 791–798. [CrossRef]
18. Yang, C.Y. Calcium and Magnesium in Drinking Water and Risk of Death from Cerebrovascular Disease. *Stroke* **1998**, *29*, 411–414. [CrossRef]
19. Catling, L.A.; Abubakar, I.; Lake, I.R.; Swift, L.; Hunter, P.R. A systematic review of analytical observational studies investigating the association between cardiovascular disease and drinking water hardness. *J. Water Heal.* **2008**, *6*, 433–442. [CrossRef]
20. Leurs, L.J.; Schouten, L.J.; Mons, M.N.; Goldbohm, R.A.; van den Brandt, P.A. Relationship between tap water hardness, magnesium, and calcium concentration and mortality due to ischemic heart disease or stroke in The Netherlands. *Environ. Health Perspect* **2010**, *118*, 414–420. [CrossRef]
21. Barooti, A.; Kamran, M.; Kharazmi, F.; Eftakhar, E.; Malekzadeh, K.; Talebi, A.; Soltani, N. Effect of oral magnesium sulfate administration on blood glucose hemostasis via inhibition of gluconeogenesis and FOXO1 gene expression in liver and muscle in diabetic rats. *Biomed. Pharmacother.* **2019**, *109*, 1819–1825. [CrossRef]
22. Kim, M.-J.; Kim, H.K. Anti-diabetic effects of electrolyzed reduced water in streptozotocin-induced and genetic diabetic mice. *Life Sci.* **2006**, *79*, 2288–2292. [CrossRef]
23. Kajiyama, S.; Hasegawa, G.; Asano, M.; Hosoda, H.; Fukui, M.; Nakamura, N.; Kitawaki, J.; Imai, S.; Nakano, K.; Ohta, M.; et al. Supplementation of hydrogen-rich water improves lipid and glucose metabolism in patients with type 2 diabetes or impaired glucose tolerance. *Nutr. Res.* **2008**, *28*, 137–143. [CrossRef]
24. Yao, H.T.; Hsu, Y.R.; Lii, C.K.; Lin, A.H.; Chang, K.H.; Yang, H.T. Effect of commercially available green and black tea beverages on drug-metabolizing enzymes and oxidative stress in Wistar rats. *Food Chem. Toxicol.* **2014**, *70*, 120–127. [CrossRef]
25. Hanioka, N.; Jinno, H.; Tanaka-Kagawa, T.; Nishimura, T.; Ando, M. Determination of UDP-glucuronosyltransferase UGT1A6 activity in human and rat liver microsomes by HPLC with UV detection. *J. Pharm. Biomed. Anal.* **2001**, *25*, 65–75. [CrossRef]
26. Habig, W.H.; Jakoby, W.B. Assays for differentiation of glutathione S-Transferases. *Methods Enzymol.* **1981**, *77*, 398–405.

27. Guan, X.; Hoffman, B.; Dwivedi, C.; Matthees, D.P. A simultaneous liquid chromatography/mass spectrometric assay of glutathione, cysteine, homocysteine and their disulfides in biological samples. *J. Pharm. Biomed. Anal.* **2003**, *31*, 251–261. [CrossRef]
28. Mohandas, J.; Marshall, J.J.; Duggin, G.G.; Horvath, J.S.; Tiller, D.J. Low activities of glutathione-related enzymes as factors in the genesis of urinary bladder cancer. *Cancer Res.* **1984**, *44*, 5086–5091.
29. Uchiyama, M.; Mihara, M. Determination of malonaldehyde precursor in tissues by thiobarbituric acid test. *Anal. Biochem.* **1978**, *86*, 271–278. [CrossRef]
30. Aleksunes, L.M.; Scheffer, G.L.; Jakowski, A.B.; Pruimboom-Brees, I.M.; Manautou, J.E. Coordinated expression of multidrug resistance-associated proteins (Mrps) in mouse liver during toxicant-induced injury. *Toxicol. Sci.* **2006**, *89*, 370–379. [CrossRef]

Sample Availability: Samples of the compounds are available from the authors.

molecules

MDPI

Article

Screening of the Active Component Promoting Leydig Cell Proliferation from *Lepidium meyenii* Using HPLC-ESI-MS/MS Coupled with Multivariate Statistical Analysis

Xiao-chen Gao [1], Jing-wei Lv [1], Chun-nan Li [1], Nan-xi Zhang [1], Lin-lin Tian [1], Xi-ying Han [2], Hui Zhang [1,*] and Jia-ming Sun [1,*]

[1] Jilin Ginseng Academy, Changchun University of Chinese Medicine, Changchun 130117, Jilin, China; gao_xiaochen@hotmail.com (X.-c.G.); jingwei-lv@hotmail.com (J.-w.L.); lcn1013@hotmail.com (C.-n.L.); nancy_8080@outlook.com (N.-x.Z.); tll19900312@163.com (L.-l.T.)

[2] College of traditional Chinese Medicine, Changchun University of Chinese Medicine, Changchun 130117, Jilin, China; hanxy385@nenu.edu.cn

* Correspondence: zhanghui-8080@163.com (H.Z.); sun_jiaming2000@163.com (J.-m.S.)

Received: 9 May 2019; Accepted: 30 May 2019; Published: 3 June 2019

Abstract: *Lepidium meyenii* is now widely consumed as a functional food and medicinal product, which is known as an enhancer of reproductive health. However, the specific chemical composition and mechanism of action for improving sexual function are unclear. The present study aims at screening and determining the potential compounds, which promote mouse leydig cells (TM3) proliferation. The partial least squares analysis (PLS) was employed to reveal the correlation between common peaks of high performance liquid chromatography (HPLC) fingerprint of *L. meyenii* and the proliferation activity of TM3. The results suggested that three compounds had good activities on the proliferation of TM3 and promoting testosterone secretion, there were *N*-benzyl-hexadecanamide, *N*-benzyl-(9z,12z)-octadecadienamide and *N*-benzyl-(9z,12z,15z)-octadecatrienamide which might be the potential bioactive markers related to the enhancing sexual ability functions of *L. meyenii*. The first step in testosterone synthesis is the transport of cholesterol into the mitochondria, and the homeostasis of mitochondrial function is related to cyclophilin D (CypD). In order to expound how bioactive ingredients lead to promoting testosterone secretion, a molecular docking simulation was used for further illustration in the active sites and binding degree of the ligands on CypD. The results indicated there was a positive correlation between the binding energy absolute value and testosterone secretion activity. In addition, in this study it also provided the reference for a simple, quick method to screen the promoting leydig cell proliferation active components in traditional Chinese medicine (TCM).

Keywords: *Lepidium meyenii*; high-performance liquid chromatography-electrospray ionization/mass spectrometry; partial least squares; ultrafiltration affinity; molecular docking

1. Introduction

Lepidium meyenii (Maca), is a Brassicaceae *Lepidium* plant native to the Andes Mountains of South America. It has been traditionally used as a food and machine over 5000 years [1]. As is usual with many traditional folk medicines, many claims have been made regarding the efficacy of Maca in treating a wide range of illnesses and medical conditions [2,3]. However, in the 20th century most of the scientific attention has been focused in the areas where the pharmacological actions of Maca seem most strongly attested, these include, enhancement of sexual drive in humans, increasing overall vigour and energy levels, and increasing sexual fertility in humans and domestic livestock [3]. *Lepidium meyenii* is rich in nutrients and secondary metabolites with a variety of biological activities. Its main chemical compositions are polysaccharide, flavone, saponin, microelement and amino acid [4]. Low polarity magamide is considered to be its unique iconic ingredient, at present, the method of solvent reflux, ultrasonic extraction, high performance liquid chromatography (HPLC) and liquid chromatography mass spectrometry were used to detect it [5].

The present studies abroad have been studying the pharmacological effects of *L. meyenii*, they focus mainly on the effect of sexual function in mice. However all of these studies are segmentary, and lack of a comprehensive and systemic assessment, as well as the effect mechanism of improving sexual function is not yet clear. Especially, there is no research on the *L. meyenii* active monomers in promoting the mechanism of sexual function [6–8].

Testosterone is a prerequisite for normal spermatogenesis. Leydig cells are the main cells responsible for the production and secretion of the testosterone hormone [9]. The raw material for testosterone synthesis is cholesterol. The rate-limiting enzyme steroidogenic acute regulatory protein (StAR) in testosterone synthesis is responsible for accelerating the transport of cholesterol to the mitochondria, which is the first step in testosterone biosynthesis. For the maintenance of the StAR function, the homeostasis of the mitochondrial function is indispensable. In the process of maintaining mitochondrial function homeostasis, CypD plays an important regulatory role. Activation of CypD leads to opening of the mitochondrial permeability transition pore (mPTP) on the outer membrane of mitochondria which causes mitochondrial damage [10,11]. Mitochondrial dysfunction results in the inhibition of StAR expression, hindering cholesterol from entering the mitochondrial stromal membrane and inhibiting testosterone secretion; the CypD inhibitor can effectively bind CypD and inhibit the *cis-trans* isomerase activity of CypD, making the StAR expression stable, ultimately promoting testosterone secretion. Although the complete mechanism of the mPTP opening remains unclear, cyclosporine A (CsA), a high-affinity cyclophilin inhibitor, blocks the mPTP opening by binding to the CypD [12–16].

Inspired by the applications mentioned above, in order to find out the bioactive markers reflecting the traditional efficacy, an effective strategy on the high-performance liquid chromatography-electrospray ionization/mass spectrometry (HPLC-ESI-MS/MS) coupling with multivariate statistical analysis was developed to screen and identify the bioactive ingredients in *L. meyenii* [17]. Molecular docking was used to investigate the mechanism of bioactive compounds for improving sexual function, as depicted in Figure 1. The present study illustrated and explained the underlying correlations between active constituents and mechanisms of action [18].

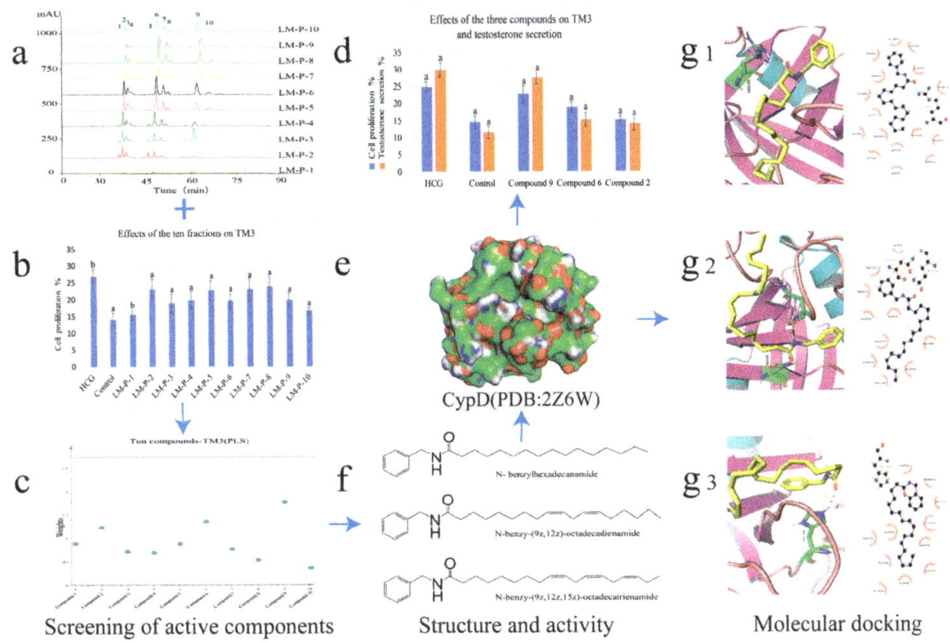

Figure 1. Strategy based on high-performance liquid chromatography-electrospray ionization/mass spectrometry (HPLC-ESI-MS/MS) coupling with the multivariate statistical analysis method to screen and identify the bioactive ingredients for the proliferation of mouse leydig cells (TM3) and promoting testosterone secretion in *Lepidium meyenii*. Molecular docking was used to investigate the mechanism of bioactive compounds. (**a**) The HPLC fingerprints of ten fractions. (**b**) Effects of the ten fractions on TM3 (a $p < 0.01$, b $p < 0.05$). (**c**) Model effect weights of the ten compounds on TM3. (**d**) Effects of the three compounds on TM3 and testosterone secretion (a $p < 0.01$). (**e**) The crystal structure of human cyclophilin D (PDB ID: 2Z6W). (**f**) The chemical structures of three bioactive markers. (**g_1**) Molecular docking of compound (**9**) with CypD showed in three-dimensional (3D) and two-dimensional (2D). (**g_2**) Molecular docking of compound (**6**) with CypD showed in 3D and 2D. (**g_3**) Molecular docking of compound (**6**) with CypD showed in 3D and 2D.

2. Results and Discussion

2.1. High-Performance Liquid Chromatography-Photodiode Array Detector-Electrospray Ionization/Mass Spectrometry Method Analysis of Ten Common Peeks

Ten compounds (**1**)–(**10**) were found in fraction LM-P-1 to fraction LM-P-10 at the characteristic wavelength of 210 nm. All the constituents of 10 fractions were separated and detected within 80 min and their MS^2 data were detected in a positive ion mode (Figure 2) according to the research of fragmentation pathway for compositions of 10 fractions in electrospray ionization using MS2 ion trap mass spectrometry and comparing retention time. Their structures were elucidated based on the analyses of ultraviolet (UV) spectra and ESI–MS2 fragmentation patterns with those of standards and the corresponding spectroscopic data given in the literatures.

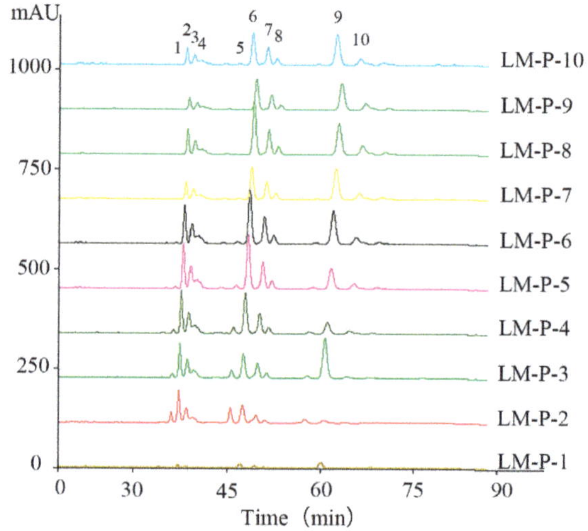

Figure 2. The HPLC fingerprints of ten fractions.

An overview of identified compounds was shown in Table 1. Meanwhile, the detailed structural analysis of common peaks was taken as an example to illustrate that our paradigm in this part bore out the correctness of structural presumption by using MS2. The Fragmentation mode were consistent with the previously reported, it suggested that Compound (**2**) of m/z 368 $[M+H]^+$ was *N*-benzyl-(9z,12z,15z)-octadecatrienamide. The collision-induced dissociation (CID) spectra of (Figure 3) were displayed as examples for the illustration of fragmentation patterns of macamides.

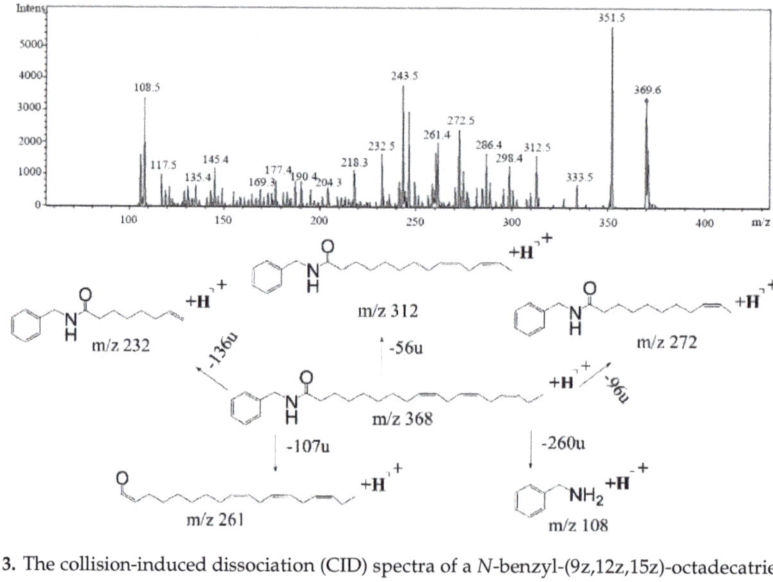

Figure 3. The collision-induced dissociation (CID) spectra of a *N*-benzyl-(9z,12z,15z)-octadecatrienamide.

The five fragment ions (m/z 56, 96, 107, 136 and 260) which corresponded to butylene, (1Z, 4Z)-heptadecadiene, benzylamine, (1Z, 4Z, 7Z)-decatriene and (9Z, 12Z, 15Z)-octadecane-triene-ketone, respectively, through a classic α-cleavage in amide linkage, were detected in all standards and were considered as the diagnostic ions of macamides [2].

Table 1. HPLC- ESI-MS data of 10 common peeks.

No.	Retention Time (min)	UV Absorption Characteristics λmax (nm)	Observed m/z	Fragment Ion	Compound Structure	Component Name
1	35.3	210	398	138,261	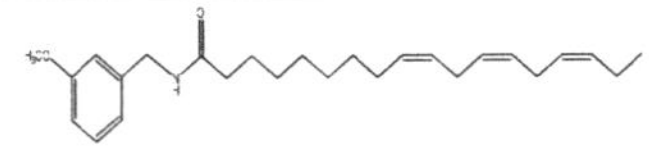	*N*-(3-methoxybenzyl)-(9zN,12zN,15z)-octadecatrienamide
2	36.4	210	368	108,232, 261,272	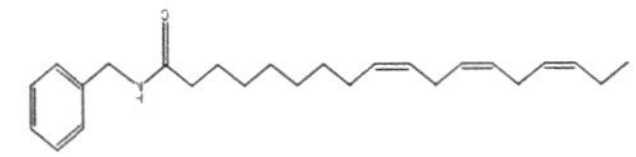	*N*-benzyl-(9z,12z,15z)-octadecatrienamide
3	37.7	210	368	108,261, 285	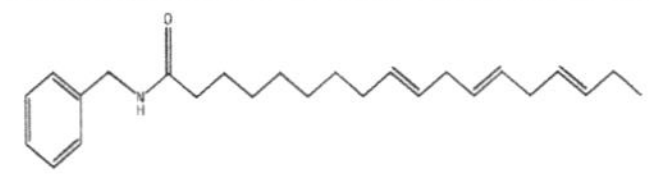	*N*-benzyl-(9E,12E,15E)-octadecatrienamide
4	38.7	210	368	108,261	Unknown	Unknown
5	44.8	210	400	138,263, 302	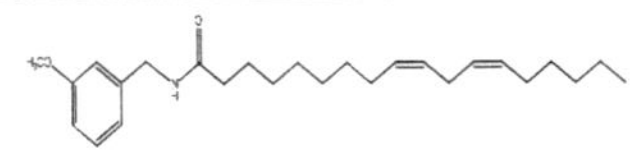	*N*-(3-methoxybenzyl)-(9Z,12Z)-octadecadienamide
6	46.6	210	370	108,232, 263,272	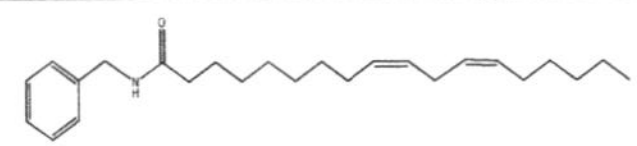	*N*-benzyl-(9z,12z)-octadecadienamide
7	48.9	210	370	108,263	Unknown	Unknown
8	50.3	210	370	108,263, 285	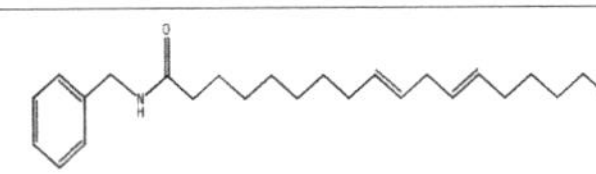	*N*-benzyl-(9E,12E)-octadecadienamide
9	59.6	210	346	108,239, 268,268	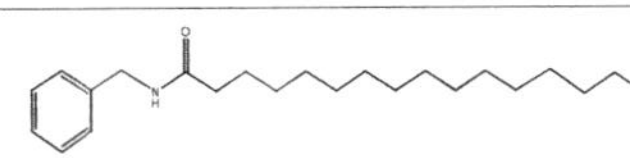	*N*- benzyl- hexadecanamide
10	63.0	210	372	108,165	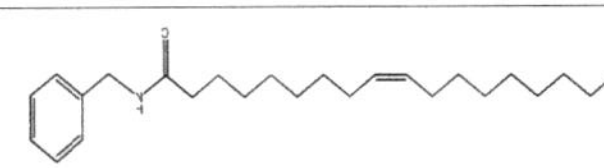	*N*-benzyl-9Z-octadecenamide

2.2. Mouse Leydig Cells Proliferation Activity of Ten Fractions

The proliferation activity of these 10 fractions was assessed using the MTT assay in TM3. All the 10 fractions were found to possess the proliferation activity (Figure 4).

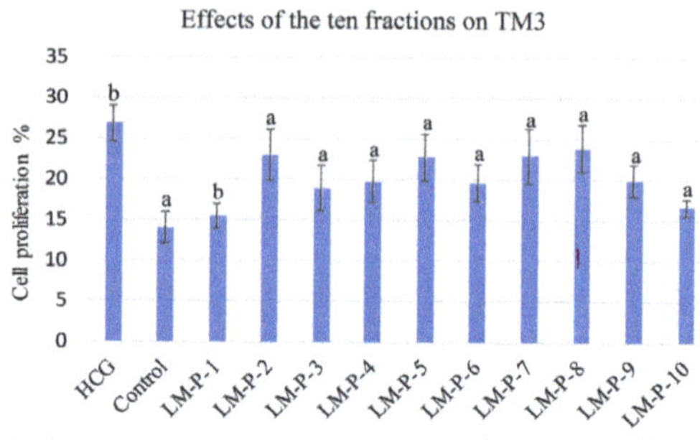

Figure 4. Effects of the ten fractions on TM3 (a $p < 0.01$, b $p < 0.05$).

2.3. Screening of Active Compounds by Using Partial Least Squares

The selected initial data was further processed by PLS in order to establish a model for predicting the potential active components in *L. meyenii*. Parameters were set as follows: Confidence level was 95%, R2 = (0.0, 0.794), Q2 = (0.0, −0.285), and the parameters showed that the established PLS model was effective. We could use the PLS to carry on the weights analysis about the impact of the common peeks area exported from ten HPLC spectra of 10 fractions (x-axis) to the proliferative activity of TM3 (y-axis) and screening of major compounds which influenced bioactivity.

In our data set, the weights plot summarized the variables both to explain X and to correlate to Y. The results were shown in (Figure 5). The weights greater than one indicated important variables, and three potential biological markers of *N*-benzyl-hexadecanamide (**9**), *N*-benzyl-(9z,12z)-octadecadienamide (**6**) and *N*-benzyl-(9z,12z,15z)-octadecatrienamide (**2**) had high contributions to the proliferation activity of TM3. Meanwhile, they were considered to be potential active compounds for further study [19].

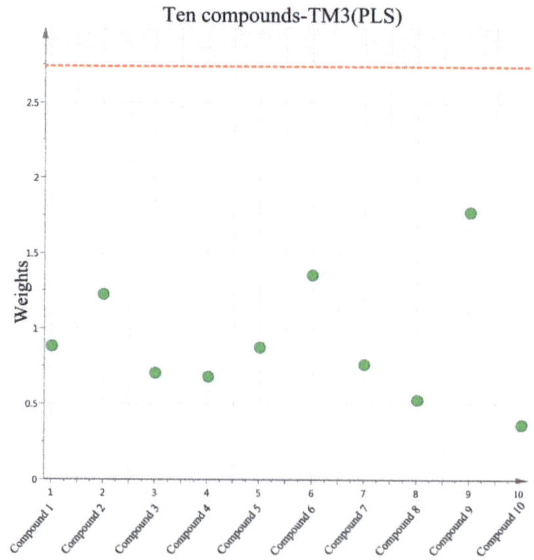

Figure 5. Model effect weights of the ten compounds on TM3.

2.4. Activity Evaluation of Active Components

The proliferation activity of the three compounds (**9**), (**6**), (**2**), were assessed using the MTT assay in TM3. All the compounds were found to possess proliferation activity.

To better evaluate the improving sexual function of the three compounds (**9**), (**6**), (**2**), the testosterone secretion assay was tested, and the results were presented in a strong correlation between the values determined by the HPLC-DAD-MS2 method and that predicted by the testosterone secretion tested data was observed [20] (Figure 6).

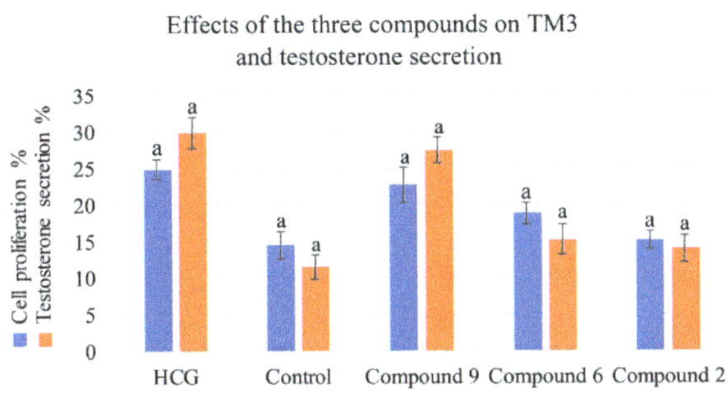

Figure 6. Effects of the three compounds on TM3 and testosterone secretion (a $p < 0.01$).

2.5. Analysis of Molecular Docking

The molecular docking study further elucidated the binding mode of the three compounds at the active site of CypD. The binding pocket of CypD was large and shallow, consisting of residues Arg55, Ile57, Phe60, Met61, Gln63, Gly72, Thr73, Gly74, Ala101, Asn102, Phe113, Trp121, Leu122, and His126 etc [21]. In which it was known, four specific residues (Arg55, Gln63, Asn102 and Trp121) were involved in hydrogen bond interactions with CsA [12]. Molecular docking simulation revealed that ligands interacted with important amino acid residues surrounding the active site through plenty of interactions including hydrogen bond acceptor, hydrogen bond donor, hydrophobic interactions. The docked molecules interacted with essential amino acid forming proteins' binding site. Unlike the case of full occupation by CSA, CypD- macamide complexes occupied only part of the binding pocket and might swing in the pocket [21]. (Figures 7a–c and 8a–c). The lowest binding energy were found: −4.79 kcal/mol for (**9**), −4.55 kcal/mol for (**6**) and −4.18 kcal/mol for (**2**). The negative binding energy ($G < 0$) indicated that there were good binding affinity between the three compounds and CypD.

Normally, the interactions between CypD and the macamide were dependent on the structures of the macamides, as the number of hydrogen bonds and hydrophobic interactions increased, the affinity degree might increase, it was shown between (**6**) and (**2**). It was interesting that, a hydrogen bond was formed between residual Arg55 and N atoms of macamide, causing the electrons of the N atom to form a regular tetrahedron of Sp3 hybrid, with single-button rotation. A mutant CypD with a single amino acid substitution (Arg to Ala at position 55) that was predicted to produce a 1000-fold attenuation in isomerase activity failed to reverse the CsA effect [22]. Therefore, the lowest binding energy were found in (**9**).

The results showed there was a specific ligand-binding ability of macamide for CypD, which could be used in the inhibition of MPT pore opening, which caused mitochondrial damage. The homeostasis of the mitochondrial function ensured the maintenance of the StAR function, which was the first step in testosterone biosynthesis. CypD inhibitor could effectively bind CypD and inhibit the cis-trans isomerase activity of CypD, making the StAR expression stable, ultimately promoting testosterone secretion [23,24]. Therefore, one of the possible mechanisms of promoting testosterone secretion for thee compounds, which could be the bioactive markers of *L. meyenii*.

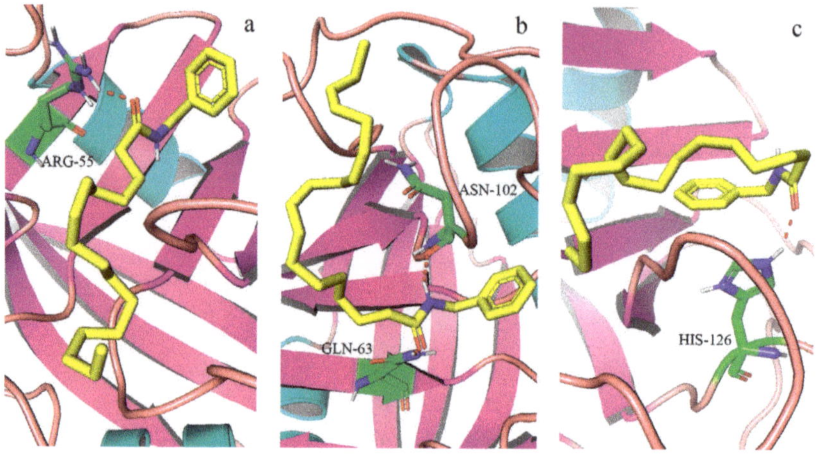

Figure 7. (**a**–**c**) Molecular docking of the three bioactive ligands with CypD, respectively. Ligands were shown in stick form and gray dashed lines were hydrogen bonds. The figure was prepared with PyMol. The interactions between bioactive ligands and binding sites were detailed in the article. (Figure 7a-(**9**), Figure 7b-(**6**), Figure 7c-(**2**)).

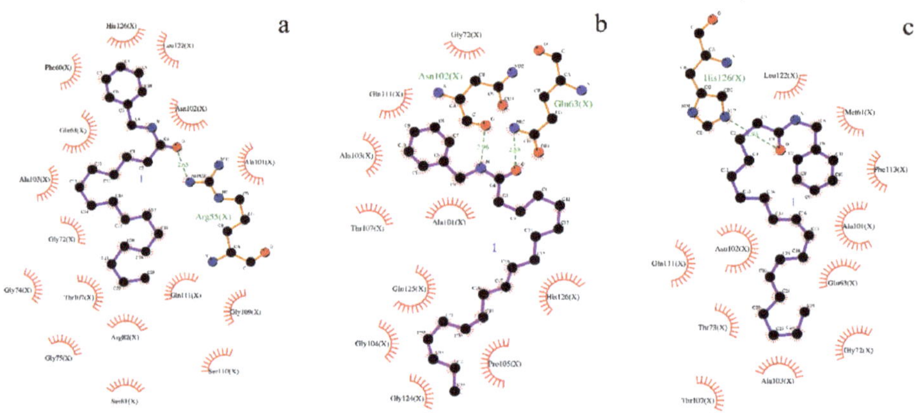

Figure 8. (**a**–**c**) Two-dimensional (2D) representation of hydrogen bond and hydrophobic interactions. Dashed lines represent hydrogen bonds, and spiked residues form hydrophobic interactions. (Figure 8a-(**9**), Figure 8b-(**6**), Figure 8c-(**2**)).

3. Materials and Methods

3.1. Materials

Lepidium meyenii was provided by Changchun University of Chinese Medicine and a voucher specimen (No. 201710) was deposited at the laboratory of Jilin Ginseng Academy, Changchun University of Chinese Medicine, P.R. China. Mouse leydig cells (TM3) were purchased from the Cell Bank of Type Culture Collection Chinese Academy of Sciences (Shanghai, China; cat. no. GNM24).

Standard compounds *N*-benzyl-hexadecanamide, *N*-benzyl-(9z,12z)-octadecadienamide and *N*-benzyl-(9z,12z,15z)-octadecatrienamide were provided by Yunnan Technical Center for Quality of Chinese Materia Medica (Yunnan, China). All standards were of purity greater than 98% and suitable for HPLC/MS/MS analysis.

3.2. Sample Preparation

The dried roots and rhizomes of *L. meyenii* (1000 g) were pulverized then sieved through a 20-mesh. The powder was extracted two times with 10 volumes of 95% ethanol (*v*/*v*) at 60 °C for 2 h. The filtrate was evaporated by a rotavapor at 60 °C and concentrated in vacuo to yield 24.32 g of brown residue. The residue (0.5 g) was further subjected to liquid-liquid partitioning to afford petroleum ether and water soluble extracts [25]. The resulting petroleum ether-soluble extract was applied to a silica gel column, and eluted with dichloromethane followed by (10:0 to 9:1, *v*/*v*) to give ten fractions (LM-P-1 to LM-P-10) [25]. The fraction was dissolved into 1 mL with acetonitrile and filtered with 0.22 μm filter membrane. The filtrate was used for HPLC analysis and testing of the proliferation of TM3.

3.3. High-Performance Liquid Chromatography-Photodiode Array Detector-Electrospray Ionization/Mass Spectrometry Method

An Agilent Technology 1100 Series HPLC system equipped with a quaternary pump, a degasser, a thermostatic auto-sampler and a photodiode array detector (DAD), was used for analysis (Agilent Technologies, Palo Alto, CA, USA). Chromatographic separations were carried out on a C18 analytical column Agilent Eclipse Plus-C18 (4.6 mm × 250 mm, 5μm) supplied by Agilent. The acetonitrile and water were used as the mobile phases (A) and (B), respectively, the optimized HPLC elution procedures were conducted as follows: 0–25 min, 80–90% (A); 25–70 min, 90–90% (A); 70–75 min, 90–100% (A). The flow-rate was 0.3 mL/min and the column temperature was maintained at 30 °C. The chromatogram was recorded at 210 nm. The injection volume of samples was 3.0 μL.

Agilent 1100 HPLC/MSD Trap mass spectrometer 6320 (Agilent) equipped with an electrospray ionization source was used in both positive and negative ion mode. An HPLC system coupled with DAD was controlled by an HPLC-MSD ChemStation software system. Auto MS2 mode of mass spectrometer was chosen to analyze the sample. The following operation parameters were used: capillary voltage: 4000 V; nebulizer pressure: 35 psi; drying gas: 9.0 L/min; gas temperature: 350 °C; skimmer voltage: 60 V. Liquid chromatography-electrospray ionization-mass spectrometry (LC-ESI-MS) accurate mass spectra were recorded across the range from 50 to 1200 m/z. The data recorded was processed with the Applied HPLC-MSD ChemStation software system [26] (1200, Agilent Technologies).

3.4. Cell Culture and Viability Assay

Mouse leydig cell (TM3) line is a mouse epithelial Leydig cell line. The TM3 cell line were grown in Dulbecco's modified Eagle's medium/F-12 nutrient mixture (DMEM/F-12) supplemented with 10% fetal bovine serum (Gibco; Thermo Fisher Scientific, Inc., Waltham, MA, USA), 1% penicillin (100 U/mL) and streptomycin (100 μg/mL) [27]. The human chorionic gonadotropin anhydrous (hCG) was obtained from Suolaibao Technology Co., Ltd. (Beijing, China).

Standard compounds *N*-benzyl hexadecanamide, *N*-benzyl-(9z,12z)-octadecadienamide and *N*-benzyl-(9z,12z,15z)-octadecatrienamide were dissolved in a culture medium containing a stock solution of 200 mg/L and further diluted to 62.5 μg/mL, 125 μg/mL, 250 μg/mL concentrations with culture medium containing 10% fetal bovine serum for 24 h. The DMEM/F12 concentrations (100 μL) was prepared as a control, and hCG concentrations (1 U/mL, 100 μL) were used as a positive control. Cells were cultured in a 37 °C incubator with 5% CO_2 and 95% air [20]. The effects of fractions on Leydig cell viability were assessed by MTT (3-(4,5-dimethylthiazol-2-yl)-2,5-diphenyltetrazolium bromide) (Roche, Basle, Switzerland). The supernatant was collected to determine the testosterone levels using the Mouse Testosterone (T) ELISA kit (cat. no. JL10895; Shanghai Yuanye Biotechnology Co., Ltd., Shanghai, China), according to the manufacturer's protocol.

3.5. Partial Least Squares Analysis and Statistical Analysis

The multivariate analysis of the acquired data was carried out by PLS using the SIMCA 11 software (Umetrics, Umea, Sweden). All assays were performed at least in triplicate and the results were expressed as a mean ± standard deviation (SD). The significant difference analysis was evaluated by one-way analysis of variance (ANOVA) test completed by the software of IBM SPSS Statistics 19 (International Business Machines Corp., New York, NY, USA). Significance was accepted at $p < 0.05$ [28].

3.6. Molecular Docking Studies

To further study the probable mechanism of the bioactive compounds with CypD, a molecular docking study which could conjecture the interactions of ligands within the constraint of receptors binding sites was performed in silico.

In the prediction, The X-ray crystal structure of CypD in the complex with its inhibitor CsA, 0.96 Å, was obtained from the Protein Data Bank (PDB ID: 2Z6W). The three-dimensional (3D) structures of the ligands were drawn and converted using ChemBioDraw Ultra and ChemBio 3D Ultra [29] (Cambridgesoft Corp., Waltham, MA, USA). The ligands and water molecules were removed from the crystal structure and the polar hydrogen was added by using AutoDock [30] (4.2.6, Department of Molecular Biology, The Scripps Research Institute, La Jolla, CA, USA).

Each grid computation was set up covering all the active sites where CsA was bounded. The grid was then concentrated on the center (80 Å, 40 Å, 80 Å, 0.375 Å, central coordinates $x = -20.347$, $y = 13.119$, and $z = 11.232$), respectively. The calculation of the docking score was repeated three times for each ligand. Fifty ligand–receptor complex conformations were generated for each test compound, in which the least building energy was considered for further analysis. Finally, PyMOL and LigPlot were used to present the docking results [31,32].

4. Conclusions

This work used the multivariate analysis to reveal some potential components, which improved sexual function from *L. meyenii*. We established an effective strategy based on HPLC-ESI-MS/MS with the PLS analysis for screening and determining the bioactive compounds which promote leydig cells proliferation and testosterone secretion. The 10 fractions were fractionated and their promoting activities on TM3 were demonstrated. With the aid of HPLC-ESI-MS/MS and the multivariate statistical software, the three potential improving sexual function markers were identified. Molecular docking was employed for further illustration in the mechanism of action for bioactivity.

In this study, correlation analysis was studied to explore the internal relationship between chemical constituents and pharmacological effects and discover the bioactive markers reflecting the traditional efficacy of *L. meyenii*. The results specified the three compounds as potential bioactive markers could lay a foundation for the improvement of quality standard of *L. meyenii*.

Author Contributions: Conceptualization, J.-m.S. and X.-c.G.; methodology, J.-m.S.; software, J.-w.L.; validation, L.-l.T., C.-n.L. and N.Z.; formal analysis, X.-y.H.; investigation, L.-l.T.; resources, H.Z.; data curation, J.-w.L.; writing—original draft preparation, X.-c.G.; writing—review and editing, J.-m.S.; visualization, J.-w.L.; supervision, J.-m.S.; project administration, H.Z.; funding acquisition, J.-m.S.

Funding: This research was funded by the National Natural Science Foundation of China, grant number 31570347 and Science and Technology funds of Administration of Traditional Chinese Medicine of Jilin Province, grant number 2019050.

Conflicts of Interest: The authors declare no conflict of interest.

References

1. Toledo, J.; Dehal, P.; Jarrin, F.; Hu, J.; Hermann, M.; Al-Shehbaz, I.; Quiros, C.F. Genetic Variability ofLepidium meyeniiand other AndeanLepidiumSpecies (Brassicaceae) Assessed by Molecular Markers. *Ann. Bot.* **1998**, *82*, 523–530. [CrossRef]

2. Pan, Y.; Zhang, J.; Li, H.; Wang, Y.Z.; Li, W.Y. Simultaneous Analysis of Macamides in Maca (Lepidium meyenii) with Different Drying Process by Liquid Chromatography Tandem Mass Spectrometry. *Food Anal. Methods* **2016**, *9*, 1686–1695. [CrossRef]
3. Wang, S.; Zhu, F. Chemical composition and health effects of maca (Lepidium meyenii). *Food Chem.* **2019**, *288*, 422–443. [CrossRef]
4. Zhou, Y.; Li, P.; Brantner, A.; Wang, H.; Shu, X.; Yang, J.; Si, N.; Han, L.; Zhao, H.; Bian, B. Chemical profiling analysis of Maca using UHPLC-ESI-Orbitrap MS coupled with UHPLC-ESI-QqQ MS and the neuroprotective study on its active ingredients. *Sci. Rep.* **2017**, *7*, 44660. [CrossRef] [PubMed]
5. Mccollom, M.M.; Villinski, J.R.; Mcphail, K.L.; Craker, L.E.; Gafner, S. Analysis of macamides in samples of Maca (Lepidium meyenii) by HPLC-UV-MS/MS. *Phytochem. Anal. Int. J. Plant Chem. Biochem. Tech.* **2010**, *16*, 463–469. [CrossRef]
6. Shin, B.C.; Lee, M.S.; Yang, E.J.; Lim, H.S.; Ernst, E. Maca (L. meyenii) for improving sexual function: A systematic review. *BMC Complement. Altern. Med.* **2010**, *10*, 44. [CrossRef]
7. Ruiz-Luna, A.C.; Salazar, S.; Aspajo, N.J.; Rubio, J.; Gasco, M.; Gonzales, G.F. Lepidium meyenii (Maca) increases litter size in normal adult female mice. *Reprod. Biol. Endocrinol.* **2005**, *3*, 16. [CrossRef]
8. Rubio, J.; Caldas, M.; Dávila, S.; Gasco, M.; Gonzales, G.F. Effect of three different cultivars of Lepidium meyenii (Maca) on learning and depression in ovariectomized mice. *BMC Complement. Altern. Med.* **2006**, *6*, 23. [CrossRef]
9. Payne, A.H.; Youngblood, G.L. Regulation of expression of steroidogenic enzymes in Leydig cells. *Biol. Reprod.* **1995**, *52*, 217–225. [CrossRef]
10. Zhao, W.H.; Gao, C.C.; Ma, X.F.; Bai, X.Y.; Zhang, Y.X. The isolation of 1,2,3,4,6-penta-O-galloyl-beta-d-glucose from Acer truncatum Bunge by high-speed counter-current chromatography. *J. Chromatogr. B* **2007**, *850*, 523–527. [CrossRef]
11. Bernardi, P. Mitochondrial transport of cations: Channels, exchangers, and permeability transition. *Physiol. Rev.* **1999**, *79*, 1127–1155. [CrossRef] [PubMed]
12. Park, I.; Londhe, A.M.; Lim, J.W.; Park, B.G.; Jung, S.Y.; Lee, J.Y.; Lim, S.M.; No, K.T.; Lee, J.; Pae, A.N. Discovery of non-peptidic small molecule inhibitors of cyclophilin D as neuroprotective agents in Aβ-induced mitochondrial dysfunction. *J. Comput.-Aided Mol. Des.* **2017**, *31*, 929–941. [CrossRef]
13. Valasani, K.R.; Vangavaragu, J.R.; Day, V.W.; Yan, S.S. Structure based design, synthesis, pharmacophore modeling, virtual screening, and molecular docking studies for identification of novel cyclophilin D inhibitors. *J. Chem. Inf. Model.* **2014**, *54*, 902–912. [CrossRef]
14. Shore, E.R.; Awais, M.; Kershaw, N.M.; Gibson, R.R.; Pandalaneni, S.; Latawiec, D.; Wen, L.; Javed, M.A.; Criddle, D.N.; Berry, N. Small molecule inhibitors of cyclophilin D to protect mitochondrial function as a potential treatment for acute pancreatitis. *J. Med. Chem.* **2016**, *59*, 2596–2611. [CrossRef] [PubMed]
15. Elkamhawy, A.; Lee, J.; Park, B.G.; Park, I.; Pae, A.N.; Roh, E.J. Novel quinazoline-urea analogues as modulators for Aβ-induced mitochondrial dysfunction: Design, synthesis, and molecular docking study. *Eur. J. Med. Chem.* **2014**, *84*, 466–475. [CrossRef] [PubMed]
16. Zhang, L.; He, D.; Li, K.; Liu, H.; Wang, B.; Zheng, L.; Li, J. Emodin targets mitochondrial cyclophilin D to induce apoptosis in HepG2 cells. *Biomed. Pharmacother.* **2017**, *90*, 222. [CrossRef] [PubMed]
17. Gao, S.; Liu, J.; Wang, M.; Liu, Y.; Meng, X.; Zhang, T.; Qi, Y.; Zhang, B.; Liu, H.; Sun, X. Exploring on the bioactive markers of Codonopsis Radix by correlation analysis between chemical constituents and pharmacological effects. *J. Ethnopharmacol.* **2019**, *236*, 31–41. [CrossRef] [PubMed]
18. He, Y.; Cheng, P.; Wang, W.; Yan, S.; Tang, Q.; Liu, D.; Xie, H. Rapid investigation and screening of bioactive components in simo decoction via LC-Q-TOF-MS and UF-HPLC-MD methods. *Molecules* **2018**, *23*, 1792. [CrossRef]
19. Zhou, B.; Zhang, Q.; Sun, J.; Xingliang, L.I.; Wei, Q. Study and application of partial least squares regression on relationship between soil nutrient and fruit quality. *Agric. Sci. Technol.* **2016**, *17*, 362–384.
20. Ni, C.; Fang, Y.; Chen, X.; Wu, K.; Li, H.; Wang, Y.; Zhenkun, L.; Lian, Q.; Ge, R.S. Stem Leydig cell regeneration in the adult rat testis is inhibited after a short-term triphenyltin exposure. *Toxicol. Lett.* **2019**, *306*, 80–89. [CrossRef]
21. Guo, H.X.; Wang, F.; Yu, K.Q.; Chen, J.; Bai, D.L.; Chen, K.X.; Shen, X.; Jiang, H.L. Novel cyclophilin D inhibitors derived from quinoxaline exhibit highly inhibitory activity against rat mitochondrial swelling and Ca^{2+} uptake/release. *Acta Pharmacol. Sin.* **2005**, *26*, 1201–1211. [CrossRef] [PubMed]

22. Helekar, S.A.; Patrick, J. Peptidyl prolyl cis-trans isomerase activity of cyclophilin A in functional homo-oligomeric receptor expression. *Proc. Natl. Acad. Sci. USA* **1997**, *94*, 5432–5437. [CrossRef]
23. Crompton, M. Mitochondrial intermembrane junctional complexes and their role in cell death. *J. Physiol.* **2000**, *529*, 11–21. [CrossRef]
24. Crompton, M.; Virji, S.; Ward, J.M. Cyclophilin-D binds strongly to complexes of the voltage-dependent anion channel and the adenine nucleotide translocase to form the permeability transition pore. *Eur. J. Biochem.* **1998**, *258*, 729–735. [CrossRef] [PubMed]
25. Chan, K.M.; Yue, G.L.; Li, P.; Wong, C.W.; Lee, K.M.; Kennelly, E.J.; Lau, B.S. Screening and analysis of potential anti-tumor components from the stipe of Ganoderma sinense using high-performance liquid chromatography/time-of-flight mass spectrometry with multivariate statistical tool. *J. Chromatogr. A* **2017**, *1487*, 162–167. [CrossRef] [PubMed]
26. Yang, Y.; Liang, X.; Jin, P.; Li, N.; Zhang, Q.; Yan, W.; Zhang, H.; Sun, J. Screening and determination for potential acetylcholinesterase inhibitory constituents from ginseng stem–leaf saponins using ultrafiltration (UF)-LC-ESI-MS2. *Phytochem. Anal.* **2019**, *30*, 26–33. [CrossRef]
27. Xie, T.; Chen, H.; Shen, S.; Huang, T.; Huang, B.; Hu, G.; Li, L.; Xu, Y. Proteasome activator REGγ promotes inflammation in Leydig cells via IkBε signaling. *Int. J. Mol. Med.* **2019**, 1961–1968. [CrossRef]
28. Chen, S.; Shi, J.; Zou, L.; Liu, X.; Tang, R.; Ma, J.; Wang, C.; Tan, M.; Chen, J. Quality Evaluation of Wild and Cultivated Schisandrae Chinensis Fructus Based on Simultaneous Determination of Multiple Bioactive Constituents Combined with Multivariate Statistical Analysis. *Molecules* **2019**, *24*, 1335. [CrossRef]
29. Milne, G.W.A. Software Review of ChemBioDraw 12.0. *J. Chem. Inf. Model.* **2010**, *50*, 2053. [CrossRef]
30. Sanner, M.F. Python: A programming language for software integration and development. *J. Mol. Gr. Model.* **1999**, *17*, 57–61.
31. Mura, C.; Mccrimmon, C.M.; Vertrees, J.; Sawaya, M.R. An introduction to biomolecular graphics. *Plos Comput. Biol.* **2010**, *6*. [CrossRef] [PubMed]
32. Laskowski, R.A.; Swindells, M.B. LigPlot+: Multiple ligand-protein interaction diagrams for drug discovery. *J. Chem. Inf. Model.* **2011**, *51*, 2778–2786. [CrossRef] [PubMed]

molecules

MDPI

Review

Advances in the Analysis of Veterinary Drug Residues in Food Matrices by Capillary Electrophoresis Techniques

Raffaella Colombo and Adele Papetti *

Department of Drug Sciences, University of Pavia, V.le Taramelli 12, 27100 Pavia, Italy; raffaella.colombo@unipv.it

* Correspondence: adele.papetti@unipv.it; Tel.: +39-0382987863; Fax: +39-0382422975

Received: 27 November 2019; Accepted: 14 December 2019; Published: 17 December 2019

Abstract: In the last years, the European Commission has adopted restrictive directives on food quality and safety in order to protect animal and human health. Veterinary drugs represent an important risk and the need to have sensitive and fast analytical techniques to detect and quantify them has become mandatory. Over the years, the availability of different modes, interfaces, and formats has improved the versatility, sensitivity, and speed of capillary electrophoresis (CE) techniques. Thus, CE represents a powerful tool for the analysis of a large variety of food matrices and food-related molecules with important applications in food quality and safety. This review focuses the attention of CE applications over the last decade on the detection of different classes of drugs (used as additives in animal food or present as contaminants in food products) with a potential risk for animal and human health. In addition, considering that the different sample preparation procedures have strongly contributed to CE sensitivity and versatility, the most advanced sample pre-concentration techniques are discussed here.

Keywords: food quality; food safety; veterinary drugs; animal food; food contaminants; capillary electrophoresis; residue analysis; pre-concentration techniques

1. Introduction

The addition of drugs (abuse or illegal use) to animal food to promote growth and protect animals can represent a potential risk of contamination of food matrices. Drugs, such as antibiotics, estrogens, non-steroidal anti-inflammatory drugs (NSAIDs), and β-agonists, which are usually used in feedstuffs, can contaminate different food products, mainly meat, milk, and dairy products, causing health problems and also serious diseases [1]. In particular, the abuse of antibiotics in food-producing animals, which contributes to the increase in risk of the transfer of antibiotic resistance from animals to humans, is a very important issue for human health. For this reason, the European Union (EU) and the Food and Drug Administration (FDA) established restrictive regulations for the control of pharmacologically active substance residues and fixed maximum residue limits (MRLs) in edible animal tissues to preserve foodstuff of animal origin and consumers [2–4].

The need of sensitive and rapid analytical techniques to detect and quantify pharmacologically active compounds, unauthorized drugs included, in animal food and foodstuff of animal origin has become mandatory for food security. Therefore, over the years, the necessity to develop and validate new analytical methods has increased [5]. The Commission Decision 2002/657/EC reported the technical guidelines and performance criteria for method validation for the control of the different residues [6]. In addition, there is a lack of regulation for veterinary drug residues (for example, for fluoroquinolones used as antimicrobials) in many foods, including baby foods [7].

EU guidelines suggest the use of the liquid chromatography (LC) technique; in particular, LC coupled with mass spectrometry (MS) was the most used approach to detect and analyze drug residues in complex matrices, such as milk and dietary products [8–12]. In addition, ion mobility spectrometry (IMS) coupled with MS represented a very promising powerful tool to detect analytes in traces [13].

Capillary electrophoresis (CE) with its well-known advantages, such as high efficiency, low consumption of sample and buffer, and rapidity, represents a potential alternative to LC methods in the analysis of drugs in different fields, including food analysis [14–16]. Another important advantage of CE rests in the versatility of applications thanks to the development of different CE separation modes. The simple addition of different molecules (surfactants, chiral selectors, polymers, particular electrolytes, and organic modifiers) to the buffer or the modification of the capillary inner wall with new packaging materials gave origin to different separation mechanisms and selectivity, increasing CE versatility and potential applications [16–19].

In particular, the use of electrospray ionization (ESI), matrix-assisted desorption/ionization (MALDI), and inductively coupled plasma (ICP) as CE-MS interfaces improved food analysis by increasing CE sensitivity [20,21]. In fact, CE-MS represents the ideal technique to detect analytes in traces with important implications in food contaminants and residue analysis [15,16].

In addition, advances in electrochemical detectors, such as CE-contactless coupled detection (CE-CCD) and CE-capacitively coupled contactless conductivity detection (CE-C^{4}D), offered very sensitive methods [22,23]. Furthermore, the development of miniaturized CE systems (microchip-CE devices) allowed the monitoring of food analytes with rapidity and sensitivity, and their use was particularly important in the detection of frauds or contaminations [24–26].

Due to the complexity of food matrices, which are mainly rich in lipids, carbohydrates, and proteins, a pre-concentration step was necessary to detect drug residues in trace amounts. This became mandatory because of the intrinsic poor sensitivity of CE [27]. Solid phase extraction (SPE) and miniaturized SPE are the most used procedures, not only for the pre-concentration step, but also for the sample clean-up. New SPE sorbents with high adsorption capacity and high resistance were studied, giving origin to selective materials for some drugs and also generating advanced high-throughput procedures able to extract different drug classes [28,29]. In addition, advances in on-chip SPE-CE procedures also allowed low-abundance analytes with high sensitivity to be analyzed [30]. Recently, the combination of traditional liquid-liquid extraction (LLE) and SPE procedures or advanced liquid extraction techniques, such as dispersive liquid-liquid microextraction (DLLME), was successfully applied to detect analytes in trace, increasing CE sensitivity [31–33].

Finally, the development of on-line procedures in which pre-concentration techniques were integrated with the CE instrumentation had many advantages, such as minimal sample loss, low cost, and rapidity [34].

In this review, we focused the attention on the potential of CE in veterinary drug residue analysis, considering the versatility of different CE-modes (mainly capillary zone electrophoresis, CZE; capillary electrochromatography, CEC; micellar electrokinetic chromatography, MEKC; nonaqueous capillary electrophoresis, NACE). CE methods and advanced sample preparation procedures combined with CE techniques in the last decade were also discussed. The main CE-modes were summarized, subdividing drugs into antibiotics (classified according to different molecular structures) and other drugs (estrogens, non-steroidal anti-inflammatory drugs, NSAIDs, and β-agonists).

2. Antibiotics

2.1. Nitroimidazoles

5-Nitroimidazoles (5-NDZs) are mainly active against Gram-negative and Gram-positive anaerobic bacteria. Some benzimidazoles (BZs) are also used as additives in stored fruit and vegetables, because of their fungicidal properties [35]. Regarding the veterinary use, these substances are prohibited [2,36], as they could be a potential risk for human health because of their genotoxicity and mutagenicity [37]. Therefore, no MRLs were established.

A rapid CEC-UV method for the analysis of 5-NDZ residues in bovine milk samples was set-up by Hernández-Mesa et al. The characteristic speed and high efficiency of CEC, which combines mobility and partition principles, are well-known to be affected by the complexity in fritting fabrication, which gives origin to pressure drops and bubble formation with loss of efficiency and reproducibility. In this work, the authors overcame this problem by proposing a simple capillary packing procedure and optimizing a sintering process (parameters: Time and temperature), thus obtaining a reproducible frit fabrication. A proper set-up of buffer composition and concentration, capillary length, and voltage allowed a reproducible analysis of eight 5-NDZs in milk samples in 15 min. In addition, a sample pretreatment using LLE and SPE methods highly increased the method sensitivity (LOQ range for all the 5-NDZs analyzed was 19–96 μg/L) [38]. This method, combining the advantages of CE and HPLC, was more selective and rapid than the simple HPLC [39] or CZE [40] methods.

The same technique (CEC-UV), but coupled with DLLME as a pre-concentration step, was used to analyze BZs in environmental and farm water [41]. DLLME, which belongs to liquid-phase microextraction (LPME) techniques, is based on the formation of a dispersion, created by adding an organic solvent mixture to an aqueous sample. It represented an ideal procedure to efficiently extract environmental pollutants and an interesting greener approach in their analysis [42].

This procedure was also used by Hernández-Mesa et al. to concentrate different 5-NDZ compounds in river water samples before the analysis with a cation-selective exhaustive injection and sweep (CSEI-sweep)-MEKC-UV method. DLLME and CSEI-sweep approaches combine an electrokinetic injection of charged cations and a sweeping in which the formation of micelles is promoted to focus the analyte. Thus, the MEKC method was more rapid, allowing the separation of six BZs in about 10 min [43]. The DLLME procedure coupled with MS detection could also improve the sensitivity of CZE. An example was the study of Tejada-Casado et al., who set-up a CZE-MS/MS method able to rapidly (about 30 min) detect and quantify twelve BZs in meat samples, as a valid alternative to HPLC methods [44].

Finally, SPE combined with CSEI-sweep-MEKC-UV allowed six nitroimidazole residues in egg samples to be quantified with LOQ values in the range of a few ng/g [45].

2.2. Fluoroquinolones

Quinolones represent a class of drugs commonly used in veterinary fields, whose MRLs have been established in the range level of 100–500 μg/kg by the EU Council Regulation and FDA [2,3]. Fluoroquinolones (FQs) are synthetic quinolones, related to nalidixic acid, which act against Gram-negative and Gram-positive bacteria by inhibiting their DNA synthesis. FQs are commonly used for livestock growth and aquaculture and are toxic to human health because they can exhibit a direct toxicity, responsible for muscular and neuronal dysfunctions, or cause antibiotic resistance or allergies [46–48].

CE techniques also represented interesting platforms in the analysis of the third and fourth generation of FQs in different food matrices (water, milk, and animal muscle), but appropriate sample pre-treatments and detection systems (LIF and MS) became mandatory to obtain a high level of sensitivity [49]. To extract milk FQs, LLE and SPE were the most used procedures, but in the literature, examples of protein precipitation (PPT) followed by SPE are also present as a good alternative [50]. In particular, molecularly imprinted polymers (MIPs), which are selective and stable sorbent materials for SPE (MISPE technique) [51], allowed promising CZE-LIF or CZE-MS methods to be obtained in the animal foodstuff analyses. MIPs can be used both before CZE analysis and as an in-line-MISPE strategy, thus obtaining sensitive and selective methods for the analysis of different complex matrices (for example, pig kidney and bovine milk) [52,53]. In fact, by using advanced MIP technologies, the low CZE-UV sensitivity can also be overcome. Magnetic molecular imprinted polymers (MMMIPs) were prepared by combining ferroferric-oxide nanoparticles, and MIPs and had a rapid and efficient binding capacity. These materials allowed a rapid and selective CZE-UV method to be obtained to separate fleroxacin, gatifloxacin, lomefloxacin (LOM), and norfloxacin in bovine milk samples [29]. The method sensitivity (LOD range: 12.9–18.8 μg/L) was slightly lower than, but comparable to, that of Springer et al. (7.5–11.6 μg/L), who separated ciprofloxacin (CIP), norfloxacin, and ofloxacin (OFL) in milk samples using a miniaturized SPE (made of carbon nanotubes as sorbents with high adsorption capacity and stability) to prepare samples [28].

Lara et al. set-up a CZE-MS/MS method for the simultaneous quantification (ng/kg level) of eight FQs, i.e., danofloxacin, sarafloxacin, CIP, marbofloxacin, enrofloxacin (ENRO), difloxacin, oxolinic acid, and flumequine, in chicken muscle samples. Two different sample preparation approaches were set-up; the first one consisted of pressurized liquid extraction (PLE), which was performed in an accelerated solvent extraction, followed by centrifugation, percolation, concentration, reconstitution, and filtration steps prior to the injection into the CE instrument; the second one was an in-line SPE, using a mixed-mode sorbent (RP and ion-exchange sorbents), in which different parameters, such as sample pH, volume, elution plug composition, and injection time, were optimized. The combination of in-line SPE-CE-MS/MS with PLE improved the selectivity and sensitivity, and this was particularly useful in multiresidue analysis [54].

Field-amplified sample stacking (FASS) with sweeping represented another on-line pre-concentration procedure used in combination with CE to rapidly quantify ENRO and CIP (LODs ng/mL) in milk and animal tissues. The FASS procedure consists of an electrokinetic injection of sample in a run buffer with high conductivity, creating an interface in which the difference in the electric field between the sample matrix and run buffer promotes sample stacking. The use of gamma-cyclodextrin in the sample matrix (sweeping procedure) could increase the FASS sensitivity, because it gave origin to micelles as in the MEKC technique, in which cyclodextrins are added to the background electrolyte (BGE) [55].

A different on-line preconcentration procedure, named field-enhanced sample injection (FESI), could increase CZE-UV sensitivity. FESI is useful for samples with low conductivity and consists of a careful optimization of the injection (pressure and time) of a water plug into the capillary in order to increase the capillary electric field by creating a sample stacking effect. The use of FESI in addition to a bubble cell capillary (with a longer window pathway than the traditional capillary) increased the method sensitivity for five different FQs (ENRO, CIP, LOM, fleroxacin, and OFL) in bovine milk samples [56].

Another CE mode, i.e., nonaqueous capillary electrophoresis (NACE)-UV, combined with a DLLME approach allowed the set-up of a selective method for the separation and quantification of eight FQs (omefloxacin, levofloxacin, marbofloxacin, CIP, sarafloxacin, ENRO, danofloxacin, and difloxacin) in water samples [57]. In NACE, BGE is added with organic solvents, mainly methanol or acetonitrile, and this promotes the separation of low-water-soluble molecules. In fact, even if the use of organic solvents could induce changes in the pKa values and mobility, the advantage consisting of an increased selectivity becomes fundamental [16]. The same CE-mode coupled with an in-line single-drop liquid-liquid-liquid microextraction (SD-LLLME) was used to separate CIP and ENRO in surface and groundwater samples. This setup allowed a reduction in the analysis time (as no sample pre-treatment was necessary) and in the volume of extraction solvent, as a buffer was used as a pH donor, a drop of NaOH as a high pH acceptor, and an organic solvent as a medium in which the analytes diffused [58].

2.3. Tetracyclines

Tetracyclines are widely used as economic broad-spectrum antibiotics against both Gram-positive and Gram-negative bacteria. MRLs were established by the Commission Regulation (EU) and ranged from 100 to 600 μg/kg, depending on animal tissues or the food sample [2]. In the literature, many works aimed to analyze and quantify these antibiotics. A simple SPE procedure with a NACE-LIF method allowed a very sensitive separation of chlortetracycline (CTC), tetracycline (TC), oxytetracycline (OTC), and doxycycline (DC) in feeds and milk with pg/mL LOD values [59]. In the same year, Deng et al.

proposed a simple CZE-enhanced chemiluminescence (ECL) method to monitor TC residues over time in crucian carp muscle of fish samples, with a sensitive detection under MRL values [60].

Recently, a field-amplified sample injection (FASI) procedure in CZE-UV was set-up for the detection of four tetracyclines in pig farms' wastewaters with results comparable to those obtained by HPLC-UV methods [61]. FASI pre-treatment is particularly suitable for large amounts of water samples and is based on a difference in electrical conductivity between the sample and the background electrolyte, which causes a stacking effect responsible for the increase in peak efficiency and method sensitivity.

The combination of two pre-concentration procedures could be useful to determine drug residues, even when present at a concentration below MRL limits. Islas et al. recently set-up a SPE step followed by a large-volume sample stacking (LVSS) approach before a CZE-UV analysis of TC. LVSS consisted of a stacking procedure with an on-line series of polarity switches (PS) and enabled an improvement in sensitivity and reproducibility, particularly useful for low-concentration analytes in complex matrices, such as milk samples (Figure 1) [62]. The same pre-concentration approach (SPE-LVSS-PS) was previously used to determine five TC residues (metacycline-MTC, OTC, TC, CTC, and DC) in water samples, reaching a high sensitivity (ng/L) in a short time (10 min) [63]. The LVSS approach alone, after a careful optimization of stacking conditions (sample zone length and stacking time), allowed a CZE-UV method to be obtained with a sensitivity around 10 ppb, a sensitivity level very similar to that obtained by using ED or LIF detectors. This method was used to detect TC, CTC, OTC, and DC in tap water samples [64].

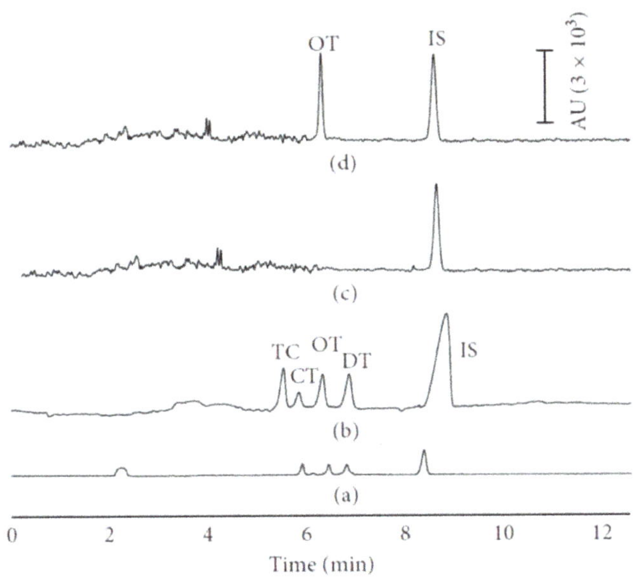

Figure 1. Solid phase extraction (SPE)-large-volume sample stacking (LVSS)-capillary electrophoresis (CE) method applied to milk sample for the detection of tetracyclines (TCs). Electropherograms of (**a**) standard TC sample (10 mg/L) analyzed by CE, (**b**) standard TC sample (1 mg/L) analyzed by LVSS-CE, (**c**) blank milk sample, and (**d**) real milk sample analyzed by SPE-LVSS-CE method [62].

Another promising extraction approach consisted of matrix solid-phase dispersion (MSPD), which required the addition of sorbents to the sample with a consequent elution step before the analysis. Mu et al. set-up an economic MSPD-CZE-UV method to rapidly (about 6 min) separate TC, OTC, and DC in milk samples [65]. In order to improve CE sensitivity, a functionalized β-cyclodextrin-ionic liquid was added in-line; unlike the conventional β-cyclodextrin, it acted as an additive able to form a complex with TCs and as a capillary coating agent. The method proposed by Zhou et al. was developed to separate four TCs in about 30 min by using CE with amperometric detection (AD) [66].

Tetracyclines can also be present in honey, representing a serious problem, as the EU Commission did not admit to the use of antibiotics in honey and did not establish MRLs for bee products [2]. Casado-Terrones et al. set-up an SPE procedure followed by a CZE-UV method to simultaneously and rapidly (16 min) determine eight tetracyclines in honey with LOD values of 23.9–49.3 μg/kg [67].

2.4. *Sulfonamides*

These molecules are widely used for the treatment of bacterial and protozoan diseases (i.e., malaria) and as growth-promoters in farm animals. The EU fixed MRLs to 100 μg/kg in different animal tissues and milk [2]. Sulfonamides could be dangerous for human health. For example, studies on sulfamethazine carcinogenesis are very controversial, but nowadays, it is not yet classified as carcinogenic by the U.S. National Toxicology Program 14th Report on Carcinogens (2016).

For the antimalarials sulfadoxine and quinine, the old literature proposed CZE-UV as the main CE-mode in meat and water analysis with sensitivity values of μg/L and mg/L for sulfadoxine and quinine, respectively [68]. More recently, Mikus et al. set-up a capillary isotachophoresis (CITP) coupled on-line with the CZE method to detect quinine in commercial beverages. This approach could be potentially used to detect drug residues in food, as CITP combined on-line with CZE-UV significantly increased the CZE sensitivity. In fact, in CITP, leading and terminating electrolytes are used to create separated zones in which the ions migrate at the same velocity, giving origin to an on-line stacking effect [69]. This combination allowed the direct analysis of samples with a lower LOD than CZE-UV methods and with sensitivity values comparable to those obtained with HPLC-UV methods [70].

Nine sulfonamides in meat samples were efficiently separated by CEC-ESI-MS. The use of a poly(divinylbenzene-octyl methacrylate) (poly-DVB-OMA) monolithic stationary phase and an on-line concentration (obtained increasing sample injection time) provided a sensitive method for detecting trace residues that needed only a simple sample pretreatment (i.e., SPE) (Figure 2) [71].

Wang et al. set-up a microfluidic CE system with LIF detection, which was able to separate four sulfonamides (sulfamethazine, sulfamethoxazole, sulfaquinoxaline, and sulphanilamide) in milk and chicken muscle extracts in 1 min with LOD values of a few μg/L. The short time required for the analysis and the fact that the plastic chips proposed were cheap made this method very useful for a rapid screening of sulfonamide residues in food samples [72].

Recently, Dai et al. proposed a CZE method with on-line ECL detection to quantify sulfadimidine, sulfadiazine, and sulfathiazole in pork and chicken meat samples. Chemiluminescence (CL) emissions were generated by the oxidation of luminol in the Ag(III)-luminol system, and sulfonamides exhibited an inhibitory effect on CL signals. The careful optimization of different parameters (buffer type and pH, voltage, and injection time) allowed a promising selective and sensitive method to be obtained for the analysis of veterinary drug residuals in animal-derived food [73].

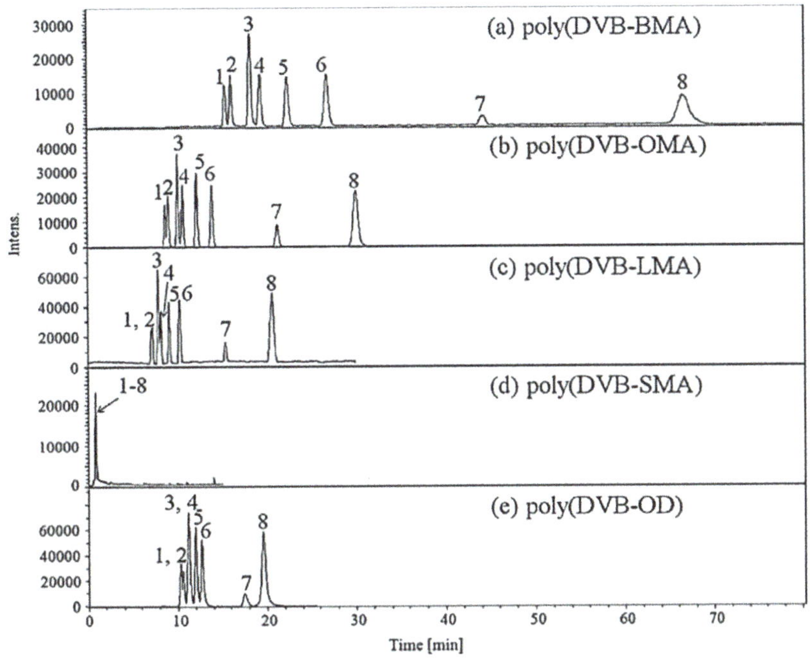

Figure 2. Capillary electrochromatography (CEC)-electrospray ionization (ESI)-mass spectrometry (MS) electropherograms obtained by using different monolithic stationary phases in the analysis of standard sulfonamides (**a–e**). The use of poly(DVB-OMA) capillary (**b**) allowed the best compromise to be obtained between resolution, efficiency, and analysis time [71].

2.5. Aminoglycosides

These antibiotics (AGs) are widely used in veterinary medicine for bacterial and protozoan infections. In addition, AGs are frequent honey contaminants from non-EU countries, which is a big issue. As mentioned above, the use of antibiotics in beekeeping is not authorized in the EU [2] and, therefore, there are no European Community regulations about MRLs for these drugs in this product. To detect these compounds in other complex matrices, i.e., biological fluids and pharmaceutical samples, CE with LIF, C^4D, or indirect detections, are requested [74]. Recently, Moreno-Gonzalez et al. detected nine AGs (three gentamicins, neomycin, apramycin, paromomycin, dihydrostreptomycin, spectinomycin, and streptomycin) in different types of honey using MISPE and FASS procedures as purification and pre-concentration steps, and a CZE-MS/MS analysis method [75]. MISPE also represented an interesting tool for the extraction and analysis of AGs in different animal-derived food (meat) [76].

In 2019, a microchip-CE-C^4D method was developed using standard solution and was proposed as a good alternative to separate in a very short time (less than 1 min) AG sulfates (gentamicin sulfate, kanamycin sulfate, and streptomycin sulfate) in foodstuff with LOD values of 1–3 μg/mL [77].

2.6. Macrolides

Macrolides are bacteriostatic antibiotics with a broad spectrum of activity against Gram-positive and Gram-negative, particularly used in the treatment of respiratory diseases in bovines and pigs, but generally also added to animal feed. Like aminoglycosides, they are listed in Group B1-antibacterial substances under the Council Directive 96/23/EC [78] and possess the same mechanism of action consisting of the inhibition of bacterial protein synthesis, but they bind the 50S instead of 30S ribosomal unit. Their presence in food could cause problems to human health, in particular, to intestinal bacterial flora, and, in the last years, their use for long periods contributed to cause the development of antibiotic resistance [79,80]. An EU directive established MRL values in the range 40–200 μg/kg for erythromycin, tylosin (TYL), tilmicosin (TIL), and spiramycin [2,80].

For the evaluation of macrolides in food, HPLC-UV or HPLC-ESI/MS were the most frequently used methods [81,82]. However, off-line or on-line pre-concentration steps were the ideal approach to overcome the derivatization step required to obtain a high sensitivity by CE. Among the on-line pre-concentration techniques, FASS represented a promising solution, as demonstrated in the analysis of TIL, erythromycin, clarithromycin, roxithromycin, and tylosin residues in milk samples by a FASS-MEKC-UV method (LOD range of 0.002–0.004 mg/kg) [83]. CZE-DAD was recently proposed to analyze macrolides (TYL and TIL) in chicken fat samples, reaching LOQ values of a few μg/kg thanks to a preconcentration step that consisted of an evolution of the DLLME procedure: In fact, the extraction and the preconcentration occurred in an ionic liquid by using ultrasound instead of dispersive solvents (this procedure is named reverse ultrasound-assisted emulsification-microextraction, R-USAEME). Organic salts, which are more stable and less toxic than the organic solvents commonly used in DLLME, were used to prepare the ionic solution [84].

2.7. *β-Lactam Antibiotics*

Penicillins and cephalosporines are classified as β-lactam antibiotics. They have the same action mechanism, but cephalosporines have a more extended spectrum. In particular, in veterinary medicine, ceftiofur and cefquinome are specifically used to treat respiratory diseases and exudative epidermitis, and meningitis, respectively. Cefquinome had MRL values more restrictive than those of ceftiofur (50–200 μg/kg vs. 100–6000 μg/kg) [2], but the use of ceftiofur must be carefully evaluated as its use could develop E. coli resistance [85,86].

CZE and MEKC modes were the best choice for the analysis of penicillins and cephalosporins in complex matrices [87,88]. For example, penicillin acid and penicillin G could be easily detected in milk by CZE-UV with a simple sample pre-treatment, which consisted of the deproteinization, extraction, and precipitation of milk protein with acetonitrile [89]. To achieve the best resolution, an on-line sample concentration step coupled with a CEC-MS method with a polymeric monolithic column was proposed by Liu et al. The injection in CE consisted of an anion selective injection (ASEI) performed by solubilizing the analytes in buffers with different pH values and promoting a stacking effect. This method was applied to analyze milk samples, obtaining a high sensitivity [90].

Regarding cephalosporins, Hancu et al. set-up a simple and rapid CZE-UV method, simply optimizing BGE composition and pH. The method was able to separate seven cephalosporins in 6 min and it was proposed for the analysis of pharmaceutical products and different complex matrices, obtaining very low LOD values; therefore, it could be useful for residue analysis [88].

2.8. *Simultaneous Analysis of Different Antibiotics*

The separation and quantification of different drugs (macrolides and tetracycline antibiotics) in feedstuffs could be simultaneously carried out by CZE-UV, as demonstrated for TIL, TYL, TC, OTC, and DC with LOD values of 0.5–1 mg/kg [91]. The method sensitivity was good considering that OTC and TYL were added in feeds for growth promotion in a range from a few to 50 mg/kg, and that TIL was also often added as an antimicrobial and respiratory diseases agent at a concentration of a few hundreds of mg/kg [92]. The separation was performed in 15 min and, thus, this CE-method could be considered useful for rapid routine analysis.

In addition, β-lactams, tetracyclines, quinolones, amphenicols, and sulphonamides were simultaneously and rapidly (8 min) separated by CZE-UV in bovine raw milk. The combination of LLE and SPE extraction procedures before CE-analysis allowed LOQ values lower than MRLs to be obtained [93].

Another example of a simultaneous separation of different antibiotics (fluoroquinolones, tetracyclines, and β-lactams) in milk was proposed by Long et al. The use of ECL detector allowed trace to be detected, reaching LOD values of cents and thousands of μg/mL [23].

If carefully optimized, CE methods could be very promising in routine screening as good alternatives to LC-MS methods. For example, Kowalski et al. set-up a SPE procedure combined with a MEKC-UV method able to selectively and successfully separate sulfonamides (sulfamethazine, sulfamerazine, sulfathiazole, sulfachloropyridazine, sulfamethoxazole, sulfacarbamide, and sulfaguanidine) and amphenicol-type antibiotics (chloramphenicol, thiamphenicol, and florfenicol) in commercial poultry samples (muscle, liver, and skin with fat) [94].

More recently, the use of FASS combined with a micelle to the solvent stacking (MSS) approach allowed the set-up of a very sensitive CZE-UV method to detect sulfamethoxazole and trimethoprim (an antibacterial agent frequently used in combination with sulfamethoxazole for respiratory and urinary infections) in dairy products, chicken eggs, and honey. MSS consisted of an injection of a micellar solution plug prior to FASS, in order to obtain a focused sample zone with an increase in CE sensitivity [95].

In 2019, a microchip-CE with an LED-induced fluorescence detector was used as a promising platform to simultaneously analyze antibiotics in food. The chip was tested for chloramphenicol (CAP) and kanamycin (Kana) quantification in milk and fish samples, obtaining rapid analysis (3 min) with LODs of pg/mL. Unlike other microchip-CE methods, which had the disadvantages of complexity and low versatility, this platform used a simple strategy, named stir-bar assisted DNA multi-arm junctions recycling, which exploited the capacity of a gold bar with the DNA probe to capture antibiotics, allowing a multiplexed detection and increasing the method sensitivity without matrix interference [26].

To analyze fluoroquinolones and sulfonamides in environmental water, He et al. set-up an on-line preconcentration procedure with a pressure-assisted electrokinetic injection (PAEKI) [96]. PAEKI resolves FASI limits in the analysis of anionic molecules, which can be depleted when the voltage is applied in reverse mode. PAEKI parameters (pressure and voltage) were optimized, obtaining LOD values of a few μg/L and improving the results obtained with hydrodynamic or electrokinetic injections [96].

3. Other Drugs

3.1. Estrogens

Estrogens are widely used in intensive farming worldwide. In EU and United States livestock, the discharge of estrogens is about 83,000 kg/year, representing a very dangerous environmental pollutant. Their low water solubility and the fact that they can easily be degraded and transformed contribute to make them important water contaminants [97].

Wu et al. proposed the use of SPE and pressurized CEC (*p*-CEC)-AD to separate five estrogens (bisphenol-A, 4-tert-octylphenol, 4-*n*-nonylphenol, 2,4-dichlorophenol, and pentachlorophenol) in chicken eggs and milk powder samples. *p*-CEC is an advancement of CEC, in which the formation of typical air bubbles of CEC is prevented by setting a micro-HPLC pump at the inlet of the CE capillary. The set-up method was selective and exhibited a 100- to 500-fold higher sensitivity than CE-UV methods, with values comparable to GC-MS methods [98].

The first DLLME approach combined with MEKC-ESI/MS to extract and analyze estrogens (estriol, 17α-estradiol, 17β-estradiol, estrone, 17β-ethinylestradiol, and their main metabolites) in different milk samples and milk derivatives was set-up by D'Orazio et al. The method was simple, less expensive, and allowed a μg/L LOD level to be obtained [32]. DLLME and MEKC-UV were also used in the analysis of hexestrol, bisphenol A, diethylstilbestrol, and dienestrol, which are phenolic environmental estrogens (PEEs) frequently present in water as contaminants. In this case, after an optimization of DLLME procedure (type and volume of extraction solvent; volume of dispersive solvent; extraction time; salt concentration), LOD values (0.3–0.6 μg/L) within the requirements of trace analysis in environmental water were obtained [33].

3.2. Non-Steroidal Anti-Inflammatory Drugs (NSAIDs)

NSAIDs are added to animal feed mainly to treat respiratory diseases and allergies, generally in association with antibiotics. In addition, they could also improve animal meat quality by reducing fats [99]. They were classified as group B substances and, for many of them, MRLs were established by the European Council in relation to different animals and food matrices [2].

Diclofenac was banned in dairy animals, and other commonly used molecules (ketoprofen, salicylic acid and salicylates, acetylsalicylic acid, and acetylsalicylates) were approved only for non-dairy animals or for animals not involved in egg production [99]. For ibuprofen and flurbiprofen, no MRL value was established, and this could represent a serious problem as NSAIDs cause important side effects (from gastrointestinal problems to cancer) in humans [100].

Regarding the analysis of NSAIDs in animal feed, LC-MS was the most widely used approach, reaching LOQ values around ng/mL [10,101]. Nevertheless, off-line and on-line stacking procedures could lead to promising sensitive CE methods. Alshana et al. set-up a rapid DLLME-FASS-CZE method to detect five main NSAIDs (etodolac, naproxen, ketoprofen, flurbiprofen, and diclofenac) and their derivatives in bovine milk with results similar to those obtained using conventional SPE-LC techniques. In fact, DLLME and FASS combined extraction and stacking procedures are able to develop a sensitive method with a 1000-fold decrease in LOQ values (μg/kg), in comparison to conventional CZE techniques [99]. Among NSAIDs, naproxen, ketoprofen, and clofibric acid are widely used in veterinary medicine and represent contaminants of emerging concern (CECs) for the aquatic environment. SPME is a rapid off-line pre-concentration step ideal for the analysis of highly polar compounds, as NSAIDs, in water samples, with high sensitivity. This was demonstrated by Espina-Benitez who set-up a rapid and economic synthesis of new coated fibers for SPME, developing a sensitive CZE-UV method [102].

3.3. β-Agonists

β-agonists are widely used not only as bronchodilators, but also as muscle growth promoters to increase bovine, lamb, and pork meat production, mainly in Asian countries. They can be toxic for human health, particularly for their effect on the cardiovascular system, and, for this reason, the EU set a very low MRL for clenbuterol (0.05–0.5 μg/kg) in meat and milk samples [2].

NACE coupled with MS (NACE-MS) allowed trace amounts of β-agonists (clenbuterol, salbutamol and terbutaline, TER) to be detected. Anurukvorakun et al. developed a NACE-MS set-up, obtaining results comparable to those obtained by an HPLC-MS/MS method. In fact, the sensitivity is very high thanks to the combination of an SPE using mixed-mode reversed phase/cation exchange cartridges and hydrodynamic and electrokinetic injections in the CE-system [103].

For the quantification of ractopamine and dehydroxyractopamine in porcine meat, Wang et al. proposed an interesting MEKC-UV method with results totally in agreement with those obtained by MS techniques. The method was an on-line stacking method, including exhaustive CSEI and sweeping with a 900 fold higher sensitivity (ng/g level) in comparison to CZE-UV. The capillary was filled with a long plug of a higher conductivity buffer and an electrokinetic injection of the sample was then performed; at the end, the compounds were separated by a sweeping buffer with SDS, whose electrostatic forces contributed to the mobility and resolution of the two analytes [104]. Similarly, FASI-sweep-MEKC improved the MEKC sensitivity by 400–2000 times (ng/mL), allowing the analysis of eight β-agonists (R-albuterol hydrochloride, cimaterol, clenbuterolhydrochloride, colterol, TER, tulobuterol, ractopamine hydrochloride, and zilpaterol) in animal feed samples. A very efficient method was proposed by combining the sensitivity derived from the stacking effect of the FASI-sweeping procedure with the particular selectivity of a dialkyl-chain anionic surfactant in place of SDS [105].

The combination of FESI and CE-C^4D represented another important alternative in the analysis of these drugs in animal feed. In 2014, Gao et al. proposed the first method able to detect TER, procaterol, formoterol, and bambuterol in pig feed. FESI as a pre-concentration step allowed an improvement in LOD values (0.02 mg/L) to be obtained in comparison to UV detection (Figure 3) [106].

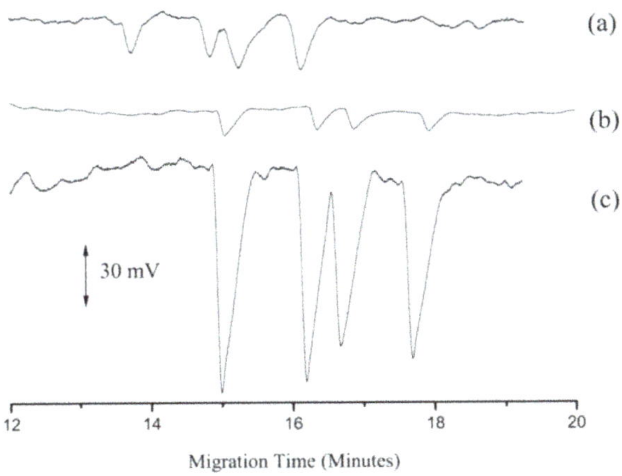

Figure 3. CE-capacitively coupled contactless conductivity detection (CE-C^{4}D) electropherograms of four β2-agonists in pig feed. (**a**) Samples (10 mg/L in background electrolyte (BGE)) without pre-concentration step; (**b**) samples (0.1 mg/L in methanol) with field-enhanced sample injection (FESI) preconcentration step; (**c**) samples (5 mg/L in methanol) with FESI preconcentration step [106].

4. Conclusions

In the last years, the use and abuse of veterinary drugs has represented a very important issue for animal and human health. Consequently, the need of sensitive and rapid analytical techniques for the analysis of veterinary drug residues in food products has become a challenge. This review highlighted the potential role of CE in this topic, as summarized in Table 1. In fact, it is evident that the recent progress in sample pre-concentration methodologies coupled to CE allowed the sensitivity limits of this technique to be overcome and to improve its versatility in the analysis of different types of molecules, in complex food matrices, too.

Table 1. CE-modes in the analysis of the mainly used veterinary drugs in different food matrices. Method sensitivities in terms of LOD or LOQ values are reported.

CE-Technique	Food Matrix	Sensitivity	Ref.
	Nitroimidazoles		
LLE-SPE-CEC-UV	bovine milk	LOQ: 19–96 (µg/L)	[38]
DLLME-CEC-UV	water	LOQ: 5.7–9.3 (µg/L)	[41]
DLLME-CZE-MS/MS	poultry and porcine meat	LOQ: 4–16 (µg/kg)	[44]
DLLME-CSEI-sweep-MEKC-UV	water	LOQ: 2.05–8.14 (ng/mL)	[43]
SPE-CSEI-sweep-MEKC-UV	egg	LOQ: 6.99–16.8 (ng/g)	[45]
	Fluoroquinolones		
SPE-CZE-UV	bovine milk	LOD: 7.5–11.6 (µg/L)	[28]
MMMIPs-CZE-UV	bovine milk	LOD: 12.9–18.8 (µg/L)	[29]
PPT/SPE-CZE-UV	bovine milk	LOQ: 0.06–0.1 (mg/kg)	[50]
MISPE-CZE-LIF	bovine milk, pig kidney	LOQ: 0.55–35 (µg/kg)	[52]
MISPE-CZE-MS/MS	bovine milk	LOQ: 3.2–4.7 (µg/kg)	[53]
PLE-SPE-CZE-MS/MS	meat	LOQ: 130–470 (ng/kg)	[54]
FASS-sweep-CZE-UV	milk, meat	LOD: 5.70, 7.39 (ng/mL)	[55]
FESI-CZE-UV and CZE-MS	bovine milk	LOQ: 2.3–8.3 (µg/kg)	[56]
DLLME-NACE-UV	water	LOQ: 5.43–461 (µg/L)	[57]
SD-LLLME-NACE-UV	water	LOD: 10.1, 55.3 (µg/L)	[58]

Table 1. *Cont.*

CE-Technique	Food Matrix	Sensitivity	Ref.
		Tetracyclines	
CZE-ECL	fish	LOD: 1.8 ng/mL	[60]
FASI-CZE-UV	water	LOQ: 23–59 μg/L	[61]
SPE-LVSS-PS-CZE-UV	milk	LOD: 18.60–23.83 (μg/L)	[62]
MSPD-CZE-UV	milk	LOD: 0.0745–0.0808 (μg/mL)	[65]
SPE-LVSS-PS-CZE-UV	water	LOQ: 67–167 (ng/L)	[63]
LVSS-CZE-UV	water	LOD: 8.1–14.5 (μg/L)	[64]
CZE-AD	water	LOD: 0.33–0.67 (μM)	[66]
SPE- CZE-UV	honey	LOD: 23.9–49.3 (μg/kg)	[67]
		Sulfonamides	
SPE-CEC-MS	meat	LOD: 0.01–0.14 (μg/L)	[71]
microchip-CE-LIF	milk, meat	LOQ: 0.6–7.7 (μg/L)	[72]
SPE-CZE-ECL	milk, meat	LOD: 0.62–3.14 (μg/mL)	[73]
CZE-UV and CZE-MS	meat, water	LOD: 0.33–180 (μg/L)	[68]
CITP-CZE-UV	beverages, water	LOD: 2.29 (ng/mL)	[69]
	Aminoglycosides		
MISPE-FASS-CZE-MS/MS	honey	LOQ: 1.4–94.8 (μg/kg)	[75]
microchip-CE-CCD	standard solutions	LODs: 0.89–3.1 (μg/mL)	[77]
		Macrolides	
FASS-MEKC-UV	milk	LOD: 0.002–0.004 (mg/kg)	[83]
R-USAEME-CZE-DAD	chicken fat	LOQ: 22.1–47.0 (μg/kg)	[84]
	β-lactam antibiotics (penicillins)		
PPT/CZE-UV	milk	LOQ: 0.04–1.7 (μg/mL)	[89]
ASEI-CEC-MS	milk	LOD: 0.05–0.2 (μg/L)	[90]
	β-lactam antibiotics (cephalosporins)		
CZE-UV	complex matrices	LOQ: 4.33–8.00 (μg/mL)	[88]
	Estrogens		
SPE-p-CEC-AD	bovine milk, diary products	LOD: 2–50 (ng/mL)	[98]
DLLME-MEKC-ESI-MS/MS	bovine and goat milk, diary products	LOD: 1–61 (μg/L)	[32]
DLLME-MEKC-UV	water	LOD: 0.3–0.6 (μg/L)	[33]
		NSAIDs	
DLLME-FASS-CZE-UV	bovine milk, dairy products	LOQ: 10–43.7 (μg/kg)	[99]
SPME-CZE-UV	water	LOQ: 2.91–3.86 (μg/L)	[102]
	β-agonists		
SPE-NACE-MS	meat	LOD: 0.3 (ppb)	[103]
CSEI-sweep-MEKC-UV	meat	LOD: 3–5 (ng/g)	[104]
FASI-sweep-MEKC-UV	commercial animal feeds	LOD: 5–20 (ng/mL)	[105]
FESI-CE-C^4D	pig feed	LOD: 0.02 (mg/L)	[106]

Author Contributions: Conceptualization, R.C., A.P.; Writing-Original Draft Preparation, R.C., A.P.; Writing-Review & Editing, R.C., A.P.

Funding: This research received no external funding.

Conflicts of Interest: The authors declare no conflict of interest.

References and Notes

1. Raza, N.; Kim, K.-H. Quantification techniques for important environmental contaminants in milk and dairy products. *TrAC Trends Anal. Chem.* **2018**, *98*, 79–94. [CrossRef]
2. Commission Regulation (EU). On pharmacologically active substances and their classification regarding maximum residue limits in foodstuffs of animal origin. *Off. J. Eur. Commun.* **2010**, *L15*, 1–72.
3. U.S. Food and Drug Administration. Chapter I: Food and Drugs Administration, Department of Health and Human Services, Subchapter E: Animal Drugs, Feed, and Related Products, Part 556: Tolerances for residues of New Animal Drugs in food, 2019.
4. Codex Alimentarius. Maximum residue limits (MRLs) and risk management recommendations (RMRs) for residues of veterinary drugs in foods, CX/MRL 2-2018.
5. Lehotay, S.J.; Chen, Y. Hits and misses in research trends to monitor contaminants in foods. *Anal. Bioanal. Chem.* **2018**, *410*, 5331–5351. [CrossRef] [PubMed]
6. Commission Regulation (EU). Concerning the performance of analytical methods and interpretation of results. *Off. J. Eur. Commun.* **2002**, *L221*, 8–36.
7. Rodriguez, E.; Moreno-Bondi, M.C.; Marazuela, M.D. Multiresidue determination of fluoroquinolone antimicrobials in baby foods by liquid chromatography. *Food Chem.* **2011**, *127*, 1354–1360. [CrossRef]
8. Robert, C.; Brasseur, P.-Y.; Dubois, M.; Delahaut, P.; Gillard, N. Development and validation of rapid multiresidue and multi-class analysis for antibiotics and anthelmintics in feed by ultra-high-performance liquid chromatography coupled to tandem mass spectrometry. *Food Addit. Contam.- Part A Chem. Anal. Control Expo. Risk Assess.* **2016**, *33*, 1312–1323. [CrossRef]
9. Berendsen, B.J.A.; Meijer, T.; Mol, H.G.J.; van Ginkel, L.; Nielen, M.W.F. A global inter-laboratory study to assess acquisition modes for multi-compound confirmatory analysis of veterinary drugs using liquid chromatography coupled to triple quadrupole, time of flight and orbitrap mass spectrometry. *Anal. Chim. Acta* **2017**, *962*, 60–72. [CrossRef]
10. Mainero Rocca, L. Veterinary drugs residues: A review of the latest analytical research on sample preparation and LC-MS based methods. *Food Addit. Contam. Part A Chem. Anal. Control Expo. Risk Assess.* **2017**, *34*, 766–784. [CrossRef]
11. Marazuela, M.D. Determination of veterinary drug residues in foods by liquid chromatography-mass spectrometry: Basic and cutting-edge applications. In *Liquid Chromatography: Applications*, 2nd ed.; Fanali, S., Haddad, P.R., Poole, C., Riekkola, M.-L., Eds.; Elsevier: Amsterdam, The Netherlands, 2017; pp. 539–570.
12. Baglai, A. Enhancing detectability of anabolic-steroid residues in bovine urine by actively modulated online comprehensive two-dimensional liquid chromatography – high-resolution mass spectrometry. *Anal. Chim. Acta* **2018**, *1013*, 87–97. [CrossRef]
13. Hernández-Mesa, M.; Escourrou, A.; Monteau, F.; Le Bizec, B.; Dervilly-Pinel, G. Current applications and perspectives of ion mobility spectrometry to answer chemical food safety issues. *TrAC Trends Anal. Chem.* **2017**, *94*, 39–53.
14. Piňero, M.-Y.; Bauza, R.; Arce, L. Thirty years of capillary electrophoresis in food analysis laboratories: Potential applications. *Electrophoresis* **2011**, *32*, 1379–1393. [CrossRef] [PubMed]
15. Álvarez, G.; Montero, L.; Llorens, L.; Castro-Puyana, M.; Cifuentes, A. Recent advances in the application of capillary electromigration methods for food analysis and Foodomics. *Electrophoresis* **2018**, *39*, 136–159. [CrossRef] [PubMed]
16. Papetti, A.; Colombo, R. High-performance capillary electrophoresis for food quality evaluation. In *Evaluation Technologies for Food Quality*; Zhong, J., Wang, X., Eds.; Elsevier: Cambridge, UK, 2019; pp. 301–377.
17. Stavrou, I.J.; Agathokleous, E.A.; Kapnissi-Christodoulou, C.P. Chiral selectors in CE: Recent development and applications (mid-2014 to mid-2016). *Electrophoresis* **2017**, *38*, 786–819. [CrossRef] [PubMed]
18. Iacob, B.C.; Bodoki, E.; Oprean, R. Recent advances in capillary electrochromatography using molecularly imprinted polymers. *Electrophoresis* **2014**, *35*, 2722–2732. [CrossRef] [PubMed]
19. Tarongoy, F.M., Jr.; Haddad, P.R.; Quirino, J.P. Recent developments in open tubular capillary electrochromatography from 2016 to 2017. *Electrophoresis* **2018**, *39*, 34–52. [CrossRef] [PubMed]
20. Týčová, A.; Ledvina, V.; Klepárník, K. Recent advances in CE-MS coupling: Instrumentation, methodology, and applications. *Electrophoresis* **2017**, *38*, 115–134. [CrossRef] [PubMed]

21. Klepárník, K. Recent advances in combination of capillary electrophoresis with mass spectrometry: methodology and theory. *Electrophoresis* **2015**, *36*, 159–178. [CrossRef] [PubMed]
22. Elbashir, A.; Schmitz, O.J.; Aboul-Enein, H.Y. Application of capillary electrophoresis with capacitively coupled contactless conductivity detection (CE-C4D): An update. *Biomed. Chromatogr.* **2017**, *31*, e3945. [CrossRef]
23. Long, C.; Deng, B.; Sun, S.; Meng, S. Simultaneous determination of chlortetracycline, ampicillin and sarafloxacin in milk using capillary electrophoresis with electrochemiluminescence detection. *Food Addit. Contam. Part A Chem. Anal. Control Expo. Risk Assess.* **2017**, *34*, 24–31. [CrossRef]
24. Ferey, L.; Delaunay, N. Food Analysis on Electrophoretic Microchips. *Sep. Purif. Rev.* **2016**, *45*, 193–226. [CrossRef]
25. Breadmore, M.C.; Wuethrich, A.; Li, F.; Phung, S.C.; Kalsoom, U.; Cabot, J.M.; Tehranirokh, M.; Shallan, A.I.; Abdul Keyon, A.S.; See, H.H.; et al. Recent advances in enhancing the sensitivity of electrophoresis and electrochromatography in capillaries and microchips (2014–2016). *Electrophoresis* **2017**, *38*, 33–59. [CrossRef] [PubMed]
26. Zhang, K.; Gan, N.; Shen, Z.; Cao, J.; Hu, F.; Li, T. Microchip electrophoresis based aptasensor for multiplexed detection of antibiotics in foods via a stir-bar assisted multi-arm junctions recycling for signal amplification. *Biosens. Bioelectron.* **2019**, *130*, 139–146.
27. Šlampová, A.; Malá, Z.; Gebauer, P.E.; Boček, P. Recent progress of sample stacking in capillary electrophoresis (2014–2016). *Electrophoresis* **2017**, *38*, 20–32. [CrossRef] [PubMed]
28. Springer, V.; Jacksén, J.; Ek, P.; Lista, A.G.; Emmer, A. Determination of fluoroquinolones in bovine milk samples using a pipette-tip SPE step based on multiwalled carbon nanotubes prior to CE separation. *J. Sep. Sci.* **2014**, *37*, 158–164. [CrossRef] [PubMed]
29. Wang, H.; Liu, Y.; Wei, S.; Yao, S.; Zhang, J.; Huang, H. Selective extraction and determination of fluoroquinolones in bovine milk samples with montmorillonite magnetic molecularly imprinted polymer and capillary electrophoresis. *Anal. Bioanal. Chem.* **2016**, *408*, 589–598. [CrossRef]
30. Ramautar, R.; Somsen, G.W.; de Jong, G.J. Developments in coupled solid-phase extraction–capillary electrophoresis 2013–2015. *Electrophoresis* **2016**, *37*, 35–44. [CrossRef]
31. Pérez-Rodríguez, M.; Pellerano, R.G.; Pezza, L.; Redigolo Pezza, H. An overview of the main foodstuff sample preparation technologies for tetracycline residue determination. *Talanta* **2018**, *182*, 1–21. [CrossRef]
32. D'Orazio, G.; Asensio-Ramos, M.; Hernández-Borges, J.; Rodríguez-Delgado, M.Á.; Fanali, S. Evaluation of the combination of a dispersive liquid-liquid microextraction method with micellar electrokinetic chromatography coupled to mass spectrometry for the determination of estrogenic compounds in milk and yogurt. *Electrophoresis* **2015**, *36*, 615–625. [CrossRef]
33. Liu, J.; Lu, W.; Liu, H.; Wu, X.; Li, J.; Chen, L. Dispersive liquid-liquid microextraction for four phenolic environmental estrogens in water samples followed by determination using capillary electrophoresis. *Electrophoresis* **2016**, *37*, 2502–2508. [CrossRef]
34. Kitagawa, F.; Otsuka, K. Recent applications of on-line sample preconcentration techniques in capillary electrophoresis. *J. Chromatogr. A* **2014**, *1335*, 43–60. [CrossRef]
35. Tadeo, J.L.; Sánchez-Brunete, C.; González, L. Pesticides: classification and properties. In *Analysis of Pesticides in Food and Environmental Samples*; Tadeo, J.L., Ed.; CRC Press: Boca Raton, FL, USA, 2008; pp. 1–34.
36. U.S. Food and Drug Administration. Chapter I: Food and Drugs Administration, Department of Health and Human Services, Subchapter E: Animal Drugs, Feed, and Related Products, Part 530: Extralabel drugs use in animals, 2019.
37. Mudry, M.D.; Martinez, R.A.; Nieves, M.; Carballo, M.A. Biomarkers of geno- toxicity and genomicin stability in a non-human primate, Cebus libidinosus (Cebidae, Platyrrhini), exposed to nitroimidazole derivatives. *Mutat. Res. Genet. Toxicol. Environ.* **2011**, *721*, 108–113. [CrossRef] [PubMed]
38. Hernández-Mesa, M.; Lara, F.J.; Cruces-Blanco, C.; García-Campaña, A.M. Determination of 5-nitroimidazole residues in milk by capillary electrochromatography with packed C18 silica beds. *Talanta* **2015**, *144*, 542–550. [CrossRef] [PubMed]
39. Sun, H.; Wang, F.; Ai, L. Simultaneous determination of seven nitroimidazole residues in meat byusing HPLC-UV detection with solid-phase extraction. *J. Chromatogr. B* **2007**, *857*, 296–300. [CrossRef] [PubMed]
40. Lin, Y.; Su, Y.; Liao, X.; Yang, N.; Yang, X.; Choi, M.M.F. Determination of five nitroimidazole residues in artificial porcine muscle tissue samples by capillary electrophoresis. *Talanta* **2012**, *88*, 646–652. [CrossRef]

41. Tejada-Casado, C.; Hernández-Mesa, M.; del Olmo-Iruela, M.; García-Campaña, A.M. Capillary electrochromatography coupled with dispersive liquid-liquid microextraction for the analysis of benzimidazole residues in water samples. *Talanta* **2016**, *161*, 8–14. [CrossRef]
42. Hashemi, B.; Zohrabi, P.; Kim, K.-H.; Shamsipur, M.; Deep, A.; Hong, J. Recent advances in liquid-phase microextraction techniques for the analysis of environmental pollutants. *TrAC Trends Anal. Chem.* **2017**, *97*, 83–95. [CrossRef]
43. Hernández-Mesa, M.; Airado-Rodríguez, D.; Cruces-Blanco, C.; García-Campaña, A.M. Novel cation selective exhaustive injection-sweeping procedure for 5-nitroimidazole determination in waters by micellar electrokinetic chromatography using dispersive liquid-liquid microextraction. *J. Chromatogr. A* **2014**, *9*, 65–72. [CrossRef]
44. Tejada-Casado, C.; Moreno-González, D.; Lara, F.J.; García-Campana, A.M. Monsalud del Olmo-Iruela, Determination of benzimidazoles in meat samples by capillary zoneelectrophoresis tandem mass spectrometry following dispersiveliquid–liquid microextraction. *J. Chromatogr. A* **2017**, *1490*, 212–219. [CrossRef]
45. Airado-Rodríguez, D.; Hernández-Mesa, M.; García-Campaña, A.M.; Cruces-Blanco, C. Evaluation of the combination of micellar electrokinetic capillary chromatography with sweeping and cation selective exhaustive injection for the determination of 5-nitroimidazoles in egg samples. *Food Chem.* **2016**, *213*, 215–222. [CrossRef]
46. Riaz, L.; Mahmood, T.; Khalid, A.; Rashid, A.; Ahmed Siddique, M.B.; Kamal, A.; Coyne, M.S. Fluoroquinolones (FQs) in the environment: A review on their abundance, sorption and toxicity in soil. *Chemosphere* **2018**, *191*, 704–720. [CrossRef]
47. Golomb, B.A.; Koslik, H.J.; Redd, A.J. Fluoroquinolone-induced serious, persistent, multisymptom adverse effects. *BMJ Case Rep* **2015**, *2015*, bcr2015209821. [CrossRef] [PubMed]
48. Wall, G.C.; Taylor, M.J.; Smith, H.L. Prevalence and characteristics of hospital inpatients with reported fuoroquinolone allergy. *Int. J. Clin. Pharm.* **2018**, *40*, 890–894. [CrossRef] [PubMed]
49. Czyrski, A. Analytical Methods for Determining Third and Fourth Generation Fluoroquinolones: A Review. *Chromatographia* **2017**, *80*, 181–200. [CrossRef] [PubMed]
50. Piňero, M.-Y.; Garrido-Delgado, R.; Bauza, R.; Arce, L.; Valcárcel, M. Easy sample treatment for the determination of enrofloxacin and ciprofloxacin residues in raw bovine milk by capillary electrophoresis. *Electrophoresis* **2012**, *33*, 2978–2986. [CrossRef] [PubMed]
51. De Quirós, A.R.-B.; Sendón, R. Molecularly imprinted polymers: Applications in food science. In *Molecularly Imprinted Polymers (MIPs): Challenges, Uses and Prospects*; Queen, T., Ed.; Nova Science Publishers, Inc.: New York, NY, USA, 2017; pp. 1–42.
52. Lombardo-Agüí, M.; García-Campaña, A.M.; Gámiz-Gracia, L.; Cruces Blanco, C. Laser induced fluorescence coupled to capillary electrophoresis for the determination of fluoroquinolones in foods of animal origin using molecularly imprinted polymers. *J. Chromatogr. A* **2010**, *1217*, 2237–2242. [CrossRef]
53. Moreno-González, D.; Lara, F.J.; Gámiz-Gracia, L.; García-Campaña, A.M. Molecularly imprinted polymer as in-line concentrator in capillary electrophoresis coupled with mass spectrometry for the determination of quinolones in bovine milk samples. *J. Chromatogr. A* **2014**, *1360*, 1–8. [CrossRef]
54. Lara, F.J.; García-Capaña, A.M.; Alés-Barrero, F.; Bosque-Sendra, J.M. In-line solid-phase extraction preconcentration in capillary electrophoresis-tandem mass spectrometry for the multiresidue detection íof quinolones in meat by pressurized liquid extraction. *Electrophoresis* **2008**, *29*, 2117–2125. [CrossRef]
55. Xu, X.; Liu, L.; Jia, Z.; Shu, Y. Determination of enrofloxacin and ciprofloxacin in foods of animal origin by capillary electrophoresis with field amplified sample stacking–sweeping technique. *Food Chem.* **2015**, *176*, 219–225. [CrossRef] [PubMed]
56. Deng, Y.; Gasilova, N.; Qiao, L.; Zhou, Y.-L.; Zhang, X.-X.; Girault, H.H. Highly sensitive detection of five typical fluoroquinolones in low-fat milk byfield-enhanced sample injection-based CE in bubble cell capillary. *Electrophoresis* **2014**, *35*, 3355–3362. [CrossRef] [PubMed]
57. Herrera-Herrera, A.V.; Hernández-Borges, J.; Borges-Miquel, T.M.; Rodríguez-Delgado, M.A. Dispersive liquid-liquid microextraction combined with nonaqueous capillary electrophoresis for the determination of fluoroquinolone antibiotics in waters. *Electrophoresis* **2010**, *31*, 3457–3465. [CrossRef] [PubMed]
58. Springer, V.H.; Lista, A.G. In-line coupled single drop liquid–liquid–liquid microextraction with capillary electrophoresis for determining fluoroquinolones in water samples. *Electrophoresis* **2015**, *36*, 1572–1579. [CrossRef]

59. Ma, T.Y.; Vickroy, T.W.; Shien, J.H.; Chou, C.C. Improved nonaqueous capillary electrophoresis for tetracyclines at subparts per billion level. *Electrophoresis* **2012**, *33*, 1679–1682. [CrossRef]
60. Deng, B.; Xu, Q.; Lu, H.; Ye, L.; Wang, Y. Pharmacokinetics and residues of tetracycline in crucian carp muscle using capillary electrophoresis on-line coupled with electrochemiluminescence detection. *Food Chem.* **2012**, *134*, 2350–2354. [CrossRef] [PubMed]
61. Díaz-Quiroz, C.A.; Hernández-Chávez, J.F.; Ulloa-Mercado, G.; Gortáres-Moroyoqui, P.; Martínez-Macías, R.; Meza-Escalante, E.; Serrano-Palacios, D. Simultaneous quantification of antibiotics in wastewater from pig farms by capillary electrophoresis. *J. Chromatogr. B* **2018**, *1092*, 386–392. [CrossRef] [PubMed]
62. Islas, G.; Rodriguez, J.A.; Perez-Silva, I.; Miranda, J.M.; Ibarra, I.S. Solid-Phase Extraction and Large-Volume Sample Stacking-Capillary Electrophoresis for Determination of Tetracycline Residues in Milk. *J. Anal. Methods Chem.* **2018**, *2018*, 5394527. [CrossRef] [PubMed]
63. Moreno-González, D.; Lupión-Enríquez, I.; Garcia-Campaña, A.M. Trace determination of tetracyclines in water samples by capillary zone electrophoresis combining off-line and on-line sample preconcentratio. *Electrophoresis* **2016**, *37*, 1212–1219. [CrossRef]
64. Wu, X.; Xu, Z.; Huang, Z.; Shao, C. Large volume sample stacking of cationic tetracycline antibiotics toward 10 ppb level analysis by capillary electrophoresis with UV detection. *Electrophoresis* **2016**, *37*, 2963–2969. [CrossRef]
65. Mu, G.; Liu, H.; Xu, L.; Tian, L.; Luan, F. Matrix Solid-Phase Dispersion Extraction and Capillary Electrophoresis Determination of Tetracycline Residues in Milk. *Food Anal. Methods* **2012**, *5*, 148–153. [CrossRef]
66. Zhou, C.; Deng, J.; Shi, G.; Zhou, T. β-cyclodextrin-ionic liquid polymer based dynamically coating for simultaneous determination of tetracyclines by capillary electrophoresis. *Electrophoresis* **2017**, *38*, 1060–1067. [CrossRef]
67. Casado-Terrones, S.; Segura-Carretero, A.; Busi, S.; Dinelli, G.; Fernández-Gutiérrez, A. Determination of tetracycline residues in honey by CZE with ultraviolet absorbance detection. *Electrophoresis* **2007**, *28*, 2882–2887. [CrossRef]
68. Amin, N.C.; Blanchin, M.D.; Aké, M.; Fabre, H. Capillary electrophoresis methods for the analysis of antimalarials. Part II. Achiral separative methods. *J. Chromatogr. A* **2013**, *1276*, 1–11. [CrossRef]
69. Mikus, P.; Maráková, K.; Veizerová, L.; Piešť anský, J. Determination of quinine in beverages by online coupling capillary isotachophoresis to capillary zone electrophoresis with UV spectrophotometric detection. *J. Sep. Sci.* **2011**, *34*, 3392–3398. [CrossRef]
70. Kluska, M.; Marciniuk-Kluska, A.; Prukala, D.; Prukala, W. Analytics of quinine and its derivatives. *Crit. Rev. Anal. Chem.* **2016**, *46*, 139–145. [CrossRef] [PubMed]
71. Cheng, Y.J.; Huang, S.H.; Singco, B.; Huang, H.-Y. Analyses of sulfonamide antibiotics in meat samples by on-line concentration capillary electrochromatography-mass spectrometry. *J. Chromatogr. A* **2011**, *1218*, 7640–7647. [CrossRef]
72. Wang, L.; Wu, J.; Wang, Q.; He, C.; Zhou, L.; Wang, J.; Pu, Q. Rapid and sensitive determination of sulfonamide residues in milk and chicken muscle by microfluidic chip electrophoresis. *J. Agric. Food Chem.* **2012**, *60*, 1613–1618. [CrossRef]
73. Dai, T.; Duan, J.; Li, X.; Xu, X.; Shi, H.; Kang, W. Determination of Sulfonamide Residues in Food by Capillary Zone Electrophoresis with On-Line Chemiluminescence Detection Based on an Ag(III) Complex. *Int. J. Mol. Sci.* **2017**, *18*, 1286. [CrossRef] [PubMed]
74. Farouk, F.; Azzazy, H.M.E.; Niessen, W.M.A. Challenges in the determination of aminoglycoside antibiotics, a review. *Anal. Chim. Acta* **2015**, *890*, 21–43. [CrossRef] [PubMed]
75. Moreno-González, D.; Lara, F.J.; Jurgovská, N.; Gámiz-Gracia, L.; García-Campana, A.M. Determination of aminoglycosides in honey by capillary electrophoresis tandem mass spectrometry and extraction with molecularly imprinted polymers. *Anal. Chim. Acta* **2015**, *891*, 321–328. [CrossRef]
76. Liu, H.; Li, N.; Liu, X.; Qian, Y.; Qiu, J.; Wang, X. Poly(N-acryloyl-glucosamine-co-methylenebisacrylamide)-based hydrophilic magnetic nanoparticles for the extraction of aminoglycosides in meat samples. *J. Chromatogr. A* **2019**, in press. [CrossRef]
77. Zhu, G.; Bao, C.; Liu, W.; Yan, X.; Liu, L.; Xiao, J.; Chen, C. Rapid detection of ags using microchip capillary electrophoresis contactless conductivity detection. *Curr. Pharm. Anal.* **2019**, *15*, 9–16. [CrossRef]

78. Commission Regulation (EU). On measures to monitor certain substances and residues thereof in live animals and animal products and repealing Directives 85/358/EEC and 86/469/EEC and Decisions 89/187/EEC and 91/664/EEC. *Off. J. Eur. Commun.* **1996**, *125*, 10–32.
79. Dinos, G.P. The macrolide antibiotic renaissance. *Br. J. Pharmacol.* **2017**, *174*, 2967–2983. [CrossRef] [PubMed]
80. Pyörälä, S.; Baptiste, K.E.; Catry, B.; van Duijkeren, E.; Greko, C.; Moreno, M.A.; Pomba, M.C.; Rantala, M.; Ružauskas, M.; Sanders, P.; et al. Macrolides and lincosamides in cattle and pigs: Use and development of antimicrobial resistance. *Vet. J.* **2014**, *200*, 230–239. [CrossRef] [PubMed]
81. Şanli, S.; Palabiyik, I.M.; Şanli, N.; Guzel-Seydim, Z.B.; Alsancak, G. Optimization of the experimental conditions for macrolide antibiotics in high performance liquid chromatography by using response surface methodology and determination of tylosin in milk samples. *J. Anal. Chem.* **2011**, *66*, 838–847. [CrossRef]
82. Dickson, L.C.; O'Byrne, C.; Chan, W. A quantitative method for residues of macrolide antibiotics in porcine kidney by liquid chromatography/ tandem mass spectrometry. *J. AOAC Int.* **2012**, *9*, 567–575. [CrossRef] [PubMed]
83. Hong, Y.-Q.; Guo, X.; Chen, G.-H.; Zhou, J.-W.; Zou, X.-M.; Liao, X.; Hou, T. Determination of five macrolide antibiotic residues in milk by micellar electrokinetic capillary chromatography with field amplified sample stacking. *J. Food Saf.* **2018**, *38*, e12382. [CrossRef]
84. Lorenzetti, A.S.; Lista, A.G.; Domini, C.E. Reverse ultrasound-assisted emulsification-microextraction of macrolides from chicken fat followed by electrophoretic determination. *LWT-Food Sci. Technol.* **2019**, *113*, 108334. [CrossRef]
85. Cameron-Veas, K.; Solà-Ginés, M.; Moreno, M.A.; Fraile, L.; Migura-Garcia, L. Impact of the use of β-lactam antimicrobials on the emergence of Escherichia coli isolates resistant to cephalosporins under standard pig-rearing conditions. *Appl. Environ. Microbiol.* **2015**, *81*, 1782–1787. [CrossRef]
86. Cameron-Veas, K.; Moreno, M.A.; Fraile, L.; Migura-Garcia, L. Shedding of cephalosporin resistant Escherichia coli in pigs from conventional farms after early treatment with antimicrobials. *Vet. J.* **2016**, *211*, 21–25. [CrossRef] [PubMed]
87. Piñero, M.-Y.; Bauza, R.; Arce, L.; Valcárcel, M. Determination of penicillins in milk of animal origin by capillary electrophoresis: Is sample treatment the bottleneck for routine laboratories. *Talanta* **2014**, *119*, 75–82. [CrossRef]
88. Hancu, G.; Sasebeşi, A.; Rusu, A.; Kelemen, H.; Ciurba, A. Study of the electrophoretic behavior of cephalosporins by Capillary Zone Electrophoresis. *Adv. Pharm. Bull.* **2015**, *5*, 223–229. [CrossRef] [PubMed]
89. Li, M.H.; He, W.; Zhang, L.; Duan, C. Analysis of penicillin and its β-lactamase hydrolysis products in milk using capillary zone electrophoresis. *Anal. Methods* **2015**, *7*, 4602–4607. [CrossRef]
90. Liu, W.-L.; Wu, C.-Y.; Li, Y.-T.; Huang, H.-Y. Penicillin analyses by capillary electrochromatography-mass spectrometry with different charged poly(stearylmethacrylate–divinylbenzene) monoliths as stationary phases. *Talanta* **2012**, *101*, 71–77. [CrossRef] [PubMed]
91. Tong, J.; Rao, Q.; Zhu, K.; Jang, Z.; Ding, S. Simultaneous determination of five tetracycline and macrolide antiobiotics in feeds using HPCE. *J. Sep. Sci.* **2009**, *32*, 4254–4260. [CrossRef]
92. Regulation (EC) No 1831/2003, Annex I: list of additives, Edition 11/2019 (277).
93. Vera-Candioti, L.; Olivieri, A.C.; Goicoechea, H.C. Development of a novel strategy for preconcentration of antibiotic residues in milk and their quantitation by capillary electrophoresis. *Talanta* **2010**, *82*, 213–221. [CrossRef]
94. Kowalski, P.; Plenis, A.; Oledzka, I.; Konieczna, L. Optimization and validation of the micellar electrokinetic capillary chromatographic method for simultaneous determination of sulfonamide and amphenicol-type drugs in poultry tissue. *J. Pharm. Biomed. Anal.* **2011**, *54*, 160–170. [CrossRef]
95. Liu, L.; Wan, Q.; Xu, X.; Duan, S.; Yang, C. Combination of micelle collapse and field-amplified sample stacking in capillary electrophoresis for determination of trimethoprim and sulfamethoxazole in animal-originated foodstuffs. *Food Chem.* **2017**, *219*, 7–12. [CrossRef]
96. He, T.; Xu, Z.; Ren, J. Pressure-assisted electrokinetic injection stacking for seven typical antibiotics in waters to achieve μg/L level analysis by capillary electrophoresis with UV detection. *Microchem. J.* **2019**, *146*, 1295–1300. [CrossRef]
97. Adeel, M.; Song, X.; Wang, Y.; Francis, D.; Yang, Y. Environmental impact of estrogens on human, animal and plant life: A critical review. *Environ. Int.* **2017**, *99*, 107–119. [CrossRef]

98. Wu, W.; Yuan, X.; Wu, X.; Lin, X.; Xie, Z. Analysis of phenolic xenoestrogens by pressurized CEC with amperometric detection. *Electrophoresis* **2010**, *31*, 1011–1018. [CrossRef]
99. European Commission. Council Regulation 324/2004/EC.2005. *Off. J. Eur. Commun.* **2004**, *L58*, 16.
100. Alshana, U.; Göğer, N.G.; Ertaş, N. Dispersive liquid–liquid microextraction combined with field-amplified sample stacking in capillary electrophoresis for the determination of non-steroidal anti-inflammatory drugs in milk and dairy products. *Food Chem.* **2013**, *138*, 890–897. [CrossRef] [PubMed]
101. Shishov, A.; Nechaeva, D.; Bulatov, A. HPLC-MS/MS determination of non-steroidal anti-inflammatory drugs in bovine milk based on simultaneous deep eutectic solvents formation and its solidification. *Microchem. J.* **2019**, *150*, 104080. [CrossRef]
102. Espina-Benitez, M.; Araujo, L.; Prieto, A.; Navalón, A.; Vílchez, J.L.; Valera, P.; Zambrano, A.; Dugas, V. Development of a new microextraction fiber combined to on-line sample stacking capillary electrophoresis UV detection for acidic drugs determination in real water samples. *Int. J. Environ. Res. Public Health* **2017**, *14*, 739. [CrossRef] [PubMed]
103. Anurukvorakun, O.; Buchberger, W.; Himmelsbach, M.; Klampel, C.W.; Suntornsuk, L. A sensitive non-aqueous capillary electrophoresis-mass spectrometric method for multiresidue analyses of beta-agonists in pork. *Biomed. Chromatogr.* **2010**, *24*, 588–599. [PubMed]
104. Wang, C.-C.; Lu, C.-C.; Chen, Y.-L.; Cheng, H.-L.; Wu, S.-M. Chemometric optimization of cation-selective exhaustive injection sweeping micellar electrokinetic chromatography for quantification of ractopamine in porcine meat. *J. Agric. Food Chem.* **2013**, *61*, 5914–5920. [CrossRef] [PubMed]
105. Hsieh, S.-Y.; Wang, C.-C.; Kou, H.-S.; Wu, S.M. Dialkyl anionic surfactant in field-amplified sample injection andsweeping-micellar electrokinetic chromatography for determinationof eight leanness-promoting β-agonists in animal feeds. *J. Pharm. Biomed. Anal.* **2017**, *141*, 222–228. [CrossRef]
106. Gao, F.; Wu, M.; Zhang, Y.; Wang, G.; Wang, Q.; He, P.; Fang, Y. Sensitive determination of four β2-agonists in pig feed by capillary electrophoresis using on-line sample preconcentration with contactless conductivity detection. *J. Chromatogr. B* **2014**, *973*, 29–32. [CrossRef]

molecules

MDPI

Review

Anti-Inflammatory and Anticancer Properties of Bioactive Compounds from *Sesamum indicum* L.—A Review

Ming-Shun Wu [1,2,3], Levent Bless B. Aquino [4], Marjette Ylreb U. Barbaza [4], Chieh-Lun Hsieh [5], Kathlia A. De Castro-Cruz [4], Ling-Ling Yang [6,7] and Po-Wei Tsai [8,*]

1. Division of Gastroenterology, Department of Internal Medicine, Wan Fang Hospital, Taipei Medical University, Taipei 116, Taiwan; vw1017@gmail.com
2. School of Medicine, College of Medicine, Taipei Medical University, Taipei 110, Taiwan
3. Integrative Therapy Center for Gastroenterologic Cancers, Wan Fang Hospital, Taipei Medical University, Taipei 116, Taiwan
4. School of Chemical, Biological, Materials Engineering and Sciences, Mapúa University, Manila 1002, Metro Manila, Philippines; leventbless12@gmail.com (L.B.B.A.); marjette.barbaza@gmail.com (M.Y.U.B.); kadecastro@mapua.edu.ph (K.A.D.C.-C.)
5. Department of Athletics Sports, College of Humanities and Social Sciences, Chang Jung Christian University, Tainan 711, Taiwan; lyrecojolly@gmail.com
6. School of Pharmacy, College of Pharmacy, Taipei Medical University, Taipei 110, Taiwan; llyang@tmu.edu.tw
7. College of Acupuncture and Oriental Medicine, Houston, TX 77063, USA
8. Department of Medical Sciences Industry, College of Health Sciences, Chang Jung Christian University, Tainan 711, Taiwan

* Correspondence: powei@mail.cjcu.edu.tw; Tel.: +886-6-278-5123

Academic Editor: Ping-Chung Kuo
Received: 9 November 2019; Accepted: 3 December 2019; Published: 4 December 2019

check for updates

Abstract: The use of foodstuff as natural medicines has already been established through studies demonstrating the pharmacological activities that they exhibit. Knowing the nutritional and pharmacological significance of foods enables the understanding of their role against several diseases. Among the foods that can potentially be considered as medicine, is sesame or *Sesamum indicum* L., which is part of the Pedaliaceae family and is composed of its lignans such as sesamin, sesamol, sesaminol and sesamolin. Its lignans have been widely studied and are known to possess antiaging, anticancer, antidiabetes, anti-inflammatory and antioxidant properties. Modern chronic diseases, which can transform into clinical diseases, are potential targets of these lignans. The prime example of chronic diseases is rheumatic inflammatory diseases, which affect the support structures and the organs of the body and can also develop into malignancies. In line with this, studies emphasizing the anti-inflammatory and anticancer activities of sesame have been discussed in this review.

Keywords: anti-inflammatory; anti-cancer; sesame extracts; sesame oil; *Sesamum indicum* L.

1. Introduction

Numerous studies have already been able to prove that different kinds of food are capable for acting against several types of diseases due to their various medicinal properties. *Sesamum indicum* L., more commonly referred to as sesame, is one of the foods that are known to exhibit pharmacological applications. Sesame, from the Pedaliaceae family [1], has been known way back several decades ago (1600 BC), and is originated from Indonesia according to the Hindu legends. Sesame is considered as one of the oldest condiments in the human history [2,3]. In line with this, sesame is one of the first crops to produce oil [4]. Its herbal medicinal property was first discovered in China and India during the 8th century BC, and is believed to relieve toothaches, to give energy, to prevent aging, to sooth the mind and body and to treat bites of insects [5,6]. The largest area of sesame cultivation is found in India and provides 27.9% of the world's sesame [7]. Some countries also cultivate sesame especially in the tropical countries of Africa and Asia [8].

Majority of the sesame lignans are found in the sesame seeds. Sesame seeds comprise of 50% oil and 25% protein and the rest are sugars, moisture, fibers and minerals [9,10]. There are four commonly known sesame lignans namely, sesamin, sesamol, sesaminol and sesamolin. These sesame lignans (Figure 1) are known for its various biological activities and applications. Sesamin, a furfuran lignan, plays a role in scavenging free radicals and lipid and glucose metabolism. Sesamol possess high antioxidant activity, which involves membrane protection from peroxidation of lipids [11,12]. Sesaminol exhibits inhibitory property against membrane lipid peroxidation and improves the tocopherols availability of vitamin E through enhancement of concentrations of liver and plasma [13,14]. Sesamolin upregulates the rate of peroxisomal fatty acid oxidation and hepatic mitochondrial and has a synergistic effect on pyrethrum insecticides [15–17]. Not only is sesame oil abundant in sesame seeds, but sesame oil also has excellent quality nutrition and stability [18]. Among other vegetable oils, sesame oil has many uses and has high therapeutic values, which make it interesting to study. The presence of sesame lignans such as sesamin, sesamol, sesamolin and other acylglycerols (oleic acid, linoleic acid, palmitic acid, stearic acid and arachidic acid) in sesame oil contributes to its distinct properties [19,20]. This paper presents an in-depth review on the anti-inflammatory and anti-cancer properties of sesame.

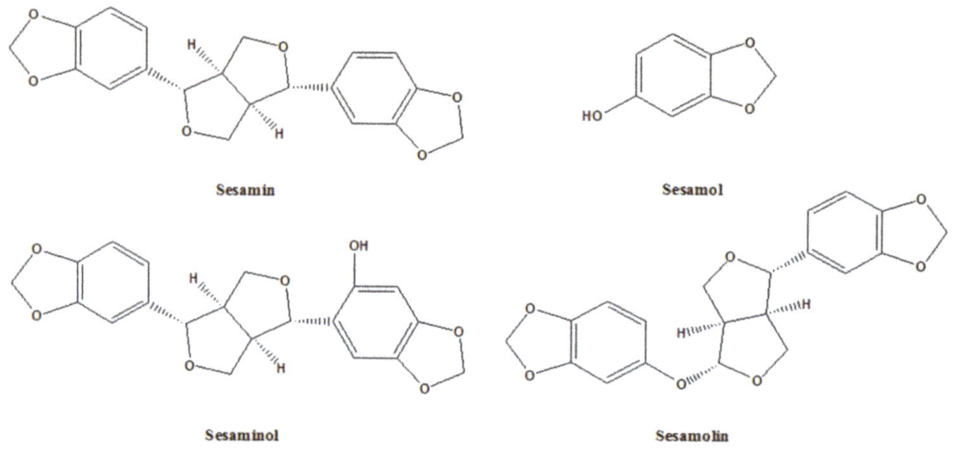

Figure 1. Chemical Structures of *Sesamum indicum* L. lignans.

2. Pharmacological Applications of Sesame

2.1. Anti-Inflammatory Activity

Inflammation has been recorded way back several centuries ago (1500 BCE–600 CE) in the ancient Indian medicine called Ayurveda [21]. In the modern era, the term 'inflammation' comes from the Latin word inflammare that has a meaning of "to set on fire" [22]. Redness, heat, pain and swelling are the characterizations of inflammation and are induced by several factors such as frostbite, infection by pathogens, burns, chemical irritants, physical injuries, oxidative stress, ischemia, toxins and hypersensitivity [23,24]. Inflammation is part of the body's defense mechanism against pathogens or pathogen-associated molecular patterns [25]. Restoring, regenerating and repairing of the damaged tissues or organ hemeostasis through cellular network and signaling pathways are involved in the inflammation process. Conversely, severe acute inflammation or prolongation of inflammation may lead to pathology, organ failure, chronic inflammatory diseases, autoimmunity and death. The active cellular components that are responsible for the processing of acute and chronic inflammation are monocytes/macrophages, endothelial cells, neutrophils, innate lymphoid cells, mucosal-associated invariant T cells, mast cells, natural killer cells, dendritic cells and other lower forms of T cells [26–29]. Tables 1–3 depicts the in vitro/in vivo models for the anti-inflammatory properties of sesame.

2.1.1. Neurodegenerative Disease

Dysregulation of microglia, a principal cell type of brain, is pivotal to recruitment of cell inflammation and expression of pro-inflammatory factors, which later results to neurodegeneration [30]. Such neurodegenerative diseases are Parkinson's disease, Alzheimer's disease and other multiple sclerosis [31]. The inflammation begins when the pattern recognition receptors (PRRs) and pathogen associated molecular patters (PAMPs) interacts [32]. A common type of PPRs that detects the lipopolysaccharide (LPS) is Toll-like receptor 4 (TLR4), which is expressed highly on the surface of the microglia when activated. Once activated, the pro-inflammatory mediator is released via mitogen-activated protein kinase (MAPK) and nuclear factor kappa-light-chain-enhancer of activated B cells (NF-κB) pathways [33] and further generation of other pro-inflammatory mediators begins (e.g., interleukin 1 beta (IL-1β), interleukin-6 (IL-6), tumor necrosis factor alpha (TNF-α), reactive oxygen species (ROS) and nitric oxide (NO)). Other pro-inflammatory molecules like prostaglandin E_2 (PGE_2) are also initiated simultaneously through phosphorylation of its subunits. There are several NF-κB subunits such as p-IKKα/β, p-IκBα and p-p65, and several MAPK subunits such as c-Jun N-terminal kinase (JNK), p38 and extracellular-signal-regulated kinase (ERK).

In the literature study [34] showed that wherein the BV-2 microglial cell line was used, sesamin shows dose-dependent decrease in expression of TLR4 in LPS-stimuli. Notably, at 50 μM concentration of sesamin, there is a significant decrease in the expression of TLR4. Furthermore, sesamin suppresses the phosphorylation of p-IkBα and p-p65 over a moderate period. In addition, sesamin decreases the phosphorylation of JNK, however, the phosphorylation of p38 is slightly reduced. Interestingly, sesamin with 50 μM indomethacin dose-dependently reduced *IL-1β* and *IL-6* mRNA gene expression and moderately reduced TNF-α before LPS exposure. Similarly, sesamin decreases the levels of cyclooxygenase-2 (*COX-2*) gene expressions and hinders the production of PGE_2. Moreover, sesamin reduced the expression of inducible nitric oxide synthase (*iNOS*) gene and as well as reduced production of NO in dose-dependent manner. Sesamin exhibits a reduced effect for neurotoxicity of LPS-mediated microglia activation, which eventually increases the viability of neuronal cells.

In another study [35], in which same microglial cell line was used, further investigation on the p38 MAPK signaling pathway was commenced to support its role in cytokine production. It has been found that 50 μM concentration of sesamin suppresses the p38 MAPK activation (40–75%) induced by LPS. The inhibitory effect of sesamin is similar to that of SB203580 (p38 MAPK inhibitor), which inhibits the production of IL-6 mRNA and protein production specifically. Similarly, sesamolin has been reported to reduce the activation of p38 MAPK induced by LPS, however it has not yet been fully studied.

One of the mental disorders associated with neuroinflammation is depression. The downregulation of norepinephrine (NE), and serotonin (5-hydroxytryptamine, 5-HT) levels and the decrease of synaptic content cause depression [36,37]. Brain derived neurotrophic factor (BDNF) controls the development of neuronal function and becomes ineffective in neurodegenerative disorders such as depression [38]. The ionized calcium binding adaptor molecule 1 (IBA-1) is responsible for the microglia activation in hippocampus and cortex [39]. Several studies already used chronic unpredictable mild stress (CUMS) for the physiological pathway elucidation of depression [40].

CUMS-induced depression in mice [41] shows relatively positive results toward forced swimming, tail suspension, elevated plus maze, Morris water maze and Y-maze tests when sesamin is administered (50 ppm/d) for 6 weeks. Furthermore, sesamin upregulates the levels of 5-HT and NE in striatum only, suggesting its beneficial effects on depressive like behaviors. Neurotrophin-3 (NT3) and BDNF shows increase expression in hippocampus when treated with sesamin. Comparatively, sesamin decreases the expression of IBA-1 expression and as a result, the production of inflammatory cytokines ceases.

Another neurodegenerative disorder is the ischemic brain stroke. Shutting off the flow of cerebral blood to thrombi results to ischemic brain stroke due to loss of oxygen and energy supply to crucial tissues of the brain. In middle cerebral artery occlusion (MCAO)-treated mice [42], sesamin (30 ppm) mitigates brain injury by suppressing the production of inflammatory mediators. Interestingly, sesamin reduces the expression levels of p-ERK1/2 together with p-p38 of ischemic brain tissue in MCAO-induced brain damage.

Sesamin at dosage 20 ppm has shown to suppress 6-hydroxydopamine (6-OHDA) that induces Parkinson's disease in rats [43] via decrease in inflammatory mediator levels in the brain. Sesamin has the capability to alleviate astrogliosis based on the lowering effect of glial fibrillary acidic protein (GFAP) immunoreactivity. Similarly, the inhibitory effect of sesamin against inflammatory agents (MAPK and *COX-2*) results to stabilizing the oxidative stress and mortality in kainic acid-induced status epilepticus [44]. This study, however, is not fully studied as the degree of inflammatory markers was not discussed.

Degradation of heme to iron, carbon monoxide and biliverdin is caused by a phase II antioxidant enzyme called heme oxygenase (HO). There are other isoforms of *HO* such as inducible HO-1 and constitutive HO-2 and HO-3 [45]. Biliverdin reductase transform biliverdin to bilirubin and bilirubin is believed to possess anti-oxidative properties [46,47].

At 100 μM concentration of sesamin, there is an increase in the (*HO-1*) protein level in RAW 264.7 macrophage cells. The *HO-1* mRNA expression, however, is not affected by sesamin. Several studies have claimed that sesamin activates Nf-κB or MAPK signaling pathway, not to mention, sesamin affects the p65 and p38 MAPK effectively in RAW 264.7 cells. The ZnPP IX, a HO-1 inhibitor, reduced the inhibitory effect of sesamin on the release of NO. The degradation of HO-1 protein through the ubiquitination pathway is partially suppressed by sesamin. The proteasome activity, however, is not affected by sesamin. Hence, the ubiquitination mechanism inhibition by sesamin is still unclear [48].

In rat pheochromocytoma PC12 cells study [49], episesamin and sesamin metabolites are investigated. Based from the luciferase reporter assays, episesamin and sesamin metabolites has the capability to activate the signaling cells nuclear factor E2-related factor 2/antioxidant response element (Nrf2/ARE), which further upgrades to phase II detoxification enzyme expression. Moreover, sesamin metabolites induced the expression of detoxification enzymes such as HO-1, y-GCSc and NQO-1 in a dose-dependent manner. Different signal transduction pathways are also affected by the sesamin metabolites through phosphorylation. In addition, sesamin metabolites increase the expression of HO-1 mRNA and protein, which comes before the nuclear translocation of Nrf2.

In the 16-week-old senescence-accelerated mouse-prone 8 (SAMP8) study [50], sesaminol exhibit a reducing effect on the inflammatory cytokines namely IL-6, IL-1β and TNF-α via real-time polymerase chain reactor (PCR) assay in the brain of the modeled mouse. This further contributes in the mitigation of the Alzheimer's disease.

2.1.2. Osteoarthritis

The degradation and the tearing down of cartilage matrix are characterized in a chronic articular disease called osteoarthritis (OA) [51,52]. Chondrocyte has become the context of pathogenesis of OA, which has an outcome of imbalance between degradation and synthesis of cartilage extracellular matrix (CEM). Inflammatory cytokine network could stimulate matrix metalloproteinases (MMPs) generation together with PGE_2 and NO in chondrocytes [53,54]. During the progression of OA, MMPs (e.g., collagenase-3/MMP-13) cleaves and denatures the type II collagen and proteoglycan at the surface of the cartilage. The degradation of the collagen and proteoglycan results to the loss of tensile strength in the matrix of cartilage due to the increase of water content [55,56]. Interleukin-1 (IL-1) and TNF-α can prompt the expression of MMPs expression via c-fos signaling pathway.

In papain-induced OA rat study [57], sesamin has been reported to exhibit chondroprotective effects for OA. Sesamin has no effect on the activity of aggrecanase, the main proteoglycan in the cartilage tissue, but has reversible effect in the expression of MMP-1, -3 and -13 in the human articular chondrocyte (HAC) culture. Not to mention, sesamin suppresses the IL-1β induced MMP expression at both protein and mRNA levels. Moreover, sesamin has been reported to reverse the synergistic effect of combined IL-1β and oncostatin M (OSM) and stops the degradation of type II collagen and proteoglycan. Having said this, sesamin can slow down the destruction of cartilage prompting to the development of OA.

Another study [58], wherein the articular cartilage of the 12 patients undergoing knee replacement surgery were acquired, suggests that sesamin activates Nrf2 signaling pathway and up-regulates HO-1 protein expression and further inhibits the inflammatory gene expressions in the OA chondrocytes. Coupled with, sesamin inhibits the activation of NF-κB during the process of Nrf2 activation.

Sesamol exhibits an inhibitory effect on MMPs expression that triggers OA [59]. The specific MMPs are MMP-1, -9 and -13 and the expression of these MMPs initiate the destruction of cartilage. The cell line used was the human chondrosarcoma cell line (SW1353) and was put into different concentrations of sesamol (5–20 μM). The study claimed that activation of MMP-9 and its expression is prompted concentration-dependently by TNF-α. MMP-1 and MMP-13, however, is not improved by TNF-α but still these are inhibited by sesamol in response to PMA (phorbol 12-myristate 13-acetate). Sesamol, moreover, has been reported to inhibit the degradation of IκB-α, which also considers the inhibition of NF-αB activation, which then propagates the TNF-α signaling. In restoring the chondrocytes via IL-1β signaling pathway, there would be production of MMP-9. The IL-Iβ induced p38 MAPK activation promotes generation of MMP-9 by chondrocytes. Having said this, sesamol has been found to alleviate the phosphorylation of p38 MAPK induced by IL-1β. Moreover, sesamol has been shown to attenuate the expression of MMP-1/-9 in MIA-induced OA in rats though the effect of sesamol in the destruction of cartilage was not studied.

One study [60] discussed the effectivity of encapsulating sesamol with micelles. Encapsulated sesamol in phosphatidyl choline micelles have no cytotoxic activity in the cells and improves its bioavailability against inflammatory. As compared to free sesamol, encapsulated sesamol has a higher percentage in decreasing the production of ROS induced by LPS intracellular. Free sesamol decreases the generation of ROS by 42.6%, whereas encapsulated sesamol decreases the generation of the same species by 74.8%. Investigation on the extent of encapsulated sesamol on other inflammatory cytokines is not yet discussed.

In a clinical trial study [61], a randomized double arm, double-blind active-controlled was designed in 104 male and female participants ages 30–70 years old. Half of the participants are the control group and half are the intervention group. The intervention and took 1.5 mL of sesame oil thrice a day for a span of 4 weeks, while the control group took diclofenac gel in the same span of time. The clinical results show that sesame oil reduces knee OA pain and other body systems related to knee OA more effectively than the control group. On the other hand, the control group was non-inferior as compared to sesame oil when it comes to knee joint stiffness. The study claimed to be the first clinical trial to conduct efficacy of sesame oil in OA patients. All things considered, the study needs extensive and deep exploration regarding other external and internal factors of OA and its response to sesame.

2.1.3. Liver Disease

Nonalcoholic fatty liver disease (NAFLD) occurs when there is generation of fat in the liver in the absence of alcohol. NAFLD is characterized by lobular hepatitis, hepatic steatosis, and liver cell injury [62]. One of the mechanisms involved in the pathogenesis of NAFLD is the inflammation in hepatocytes [63]. Nuclear receptor such as Liver X receptor α (LXR-α) and peroxisome proliferator-activated receptor α (PPAR-α) are responsible for the regulation of lipid homeostasis and are crucial [64].

In the high-fat diet (HFD) Sprague-Dawley rat model [65], sesamin administered at dose 160 ppm by weight exhibit a reduced level of steatosis through inhibition of gene tags: *TC* and *TAG* accumulation.

Equally, sesamin reversed the effect of HFD to the levels of inflammatory cytokines such as IL-6 and TNF-α. In parallel, sesamin downregulates the LXR-α prompted by HFD, while sesamin upregulates the expression of PPAR-α, which stimulates attenuation in the hepatic fat accumulation.

The life threatening of fulminant hepatic failure (FHF) is described as the immense hepatocyte necrosis and sudden decrease in the liver function. This results to multiple organ disfunction syndromes and hepatic encephalopathy [66,67]. In lipopolysaccharide/D-galactosamine (LPS/D-GalN)-induced FHF mice model [68], 100 ppm of sesamin shows improved mortality and promotes serum aminotransferases activity for 6 h. Moreover, sesamin possess hepatoprotection by reducing the regulation of expression of hepatic *TNF-α* mRNA and protein as well as intercellular adhesion molecule-1 (ICAM-1) and endothelial cell adhesion molecule-1 (ECAM-1). Similarly, sesamin downregulates the expression of TLR4 hindering the MAPK and NF-κB signaling pathway. In one small study, sesamin alleviates the synergistic effect of combined lead (Pb) and LPS causing acute hepatic injury effectively. The suppressive effect of sesamin positively affects several inflammatory signaling pathways such as *COX-2*, *iNOS*, JNK and p38 MAPKs, CHOP and GADD45b [69].

The conversion of sesamin into catechol derivatives and converted further into glucuronides or sulfates in the liver [70–73] has been reported to have anti-inflammatory effect on macrophage cell line (J774.1 cells). The said sesamin metabolites have been reported to have strong inhibitory effects on the LPS-induced NO production [74]. Through methylation, the catechol group of the sesamin metabolites was found to cause this inhibition. Though the level of the inhibitory effect of the sesamin metabolites differs, its inhibitory strength is relative to one another. One of the noticeable interactions is the deconjugation of (7α,7′α,8α,8′α)-3,4-dihydroxy-3′,4′-methylenedioxy-7,9′:7′,9-diepoxylignane (SC1) glucuronide to (7α,7′α,8α,8′α)-3-methoxy-4-hydroxy-3′,4′-methylenedioxy-7,9′:7′,9-diepoxylignane (SC1m) in macrophages. With this, the time-dependent deconjugation enhances the inhibitory effect against the LPS-induced NO production in macrophages. Correspondingly, the inhibitory effects of SC1 glucuronide are controlled by the β-glucuronidase activity and catechol-*O*-methyltransferase (COMT) activity in macrophages. β-glucuronidase activity hinders the accumulation of SC1m in macrophages and hence, withdraws its inhibitory effect. On the other hand, the COMT activity helps the inhibitory effects of SC1 glucuronide in macrophages. The sulfates of the sesamin metabolites, however, possess weak or non-inhibitory effects towards the LPS-stimulated macrophages. The conversion of sesamin to SC1 during the metabolism of macrophage is therefore the prominent form for the anti-inflammatory effects [75].

Sesame oil, at dosage 1 and 2 ppm, possess protective effect against nutritional steatohepatitis in the methionine-choline deficient (MCD) mice model [76]. Sesame oil with concentration of 4 ppm has been reported to mitigate hepatic injury, α-SMA, fibrosis and reduces the activity of MMP-2, and -9. However, at the same concentration of sesame oil, it elevates the expression of PPAR-γ and tissue inhibitor matrix metalloproteinase 1 (TIMP-1) [77]. It was mentioned, consequently, that consumption of a large dosage of sesame oil might cause steatohepatitis due to suppression of its antioxidative effect [78]. PPAR-γ, when activated, is responsible for suppressing the expression of collagen and inhibition of hepatic fibrosis by blocking the profibrogenic transforming growth factor beta (TGF-β)/Smad pathway which then down-regulates the expression of α-SMA expression [79–81]. MMPs cleave the fibrillar extracellular matrix (ECM) and aid in apoptosis, specifically MMP-2 and -9. As mentioned above, sesame oil inhibits MMP-2 and -9 as it regulates the expression of TIMP-1. This results to alleviating the injury, hepatic stellate cell (HSC) activation and degradation of ECM and utmost is reversing the fibrosis [82]. Conversely, the US Food and Drug Administration's approval for its medication for the treatment of non-alcoholic steatohepatitis (NASH) has not yet been made.

In a different rat model [83], sesame oil has been found to possess preventive expression of proteins in HFD-induced endoplasmic reticulum (ER) stress and to prevent the initiation of apoptosis. Furthermore, sesame oil has been reported to decrease the expression levels of lipogenic transcription factors (by 44.58%) and enzymes in the liver, which also increases the expression of fatty acid oxidation-related gene. Similarly, sesame oil suppresses the regulation of sterol regulatory element-binding protein 1 (SREBP-1) by 22.91% and the expression of FAS by 50.38%, however, it promotes the expression levels of PPAR-α and carnitine palmitoyl transferase 1 (CPT-1). In view of these regulations, this increases the hepatic lipid oxidation and hence prevents hepatic fat accumulation.

2.1.4. Diabetic Eye Disease

Diabetic retinopathy is a microvascular complication that affects the neurovascular of the retina as a result of neurodegeneration, neuro-inflammation, eventual fibrosis and other diabetic-related damage [84]. In general theory, pathological conditions related to inflammatory activates microglia in the retina and further initiate neuro-inflammation [85]. This eventually led to tissue ischemia, vascular occlusion and cell death causing blindness [86].

In streptozotocin (STZ)-induced diabetic retinopathy mice model, administered 30 ppm of sesamin alleviates the retinal inflammation. The treatment decreases the mRNA levels of *TNF-α* and *ICAM-1* I unlike in the diabetic group. This further elaborates the possibility of reducing the production of inflammatory cytokines in diabetic retina. Sesamin suppresses the induced diabetic retinal injury by inhibition of TNF-α and microglia Iba-1 [87].

2.1.5. Inflammatory Bowel Disease

Ulcerative colitis is an inflammatory bowel disease (IBD) that involves damaging of mucosal tissue via dysregulation of the inflammatory system. In the dinitrochlorobenzene (DNCB)-induced IBD albino rat model [88], sesamol decreases the activity of myeloperoxidase (MPO), which is considered to manifest the anti-inflammatory activities. Although sesamol has incapability to decrease the IL-6 and TNF-α cytokine levels induced by DNCB, sesamol can undergo through ROS pathway.

Aspirin, when ingested, can kill different inflammatory diseases however this can possibly cause acute gastroduodenal injury, which considers the bleeding of ulcers [89]. This happens when there is lipid peroxidation in gastric mucosal caused by ROS, which then regulates the inflammatory cytokines [90,91]. In an aspirin-induced gastric mucosal rat model [92], sesamol has been reported to have suppression effect on neutrophil activation and infiltration when aspirin induces gastric inflammation. The activation of neutrophil initiates the expression of proinflammatory mediators and eventually upregulates the production of nitric oxide, which causes cell damage and lipid peroxidation. The inhibition of sesamol on the neutrophil activation, moreover, is reported to have no effect on the physiological aspect of the aspirin-treated system.

In one study [93], sesaminol triglucoside can be metabolized to enterolignans through the walls of the large intestines such as ST-1, ST-2 and ST-3. Correspondingly, these sesaminol triglucoside metabolites are transformed further into hydroxylate metabolites when absorbed to the intestines and is excreted in urine. ST-1 and ST-2 possess catechol moiety, which considers its antioxidant activity. ST-2 has showed a remarkable inhibitory effect against the LPS stimulated TNF-α and IL-6 production in RAW 264.7 cells. STG blocks the generation of NO, which is induced by LPS and inhibits cytosolic phospholipase A2 (cPLA2), *COX-2* and *iNOS* expression [94].

In one small study of adhesive small bowel obstruction (SBO) [95], a clinical trial wherein sixty-four patients (control: 33 patients; intervention: 31 patients) were administered, is conducted in a span of three hours with 150 mL sesame oil. The results show more effectivity of sesame oil in SBO as compared to the control group. Only a few patients were required to undergo surgery in the intervention group as compared to the control group and the observed duration of stay in the hospital was shorter than the control group.

2.1.6. Cardiovascular Disease

Endothelial dysfunction ignites the start of the chronic inflammatory process called atherosclerosis. This pathogen pathway is enhanced by transforming low-density lipoprotein (LDL) to oxidized low-density lipoprotein (oxLDL) as a consequence of ROS and oxidative stress. Exposure of endothelial cells to oxLDL promotes the expression of proinflammatory cytokines [96].

In the human umbilical vein endothelial cells (HUVECs) model [97], sesamin at a dose of 100 μM has nearly 100% inhibitory effect on the activation of NF-κB induced by oxLDL. Similarly, sesamin also suppresses the discharge of interleukin IL-8, and endothelin ET-1, prompted by oxLDL. The expression of adhesion molecules on the surface has been attenuated by sesamin by half of the initial dose.

Cardiac hypertrophy is the unusual enlargement of heart muscles that leads to the changes in the extracellular matrix of the heart. Renin-angiotensin system (RAS) is crucial in the development of cardiac hypertrophy especially in the left ventricle. MAPK stimulates the hypertrophic response of Angiotensin II (Ang II), the bioactive peptide component of RAS [98–101]. In the study of DOCA/salt-induced left ventricular hypertrophy (LVH) rat model, sesame oil exhibited a down-regulation of p-p38 and p-JNK levels. However, sesame oil had no observed effect on the ERK activation in the LVH rat [102].

2.1.7. Lung Disease

Acute lung injury is caused by lung inflammation as a consequence of endotoxemia [103]. Promotion of pulmonary inflammatory cell sequestration and enhanced production of pro-inflammatory mediators affects the alveolar space resulting to lung dysfunction [104–107]. In the study of systematic endotoxin-induced acute pulmonary injury in rats [108], administered sesamol at 1–3 ppm blocks the inflammatory cells induced by LPS in infiltrating the alveolar space, suppresses the protein leakage and expression of inflammatory cytokines in bronchoalveolar lavage fluid (BALF). The NF-κB activation in alveolar macrophage is inhibited by sesamol. Sesamin at dosages 1 and 3 ppm, moreover, suppresses the generation of nitric oxide related to the alveolar macrophage.

Leukotrienes are mediators of lipid that involves the pathogenesis in asthma. The enzyme 5-lipoxygenase (5-LOX) draw in the production of inflammatory mediators through metabolism of arachidonic acid into leukotriene B_4 (LTB_4) and leukotriene C_4 (LTC_4), potent inflammatory mediators, by the presence of five lipoxygenase activating protein (FLAP). LTC_4 is claimed to possess powerful inflammatory eicosanoid, which enhances vascular permeability [109]. Sesamin has been reported to reduce the levels of LTB_4 more effectively than sesamol. On the contrary, sesamin and sesamol downregulates the serum level of LTC_4. The cumulative effects of the combined sesamin and sesamol reduces the said inflammatory mediators, although its reducible strength is equivalent to that of the individual effects of sesamol and sesamin suggesting that the synergistic interactions are absent. For the most part, sesamin and sesamol exhibits anti-leukotriene effects, which downregulates the receptors and key enzymes of leukotriene pathway and further diminishing the pro-inflammatory leukotrienes production [110].

Non-heme iron-containing enzyme called lipoxygenases (LOX) is characterized by its catalysis activity in incorporating molecular oxygen into polyunsaturated fatty acids [111]. The conversion of arachidonic acid to hydroxyeicosatetraenoic acid (HETE) via metabolism with LOX promotes expression of leukotrienes and regulates the inflammation pathway [112].

In the study of kinetic inhibition [113], sesamol competitively inhibits LOX with IC50 value of 51.84 μM and an inhibitory constant (K_i) of 4.9 μM. The competitive inhibition is happening in either the active substrate site or the active metal ion site. Hence, in the ferric reducing ability power (FRAP) assay of the same study, 55.35 μM of sesamol reduces half of the Fe^{3+}-LOX into Fe^{2+}-LOX indicating the partial interaction of sesamol to the active metal ion site.

Table 1. In vivo models for the anti-inflammatory effects of sesame lignans.

Compound	Inflammatory Disease/Disorder	Rat Model	References
Sesamin	Depression	Chronic unpredictable mild stress (CUMS) rat model	[41]
	Ischemic brain stroke	Middle cerebral artery occlusion (MCAO) rat model	[42]
	Parkinson's disease	6-hydroxydopamine (6-OHDA) rat model	[43]
	Osteoarthritis	Papain-induced osteoarthritis rat model	[57]
	Hepatic steatosis	High-fat diet rat model	[65]
	Fulminant hepatic failure	D-galactosamine (D-GalN)-sensitized rat model	[66]
	Acute hepatic injury	Lead-induced acute hepatic injury rat model	[69]
	Diabetic Retinopathy	Streptozotocin (STZ) induced rat model	[87]
	LPS-induced leukotrienes generation	ad libitum semi-synthetic diet rat model	[110]
Sesamol	Ulcerative colitis	Dinitrochlorobenzene (DNCB)—induced rat model	[88]
	Gastric ulceration	Aspirin-induced gastric mucosal rat model	[92]
	Acute lung injury	Endotoxin-induced acute pulmonary inflammation rat model	[108]
Sesaminol	Alzheimer's disease	senescence-accelerated mouse-prone 8 (SAMP8) model	[50]
Sesame Oil	Nonalcoholic steatohepatitis	Methionine-choline deficient (MCD) diet rat model	[76]
	Hepatic steatosis	High-fat diet-fed rat model	[83]
	Cardiac hypertrophy	Deoxycorticosterone/salt (DOCA/salt)-induced hypertension uninephrectomized rat model	[102]

Table 2. In vitro models for the anti-inflammatory effects of sesame lignans.

Compound	Mechanism of Action	Cell Line	References
Sesamin	Inhibition of LPS-induced TLR4 expression	BV-2 microglial cell	[34]
	Inhibition of LPS-induced IL-6 mRNA and protein	BV-2 microglial cell	[35]
	Inhibition of HO-1 protein ubiquitination	RAW 264.7 murine macrophage cells	[48]
	Activation of Nrf2/ARE	PC12 rat pheochromocytoma cells.	[49]
	Inhibition of IL-1β-stimulated human osteoarthritis chondrocytes.	Primary chondrocytes	[58]
	Inhibition of oxidized low-density lipoprotein (oxLDL)-induced endothelial dysfunction	Human umbilical vein endothelial cells (HUVECs)	[97]
Episesamin and Sesamin metabolites	Activation of Nrf2/ARE	PC12 rat pheochromocytoma cells.	[49]
Sesamin Catechol Glucuronides	Inhibition of LPS-induced NO production	J774.1 mouse macrophage-like cells	[75]
Sesamol	Inhibition of MMPs expression	SW1353 human chondrosarcoma cells	[59]
	Inhibition of inflammatory LOX	Soy LOX-1 enzyme model	[113]
Sesaminol Triglucoside	Inhibition of IL-6 and TNF-α	RAW 264.7 murine macrophage cells	[93]
Sesamolin	Reduce the activation of p38 MAPK	BV-2 microglial cell	[35]

2.1.8. Others

The extent of the effect of other sesame extracts on the inflammatory cytokines and mediators has also been investigated thoroughly. One study involves the aqueous extract of sesame oil (SOAE) [114], which has twenty-eight identified molecules that ranges from moderate to polar in nature. The SOAE-methoxyphenol derivatives (SOAE-8; i.e., vanillyl alcohol, p-hydroxyphenylacetic acid, vanillic acid, coniferyl alcohol, p-coumaric acid, ferulic acid, sinapic acid and syringic acid) are the key components for its anti-inflammatory property. The absence of the sesame lignans in SOAE paves way for the study to be engrossing. Monocyte derived macrophages (MDMs) and RAW 264.7 macrophage cells were used, and the results are positive with slight distinction. SOAE-8 successfully reduced dose-dependently the mRNA levels of the inflammatory markers (*IL-6, IL-1β* and *TNF-α*) in MDMs. Meanwhile, in RAW264.7 cells, only *TNF-α* mRNA level was not reduced dose-dependently by SOAE-8.

The ethanol extract in black sesame seeds (BSSEE) attenuates liver inflammatory response in fructose-induced NAFLD rat model [115]. Three major lignans, which are sesamin (16.33%), sesaminol (1.92%) and sesamolin (13.06%), were found in BSSEE. Three major lignans, which are sesamin (16.33%), sesaminol (1.92%) and sesamolin (13.06%), were found in BSSEE. Inflammatory cytokines are dose-dependently reduced by BSSEE when administered at 0.5–2 ppm. Infiltration of inflammatory cells is also hindered in the presence of BSSEE. Correspondingly, BSSEE with concentrations of 1 ppm and 2 ppm promotes the activation of Nrf2 and improves the levels of MAPKs and NF-κB. In another study of BSSEE, Freund's complete adjuvant (FCA)-induced arthritis rat model was used [116]. The study examined the effect of BSSEE on the inflammatory cytokines (IL-6 and TNF-α). It has been reported that at dosage 800 ppm of BSSEE, the levels of inflammatory cytokines induced by rheumatoid arthritis are reduced in the span of 28 days. Comparatively, extracts of black sesame seeds via CO_2 supercritical fluid extraction (SFE) exhibits neuroprotective activity against ischemia [117]. While its composition is mainly made up of fatty acids (caprylic acid, capric acid, lauric acid, myristic acid, palmitoleic acid, margaric acid, linolenic acid, arachidic acid, behenic acid, palmitic acid, stearic acid, linoleic acid and oleic acid) and phytosterol (cholesterol, brassicasterol, stigmasterol, Δ-5, 24 stigmastadienol, Δ-7 stigmastanol, Δ-7 avenasterol, eritrodiol, campesterol + campestanol + 24 methylene cholesterol, clerosterol + Δ-5, 23 stigmastadienol, Δ-5 avenasterol and β-Sitosterol + sitostanol), its synergistic interactions play a vital role. The infiltration of leukocyte reduces when treated with CO_2 SFE extracts of black sesame seeds. On the contrary, the study suggests further evaluation of its mediating-effect on the neuronal disorders.

The ethanol extract in the sesame coat (EESC) also manifests anti-inflammatory property [118]. It was reported that EESC contains sesamin, sesamolin, phenolic compounds and tetranortriterpenoids [119]. Using RAW 264.7 macrophage cell line, EESC (0.08 ppm) exhibited lowering effect on the levels of NO production as well as PGE_2 production. EESC, at the same dosage, inhibits both the *iNOS* and *COX-2* protein expressions by 94% and 53%. Inhibitory effect on the LPS-induced degradation of IκB-protein has been observed when EESC was applied.

Table 3. In vitro/in vivo applications of different extracts from different sesame components.

Sesame Component	Mode of Extraction/Solvent	In vitro/In vivo	References
Sesame Oil	Aqueous extract	RAW 264.7 macrophage cell line	[114]
Black Sesame Seeds	Ethanol extract	Fructose-induced NAFLD rat model	[115]
Black Sesame Seeds	Ethanol extract	Freund's complete adjuvant (FCA)-induced arthritis rat model	[116]
Black Sesame Seeds	CO_2 supercritical fluid extraction	Endothelin-1-induced focal ischemia rat model	[117]
Sesame Coat	Ethanol extract	RAW 264.7 macrophage cell line	[118]

2.2. Anti-Cancer Activity

During inflammation, the immune system of the body is tasked to the release of reactive oxygen and nitrogen species (RONS) to fight against pathogens and to protect the body. RONS are also responsible for tissue repair and regeneration [120]. However, these chemicals are able to obstruct DNA repair mechanisms, which can potentially lead to DNA damage. With damaged DNA, the chance of mutations significantly rises, promoting tumorigenesis. It was reported that chronic inflammation is part of the 15% precedencies of recorded cancer cases [121]. Millions of cancer cases are recorded every year and millions of people succumb to different types of cancer. In 2018, 9.6 million deaths in the world were caused by cancer [122]. Due to this, efforts in finding a cure for and attempts to understand cancer continue to rise [123]. Cancer is commonly defined as the abnormal growth of the cells, but there are indications that mark the development of cancer in the human body. It was suggested that six essential alterations in cell physiology are mainly responsible for the uncontrolled growth of the cells. Self-sufficient in growth signals, insensitive to growth-inhibitory signals, evasive of apoptosis, unlimited replication potential, maintenance of angiogenesis and invasive to tissues are the six characteristics of cells during tumor development [124,125]. However, in a study more than a decade later after these indications were identified, two other hallmarks of cancer have been discovered. It was reported that tumor cells are also able to reprogram energy metabolism and are evasive of immune destruction, with genome instability and inflammation as the primary causes [126]. Knowing these hallmarks allow the creation of defensive strategies against cancer and also understanding the fundamentals of these indications opens the opportunity for a better comprehension as to how neoplastic diseases emerge.

2.2.1. Lung Cancer

Oil extracted from sesame stood out among other vegetable oils because of its exceptional nutritional characteristics [2]. Sesame lignans have demonstrated several pharmacological applications [127], including anti-proliferative activity, which made them subjects of anticancer studies. Lung cancer, which is the most diagnosed cancer in 2018 [122], has been studied by Harikumar and colleagues by treating human lung adenocarcinoma cell line H1299 with sesamin solution. Sesamin was able to significantly inhibit the proliferation of H1299 cells, with a 50% inhibitory dose of 40.1 μmol/L [128]. Sesaminol has also been proven to exhibit anticancer property against lung carcinoma A549 cells at a concentration of 50 μM and 6 h of treatment [129]. The mechanism of action of sesaminol on A549 cells is tackled in the following section (2.2.2 Breast Cancer).

Another lignan of sesame, sesamol, has also been explored for its anticancer properties [130]. It has been studied for its apoptotic effect in lung adenocarcinoma SK-LU-1 cell line. After 48 h of treatment, the lignan showed selective antiproliferation effect with an IC_{50} of 2.7 mM on SK-LU-1 cells and 7.6 mM on Vero cells. The apoptotic effect of sesamol was also found to be both time and dose-dependent. The higher the concentration, the higher number of cell deaths was recorded. At 2.7 mM sesamol, evident necrosis occurred in a time-dependent manner, while at 5.4 mM sesamol, early stage apoptosis was affected [131,132]. A 24-h sesamol treatment of SK-LU-1 cells showed an increase in the activities of caspase 3/7, which largely participate in the propagation of death signals. Specifically, the increase and activation of caspase 3 causes the cleavage of a DNA repair protein known as PARP. Consequently, DNA damage occurs and cell deaths are achieved. The loss of mitochondrial membrane potential ($\Delta\psi m$) was also investigated as one of the factors of apoptosis and a 48-h treatment revealed that longer exposure to sesamol leads to a greater decrease in $\Delta\psi m$. This proposes that mitochondria are part of the apoptotic pathway that occurs in SK-LU-1 cells [131,132].

Another study revealed that sesamin also has a protective action against the down-regulation of the PI3K-Akt signaling pathway in a nickel-induced apoptosis in mouse liver. PI3K-Akt pathway is responsible for the restriction of apoptosis and for the promotion of cell survival [133,134]. This action of sesamin showed that the lignan is capable of preventing DNA damage, hence, it becomes a potential anticancer agent. In line with this, it was also reported that the ability of sesamin to reduce

COX-2 gene expressions in A549 cell line occurs through PI3K-Akt pathway [135] and inhibition of this pathway inactivates inflammatory response, which reduces restenosis [136].

2.2.2. Breast Cancer

Another common type of cancer is the female breast cancer, which accounts for 11.6% of the total cancer deaths [122]. For that reason, breast cancer is also widely studied. In 2007, sesamin was examined for its effect on the proliferation of human breast cancer cell line MCF-7 for 24 h. Results showed that the inhibition is done through G1 phase growth arrest and is dose-dependent, with a cytostatic effect at 100 μM sesamin. The lignan was also used to assess the effect of down-regulation of cyclin D1 protein expression in different types of human tumor cells including the breast cancer cell line T-47D, the lung cancer cell line A549, the transformed renal cell line 293T, the immortalized keratinocyte cell line HaCaT, the melanoma cell line UACC-62 and the osteosarcoma cell line MG63, which proved that suppression of sesamin generally occurs in the tumor cells. In line with this, the down-regulation of cyclin D1 protein expression was examined as one of the main factors that cause the growth inhibitory effect of sesamin against cyclin D1-depleted MCF-7 breast cancer cells. Results showed that the inhibitory effect of sesamin is largely dependent on the presence of cyclin D1 as cyclin D1-depleted MCF-7 cells were almost insensitive to the sesamin treatment [137]. In the study of Harikumar, the same antiproliferation activity of sesamin was also observed against breast cancer cells MDA-MB-231 with an IC_{50} value of 51.1 μmol/L [128].

The anticancer function of sesamin also manifests even with a different mechanism of action. Sesamin was also able to impede the proangiogenic activity of MCF-7 cells. A 24-h pretreatment of MCF-7 cells and macrophages were carried out with 50 μM of sesamin, which resulted to the suppression of angiogenesis upon observation of the endothelial capillary tube assay and the network formation of the cells. Observation done on MCF-7 cells alone, however, revealed that the sesamin treatment was not significantly cytotoxic to the breast cancer cells as it did not decrease the cells' viability even at sesamin concentration of 100 μM within a 72-h treatment period. MϕCM-treated MCF-7 cells were also examined for the same activity and sesamin showed a more promising inhibition performance than when MCF-7 cells were treated with the control medium. The same suppressive effect of sesamin was also exhibited against a more malignant cell line, MDA-MB-231 cells, treated with MϕCM. To properly investigate the mechanism of action of sesamin on the tumor cells, vascular endothelial growth factor (VEGF) and matrix metallopeptidase 9 (MMP-9), which are essential factors of angiogenesis, were evaluated. MϕCM treatment was found to induce *VEGF* and *MMP-9* mRNA expression in MCF-7 cells, but treatment with sesamin drastically hindered this. The same results were obtained for MDA-MB-231 cells [138].

A combined treatment with γ-tocotrienol also prevents the proliferation of MCF-7 and MDA-MB-231 cells. It was suspected that sesamin could increase the antiproliferative activity of γ-tocotrienol by inhibiting its metabolic degradation but the study revealed that the synergistic inhibiting effect of the two compounds is the result of the induction of G1 cell cycle arrest and the reduction of protein expression levels involved in the cell cycle. The synergistic inhibitory performance of sesamin and γ-tocotrienol is also effective against the murine +SA mammary epithelial cell line of a mouse [139]. An in vivo study on the proliferation of MCF-7 cells in athymic mice presented a comparison of the apoptotic activity of sesamin and of a lignan of flaxseed, secoisolariciresinol diglucoside (SDG), in which sesamin showed a more promising activity [140]. Tumor cell proliferation declined by 38% with sesamin treatment and cell apoptosis rose by 91% as opposed to the 37% reduction of proliferation with SDG treatment without exhibition apoptotic property. Their ability to down-regulate growth factor receptors known as EGFR and HER2 explains the reduced tumor growth and sesamin's ability to reduce pMAPK expression causes it to be more effective than SDG.

Just like with sesamin, sesamol was also used against the human breast cancer MCF-7 cells. Endothelial cell line EA.hy 926 was used to observe morphological changes that are caused by the exposure to sesamol. Endothelial cells were treated with sesamol for 72 h with a concentration range of

0.05–1.0 mM. It was found that lower doses of sesamol caused the cells to swell, while a concentration of 1 mM caused apparent cell death. Fragmented nuclei were present in the cells that were treated with 1 mM sesamol, indicating apoptosis. To check whether sesamol is capable of inhibiting the growth of MCF-7 cells, the cells were subjected to a 3-day treatment at 0.10 mM. This revealed that cell numbers are indeed lower than the controls. Combining PI labeling with a TUNEL assay led to the information that cell deaths occur in both S and G_2/M phases [141]. Whether the inhibitory effect of sesamol on MCF-7 cells was solely caused by the apoptosis was disclosed in the study. No other pathway of inhibition was presented.

Aside from this, sesaminol has also been utilized against breast cancer cell lines MDA-MB-231 and MCF-7, along with lung carcinoma cell line A549 stated in the previous section. As mentioned, restriction of cyclin D1 expression plays a major role in some of the inhibitory pathways induced by sesame lignans. It was reported that the same statement applies to sesaminol, which was verified when a 6-h treatment at 50 μM sesaminol reduced cyclin D1 expression in MCF-7 cells. Although the same conditions did not produce the same result in MDA-MB-231 cells, it was stated that 100 μM sesaminol with 24 h to 48 h of sesaminol treatment achieved the desired result. To investigate how sesaminol manifests the same activity, sesaminol-immobilized FG beads were incubated with MCF-7 cells to identify, which proteins would bind to the beads. Among the proteins was an inner mitochondrial membrane protein known as adenine nucleotide translocase 2 or ANT2. This protein is said to be overexpressed in different malignant tumors and is recognized as an oncoprotein. To confirm whether ANT2 is involved in the anticancer activity of sesaminol, a knockdown of ANT2 in MCF-7 cells was performed. This resulted to growth inhibition and accumulation of cells in the G1 phase [129].

2.2.3. Prostate Cancer

Following breast cancer is the prostate cancer, which is responsible for 7.1% of the total cancer deaths. Similar with the findings on γ-tocotrienol, γ-tocopherol was also found to have antiproliferation effect on the human prostate cancer cell line PC-3 [142] and that intake of sesamin, sesamolin or sesame seed oil in general [143] has a significant impact on the plasma levels of tocopherols in humans [144]. The presence of γ-tocopherol becomes more evident with sesamin intake and at low levels of α-tocopherol [145]. It was reported that sesamin prevents the tocopherol from being metabolized by inhibiting CYP3A-dependent carboxychroman production to which tocopherol is metabolized to [146]. The same lignan was also found effective against the solid tumor cell line DU145 and the 50% inhibitory sesamin dose was 60.2 μmol/L [128].

2.2.4. Colon Cancer

Colorectal cancer, which is next to prostate cancer as one of the leading causes of cancer deaths [122], was investigated to determine whether sesame lignans can act against its cell lines. Sesamin was capable of suppressing the growth of colon cancer cells HCT116 with an IC_{50} value of 57.2 μmol/L [128] and in the study of Watanabe, colon adenocarcinoma RKO cells were also used to observe the effectivity of sesaminol against the proliferation of these cancer cells and the same results were observed with 6-h treatment of 50 μM sesaminol [129]. Five compounds found in sesame seeds, including sesamol, were tested for their ability to suppress the transcriptional activity of COX-2 because its excessive production of prostaglandin is an essential factor of colorectal cancer. The human colon cancer cell line DLD-1 was used to measure the activity of COX-2 up to 100 μM of the test compounds. At 100 μM, the inhibiting effect of sesamol was 50%, while ferulic acid, sesamin, sesamolin and syringic acid did not manifest any favorable results. Due to this, sesamol was further examined in terms of its capability to hinder intestinal polyp formation in Min mice. The effects of administering 500 ppm sesamol to Min mice for eight weeks were observed. It was stated that sesamol administration did not affect the mice in any aspect abnormally and that it successfully lessened the number of polyps in the small intestine and colon by 75% of the untreated group [147].

Human colorectal carcinoma cell line HCT116 was also evaluated as the target of sesamol, to which it caused cell death with an IC_{50} value of 2.59 mM. Concentration range 0.5–5 mM showed exceptional inhibition of the survival of HCT116, but not on the viability of the normal Vero cells. The cell cycle arrest ability of sesamol was then compared to cisplatin with an exposure of 48 h. Sesamol was able to induce G_0/G_1 cell cycle arrest at a low concentration (0.05 mM) as opposed to 100 μM of cisplatin. At both low and high concentrations, sesamol was also able to arrest the S phase, with the highest cell cycle arrest peaking at 1 mM [148]. Treatment beyond this concentration revealed a relationship with the S phase arrest that is inversely proportional. Sesamol increases the cells in the S phase and decreases the number in the G_0/G_1 phase. Analyzing this led to the understanding that subG1 phase is also suppressed by both sesamol and cisplatin, leading to DNA fragmentation and cell death. Data gathered translate to the fact that the mitochondria are indeed a factor in the apoptotic pathway induced by sesamol as it enhanced $\Delta\psi m$. Colon cancer induced by 1,2-dimethylhydrazine (DMH) in Wistar rats was observed to whole sesame paste (WSP) and resistant starch type 2 (RS2) as anticancer agents. It was reported that both WSP and RS2 have restrictive actions against the initiation of DMH-induced colorectal cancer and they are capable of reducing the number of mucin depleted foci [149]. This result is similar to the study on azoxymethane-induced colon carcinogenesis, which proved that sesame could act against the said cancer [150].

2.2.5. Liver Cancer

The induction of cell cycle arrest was also tested on the human hepatocellular carcinoma cell line HepG2. The MTT assay was used to observe the viability of HepG2 cells under the influence of sesamin and the data indicated that the cells were inhibited after 48 h with an IC_{50} value of 98 μM, but sesamin was less cytotoxic to L02 cells. Unlike with the growth inhibition of the breast cancer cells, the antiproliferation activity on HepG2 cell was caused by the suppression of the STAT3 signaling pathway, which controls genes that participate both in cell cycle and apoptosis. This induces G2/M phase arrest and a dose-dependent early apoptosis resulting to reduced proliferation [151]. Parallel observations were made when sesamol was used against HepG2 cells, in which the antiproliferative activity of 1 mM sesamol was over 90%. Just like with other inhibitory effects of sesame lignans, this was also reported as concentration-dependent. To understand which cell death patterns occur in HepG2 cells, the deaths were observed at different concentrations. Chromosomal DNA fragmentation was noticeable at as low as 50 μM sesamol. At this concentration, characteristics of apoptosis such as nuclear shrinkage and membrane blebbing were recorded. At higher doses like 200 and 1000 μM, necrosis was observed [152,153].

The location of sesamol in cells was identified to further analyze its apoptotic effect. It was shown that sesamol undergoes nuclear localization in HepG2 cells and this phenomenon is related to sesamol's cytotoxicity because this means that sesamol can travel to and accumulate in the nuclei of cancer cells. The transportation of sesamol into the nucleus, however, was not discussed in the study. In a more recent study of the same group of researchers, it was revealed that the intrinsic apoptosis pathway was achieved by the reduction of the Bcl-2 expression, although there were no changes observed on the expression of Bax. Sesamol also enhanced cytochrome *c* release, which activates caspase-3. As discussed previously, caspase-3 activation leads to the cleavage of PARP, which is responsible for DNA repair. In contrast, the extrinsic pathway happened through the amplified protein expression of Fas/FasL and through the activation of tBid and caspase-8, caused by sesamol. Further experimentations also showed that sesamol suppressed both autophagy and mitophagy in the liver cancer cells via reduced LC3 expression, an indicator of autophagy, and via triggering the loss of $\Delta\psi m$ [152,153]. Phenolic extract from black and white sesame seeds have also been used to study the same activity and have been proven to exhibit significant antiproliferative property [154].

2.2.6. Cervical Cancer

Cervical cancer, ranked fourth as the leading cause of cancer death among females [125], was studied by observing the effect of sesamin on the viability and the migration of the HeLa human cervical cancer cell line. The CCK-8 assay was used to determine the cells' viability, while scratch wound assays were used for the migration test. Similar with other inhibitory actions of sesamin, its effect on the HeLa cells was also found to be dose-dependent. The apoptosis rate of HeLa cells also increased with 50 μM sesamin treatment within 48 h as compared to the HeLa cells without sesamin treatment and this occurrence was caused by the increased ratio of Bax, a pro-apoptotic protein, to Bcl-2, an anti-apoptotic protein [155]. It was also revealed that sesamin treatment increased the injured endoplasmic reticulum leading to programmed cell deaths. There was also up-regulation of the levels of p-IRE1α and p-JNK in HeLa cells, which were reported to be the pathway responsible for the ER-stress mediated apoptosis.

2.2.7. Blood Cancer

Sesamin can also suppress the proliferation of human leukemic cell lines, KBM-5 and K562, and of a multiple myeloma cell line, U266. The IC_{50} values of sesamin for these cells are 42.7, 48.3 and 51.7 μmol/L, respectively. While investigating the biological pathway of sesamin in these cells, it was discovered that pretreatment with the lignan allows the restriction of cyclooxygenase-2 (COX-2) and cyclin D1 expressions induced by tumor necrosis factor (TNF), which is a cell signaling protein. These expressions are known to play essential roles in the propagation of cancer. Aside from this, TNF can also induce expression of gene products involved in angiogenesis and sesamin exhibited the same inhibitory action against these expressions. It was revealed that sesamin is capable of hindering the growth of the cells and its inhibition ability depends on both its concentration and the duration of the treatment. The inhibitory action of sesamin was made possible with the induction of TNF to NF-κB activation, which is responsible for the involved cellular responses. Sesamin was able to augment the apoptotic activity of TNF by downregulating the expression of gene products [131].

Aside from the study of Harikumar on leukemic cells using sesamin treatment, study utilizing sesamol against blood cancer cells have also been carried out. Interaction of sesamol with human lymphoid leukemia Molt 4B cells resulted to growth inhibition and induced apoptosis in a concentration-dependent manner. Morphological change indicating apoptosis and DNA fragmentation were observed in the cancer cells and the fragments of DNA increased the longer the contact time. These changes were not observed in sesamol-treated normal lymphocytes, leading to the conclusion that sesamol could induce cell death to restrict the growth of the leukemic cells [156]. The study did not present a detailed mechanism of action of the DNA fragmentation nor the apoptotic pathway.

A study in 2010, both in vivo and in vitro, presented the cytotoxic activity of two oxidation products of sesamol, a trimer and a tetramer. Sesamol and its oxidation products were used to treat rat thymocytes to know how the compounds will change the lethality of the cells. $FeCl_3$ was reacted with sesamol to undergo oxidation and yield the trimer and the tetramer. A 24-h incubation with 30 μM sesamol did not affect the population of the cells exerting propidium fluorescence, which is used as an indicator for dead cells. On the other hand, trimer at the same concentration resulted to a slight growth in population, while tetramer amplified it. The biological pathway of the apoptotic effect of tetramer was similar to that of sesamol against the SK-LU-1 cell line. Tetramer managed to elevate the activity of caspases, which in turn increased DNA damage. When tetramer was tested on K562 cells as an antiproliferative agent, the results showed that K562 cells were inhibited depending on the concentration of tetramer, which ranged from 3 to 30 μM. Only the concentration above 10 μM was able to manifest significant inhibitory performance. It was also discovered that at 30 μM, tetramer already exhibits a significant increase in lethality. Unfortunately, at these specific concentrations, the cytotoxicity of tetramer on normal cells was found to be greater than the cancer cells, which makes it difficult to consider the compound as a possible anticancer agent [157]. To be able to utilize tetramer against K562 cells, a mechanism of action that protects the normal cells from its cytotoxic action must first be developed. Acute myeloid leukemia cells, HL-60 and Molt-4, were also examined and only HL-60 cells suffered from DNA fragmentation due to exposure to sesamol [158].

The idea of sesaminol acting against human lymphoid leukemia Molt 4B cell line was also investigated by the same research team that explored sesamol's effect on Molt 4B cells and parallel data were acquired as with the study on sesamol. The inhibition was also concentration-dependent and morphological changes indicating apoptosis were also reported. Apoptotic bodies were observed after three days of treatment with 45 μM sesaminol and it was noted that the growth inhibition of sesaminol is better than other sesame lignans. Specific mechanism of action of sesaminol on its induced apoptosis in Molt 4B cells was also not presented in this study [159,160].

On the other hand, along with sesamin, sesamolin was studied for its inhibitory effect against Burkitt's lymphoma cells, Raji. The study aimed to utilize the sesame lignans to improve NK cell lysis activity so identification of a cancer cell line that has low cytotoxicity against NK cells was first carried out. Human leukemia cell line K562, T cell leukemia cell line Jurkat, and human Burkitt's lymphoma cell line Raji were the three tumor cell lines used with LHD assay. Both K562 and Jurkat cells were highly sensitive to the cytolysis activity of the NK cells in contrast with its effect on Raji cells. Due to this, Raji cell line was used as the subject of study. It was shown that sesamolin decreased the cells' viability by 31% compared to the untreated cells at a concentration of 80 μg/mL. Meanwhile, sesamin showed an even greater cytotoxic activity with concentrations higher than 2.5 μg/mL. Since the goal of the study was to use the lignans as boosting agents, concentrations were set at a level that is not toxic to Raji cells. Sesamolin-treated Raji cells were proven to be more sensitive to NK cell lysis than the untreated ones, confirming that sesamol enhanced the lysis activity of the NK cells. On the other hand, no significant results were observed with the tests involving sesamin. Since NKG2D ligands are deemed as one of the key role players in the activity of NK cells against cancer cells, it was assessed whether sesamolin treatment has an impact on its expression. Below toxic concentration of sesamolin, it was confirmed that the expression indeed increased. The same amplification of expression was observed with other key role players namely ULBP-1, ULPB-2 and MIC-A/B. The latter part of the study reported that sesamolin increased the phosphorylation of the ERK pathway, which is one of the pathways responsible for NKG2D ligand expression in Raji cells [161]. Although sesamin did not affect the NK cell lysis activity, it is important to note that it exhibits a much greater cytotoxicity to Raji cells even at low concentrations. This presents the possibility of sesame acting against the growth of Raji cells in a different pathway.

2.2.8. Skin Cancer

The antimelanogenesis property of sesamin, was evaluated in comparison to recognized depigmenting agents, kojic acid and β-arbutin. Its ability to function as a sunscreen was investigated through the measurement of UV absorption. Its effects against tyrosinase activity through mushroom and cellular tyrosinase were also observed. The results showed that sesamin was able to absorb ultraviolet in the UVB range with an absorbance intensity of approximately 1.3 at 290 nm, as opposed to the absorbances of kojic acid and β-arbutin equal to 0.2 at 300 nm. Sesamin also exhibited a slight inhibitory effect against tyrosinase. In the same study above, similar antimelanogenesis findings were observed for sesamol. The lignan was also able to absorb ultraviolet in the UVB range with an absorbance intensity of 0.8 at 300 nm, four folds higher than that of kojic acid and β-arbutin. Its inhibition of mushroom tyrosinase activity in vitro was shown to be concentration-dependent just like with kojic acid. The IC_{50} values for sesamol and kojic acids were 0.33 μg/mL and 6.15 μg/mL, respectively. On the other hand, β-arbutin did not exhibit an inhibitory activity, leading to the conclusion that sesamol has the greatest effect against tyrosinase enzyme [162].

To further assess the ability of sesamol, its cytotoxicity on Vero and SK-MEL2 cell lines were examined. Cytotoxicity of sesamol to Vero cell line was 22.8% after 48 h of treatment, while at a concentration range of 600–800 μg/mL, its cytotoxicity to SK-MEL2 cell line was between 42–97%, with an IC_{50} value of 608.93 μg/mL. Sesamolin, another lignan, was also tested for its antimelanogenesis activity. Similar with the results in the inhibition of mushroom tyrosinase activity, the sesame lignans and kojic acid obstructed the activity of cellular tyrosinase, while β-arbutin was negative. In line with this, sesamolin was also found to manifest the highest inhibition equal to 50% compared to the other two sesame compounds with only 23% of inhibition. Inhibition of melanin pigment was also investigated among the test compounds and sesamolin showed the most favorable inhibition performance at 25 μg/mL. Other proteins involved in melanogenesis, TRP-1 and TRP-2, were also exposed to the sesame compounds and their levels in SK-MEL2 cells were analyzed using Western blot analysis. It was illustrated that sesamolin lowered TRP-1 and TRP-2 protein levels by 36% and 15%, respectively. Sesame oil was also tested against SK-MEL cells [162], while sesaminol was tested against the SK-MEL-28 cell line, which was found to reduce cyclin D1 expression at a concentration of 100 μM within one to two days of incubation period [132].

Studies have shown that essential polyunsaturated fatty acids such as linoleic acid are capable of exhibiting antiproliferative activity against malignant cell lines. It was also reported that sesame oil is made up of 96% triglycerides and around 90% of its esterified fatty acids are oleic and linoleic acids in an approximately equal proportion. In line with this, sesame oil was proven to manifest greater growth inhibitory effect against melanoma cells than compared to its effect on normal melanocytes. An incubation period of five days showed that the average growth rate of melanoma cells and normal cells were 2.6 and 2.2, respectively [163]. Further investigation on the pathway of inhibition, however, was not executed. In a similar study, the effect of sesamin, sesamol and sesamolin on the synthesis of melanin in mouse melanoma B16F10 cells was observed and only sesamol showed a significant inhibitory effect, which was approximately 63% of the synthesis at 100 μg/mL. The production of melanin was halted by sesamol by inhibiting the specific activities of mushroom tyrosinase, monophenolase and diphenolase. At the same concentration, sesamol was also able to disrupt the viability of the cells by 60% as it was found to exhibit an apoptotic effect [164].

The chemopreventive capability of sesamol, along with sesame oil and two other products, against two-stage skin carcinogenesis of mice was also inspected in vivo. The two-stage skin carcinogenesis initiated by 7,12-dimethylbenz(a)anthracene (DMBA) and promoted by the tumor promoter 12-*O*-tetradecanoylphorbol-13-acetate (TPA) was exposed to resveratrol, sesamol, sesame oil and sunflower oil. Ten weeks of treatment and prior to TPA promotion, the mice showed gross tumor incidence of 20%, 20%, 30% and 20%, respectively, against the 100% control group. After 20 weeks of treatment, only resveratrol and sesamol manifested more than 30% and 10% inhibitory potential, respectively. The ability of the compounds to cancel TPA tumor promotion and impede tumor latency further demonstrated their chemopreventive effects [165]. In-depth mechanisms of action of the products involved against the proliferation of skin papillomas in the mice have not been provided. To test sesamol's impact on the process of neoplastic development, an ex vivo research on the permeation of sesamol to the LACA mice skin was assessed by using sesamol solution and cream base sesamol-loaded solid lipid nanoparticles (S-SLN), while an in vivo method utilized sesamol solution, sesamol ointment and S-SLN. Ex vivo skin permeation of sesamol as a free drug solution was recorded to be much higher than that of S-SLN, which has three times higher skin retention and a 40% drop on the flux. This proposes that SLN is a potential transport service for sesamol. This is consistent with the results of the in vivo study wherein the number of papillomas on the dorsal skin of the mice was checked and there was 0% incidence of skin tumors [158].

2.2.9. Others

Attempts to apply sesamin against pancreatic cancer and skin cancer were also executed. In 1994, pancreatic carcinogenesis initiated with *N*-nitrosobis (2-oxoproyl) amine (BOP) was observed in vivo by controlled diet of Syrian golden hamsters. The effect of sesamin-supplemented diet, a strategy to lower cholesterol levels, on the progress of the pancreatic cancer was evaluated. However, within a four-month period of treatment, it was concluded that although sesamin successfully lowered the cholesterol contents, this did not have any significant effect on BOP-initiated pancreatic cancer in hamsters [166,167]. Even so, the study emphasizes that the period of observation was done on a short term, which basis and suggests that favorable results possibly occur in a time-dependent manner. In line with this, a more recent study reported that sesamin was able to suppress the growth of pancreatic cancer MiaPaCa-2 cells with an IC_{50} value of 58.3 μmol/L [129].

The antitumor effect of sesamol was also determined using MA-10 cells, a mouse Leydig tumor cell line. Morphological changes caused by sesamol were observed and cells without sesamol treatment showed signs of normal cell growth phenomena, while those treated with sesamol appeared differently, depending on the duration of the treatment. Plasma membrane blebbing was seen after 12 h of treatment. These findings suggest the apoptotic property of sesamol. To confirm this, MA-10 cells viability against sesamol treatment was assessed. It was then revealed that sesamol is capable of inducing the death of MA-10 cells and that the dose of sesamol and the time of treatment are significant factors that affect the performance of sesamol. Flow cytometry showed DNA fragmentation at the subG1 phase of the cell cycle and verified the apoptosis of MA-10 cells. Similar with the established pathways of other studies, the activation of caspase-3 expression was also observed, consequently inducing apoptosis [168].

The following table (Table 4) provides a summary of the anticancer activities of the lignans of sesame, including their mechanisms of action against each cancer cell line discussed above:

Table 4. Sesame lignans against different types of cancer and the mechanisms of action.

Compound	Cancer Type	Cell Line	Mechanism of Action	Reference
Sesame Oil	Skin Cancer			
	human malignant melanoma	SK-MEL	-	[163]
	two-stage mouse skin carcinogenesis	in vivo	protection against TPA tumor promotion	[165]
Sesame Extract	Liver Cancer			
	human hepatocellular carcinoma	HepG2	-	[154]
Sesamin	Blood Cancer			
	chronic myeloid leukemia	K562 KBM-5	inhibition of TNF-induced NF-κB activation	[128]
	myeloma	U266	inhibition of TNF-induced NF-κB activation	[128]
	Breast Cancer	MCF-7	induction of G1 cell cycle arrest; down-regulation of Cyclin D1 protein	[137]
		MCF-7 MDA-MB-231	inhibition of macrophage-induced VEGF and MMP-9 mRNA expressions	[138]
		MDA-MB-231	inhibition of TNF-induced NF-κB activation	[128]
		MCF-7 MDA-MB-231	induction of G1 cell cycle arrest and reduction of protein expression levels	[139]
		MCF-7	down regulation of growth factor receptors EGFR, HER2, and pMAPK expression	[140]
	Cervical Cancer	HeLa	favored apoptosis through the increase of Bax/Bcl-2 ratio; ER stress-mediated apoptosis by IRE1α/JNK pathway	[155]
	Colon Cancer			
	colon carcinoma	HCT116	inhibition of TNF-induced NF-κB activation	[128]
	Liver Cancer			
	human hepatocellular carcinoma	HepG2	suppression of the STAT3 signaling pathway	[151]
	Lung Cancer			
	human lung adenocarcinoma	H1299	inhibition of TNF-induced NF-κB activation	[128]
	Pancreatic cancer	MiaPaCa-2	inhibition of TNF-induced NF-κB activation	[128]
	Prostate Cancer	PC-3	degradation of γ-tocopherol metabolism	[146]
		DU145	inhibition of TNF-induced NF-κB activation	[128]
	Skin Cancer			
	malignant melanoma	SK-MEL2	absorption of ultraviolet in the UV range and inhibition of mushroom and cellular tyrosinase	[162]

Table 4. *Cont.*

Compound	Cancer Type	Cell Line	Mechanism of Action	Reference
Sesaminol	Blood Cancer			
	lymphoid leukemia	Molt 4B	DNA fragmentation leading to apoptosis	[159,160]
	Breast Cancer	MCF-7 MDA-MB-231	Reduction of cyclin D1 expression by binding to ANT2 protein	[129]
	Colon Cancer			
	colon carcinoma	RKO	Reduction of cyclin D1 expression by binding to ANT2 protein	[129]
	Lung Cancer			
	lung adenocarcinoma	A549	Reduction of cyclin D1 expression by binding to ANT2 protein	[129]
	Skin Cancer			
	melanoma	SK-MEL-28	Reduction of cyclin D1 expression by binding to ANT2 protein	[129]
Sesamol	Blood Cancer			
	acute myeloid leukemia	HL-60	DNA fragmentation leading to apoptosis	[158]
	chronic myeloid leukemia	K562	oxidation to tetramer; increased caspases activity leading to DNA damage	[157]
	lymphoid leukemia	Molt 4B	DNA fragmentation leading to apoptosis	[156]
	Breast Cancer	MCF-7	growth inhibition and apoptosis in S and G_2/M phases	[141]
	Colon Cancer			
	colon adenocarcinoma	DLD-1	suppression of cyclooxygenase-2 transcriptional activity	[147]
	colon carcinoma	HCT116	subG1 phase cell cycle arrest causing cell death	[148]
	Liver Cancer			
	human hepatocellular carcinoma	HepG2	induced apoptosis and necrosis via DNA fragmentation and induced apoptosis via suppression of autophagy	[152,153]
	Lung Cancer			
	lung adenocarcinoma	SK-LU-1	increased activity of caspase 3 leading to DNA damage	[131,132]
	Skin Cancer			
	human malignant melanoma	SK-MEL2	absorption of ultraviolet in the UV range and inhibition of mushroom and cellular tyrosinase	[162]
	mouse melanoma	B16F10	inhibition of monophenolase and diphenolase activities and promotion of apoptosis	[164]
	two-stage mouse skin carcinogenesis	in vivo	protection against TPA tumor promotion	[165]
	Others			
	Mouse Leydig tumor	MA-10	increased activity of caspase 3 leading to DNA damage at subG1 phase	[168]
Sesamolin	Blood Cancer			
	Burkitt's lymphoma	Raji	enhancement of NK cell lysis activity via escalated NKG2D ligand expression	[161]
	Prostate Cancer	PC-3	degradation of γ-tocopherol metabolism	[146]
	Skin Cancer			
	human malignant melanoma	SK-MEL2	absorption of ultraviolet in the UV range and inhibition of mushroom and cellular tyrosinase	[162]

3. Conclusions

The bioactive components of *Sesamum indicum* L., such as sesamin, sesaminol, sesamol and sesamolin, play essential roles in combating different types of biological and pharmacological concerns and are able to exhibit promising medicinal properties against the diseases. One of the notable properties of sesame lignans is anti-inflammation. Inflammation is a defense mechanism of a body against foreign substances and chronic inflammation persist to different kind of diseases. Sesame lignans hinder the propagation of inflammatory cytokines and inflammatory mediators, which further leads to alleviating inflammatory-related diseases such as osteoarthritis, cardiovascular disease, neurodegenerative disease, inflammatory bowel disease, diabetic eye disease, lung disease and liver disease. Other sesame extracts apart from the sesame lignans also exhibit mitigation of inflammatory-related pathways. These are proven by the in vivo and in vitro models of inflammatory-induced systems. Another alarming complication that can be instigated by inflammation is cancer. Cancer, which is the abnormal growth of the cells, may develop when the body suffers from DNA damage caused by the chemicals released during inflammation. Interestingly, it was also revealed that the lignans of sesame manifest anticancer activities against different cancer cell lines with different mechanisms of action. Both in vitro and in vivo studies have presented that the lignans are capable of inhibiting the growth of the cancer cells by down-regulating protein expressions, by suppressing the production of gene products, and through cell cycle arrest. Consequently, the lignans also induce either necrosis or apoptosis to the cells, inflicting an antiproliferation effect. Sesame lignans have been proven to manifest these anticancer effects against the tumor cells of lung cancer, breast cancer, colon cancer, prostate cancer, cervical cancer, blood cancer, skin cancer and even pancreatic cancer. With all things considered, sesame proves that food can indeed become a medicine and that foods do not only possess nutritional value, but they also have medicinal worth. The lignans of sesame that manifest anti-inflammatory and anticancer properties and the specific diseases that they act against are summarized in Figure 2 below:

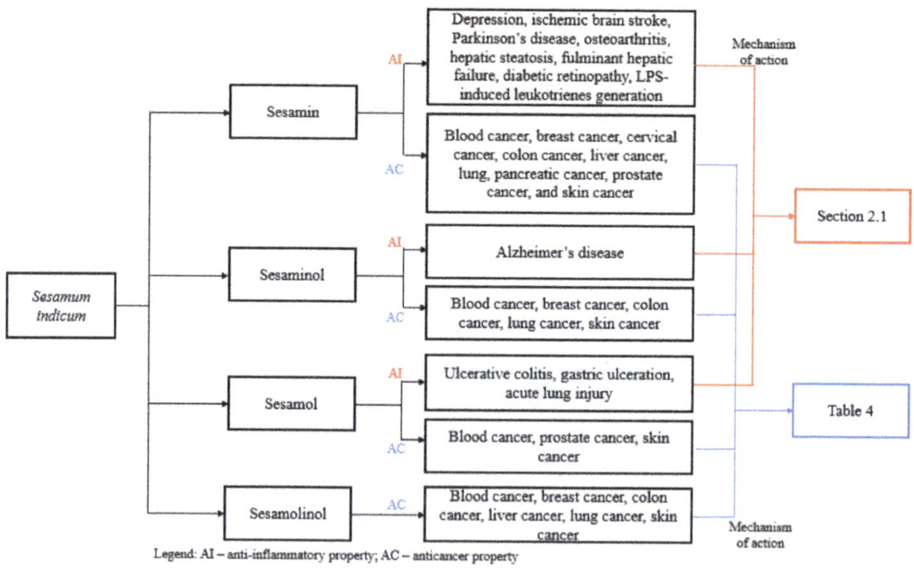

Figure 2. Anti-inflammatory and anti-cancer activities of major sesame lignans.

Author Contributions: M.-S.W., L.B.B.A., M.Y.U.B.: developed the concept for the review, wrote manuscript and original draft preparation; K.A.D.C.-C., C.-L.H, L.-L.Y.: contributed literature search and data curation; P.-W.T.: supervision and final revision.

Funding: This research received no external funding.

Conflicts of Interest: The authors declare no conflict of interest.

References

1. Grougnet, R.; Magiatis, P.; Laborie, H.; Lazarou, D.; Papadopoulos, A.; Skaltsounis, A.-L. Sesamolinol Glucoside, Disaminyl Ether, and Other Lignans from Sesame Seeds. *J. Agric. Food Chem.* **2012**, *60*, 108–111. [CrossRef] [PubMed]
2. Nagendra Prasad, M.N.; Sanjay, K.R.; Prasad, D.S.; Vijay, N.; Kothari, R.; Nanjunda Swamy, S. A Review on Nutritional and Nutraceutical Properties of Sesame. *J. Nutr. Food Sci.* **2012**, *2*, 1–6.
3. Anilakumar, K.R.; Pal, A.; Khanum, F.; Bawa, A.S. Nutritional, Medicinal and Industrial Uses of Sesame (*Sesame indicum* L.) Seeds - A Overview. *Agric. Conspec. Sci.* **2010**, *75*, 159–168.
4. Muta, N.; M, P.N. A review on sesame—An ethno medicinally significant oil crop. *Int. J. Life Sci. Pharm. Res.* **2017**, *7*, 58–63.
5. Bedigian, D. History and lore of sesame in Southwest Asia. *Econ. Bot.* **2004**, *58*, 329–353. [CrossRef]
6. Bedigian, D.; Harlan, J.R. Evidence for cultivation of sesame in the ancient world. *Econ. Bot.* **1986**, *40*, 137–154. [CrossRef]
7. Khan, M.S.; Raj, S.K.; Snehi, S.K. First report of 'Candidatus phytoplasma asteris' affecting sesame cultivation in India. *J. Plant Pathol.* **2007**, *89*, 301–305.
8. Saydut, A.; Duz, M.Z.; Kaya, C.; Kafadar, A.B.; Hamamci, C. Transesterified sesame (*Sesamum indicum* L.) seed oil as a biodiesel fuel. *Bioresour. Technol.* **2008**, *99*, 6656–6660. [CrossRef]
9. Moazzami, A. Sesame seed lignas: diversity, human metabolism and bioactivities. Doctoral Thesis, Swedish University of Agricultural Sciences, Uppsala, Sweden, 2006.
10. Pathak, N.; Rai, A.K.; Kumari, R.; Bhat, K.V. Value addition in sesame: A perspective on bioactive components for enhancing utility and profitability. *Pharmacogn. Rev.* **2014**, *8*, 147–155.
11. Unnikrishnan, M.K.; Kumar, M.S.; Satyamoorthy, K.; Joshi, R. Free radical reactions and antioxidant activities of sesamol: Pulse radiolytic and biochemical studies. *J. Agric. Food Chem.* **2005**, *53*, 2696–2703.
12. Juan, S.; Chen, S.; Qiuhui, H. Antioxidant activity of brown pigment and extracts from black sesame seed (*Sesamum indicum* L.). *Food Chem.* **2005**, *91*, 79–83.
13. Osawa, T. Protective role of dietary polyphenols in oxidative stress. *Mech. Age. Develop.* **1999**, *111*, 133–139. [CrossRef]
14. Yamashita, K.; Nohara, Y.; Katayama, K.; Namiki, M. Sesame Seed Lignans and γ-Tocopherol Act Synergistically to Produce Vitamin E Activity in Rats. *J. Nutr.* **1992**, *122*, 2440–2446. [CrossRef] [PubMed]
15. Morris, J.B. Food, industrial, nutraceutical, and pharmaceutical uses of sesame genetic resources. In *Trends in New Crops And New Uses*; ASHS Press: Alexandria, VA, USA, 2002; pp. 153–156.
16. Simon, J.E.; Chadwick, A.F.; Craker, L.E. *Herbs: An Indexed Bibliography, 1971–1980. The Scientific Literature on Selected Herbs, and Aromatic and Medicinal Plants of the Temperate Zone*; Archon Books: Hamden, CT, USA, 1984.
17. Beckstrom-Sternberg, S.M.; Duke, J.A.; Wain, K.K. The Ethnobotany Database. 1994. Available online: http://arsgenome.cornell.edu/cgibin/WebAce/webace?db=ethnobotdb.
18. Nayar, N.M.; Mehra, K.L. Sesame: Its uses, botany, cytogenetics, and origin. *Econ. Bot.* **2002**, *24*, 20–31. [CrossRef]
19. Shahidi, F.; Amarowi, R.; Abou-Gharbia, H.A.; Shehata, A.A.Y. Endogenous antioxidants and stability of sesame oil as affected by processing and storage. *J. Am. Oil Chem. Soc.* **1997**, *74*, 143–148. [CrossRef]
20. Jamieson, G.S.; Baughman, W.F. The chemical composition of sesame oil. *J. Am. Chem. Soc.* **1924**, *46*, 775–778. [CrossRef]
21. Ballakur, V. Inflammation in ayurveda and modern medicine. *Int. Ayurvedic Med. J.* **2013**, *1*, 1–7.
22. Scott, A.; Khan, K.M.; Cook, J.L.; Duronio, V. What is "inflammation"? Are we ready to move beyond Celsus? *Br. J. Sports Med.* **2004**, *38*, 248–249. [CrossRef]
23. Janeway, C.; Travers, P.; Walport, M.; Shlomchik, M. Inflammation. In *Immunobiology*; Garland Science: New York, NY, USA, 2001; pp. 41–43.
24. Hsu, D.-Z.; Chu, P.-Y.; Liu, M.-Y. Sesame Seed (*Sesamum indicum* L.) Extracts and Their Anti-Inflammatory Effect. *Emerg. Trends Diet. Compon. Prev. Combat. Dis.* **2012**, *1093*, 335–341.
25. Fullerton, J.N.; Gilroy, D.W. Resolution of inflammation: a new therapeutic frontier. *Nat. Rev. Drug Dis.* **2016**, *15*, 551. [CrossRef]
26. Kumar, V.; Ahmad, A. Role of MAIT cells in the immunopathogenesis of inflammatory diseases: new players in old game. *Int. Rev. Immunol.* **2017**, *37*, 1–21. [CrossRef] [PubMed]

27. Kumar, V. Innate lymphoid cells: new paradigm in immunology of inflammation. *Immunol. Lett.* **2014**, *157*, 23–37. [CrossRef] [PubMed]
28. Kumar, V.; Sharma, A. Neutrophils: Cinderella of innate immune system. *Int. Immunopharmacol.* **2010**, *10*, 1325–1334. [CrossRef] [PubMed]
29. Kumar, V.; Sharma, A. Mast cells: emerging sentinel innate immune cells with diverse role in immunity. *Mol. Immunol.* **2010**, *48*, 14–25. [CrossRef]
30. Kempuraj, D.; Thangavel, R.; Natteru, P.; Selvakumar, G.; Saeed, D.; Zahoor, H.; Zaheer, A. Neuroinflammation Induces Neurodegeneration. *J. Neurol. Neurosurg. Spine* **2017**, *1*, 1003.
31. Chen, W.-W.; Zhang, X.; Huang, W.-J. Role of neuroinflammation in neurodegenerative diseases (Review). *Mol. Med. Rep.* **2016**, *13*, 3391–3396. [CrossRef]
32. Akira, S.; Takeuchi, O. Pattern recognition receptors and inflammation. *Cell* **2010**, *140*, 805–820.
33. Kacimi, R.; Giffard, R.; Yenari, M. Endotoxin-activated microglia injure brain derived endothelial cells via NF-kB, JAK-STAT and JNK stress kinase pathways. *J. Inflamm.* **2011**, *8*, 1–15. [CrossRef]
34. Udomruk, S.; Kaewmool, C.; Pothacharoen, P.; Phitak, T.; Kongtawelert, P. Sesamin suppresses LPS-induced microglial activation via regulation of TLR4 expression. *J. Funct. Foods* **2018**, *49*, 32–43. [CrossRef]
35. Jeng, K.-G.; Hou, R.C.W.; Wang, J.; Ping, L.-I. Sesamin inhibits lipopolysaccharide-induced cytokine production by suppression of p38 mitogen-activated protein kinase and nuclear factor-kB. *Immunol. Lett.* **2005**, *97*, 101–106. [CrossRef]
36. Hamon, M.; Blier, P. Monoamine neurocircuitry in depression and strategies for new treatments. *Prog. Neuropsychopharmacol. Biol. Psychiatry* **2013**, *45*, 54–63. [CrossRef] [PubMed]
37. Root, D.H.; Hoffman, A.F. Norepinephrine activates dopamine D4 receptors in the rat lateral habenula. *J. Neurosci.* **2015**, *35*, 3460–3469. [CrossRef] [PubMed]
38. Yan, T.; He, B.; Wan, S.; Xu, M.; Yang, H.; Xiao, F.; Bi, K.; Jia, Y. Antidepressant-like effects and cognitive enhancement of Schisandra chinensis in chronic unpredictable mild stress mice and its related mechanism. *Sci. Rep.* **2017**, *7*, 6903. [CrossRef] [PubMed]
39. Ohsawa, K.; Imai, Y.; Sasaki, Y.; Kohsaka, S. Microglia/macrophage-specific protein Iba1 binds to fimbrin and enhances its actin-bundling activity. *J. Neurochem.* **2004**, *88*, 844–856. [CrossRef] [PubMed]
40. Farooq, R.K.; Isingrini, E.; Tanti, A.; Le Guisquet, A.M.; Arlicot, N.; Minier, F.; Leman, S.; Chalon, S.; Belzung, C.; Camus, V. Is unpredictable chronic mild stress (UCMS) a reliable model to study depression induced neuroinflammation? *Behav. Brain Res.* **2012**, *231*, 130–137. [CrossRef] [PubMed]
41. Zhao, Y.; Wang, Q.; Jia, M.; Fu, S.; Pan, J.; Chu, C.; Liu, X.; Liu, X.; Liu, Z. (+)-Sesamin attenuates chronic unpredictable mild stress-induced depressive-like behaviors and memory deficits via suppression of neuroinflammation. *J. Nutr. Biochem.* **2019**, *64*, 61–71. [CrossRef]
42. Ahmad, S.; Elsherbiny, N.M.; Haque, R.; Khan, M.B.; Ishrat, T.; Shah, Z.A.; Khan, M.M.; Ali, M.; Jamal, A.; Katare, D.P.; et al. Sesamin attenuates neurotoxicity in mouse model of ischemic brain stroke. *NeuroToxicology* **2014**, *45*, 100–110. [CrossRef]
43. Baluchnejadmojarad, T.; Mansouri, M.; Ghalami, J.; Mokhtari, Z.; Roghani, M. Sesamin imparts neuroprotection against intrastriatal 6-hydroxydopamine toxicity by inhibition of astroglial activation, apoptosis, and oxidative stress. *Biomed. Pharmacol.* **2017**, *88*, 754–761. [CrossRef]
44. Hsieh, P.F.; Hou, C.-W.; Yao, P.-W.; Wu, S.-P.; Peng, Y.-F.; Shen, M.-L.; Lin, C.-H.; Chao, Y.-Y.; Chang, M.-H.; Jeng, K.-C. Sesamin ameliorates oxidative stress and mortality in kainic acid-induced status epilepticus by inhibition of MAPK and COX-2 activation. *J. Neuroinflammation* **2011**, *8*, 57. [CrossRef]
45. Maines, M.D. The heme oxygenase system: a regulator of second messenger gases. *Annu. Rev. Pharmacol. Toxicol.* **1997**, *37*, 517–554. [CrossRef]
46. Tenhunen, R.; Marver, H.S.; Schmid, R. The enzymatic conversion of heme to bilirubin by microsomal heme oxygenase. *Proc. Natl. Acad. Sci. USA* **1968**, *61*, 748–755. [CrossRef] [PubMed]
47. Kirkby, K.A.; Adin, C.A. Products of heme oxygenase and their potential therapeutic applications. *Am. J. Physiol. Renal. Physiol.* **2006**, *290*, F563–F571. [CrossRef] [PubMed]
48. Fukunaga, M.; Ohnishi, M.; Shiratsuchi, A.; Kawakami, T.; Takahashi, M.; Motomura, M.; Egusa, K.; Urasaki, T.; Inoue, A. Sesamin increases heme oxygenase-1 protein in RAW 264.7 macrophages through inhibiting its ubiquitination process. *Euro. J. Pharmacol.* **2014**, *741*, 214–221. [CrossRef] [PubMed]

49. Hamada, N.; Tanaka, A.; Fujita, Y.; Itoh, T.; Ono, Y.; Kitagawa, Y.; Tomimori, N.; Kiso, Y.; Akao, Y.; Nozawa, Y.; et al. Involvement of heme oxygenase-1 induction via Nrf2/ARE activation in protection against H_2O_2-induced PC12 cell death by a metabolite of sesamin contained in sesame seeds. *Bioorg. Med. Chem.* **2011**, *19*, 1959–1965. [CrossRef] [PubMed]
50. Katayama, S.; Sugiyama, H.; Kushimoto, S.; Uchiyama, Y.; Hirano, M.; Nakamura, S. Effects of Sesaminol Feeding on Brain Aβ Accumulation in a Senescence-Accelerated Mouse-Prone 8. *J. Agric. Food Chem.* **2016**, *64*, 4908–4913. [CrossRef]
51. Hayami, T.; Pickarski, M.; Zhuo, Y.; Wesolowski, G.A.; Rodan, G.A.; Duong, L.T. Characterization of articular cartilage and subchondral bone changes in the rat anterior cruciate ligament transection and meniscectomized models of osteoarthritis. *Bone* **2006**, *38*, 234–243. [CrossRef]
52. Pearle, A.D.; Warren, R.F.; Rodeo, S.A. Basic science of articular cartilage and osteoarthritis. *Clin. Sports Med.* **2005**, *24*, 1–12. [CrossRef]
53. Goldring, M.B. Osteoarthritis and cartilage: the role of cytokines. *Curr. Rheumatol. Rep.* **2000**, *2*, 459–465. [CrossRef]
54. Daheshia, M.; Yao, J.Q. The interleukin-1β pathway in the pathogenesis of osteoarthritis. *J. Rheumatol.* **2008**, *35*, 2306–2312. [CrossRef]
55. Struglics, A.; Larsson, S.; Pratta, M.A.; Kumar, S.; Lark, M.W.; Lohmander, L.S. Human osteoarthritis synovial fluid and joint cartilage contain both aggrecanase- and matrix metalloproteinase-generated aggrecan fragments. *Osteoarthr. Cartil.* **2006**, *14*, 101–113. [CrossRef]
56. Mitchell, P.G.; Magna, H.A.; Reeves, L.M.; Lopresti-Morrow, L.L.; Yocum, S.A.; Rosner, P.J.; Geoghegan, K.F.; Hambor, J.E. Cloning, expression, and type II collagenolytic activity of matrix metalloproteinase-13 from human osteoarthritic cartilage. *J. Clin. Invest.* **1996**, *97*, 761–768. [CrossRef] [PubMed]
57. Phitak, T.; Pothacharoen, P.; Settakorn, J.; Poompimol, W.; Caterson, B.; Kongtawelert, P. Chondroprotective and anti-inflammatory effects of sesamin. *Phytochemistry* **2012**, *80*, 77–88. [CrossRef] [PubMed]
58. Kong, P.; Chen, G.; Jiang, A.; Wang, Y.; Song, C.; Zhuang, J.; Xi, C.; Wang, G.; Ji, Y.; Yan, J. Sesamin inhibits IL-1β-stimulated inflammatory response in human osteoarthritis chondrocytes by activating Nrf2 signalling pathway. *Oncotarget* **2016**, *7*, 83720–83726. [CrossRef] [PubMed]
59. Lu, Y.-C.; Jayakumar, T.; Duann, Y.-F.; Chou, Y.-C.; Hsieh, C.-Y.; Yu, S.-Y.; Sheu, J.-R.; Hsiao, G. Chondroprotective Role of Sesamol by Inhibiting MMPs Expression via Retaining NF-κB Signalling in Activated SW1353 Cells. *J. Agric. Food Chem.* **2011**, *59*, 4969–4978. [CrossRef] [PubMed]
60. Yashaswini, P.S.; Kurrey, N.K.; Singh, S.A. Encapsulation of sesamol in phosphatidyl choline micelles: Enhanced bioavailability and anti-inflammatory activity. *Food Chem.* **2017**, *228*, 330–337. [CrossRef] [PubMed]
61. Askari, A.; Ravansalar, S.A.; Naghizadeh, M.M.; Mosavat, S.Y.; Khodadoost, M.; Jazani, A.M.; Hashempur, M.H. The efficacy of topical sesame oil in patients with knee osteoarthritis: A randomized double-blinded active-controlled non-inferiority clinical trial. *Complement. Ther. Med.* **2019**, *47*, 102183. [CrossRef] [PubMed]
62. Clark, J.M.; Diehl, A.M. Defining nonalcoholic fatty liver disease: implications for epidemiologic studies. *Gastroenterology* **2003**, *124*, 248–250. [CrossRef]
63. Miller, A.M.; Wang, H.; Bertola, A.; Park, O.; Horiguchi, N.; Ki, S.H.; Yin, S.; Lafdil, F.; Gao, B. Inflammation-associated interleukin-6/signal transducer and activator of transcription 3 activation ameliorates alcoholic and nonalcoholic fatty liver diseases in interleukin-10-deficient mice. *Hepatology* **2011**, *54*, 846–856. [CrossRef]
64. Yoshikawa, T.; Ide, T.; Shimano, H.; Yahagi, N.; Amemiya-Kudo, M.; Matsuzaka, T.; Yatoh, S.; Kitamine, T.; Okazaki, H.; Tamura, Y.; et al. Cross-Talk between Peroxisome Proliferator-Activated Receptor (PPAR) α and Liver X Receptor (LXR) in Nutritional Regulation of Fatty Acid Metabolism. I. PPARs Suppress Sterol Regulatory Element Binding Protein-1c Promoter through Inhibition of LXR Signalling. *Mol. Endocrinol.* **2003**, *17*, 1240–1254.
65. Zhang, R.; Yu, Y.; Hu, S.; Zhang, J.; Yang, H.; Han, B.; Cheng, Y.; Luo, X. Sesamin ameliorates hepatic steatosis and inflammation in rats on a high-fat diet via LXRα and PPARα. *Nutr. Res.* **2016**, *36*, 1022–1030. [CrossRef]
66. Bernal, W.; Auzinger, G.; Dhawan, A.; Wendon, J. Acute liver failure. *Lancet* **2010**, *376*, 190–201. [CrossRef]
67. Felipo, V. Hepatic encephalopathy: effects of liver failure on brain function. *Nat. Rev. Neurosci.* **2013**, *14*, 851–858. [CrossRef] [PubMed]

68. Ma, L.; Gong, X.; Kuang, G.; Jiang, R.; Chen, R.; Wan, J. Sesamin ameliorates lipopolysaccharide/D-galactosamine-induced fulminant hepatic failure by suppression of Toll-like receptor 4 signalling in mice. *Biochem. Biophys. Res. Commun.* **2015**, *461*, 230–236. [CrossRef] [PubMed]
69. Chiang, H.-M.; Chang, H.; Yao, P.-W.; Chen, Y.-S.; Jeng, K.-C.; Wang, J.-S.; Hou, C.-W. Sesamin reduces acute hepatic injury induced by lead coupled with lipopolysaccharide. *J. Chin. Med. Assoc.* **2014**, *77*, 227–233. [CrossRef]
70. Nakai, M.; Harada, M.; Nakahara, K.; Akimoto, K.; Shibata, H.; Miki, W.; Kiso, Y. Novel antioxidative metabolites in rat liver with ingested sesamin. *J. Agric. Food Chem.* **2003**, *51*, 1666–1670. [CrossRef]
71. Yasuda, K.; Ikushiro, S.; Kamakura, M.; Ohta, M.; Sakaki, T. Metabolism of sesamin by cytochrome P450 in human liver microsomes. *Drug Metab. Dispos.* **2010**, *38*, 2117–2123. [CrossRef]
72. Yasuda, K.; Ikushiro, S.; Kamakura, M.; Munetsuna, E.; Ohta, M.; Sakaki, T. Sequential metabolism of sesamin by cytochrome P450 and UDP-glucuronosyltransferase in human liver. *Drug Metab. Dispos.* **2011**, *39*, 1538–1545. [CrossRef]
73. Tomimori, N.; Tanaka, Y.; Kitagawa, Y.; Fujii, W.; Sakakibara, Y.; Shibata, H. Pharmacokinetics and safety of the sesame lignans, sesamin and episesamin, in healthy subjects. *Biopharm. Drug Dispos.* **2013**, *34*, 462–473. [CrossRef]
74. Tomimori, N.; Rogi, T.; Shibata, H. Absorption, distribution, metabolism, and excretion of [^{14}C] sesamin in rats. *Mol. Nutr. Food Res.* **2017**, *61*, 1600844. [CrossRef]
75. Abe-Kanoh, N.; Lee, Y.; Takemoto, D.; Ono, Y.; Shibata, H.; Ohnishi, K.; Kawai, Y. Sesamin catechol glucuronides exert anti-inflammatory effects by suppressing IFN-β and iNOS expression through the deconjugation in macrophage-like J774.1 cells. *J. Agric. Food Chem.* **2019**, *67*, 7640–7649. [CrossRef]
76. Periasamy, S.; Chien, S.-P.; Chang, P.-C.; Hsu, D.-Z.; Liu, M.-Y. Sesame oil mitigates nutritional steatohepatitis via attenuation of oxidative stress and inflammation: a tale of two-hit hypothesis. *J. Nutr. Biochem.* **2014**, *25*, 232–240. [CrossRef] [PubMed]
77. Periasamy, S.; Hsu, D.-Z.; Chang, P.-C.; Liu, M.-Y. Sesame oil attenuates nutritional fibrosing steatohepatitis by modulating matrix metalloproteinases-2, 9 and PPAR-γ. *J. Nutr. Biochem.* **2014**, *25*, 337–344. [CrossRef] [PubMed]
78. Hsu, D.Z.; Chien, S.P.; Li, Y.H.; Liu, M.Y. Sesame oil does not show accumulatively enhanced protection against oxidative-stress-associated hepatic injury in septic rats. *J. Parenter. Enter. Nutr.* **2008**, *32*, 276–280. [CrossRef] [PubMed]
79. Wang, Z.; Xu, J.-P.; Zheng, Y.-C.; Chen, W.; Sun, Y.-W.; Wu, Z.-Y.; Luo, M. Peroxisome proliferator-activated receptor gamma inhibits hepatic fibrosis in rats. *Hepatobiliary Pancreat. Dis. Int.* **2011**, *10*, 64–71. [CrossRef]
80. Abdelmegeed, M.A.; Yoo, S.H.; Henderson, L.E.; Gonzalez, F.J.; Woodcroft, K.J.; Song, B.J. PPAR alpha expression protects male mice from high fat-induced non-alcoholic fatty liver. *J. Nutr.* **2011**, *141*, 603–610. [CrossRef] [PubMed]
81. Zhang, F.; Lu, Y.; Zheng, S. Peroxisome proliferator-activated receptor-γ crossregulation of signalling events implicated in liver fibrogenesis. *Cell Sign.* **2012**, *24*, 596–605. [CrossRef]
82. Periasamy, S.; Yang, S.S.; Chen, S.Y.; Chang, C.C.; Liu, M.Y. Prophylactic sesame oil attenuates sinusoidal obstruction syndrome by inhibiting matrix metalloproteinase- 9 and oxidative stress. *J. Parenter. Enter. Nutr.* **2013**, *37*, 529–537. [CrossRef]
83. Kim, M.; Woo, M.; Noh, J.S.; Choe, E.; Song, Y.O. Sesame oil lignans inhibit hepatic endoplasmic reticulum stress and apoptosis in high-fat diet-fed mice. *J. Funct. Foods* **2017**, *37*, 658–665. [CrossRef]
84. Gologorsky, D.; Thanos, A.; Vavvas, D. Therapeutic interventions against inflammatory and angiogenic mediators in proliferative diabetic retinopathy. *Mediat. Inflamm.* **2012**, *2012*, 1–10. [CrossRef]
85. Saijo, K.; Glass, C.K. Microglial cell origin and phenotypes in health and disease. *Nat. Rev. Immunol.* **2011**, *11*, 775–787. [CrossRef]
86. Elsherbiny, N.M.; Ahmad, S.; Naime, M.; Elsherbini, A.M.; Fulzele, S.; Al-Gayyar, M.M.; Eissa, L.A.; El-Shishtawy, M.M.; Liou, G.I. ABT-702, an adenosine kinase inhibitor, attenuates inflammation in diabetic retinopathy. *Life Sci.* **2013**, *93*, 78–88. [CrossRef] [PubMed]
87. Ahmad, S.; ElSherbiny, N.M.; Jamal, M.S.; Alzahrani, F.A.; Haque, R.; Khan, R.; Zaidi, S.K.; AlQahtani, M.H.; Liou, G.I.; Bhatia, K. Anti-inflammatory role of sesamin in STZ induced mice model of diabetic retinopathy. *J. Neuroimmunol.* **2016**, *295–296*, 47–53. [CrossRef] [PubMed]

88. Kondamudi, P.K.; Kovelamudi, H.; Mathew, G.; Nayak, P.G.; Rao, C.M.; Shenoy, R.R. Modulatory effects of sesamol in dinitrochlorobenzene-induced inflammatory bowel disorder in albino rats. *Pharm. Rep.* **2013**, *65*, 658–665. [CrossRef]
89. Hawkey, C.J. Non-steroidal anti-inflammatory drug gastropathy: Causes and treatment. *Scand. J. Gastroenterol.* **1996**, *220*, 124–127. [CrossRef]
90. Tappel, A.L. Lipid peroxidation damage to cell components. *Fed. Proc.* **1973**, *32*, 1870–1874.
91. Szabo, C.; Cuzzocrea, S.; Zingarelli, B.; O'Connor, M.; Salzman, A.L. Endothelial dysfunction in a rat model of endotoxic shock. Importance of the activation of poly (ADPribose) synthetase by peroxynitrite. *J. Clin. Investig.* **1997**, *100*, 723–735. [CrossRef]
92. Hsu, D.-Z.; Chu, P.-Y.; Chandrasekaran, V.R.M.; Liu, M.-Y. Sesame lignan sesamol protects against aspirin-induced gastric mucosal damage in rats. *J. Funct. Foods* **2009**, *1*, 349–355. [CrossRef]
93. Jan, K.-C.; Ku, K.-L.; Chu, Y.-H.; Hwang, L.S.; Ho, C.-T. Tissue Distribution and Elimination of Estrogenic and Anti-Inflammatory Catechol Metabolites from Sesaminol Triglucoside in Rats. *J. Agric. Food Chem.* **2010**, *58*, 7693–7700. [CrossRef]
94. Lee, S.Y.; Son, D.J.; Lee, Y.K.; Lee, J.W.; Lee, H.J.; Yun, Y.W.; Ha, T.Y.; Hong, J.T. Inhibitory effect of sesaminol glucosides on lipopolysaccharide-induced NF-kappaB activation and target gene expression in cultured rat astrocytes. *Neurosci. Res.* **2006**, *56*, 204–212. [CrossRef]
95. Ji, Z.-L.; Li, J.-S.; Yuan, C.-W.; Chen, W.; Zhang, Y.-N.; Ju, X.-T.; Tang, W.-H. Therapeutic value of sesame oil in the treatment of adhesive small bowel obstruction. *Am. J. Surg.* **2010**, *199*, 160–165. [CrossRef]
96. Yeh, M.; Leitinger, N.; de Martin, R.; Onai, N.; Matsushima, K.; Vora, D.K.; Berliner, J.A.; Reddy, S.T. Increased transcription of IL-8 in endothelial cells is differentially regulated by TNF-R and oxidized phospholipids. *Arterioscler. Thromb. Vasc. Biol.* **2001**, *21*, 1585–1591. [CrossRef] [PubMed]
97. Lee, W.-J.; Ou, H.-C.; Wu, C.-M.; Lee, I.; Lin, S.-Y.; Lin, L.-Y.; Tsai, K.-L.; Lee, S.-D.; Sheu, W.H.-H. Sesamin Mitigates Inflammation and Oxidative Stress in Endothelial Cells Exposed to Oxidized Low-Density Lipoprotein. *J. Agric. Food Chem.* **2009**, *57*, 11406–11417. [CrossRef] [PubMed]
98. Shenasa, M.; Shenasa, H.; El-Sherif, N. Left ventricular hypertrophy and arrhythmogenesis. *Card. Electrophysiol. Clin.* **2015**, *7*, 207–220. [CrossRef] [PubMed]
99. Gonzalez, A.A.; Prieto, M.C. Renin and the (pro) renin receptor in the renal collecting duct: role in the pathogenesis of hypertension. *Clin. Exp. Pharmacol. Physiol.* **2015**, *42*, 14–21. [CrossRef]
100. Omura, T.; Yoshiyama, M.; Yoshida, K.; Nakamura, Y.; Kim, S.; Iwao, H.; Takeuchi, H.; Yoshikawa, J. Dominant negative mutant of c-Jun inhibits cardiomyocyte hypertrophy induced by endothelin 1 and phenylephrine. *Hypertension* **2002**, *39*, 81–86. [CrossRef]
101. Abdelhamid, G.; El-Kadi, A.O. Buthionine Sulfoximine, an inhibitor of glutathione biosynthesis, induces expression of soluble epoxide hydrolase and markers of cellular hypertrophy in a rat cardiomyoblast cell line: roles of the NF-κB and MAPK signaling pathways. *Free Radic. Bio. Med.* **2015**, *82*, 1–12. [CrossRef]
102. Liu, C.-T.; Liu, M.-Y. Daily sesame oil supplementation attenuates local renin-angiotensin system via inhibiting MAPK activation and oxidative stress in cardiac hypertrophy. *J. Nutr. Biochem.* **2017**, *42*, 108–116. [CrossRef]
103. Bone, R.C. The pathogenesis of sepsis. *Ann. Intern. Med.* **1991**, *115*, 457–469. [CrossRef]
104. Matthay, M.A.; Zimmerman, G.A. Acute lung injury and the acute respiratory distress syndrome: four decades of inquiry into pathogenesis and rational management. *Am. J. Respir. Cell Mol. Biol.* **2005**, *33*, 319–327. [CrossRef]
105. Braude, S.; Nolop, K.B.; Hughes, J.M.B.; Barnes, P.J.; Royston, D. Comparison of lung vascular and epithelial permeability indices in the adult respiratory distress syndrome. *Am. Rev. Respir. Dis.* **1986**, *133*, 1002–1005.
106. Tomashefski, J. Pulmonary pathology of the adult respiratory distress syndrome. *Clin. Chest Med.* **1990**, *11*, 593–619. [PubMed]
107. Sinclair, D.G.; Braude, S.; Haslam, P.L.; Evans, T.W. Pulmonary endothelial permeability in patients with severe lung injury. *Clin. Correl. Nat. His. Chest* **1994**, *106*, 535–539. [CrossRef] [PubMed]
108. Chu, P.-Y.; Chien, S.-P.; Hsu, D.-Z.; & Liu, M.-Y. Protective effect of sesamol on the pulmonary inflammatory response and lung injury in endotoxemic rats. *Food Chem. Toxicol.* **2010**, *48*, 1821–1826. [CrossRef] [PubMed]
109. Liu, M.; Yokomizo, T. The role of leukotrienes in allergic diseases. *Allergol. Int.* **2015**, *64*, 17–26. [CrossRef]

110. Yashaswini, P.S.; Sadashivaiah, B.; Ramaprasad, T.R.; Singh, S.A. In vivo modulation of LPS induced leukotrienes generation and oxidative stress by sesame lignans. *J. Nutr. Biochem.* **2017**, *41*, 151–157. [CrossRef]
111. Andreou, A.; Gobel, C.; Hamberg, M.; Feussner, I. A bisallylic mini-lipoxygenase from cyanobacterium Cyanothece sp. that has an iron as cofactor. *J. Biol. Chem.* **2010**, *285*, 14178–14186. [CrossRef]
112. Mahesha, H.G.; Singh, S.A.; Appu Rao, A.G. Inhibition of lipoxygenase by soy isoflavones: Evidence of isoflavones as redox inhibitors. *Arch. Biochem. Biophys.* **2007**, *461*, 176–185. [CrossRef]
113. Yashaswini, P.S.; Rao, A.G.A.; Singh, S.A. Inhibition of lipoxygenase by sesamol corroborates its potential anti-inflammatory activity. *Int. J. Biol. Macromol.* **2017**, *94*, 781–787. [CrossRef]
114. Deme, P.; Narasimhulu, C.A.; Parthasarathy, S. Identification and evaluation of anti-inflammatory properties of aqueous components extracted from sesame (*Sesamum indicum*) oil. *J. Chromatogr. B* **2018**, *1087–1088*, 61–69. [CrossRef]
115. Yang, Y.; Wang, J.; Zhang, Y.; Li, J.; Sun, W. Black Sesame Seeds Ethanol Extract Ameliorates Hepatic Lipid Accumulation, Oxidative Stress and Insulin Resistance in Fructose-induced Nonalcoholic Fatty Liver Disease. *J. Agric. Food Chem.* **2018**, *66*, 10458–10469. [CrossRef]
116. Ruckmani, A.; Meti, V.; Vijayashree, R.; Arunkumar, R.; Konda, V.R.; Prabhu, L.; Madhavi, E.; Devi, S. Anti-rheumatoid activity of ethanolic extract of Sesamum indicum seed extract in Freund's complete adjuvant induced arthritis in Wistar albino rats. *J. Tradit. Complement. Med.* **2018**, *8*, 377–386. [CrossRef] [PubMed]
117. Botelho, J.R.S.; Medeiros, N.G.; Rodrigues, A.M.C.; Araújo, M.E.; Machado, N.T.; Guimarães Santos, A.; Rogerio Santos, I.; Gomes-Leal, W.; Carvalho, R.N. Black sesame (*Sesamum indicum* L.) seeds extracts by CO2 supercritical fluid extraction: Isotherms of global yield, kinetics data, total fatty acids, phytosterols and neuroprotective effects. *J. Supercrit. Fluids* **2014**, *93*, 49–55. [CrossRef]
118. Wang, B.-S.; Chang, L.-W.; Yen, W.-J.; Duh, P.-D. Antioxidative effect of sesame coat on LDL oxidation and oxidative stress in macrophages. *Food Chem.* **2007**, *102*, 351–360. [CrossRef]
119. Chang, L.W.; Yen, W.J.; Huang, S.C.; Duh, P.D. Antioxidant activity of sesame coat. *Food Chem.* **2002**, *78*, 347–354. [CrossRef]
120. Kay, J.; Thadhani, E.; Samson, L.; Engelward, B. Inflammation-Induced DNA Damage Mutations and Cancer. *DNA Repair* **2019**, *83*, 102673. [CrossRef]
121. Greten, F.R.; Grivennikov, S.I. Inflammation and Cancer: Triggers, Mechanisms, and Consequences. *Immunity* **2019**, *51*, 27–41. [CrossRef]
122. Bray, F.; Ferlay, J.; Soerjomataram, I.; Siegel, R.L.; Torre, L.A.; Jemal, A. Global Cancer Statistics 2018: GLOBOCAN Estimates of Incidence and Mortality Worldwide for 36 Cancers in 185 Countries. *CA Cancer J. Clin.* **2018**, *68*, 394–424. [CrossRef]
123. Cabral, B.P.; Fonseca, M.G.D.; Mota, F.B. The Recent Landscape of Cancer Research Worldwide: A Bibliometric and Network Analysis. *Oncotarget* **2018**, *9*, 30474–30484. [CrossRef]
124. Gutschner, T.; Diederichs, S. The Hallmarks of Cancer: A Long Non-Coding RNA Point of View. *RNA Biol.* **2012**, *9*, 703–719. [CrossRef]
125. Hanahan, D.; Weinberg, R.A. The Hallmarks of Cancer. *Cell Press.* **2000**, *100*, 57–70. [CrossRef]
126. Hanahan, D.; Weinberg, R.A. Hallmarks of Cancer: The Next Generation. *Cell* **2011**, *144*, 646–674. [CrossRef] [PubMed]
127. Prakash, K.; Naik, S.N. Bioactive Constituents as a Potential Agent in Sesame for Functional and Nutritional Application. *J. Bioresour. Eng. Technol.* **2014**, *1*, 48–66.
128. Harikumar, K.B.; Sung, B.; Tharakan, S.T.; Pandey, M.K.; Joy, B.; Guha, S.; Krishnan, S.; Aggarwal, B.B. Sesamin Manifests Chemopreventive Effects through the Suppression of NF-κB-Regulated Cell Survival, Proliferation, Invasion, and Angiogenic Gene Products. *Mol. Cancer Res.* **2010**, *8*, 751–761. [CrossRef] [PubMed]
129. Watanabe, M.; Iizumi, Y.; Iizuka-Ohashi, M.; Sowa, Y.; Sakai, T. The Pleiotropic Regulation of Cyclin D1 by Newly Identified Sesaminol-Binding Protein ANT2. *Oncogenesis* **2017**, *6*, 1–11. [CrossRef] [PubMed]
130. Majdalawieh, A.F.; Mansour, Z.R. Sesamol, a Major Lignan in Sesame Seeds (*Sesamum indicum*): Anti-Cancer Properties and Mechanisms of Action. *Eur. J. Pharmacol.* **2019**, *855*, 75–89. [CrossRef] [PubMed]
131. Siriwarin, B.; Weerapreeyakul, N. Sesamol Induced Apoptotic Effect in Lung Adenocarcinoma Cells through Both Intrinsic and Extrinsic Pathways. *Chem.-Biol. Interact.* **2016**, *254*, 109–116. [CrossRef] [PubMed]

132. Siriwarin, B.; Weerapreeyakul, N.; Tanthanuch, W.; Thumanu, K. Biomolecular Changes and DNA Targeting Effect of Sesamol in Human Lung Adenocarcinoma (SK-LU-1) Cells by FTIR Microscopy. *Asian Pac. J. Trop. Biomed.* **2018**, *8*, 377–386.
133. Majdalawieh, A.F.; Massri, M.; Nasrallah, G.K. A Comprehensive Review on the Anti-Cancer Properties and Mechanisms of Action of Sesamin, a Lignan in Sesame Seeds (*Sesamum indicum*). *Eur. J. Pharmacol.* **2017**, *815*, 512–521. [CrossRef]
134. Liu, C.-M.; Zheng, G.-H.; Ming, Q.-L.; Chao, C.; Sun, J.-M. Sesamin Protects Mouse Liver against Nickel-Induced Oxidative DNA Damage and Apoptosis by the PI3K-Akt Pathway. *J. Agric. Food Chem.* **2013**, *61*, 1146–1154. [CrossRef]
135. Fang, Q.; Zhu, Y.; Wang, Q.; Song, M.; Gao, G.; Zhou, Z. Suppression of Cyclooxygenase-2 Increases Chemosensitivity to Sesamin through the Akt-PI3K Signalling Pathway in Lung Cancer Cells. *Int. J. Mol. Med.* **2019**, *43*, 507–516.
136. Bilancio, A.; Rinaldi, B.; Oliviero, M.A.; Donniacuo, M.; Monti, M.G.; Boscaino, A.; Marino, I.; Friedman, L.; Rossi, F.; Vanhaesebroeck, B.; et al. Inhibition of p110δ PI3K Prevents Inflammatory Response and Restenosis After Artery Injury. *Biosci. Rep.* **2017**, *37*, 1–12. [CrossRef] [PubMed]
137. Yokota, T.; Matsuzaki, Y.; Koyama, M.; Hitomi, T.; Kawanaka, M.; Enoki-Konishi, M.; Okuyama, Y.; Takayasu, J.; Nishino, H.; Nishikawa, A.; et al. Sesamin, a lignan of sesame, down-regulates cyclin D1 protein expression in human tumor cells. *Cancer Sci.* **2007**, *98*, 1447–1453. [CrossRef] [PubMed]
138. Lee, C.-C.; Liu, K.-J.; Wu, Y.-C.; Lin, S.-J.; Chang, C.-C.; Huang, T.-S. Sesamin Inhibits Macrophage-Induced Vascular Endothelial Growth Factor and Matrix Metalloproteinase-9 Expression and Proangiogenic Activity in Breast Cancer Cells. *Inflammation* **2010**, *34*, 209–221. [CrossRef] [PubMed]
139. Akl, M.R.; Ayoub, N.M.; Abuasal, B.S.; Kaddoumi, A.; Sylvester, P.W. Sesamin Synergestically Potentiates the Anticancer Effects of γ-Tocotrienol in Mammary Cancer Cell Lines. *Fitoterapia* **2013**, *84*, 347–359. [CrossRef] [PubMed]
140. Truan, J.S.; Chen, J.-M.; Thompson, L.U. Comparative Effects of Sesame Seed Lignan and Flaxseed Lignan in Reducing the Growth of Human Breast Tumors (MCF-7) at High Levels of Circulating Estrogen in Athymic Mice. *Nutr. Cancer* **2012**, *64*, 65–71. [CrossRef] [PubMed]
141. Jacklin, A.; Ratledge, C.; Welham, K.; Bilko, D.; Newton, C.J. The Sesame Seed Oil Constituent, Sesamol, Induces Growth Arrest and Apoptosis of Cancer and Cardiovascular Cells. *Ann. N. Y. Acad. Sci.* **2003**, *1010*, 374–380. [CrossRef]
142. Galli, F.; Stabile, A.M.; Betti, M.; Conte, C.; Pistilli, A.; Rende, M.; Floridi, A.; Azzi, A. The Effect of α- and γ-Tocopherol and Their Carboxyethyl Hydroxychroman Metabolites on Prostate Cancer Cell Proliferation. *Arch. Biochem. Biophys.* **2004**, *423*, 97–102. [CrossRef]
143. Isha, D.; Milind, P. Eat Til and Protect Dil. *Int. Res. J. Pharm.* **2012**, *3*, 54–57.
144. Cooney, R.V.; Custer, L.J.; Okinaka, L.; Franke, A.A. Effects of Dietary Sesame Seeds on Plasma Tocopherol Levels. *Nutr. Cancer* **2009**, *39*, 66–71. [CrossRef]
145. Yamashita, K.; Iizuka, Y.; Imai, T.; Namiki, M. Sesame Seed and Its Lignans Produce Marked Enhancement of Vitamin E Activity in Rats Fed a Low α-Tocopherol Diet. *Lipids* **1995**, *30*, 1019–1028. [CrossRef]
146. Jeng, K.C.G.; Hou, R.C.W. Sesamin and Sesamolin: Nature's Therapeutic Lignans. *Curr. Enzyme Inhib.* **2005**, *1*, 11–20. [CrossRef]
147. Shimizu, S.; Fujii, G.; Takahashi, M.; Nakanishi, R.; Komiya, M.; Shimura, M.; Noma, N.; Onuma, W.; Terasaki, M.; Yano, T.; et al. Sesamol Suppresses Cyclooxygenase-2 Transcriptional Activity in Colon Cancer Cells and Modifies Intestinal Polyp Development in $Apc^{Min/+}$ Mice. *J. Clin. Biochem. Nutr.* **2014**, *54*, 95–101. [CrossRef] [PubMed]
148. Khamphio, M.; Barusrux, S.; Weerapreeyakul, N. Sesamol Induces Mitochondrial Apoptosis Pathway in HCT116 Human Colon Cancer Cells via Pro-Oxidant Effect. *Life Sci.* **2016**, *158*, 46–56. [CrossRef] [PubMed]
149. Silva-Prado, L.; Azevedo, L.; Oliveira, J.A.C.; Moreire, A.P.M.; Schmiele, M.; Chang, Y.K.; Paula, F.B.A.; Clerici, M.T.P.S. Sesame and Resistance Starch Reduce the Colon Carcinogenesis and Oxidative Stress in 1,2-dimethylhydrazine-induced Cancer in Wistar Rats. *Food Res. Int.* **2014**, *62*, 609–617. [CrossRef]
150. Sheng, H.Q.; Hirose, Y.; Hata, K.; Zheng, Q.; Kuno, T.; Asano, N.; Yamada, Y.; Hara, A.; Osawa, T.; Mori, H. Modifying Effect of Dietary Sesaminol Glucosides on the Formation of Azoxymethane-Induced Premalignant Lesions of Rat Colon. *Cancer Lett.* **2007**, *246*, 63–68. [CrossRef]

151. Deng, P.; Wang, C.; Chen, L.; Wang, C.; Du, Y.; Yan, X.; Chen, M.; Yang, G.; He, G. Sesamin Induces Cell Cycle Arrest and Apoptosis through the Inhibition of Signal Transducer and Activator of Transcription 3 Signalling in Human Hepatocellular Carcinoma Cell Line HepG2. *Biol. Pharm. Bull.* **2013**, *36*, 1540–1548. [CrossRef]
152. Liu, Z.; Xiang, Q.; Du, L.; Song, G.; Wang, Y.; Liu, X. The Interaction of Sesamol with DNA and Cytotoxicity, Apoptosis, and Localization in HepG2 Cells. *Food Chem.* **2013**, *141*, 289–296. [CrossRef]
153. Liu, Z.; Ren, B.; Wang, Y.; Zou, C.; Qiao, Q.; Diao, Z.; Mi, Y.; Zhu, D.; Liu, X. Sesamol Induces Human Hepatocellular Carcinoma Cells Apoptosis by Impairing Mitochondrial Function and Suppression Autophagy. *Sci. Rep.* **2017**, *7*, 45728. [CrossRef]
154. Zhou, L.; Lin, X.; Abbasi, A.M.; Zheng, B. Phytochemical Contents and Antioxidant and Antiproliferative Activities of Selected Black and White Sesame Seeds. *Biomed. Res. Int.* **2016**, *2016*, 1–9. [CrossRef]
155. Dou, H.; Yang, S.; Hu, Y.; Xu, D.; Liu, L.; Li, X. Sesamin Induces ER Stress-Mediated Apoptosis and Activates Autophagy in Cervical Cancer Cells. *Life Sci.* **2018**, *200*, 87–93. [CrossRef]
156. Miyahara, Y.; Hibasami, H.; Katsuzaki, H.; Imai, K.; Komiya, T. Sesamol Induces Apoptosis in Human Lymphoid Leukemia Molt 4B Cells. *Food. Sci. Technol. Res.* **2000**, *6*, 201–203. [CrossRef]
157. Fujimoto, A.; Shingai, Y.; Oyama, T.B.; Kawanai, T.; Hashimoto, E.; Koizumi, K.; Kimura, K.; Masuda, T.; Oyama, Y. Apoptosis-Inducing Action of Two Products from Oxidation of Sesamol, an Antioxidative Constituent of Sesame Oil: A Possible Cytotoxicity of Oxidized Antioxidant. *Toxicol. Vitro* **2010**, *24*, 1720–1726. [CrossRef] [PubMed]
158. Geetha, T.; Kapila, M.; Prakash, O.; Deol, P.K.; Kakkar, V.; Kaur, I.P. Sesamol-Loaded Solid Lipid Nanoparticles for Treatment of Skin Cancer. *J. Drug Target.* **2015**, *23*, 159–169. [CrossRef] [PubMed]
159. Namiki, M. Nutraceutical Functions of Sesame: A Review. *Crit. Rev. Food Sci. Nutr.* **2007**, *47*, 651–673. [CrossRef] [PubMed]
160. Miyahara, Y.; Katsuzaki, H.; Imai, K.; Osawa, T.; Ina, K.; Komiya, T. Sesaminol from Sesame Seed Induces Apoptosis in Human Lymphoid Leukemia Molt 4B Cells. *Int. J. Mol. Med.* **2001**, *7*, 485–488. [CrossRef]
161. Kim, J.H.; Lee, J.K. Sesamolin Enhances NK Cell Lysis Activity by Increasing the Expression of NKG2D Ligands on Burkitt's Lymphoma Cells. *Int. Immunopharmacol.* **2015**, *28*, 977–984. [CrossRef] [PubMed]
162. Srisayam, M.; Weerapreeyakul, N.; Kanokmedhakul, K. Inhibition of Two Stages of Melanin Synthesis by Sesamol, Sesamin, and Sesamolin. *Asian Pac. J. Trop. Biomed.* **2017**, *7*, 886–895. [CrossRef]
163. Smith, D.E.; Salerno, J.W. Selective Growth Inhibition of a Human Malignant Melanoma Cell Line by Sesame Oil In Vitro. *Prostaglandins Leukot. Essent. Fat. Acids* **1992**, *46*, 145–150. [CrossRef]
164. Kumar, C.M.; Sathisha, U.V.; Dharmesh, S.; Rao, A.G.A.; Singh, S.A. Interaction of Sesamol (3,4-methylenedioxyphenol) with Tyrosinase and its Effect on Melanin Synthesis. *Biochimie* **2011**, *93*, 562–569. [CrossRef]
165. Kapadia, G.J.; Azuine, M.A.; Tokuda, H.; Takasaki, M.; Mukainaka, T.; Konoshima, T.; Nishino, H. Chemopreventive Effect of Resveratrol, Sesamol, Sesame Oil and Sunflower Oil in the Epstein-Barr Virus Early Antigen Activation Assay and the Mouse Skin Two-Stage Carcinogenesis. *Pharmacol. Res.* **2002**, *45*, 499–505. [CrossRef]
166. Sugano, M.; Gu, J.-Y.; Yamada, K. Sesamin, a Multifunctional Factor for Prevention of Various Disorders Including Carcinogenesis. *Food Fact. Cancer Prev.* **1997**, 245–248.
167. Ogawa, T.; Makino, T.; Hirose, N.; Sugano, M. Lack of Influence of Low Blood Cholesterol Levels on Pancreatic Carcinogenesis After Initiation with *N*-nitrosobis (2-oxopropyl)amine in Syrian golden hamsters. *Carcinogensis* **1994**, *15*, 1663–1666. [CrossRef] [PubMed]
168. Chen, Y.-H.; Leu, S.-F.; Jen, C.-Y.; Huang, B.-M. Effects of Sesamol on Apoptosis and Steroidogenesis in MA-10 Mouse Leydig Tumor Cells. *J. Agric. Food Chem.* **2011**, *59*, 9885–9891. [CrossRef] [PubMed]

MDPI
St. Alban-Anlage 66
4052 Basel
Switzerland
Tel. +41 61 683 77 34
Fax +41 61 302 89 18
www.mdpi.com

Molecules Editorial Office
E-mail: molecules@mdpi.com
www.mdpi.com/journal/molecules

www.ingramcontent.com/pod-product-compliance
Lightning Source LLC
Chambersburg PA
CBHW061831210926
51850CB00006B/859